Nanoelectronics Devices: Design, Materials, and Applications (Part I)

Edited by

Gopal Rawat

*SMIEEE, School of Computing and Electrical Engineering
(SCEE), Indian Institute
of Technology Mandi (IIT Mandi), Kamand-175075, Mandi,
Himachal Pradesh, India*

&

Aniruddh Bahadur Yadav

*Department of Electronics and Communication Engineering
Velagapudi Ramakrishna Siddhartha
Engineering College (Autonomous),
Kanuru Vijayawada Andhra Pradesh - 520007
India*

Nanoelectronics Devices: Design, Materials, and Applications (Part I)

Editors: Gopal Rawat and Aniruddh Bahadur Yadav

ISBN (Online): 978-981-5136-62-3

ISBN (Print): 978-981-5136-63-0

ISBN (Paperback): 978-981-5136-64-7

© 2023, Bentham Books imprint.

Published by Bentham Science Publishers Pte. Ltd. Singapore. All Rights Reserved.

First published in 2023.

need for a court order if at any point you breach any terms of this License Agreement. In no event will any delay or failure by Bentham Science Publishers in enforcing your compliance with this License Agreement constitute a waiver of any of its rights.

3. You acknowledge that you have read this License Agreement, and agree to be bound by its terms and conditions. To the extent that any other terms and conditions presented on any website of Bentham Science Publishers conflict with, or are inconsistent with, the terms and conditions set out in this License Agreement, you acknowledge that the terms and conditions set out in this License Agreement shall prevail.

Bentham Science Publishers Pte. Ltd.
80 Robinson Road #02-00
Singapore 068898
Singapore
Email: subscriptions@benthamscience.net

BENTHAM SCIENCE

CONTENTS

FOREWORD

The idea of studying nano-structures was established in the year 1959 when Nobel laureate and theoretical physicist Richard Feynman spoke about miniaturizing old-age computers in his famous talk entitled "There's plenty of room at the bottom". Recently, the 2016 Nobel Prize in Chemistry awarded to Prof. Jean Pierre Sauvage, Prof. Sir J. Fraser Stoddart, and Prof. Bernard Lucas Feringa "for the design and synthesis of molecular machines", established yet another key milestone in the history of nanotechnology. Nanotechnology, in recent years, has delivered innovations that have the potential to completely transform the direction of technological advances in a wide range of applications such as nano-engineering, nano-materials, nano-electronics and nano-medicine. The evolution of nano-technology is undoubtedly going to result in sweeping changes to humanity and our society. Not just limited to engineering and material sciences, the innovations in nano-technology have allowed us to come up with a logical deduction towards understanding the complex and multiple associations within biomolecules and biological systems.

This book takes us on a voyage through the realm of new discoveries and innovations in the field of nano-technology. The book consists of twenty-three chapters; each chapter addresses specific facets of nanoscience and nano-engineering involving basic concepts of physics, chemistry, biology and mathematics for the development of new products. This enthralling journey takes us through multifarious terrains of nanotechnology, such as nano-electronic products, nano-technology for electronic modules, characteristics and application of field effect transistors as sensors. In addition, it includes extensive discussion on nanomaterials applications, including but not limited to, nano-dentistry and cosmetic industry and energy harvesting using solar cells.

In light of the recent inventions, discoveries and breakthroughs in the field of nano-technology, this book promises to equally delight nano-technology researchers, students and enthusiasts.

Rahul Checker
Radiation Biology & Health Sciences Division, Bhabha Atomic Research Centre
Trombay
Mumbai 400085
India

PREFACE

In recent years, nanoelectronics devices have been finding potential applications in all fields that are easing the life of human beings, like agriculture, energy harvesting, medicine, battery, sensors, optoelectronics, pyro-photonic, food processing, bio-nanotechnology, aerospace, automobiles, pharmaceutical, paints, cosmetic, aeronautic, medical imaging (MRI, ultrasound), *etc*. Therefore, it is necessary to produce well-organized information for researchers, scholars, academicians, and scientists working on the design, synthesis, and simulation of nanoelectronics devices and simulation modeling and characterization of materials for nanoelectronics devices. This information would help those working on nanoelectronics to conduct academic classes, research, and develop new products. Industries are producing nanoelectronics devices for different applications by putting great effort into the research and development to make them more efficient, reliable, durable, and low cost, which needs systematic information on current developments in the field.

These requirements have encouraged the editors to invite proficient, learned, and expertized authors to contribute chapters and provides in-depth information and concept of present nanoelectronics devices considering material synthesis, design, and synthesis of nanoelectronics devices. The authors of chapters have worked in the field for a long time and proven themselves in their field over the globe. The chapters are classified as nanotechnology-based solar-cell, medical devices, power grid devices, agriculture devices, cosmetic biosensors, and advanced transistors. The chapters present theoretical and experimental work and broad reviews of the existing state-of-the-art. The simulation and modeling of the nanoelectronics devices and materials for nanoelectronics devices are also considered equally. This book is for a one-semester course on nanoelectronics devices. The book is classified into two volumes one and two. First volume consists fourteen chapters, those presented here.

ORGANIZATION OF PART I

Chapter 1 of the book starts with a broad introduction to nanoelectronics devices and nanomaterials, where the authors present the present, past, and future of nanoelectronics. Nanotechnology briefly introduces and applicability of MOORs law. This chapter produces the classification of the nanomaterials based on the size and different nanofabrication approaches, like top-down and bottom-up.

Chapter 2 presents the self-assembly of the monolayer for molecular electronics, an integrated part of nanoelectronics. Transport of charge carrier, electrode, and measurement of the such device covered in detail. Materials used for the fabrication of the molecular devices are also presented.

Chapter 3 presents the performance of the core cell double gate junctionless transistors for current nanoelectronics-based integrated circuits. A multigate field effect transistor properties modified by channel doping. Fermi level adjustment in junctionless transistor explained properly. In junctionless transistors, there is a core, and cell structure improves its performance.

Chapter 4 presents tunneling field effect transistors that lead to the area reduction on the integrated circuits, mostly in non-polar devices. The results are simulated for a channel length of 20 nm under varying temperatures to understand the leakage current and other transistor parameters.

Chapter 5 presents a nanowire-based field effect transistor that overcomes the limitations of the conventional MOSFET. The negative capacitance of the nanowire-based field effect transistor is explained in depth. Band bending and distribution of potential simulated and presented. The implication of transistors in integrated circuits is presented with illustrations and simulation results.

Chapter 6 discusses an electrode's effect on the linear region of a thin film transistor. The triple material double gate was applied to control current conduction in the channel, and silicon was the channel material. The three metals are arranged horizontally of varying lengths affecting the transistor performance. The transistor performance is analyzed in both analog/RF ranges.

Chapter 7 deals mechanism of gas sensing by thin film transistors whose channel is of II-IV semiconductor, an effective transistor in the current generation of technology. Novel metals improve the gas sensing properties of the bare II-IV-based semiconductor. Transistor active layers have different characteristics identified by different characterization tools: atomic force microscope, X-ray diffraction, *etc*. Electrical characteristics in hydrogen and without hydrogen are investigated.

Chapter 8 produces a wast state-of-art of fin FET important nanoelectronics. Fin FET shows high mobility of the carriers, an improvement over the double gate. Details about current conduction and control over the channel are identified. SOI further improves the properties of this transistor. New trends in Fin FET technology are also presented in detail which would help readers to do future development in existing technology. Fin FET physical structure is evaluated intensively.

Chapter 9 provides the best example of the integration of electronics and optics. Such transistors are tunneling types those explained in previous chapters. Simulation results obtained by technology computer-aided design produced. Working and geometry are explained properly.

Chapter 10 produces self-powered photodetectors, a nanoelectronics device important for optical communication applications. Single nanobelt to pyro-photonic devices are presented, and different terminology associated with these photodetectors is defined. Quantum confinement in low dimensional materials and energy level in those illustrated. The mechanism of self-powered is also appropriately explained.

Chapter 11 solar cells attracted significant attention as fossil fuel ended and the economy expanded rapidly—nanomaterials and nanostructures are producing better solar cells and increasing the energy harvesting efficiency many folds. Different crystal structures are also illustrated that decide mobility, carrier transport, band gap, and other mechanical properties of any materials—solar cells are classified in first, second, and third generation depending on material composition efficiency and structure, *etc*.

Chapter 12 introduces the lead-free solar cell based on nanomaterials. Other methods used to harvest energy are also listed and discussed. Why nanomaterials are needed in energy harvesting is also elaborated.

Chapter 13 provides information on how nanomaterials are helpful in energy harvesting. Janus Materials, Van-der-Waals Structures, Chalcogenides Materials, and Organic nanomaterials are explained. Solar cell classification is done by considering materials that will help readers to select materials for specific solar cell fabrication.

Chapter 14 detailed the hybridization potentiality of the material for energy harvesting and storage. Conversion efficiency, fill factor, short circuit current density, *etc.*, are illustrated profoundly to provide an in-depth concept of solar cells. Many roots are listed to harvest energy cost-effectively. Many single solar cells connection in a solar panel are also discussed. Emphasis on electron-hole generation at the junction and transport to load. Equivalent circuits of solar cells would further enhance the reader's knowledge.

ACKNOWLEDGMENTS

Editors special thanks to almighty god for giving light to write a book on nanoelectronics devices: design, material, and application to serve scientific society. We sincerely thank all contributors, reviewers, and colleagues who made this project successful. Editor's special thanks go to Mr. B.S. Sannakashappanavar for his valuable time compiling the book in its present form. Editors also thank their M.Tech and Ph.D. students for their support while editing the book.

Gopal Rawat
SMIEEE, School of Computing and Electrical Engineering (SCEE), Indian Institute of Technology Mandi (IIT Mandi), Kamand-175075, Mandi, Himachal Pradesh, India

&

Aniruddh Bahadur Yadav
Department of Electronics and Communication Engineering
Velagapudi Ramakrishna Siddhartha Engineering College (Autonomous),
Kanuru Vijayawada Andhra Pradesh - 520007, India

List of Contributors

Abhishek Bhattacharjee	Department of ECE, Tripura Institute of Technology Narsingarh, Agartala-799015, India
Aswin Ramesh	Department of Physics, National Institute of Technology Calicut, Calicut-673601, Kerala, India
Amit Saini	Cadre Design Systems, Ghaziabad, India
Amandeep Singh	School of Electronics and Electrical Engineering, Lovely Professional University, Phagwara, Punjab, India
Ankita Saini	Department of Chemistry, P.D.M. University Bahadurgarh, Haryana, India
Anup Shrivastava	Computational Nano-Materials Research Laboratory (CNMRL), Indian Institute of Information Technology, Allahabad, India
Brinda Bhowmick	Department of Electronics and Communication Engineering, National Institute of Technology Silchar, Assam, 788010, India
Bhavesh Vyas	Department of EEE & ECE, K.R. Mangalam University, Gurgaon, Haryana, India
C. S. Suchand Sangeeth	Department of Physics, National Institute of Technology Calicut, Calicut-673601, Kerala, India
Dharmendra Singh Yadav	National Institute of Technology, Hamirpur, Himachal Pradesh, India
Gopal Rawat	SMIEEE, School of Computing and Electrical Engineering (SCEE), Indian Institute of Technology Mandi (IIT Mandi), Kamand-175075, Mandi, Himachal Pradesh, India
Ghanshyam Singh	Department of Electronics and Communication Engineerin, Malaviya National Institute of Technology Jaipur, Rajasthan, 302017, India
Hemant Kumar	Department of Electronics and Communication Engineering, Jaypee Institute of Information Technology, Noida, Uttar Pradesh, India
Jayesh Vyas	Department of Mechanical & Chemical Engineering, Indian Institute of Technology, Jammu, India
Jaismon Francis	Department of Physics, National Institute of Technology Calicut, Calicut-673601, Kerala, India
Jyoti Kandpal	Department of ECE, Graphic Era Hill University Dehradun, Uttarakhand, India
Koushik Dutta	PDPM Indian Institute of Information Technology, Design and Manufacturing, Jabalpur, India
Lintu Rajan	National Institute of Technology, Calicut, Kerala, 673601, India
Mohit Agarwal	Thapar Institute of Engineering and technology, Patiala, Punjab 147004, India
Prajwal Roat	National Institute of Technology, Hamirpur, Himachal Pradesh, India
Prabhat Singh	National Institute of Technology, Hamirpur, Himachal Pradesh, India

Puja Acharya	Department of EEE & ECE, K.R. Mangalam University, Gurgaon, Haryana, India
Puspa Devi Pukhrambam	Department of Electronics and Communication Engineering, National Institute of Technology Silchar, Assam, 788010, India
Rahul Ghosh	Department of ECE, Tripura Institute of Technology Narsingarh, Agartala-799015, India
Rupanjal Debbarma	Department of ECE, Tripura Institute of Technology Narsingarh, Agartala-799015, India
Sukanya Ghosh	National Institute of Technology, Calicut, Kerala, 673601, India
Shekhar Verma	Lovely Professional University, Jalandhar, Punjab, India
Sweta Chander	School of Electronic and Electrical Engineering, Lovely Professional University, Phagwara, Punjab, India
Sanjeet Kumar Sinha	School of Electronic and Electrical Engineering, Lovely Professional University, Phagwara, Punjab, India
Sunil Kumar Saini	Department of Chemistry, P.D.M. University, Bahadurgarh, Haryana, India
Sumeen Dalal	Department of Chemistry, P.D.M. University, Bahadurgarh, Haryana, India
Shivani Saini	Computational Nano-Materials Research Laboratory (CNMRL), Indian Institute of Information Technology, Allahabad, India
Sanjai Singh	Computational Nano-Materials Research Laboratory (CNMRL), Indian Institute of Information Technology, Allahabad, India
Shikha Kumari	PDPM Indian Institute of Information Technology, Design and Manufacturing, Jabalpur, India
Tanmoy Majumder	Department of ECE, Tripura Institute of Technology Narsingarh, Agartala-799015, India
Talapati Akhil Sai	PDPM Indian Institute of Information Technology, Design and Manufacturing, Jabalpur, India
Vishal Narula	Lovely Professional University, Jalandhar, Punjab, India
Vandana Devi Wangkheirakpam	Department of Electronics and Communication Engineering, Indian Institute of Information Technology Senapati, Manipur, India
Varun Goel	Department of Electronics and Communication Engineering, Jaypee Institute of Information Technology, Noida, Uttar Pradesh, India
Vineet Dahiya	Department of EEE & ECE, K.R. Mangalam University, Gurgaon, Haryana, India

CHAPTER 1

Role of Nanotechnology in Nanoelectronics

Jyoti Kandpal[1,*] and **Gopal Rawat**[2]

[1] *Department of ECE, Graphic Era Hill University, Dehradun, Uttarakhand, India*

[2] *SMIEEE, School of Computing and Electrical Engineering (SCEE), Indian Institute of Technology Mandi (IIT Mandi), Kamand-175075, Mandi, Himachal Pradesh, India*

Abstract: Nanotechnology is concerned with creating and applying materials with nanoscale dimensions in various facets of life. Additional features have been introduced to the world of electronics due to advancements in nanotechnology. The development and cost-effective manufacturing of cutting-edge components that function quickly, use less power and can be packed at much higher densities is made possible by nanotechnology's new and unique features. There is a revolution in biotechnology, food, the military, and medicine using nanotechnology.

Keywords: BJT, CMOS, ENIAC, FinFET, MOSFETs.

INTRODUCTION

Digital logic, which needs to offer a technology foundation for two different device types—high-performance logic and low-power/high-density logic—takes up a significant percentage of semiconductor device manufacture. Therefore, speed, power, density, price, capacity, and time to market are important factors for this technology platform. To preserve historical patterns of increasing device performance at lower power and cost while still operating in large volumes, the More Moore roadmap offers an enablement vision for further scaling of MOSFETs.

By European Nanoelectronics Initiative Advisory Council (ENIAC), the silicon-based micro or nanoelectronics industry can be analyzed utilizing three general categories as shown in Fig. (**1**) [1, 2].

---

*Corresponding author **Jyoti Kandpal**: Department of ECE, Graphic Era Hill University Dehradun Uttarakhand, India; E-mail: jayakandpal27@gmail.com

Gopal Rawat & Aniruddh Bahadur Yadav (Eds.)

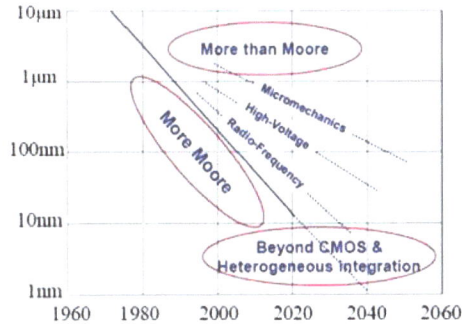

Fig. (1). ENIAC predication for microelectronics future.

1. **Advanced CMOS (More Moore)**: To continue the downsizing of transistors, particularly in the improved use of metal gates with suitable work functions, high-k oxides, and high-k oxides as insulators, to ensure an effective throughput while decreasing the leakage *via* gate stack.

More Moore targets bringing PPAC value for node scaling every 2–3 years [3]:

  a. (P)performance: >10% more operating frequency at scaled supply voltage.
  b. (P)ower: >20% less energy per switching at a given performance.
  c. (A)rea: >30% less chip area footprint.
  d. (C)ost:<30% more wafer cost 15% less die cost for scaled die.

2. **More than Moore**: Modern CMOS technology has demonstrated itself as inherently constrained. Radiofrequency, analogue circuits, switches with high-voltage, actuators, and motion sensors are non-digital functions that call for a combination of technologies customized to a particular need. To overcome these obstacles and implement new features like mechanics, optics, acoustics, ferroelectrics, *etc.*, "more than Moore" is needed.

The relevance of more-than-Moore devices, which combine performance, integration, and cost without being restricted to CMOS scaling, will continue to expand. These devices include MEMS, power electronics, CMOS image sensors, and R.F. devices. There are four standards more–than-Moore emphasizes [4].

  a. System On a Chip (SoC)
  b. System On a Package (SoP)
  c. System In a Package (SIP)
  d. Multiple Chip Module (MCM)

3. **Beyond CMOS.** New materials, whether inorganic or organic, are covered with new operating principles, such as those that replace electrons with magnetic excitation or spin and unique architectural designs. Examples of alternatives beyond CMOS include new materials for interconnects and transistors such as nanowires and carbon nanotubes, switches working with resistive change polymers for memories, the electronic properties of organic molecules, and memory and computing architectures to utilize the capabilities of these new devices fully.

The designing of electronic systems on a single chip requires optimum exploitation of "system-on-chip" devices, and it will be a vital component of the future of nanoelectronics, integrating "More Moore" and "More than Moore" with new "heterogeneous integration" technologies. Another advancement is "system-in-package", which uses different optimized process technologies for combining multiple distinct sub-systems in a single package.

WHAT IS NANOSCALE?

The word "nano" originated from the Greek word "Nanos," which means " tiny, dwarf, or exceedingly small". As per the International Systems of Units, the prefix "nano" represents one billionth or 10^{-9}. It means one nanometer is one billionth of a meter [4]. Fig. (**2**) shows the visual illustration example of nanoscale:

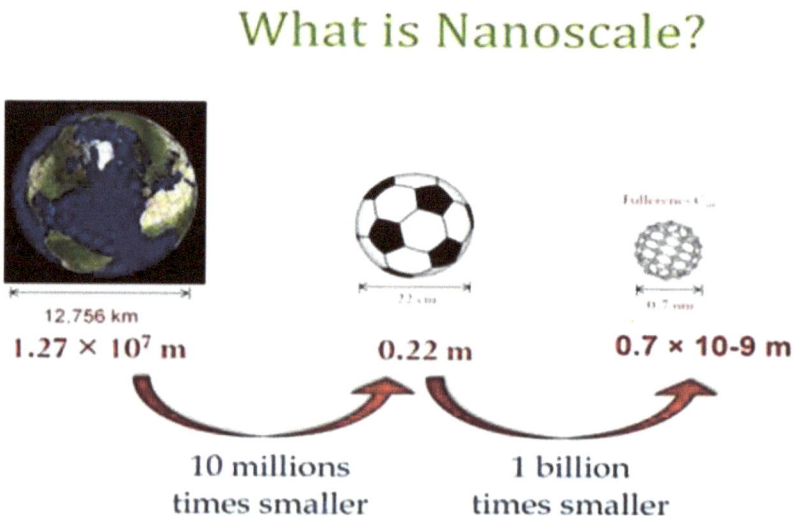

Fig. (2). A Nanometric Scale.

WHAT IS NANOTECHNOLOGY?

Nanotechnology is a research branch exploring the potential for altering matter at the nanoscale to create new materials or cutting-edge gadgets that will serve human needs in various disciplines. Nanotechnology is also defined as the study of incredibly tiny objects and the applications in which they are employed.

Further, the study of materials that display extraordinary qualities, functions and phenomena due to their small dimensions is known as nanoscience. Nanoscience and nanotechnology focus on groups of atoms smaller than 1 nm, at least in one size. As a result, many goods appropriate to a wide range of scientific fields can be created using. Additionally, nanotechnology, which deals with material measuring less than 1 mm, is termed "creation," "exploitation," and "synthesis".

Materials at the nanoscale are used in nanotechnology, which is still a relatively new field for researchers. However, commercial nanotechnology applications are overgrowing and can significantly advance humankind's ability to safeguard the environment. These technologies are applied in novel and intriguing ways in various industries, including producing computer chips, textiles, environmental cleanup, and medicine. Further, it is essential to emphasize that nanoscience and nanotechnology are not brand-new scientific discoveries. For instance, the development of immunology during the past 50 years has given rise to several new problems. With ongoing advancements, researchers were attracted to molecular biology and genomics, a branch of neuroscience and microtechnology. The current focus of nanotechnology is on materials with a size between 1 and 100 nm. Due to material advancements, researchers can manage the accuracy and precision of novel nanomaterials used at these sizes. Similarly, new uses made possible by distinct nanoscale phenomena were not possible while working with bulk materials made up of molecules or even a single atom [4 - 6].

HISTORY AND ADVANCEMENTS IN NANOTECHNOLOGY FIELD

Heinrich Rohrer is acknowledged as the originator of nanotechnology. However, American Nobel laureate and physicist Richard Feynman were the first to discuss its potential uses in 1959 at the California Institute of Technology (Caltech). As a result, this region consolidated was promoted and came into its own in the twenty-first century. It encompasses other disciplines, such as molecular biology, organic chemistry and micro-manufacturing. Towards economic expansion of this sector and to gain profit, the National Nanotechnology Initiative (NNI) invested more than 18 billion dollars in the U.S. between 2001 and 2022. Table **1** shows the development in the field of nanotechnology [7].

Table 1. Timeline showcasing the developments in the field of Nanotechnology [7].

Year	Developments
1959	Famous physicist Richard Feynman lecture on Nanotechnology - There is Plenty of Room at the Bottom.
1974	Norio Taniguchi created the concept of Nanotechnology.
1977	K. Eric Drexler presented the idea of molecular Nanotechnology.
1981	The scanning tunnelling microscope is created by IBM (STM).
1985	Bucky Ball (Fullerenes) was found.
1986	The phrase "Nanotechnology" was first coined by K. Eric Drexler in his book "Engines of Creation: The coming era of Nanotechnology" and Atomic Force Microscopy (AFM) was invented.
1991	SumioLijima found carbon nanotubes, and MITI Japan announced the bottom-up "atom factory" concept.
1997	James R. Von Her II established the first molecular Nanotechnology corporation, Zyex. Also, the nanorobotics system's initial design was implemented.
1999	R. Freitas released the first book on nanomedicine titled NanoMedicine. In addition, safety regulations for Nanotechnology were introduced.
2000	National Nanotechnology Initiative (NNI) commenced as an R&D initiative.
2001	Award of Feynman Prize for "Theory of Nanometer-scale Electronic Devices" and "The Synthesis of Carbon Nanotubes and Nanowires".
2002	Feynman Prize for DNA enabling the self-assembly utilized for the new structure and works on a model molecular machine system.
2003	Feynman Prize was given for the integration of biological molecular motors with silicon devices.
2005	Active Nanostructures like 3D transistors, amplifiers, targeted drugs, adaptive structure.
2010	We integrated Nano systems like 3D networking and new hierarchical architecture, and robotics.
2015-2020	Molecular Nanosystems Ex. Molecular Device by atomic design, emerging function.
2022	TSMC introduce the 3nm node with 250 million transistors per square millimetre of silicon.

CATEGORIES OF NANOMATERIAL

Based on their shape and size, nanomaterials are divided into four different categories - zero, one, two and three dimensions. How dimensions are used to categorize nanomaterials is explained in the subsequent paragraphs, and parameters are defined in Table **2** [6 - 9].

Table 2. Density of electrons in different degrees of freedom.

Degree of Freedom	Density of state	Effective Density of States
Three	$\rho_{DOS}^{3D} = \dfrac{1}{2\pi^2}\left(\dfrac{2m^*}{\hbar^2}\right)^{\frac{3}{2}}\sqrt{E - E_C}$	$N_c^{3D} = \dfrac{1}{\sqrt{2}}\left(\dfrac{m^* kT}{\pi\hbar^2}\right)^{\frac{3}{2}}$
Two	$\rho_{DOS}^{2D} = \dfrac{m^*}{\pi\hbar^2}\,\sigma(E - E_C)$	$N_c^{2D} = \dfrac{m^*}{\pi\hbar^2}\,kT$
One	$\rho_{DOS}^{1D} = \dfrac{m^*}{\pi\hbar}\sqrt{\dfrac{m^*}{2(E - E_C)}}$	$N_c^{1D} = \sqrt{\dfrac{m^* kT}{2\pi\hbar^2}}$
Zero	$\rho_{DOS}^{0D} = 2\delta(E - E_C)$	$N_c^{0D} = 2$

Where m* is the effective mass, E_c is energy of the conduction band, ħ is Planck constant, N_c is Electron density at the conduction band.

a. *Zero-dimensional*: The nanomaterial with all three dimensions in the nanoscale are termed zero-dimensional (0D) nanomaterials. N.P.s made of metals like palladium, gold, silver, platinum or quantum dots are examples. With a diameter of 1 to 50 nm, N.P.s can be spherical. It has been noted that some cube and polygon forms are 0D nanomaterials.
b. *One-dimensional*: These nanomaterials have three dimensions, with two being macroscale and one being in the range of 1-100 nm. One-dimensional nanomaterials (1D) include nanowires, nanotubes, nanofibers, and nanorods. In addition, quantum dots, metals like Ag, Au and Si, metal oxides such as TiO_2, ZnO, CeO_2 and other materials can produce one-dimensional nanostructures.
c. *Two-dimensional*: This class of nanomaterials has two dimensions at the nanoscale and one at the macroscale. Two-dimensional (2D) nanomaterials include nano-thin films, nanosheets, nano walls, and nanofilm multilayers. 2D nanomaterial can contain many square micrometres of surface area while having a thickness in the nanoscale range.
d. *Three-dimensional*: In three-dimensional (3D) nanomaterials, there are only macroscale and no nanoscale dimensions. Blocks, which can be as small as one nanometer or as large as one hundred nanometers or more, are the minor units that make up bulk materials, which are 3D nanomaterials.

Fig. (**3**) shows the density of electron states in bulk semiconductor structures in zero, one, two and three dimensions.

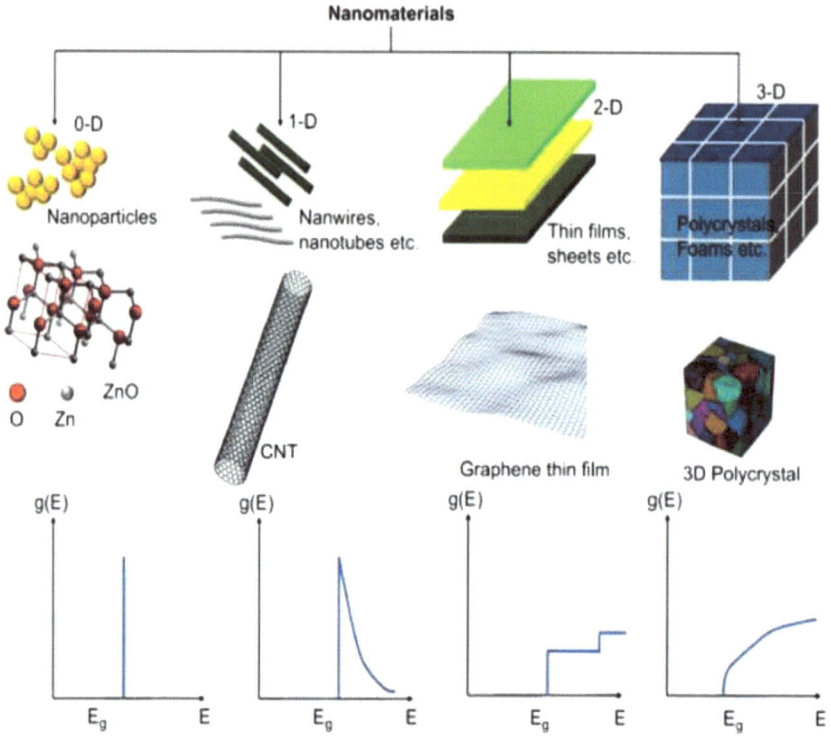

Fig. (3). Representation of zero, one, two, and three-dimensional nanomaterials & variation of electron density in the semiconductor by changing the dimension [8 - 12].

Various techniques are available for nanomaterials in different forms like nanorods, nanowires, nanotubes, nanoclusters, thin films, colloidal N.P.s *etc*. In addition, nanomaterials can be made by modifying conventional techniques. A synthesis technique can be selected based on the type of the desired nanomaterial, its size and quantity.

APPROACHES IN NANOTECHNOLOGY AND FABRICATION

Different forms of nanotechnology exist depending on how they operate (top-down or bottom-up), and the media (dry or wet) used.

Proceed Based Approach

There are two main approaches to synthesizing nanostructured material: Top-down and Bottom-up, as shown in Fig. (**4**) [13].

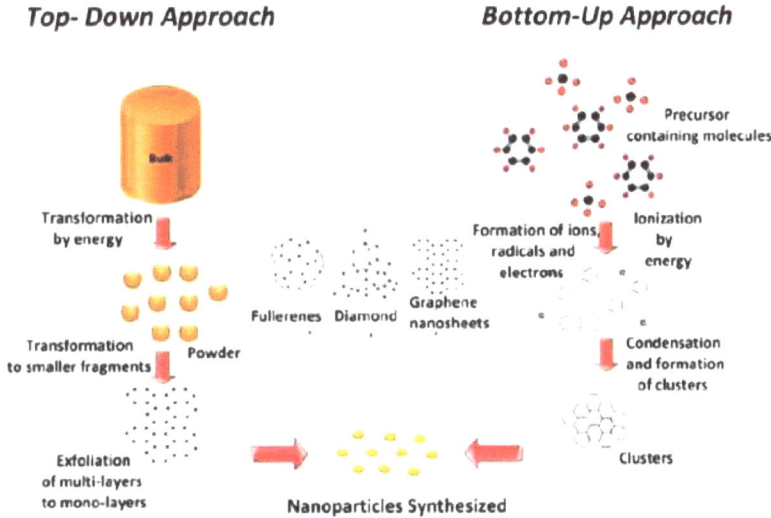

Fig. (4). Methods for the synthesis of nanostructures material.

Top-down Approach (A materials perspective - Larger to Smaller) By physically or chemically dissolving more extensive materials, one can produce nanoscale materials. As the size of the system shrinks, a variety of physical phenomena become more noticeable. Quantum mechanical effects are among them, as well as statistical mechanical effects. The nanoelectromechanical systems (NEMS), linked to microelectromechanical systems (MEMS), can also be manufactured utilizing solid-state processes. Once MEMS could be produced utilizing modified semiconductor device production techniques, typically used to create electronics, they became viable.

Bottom-Up Approach (A molecular perspective - Simple to Complex) With advancements in synthetic chemistry, any structure can be created for tiny molecules. Today, many valuable compounds, including medicines and commercial polymers, are produced using these techniques. "molecular nanotechnology," sometimes known as "molecular manufacturing," refers to molecularly scaled constructed Nanosystems. Based on the principles of mech a synthesis, a molecular assembler can build the desired structure or gadget atom by atom.

Medium Based Approaches

Depending on the medium, nanotechnologies are divided into Wet, Dry and Computational categories [14, 15].

Organic things, including tissues, membranes, enzymes, and other cellular components, are related to wet nanotechnology.

The study of physical and the creation of materials like silicon and carbon (*i.e.*, inorganic) are covered under dry nanotechnology.

Computational nanotechnology deals with simulations of structures having a size of a nanometer. For optimal functionality, wet, dry and computational technologies depend on each other. The dependency of the three nanotechnologies, as mentioned earlier, is shown in Fig. (**5**).

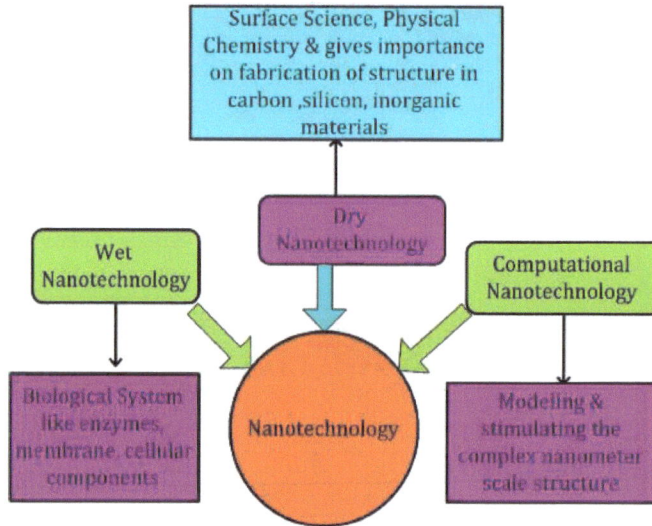

Fig. (5). Different types of Nanotechnology based on different synthesis mediums.

NANOTECHNOLOGY IN NANOELECTRONICS

The study of procedures and material handling at atomic, molecular and macromolecular scales, where a significant change in the properties takes place compared to a larger scale, is known as nanoscience. Nanotechnologies are the "practical" gadgets created by using nanoscience.

When we use nanotechnology in electronic components, it is referred to as "Nanoelectronics". It has many uses in computing and producing electronic goods like iPod Nanos' Flash memory chips, mouse, keyboards and cell phone castings with antimicrobial and antibacterial coatings.

The main areas of study in nanoelectronics technology include classification, direction and synthesis of electronic components of nanoscale size. To overcome scalability restrictions, a nanoelectronics device is a tiny device. As a result, electronics' size, weight and power consumption may be reduced while their functionality is increased, according to Nanoelectronics [16, 17]. Although "nanotechnology" refers to the use of technology fewer than 100 nanometers in size, nanoelectronics frequently refers to tiny transistors; therefore, quantum mechanical properties and interatomic interactions must be thoroughly and in-depth explored. Options for nanoelectronics include enhanced molecular electronics, hybrid molecular/semiconductor electronics, silicon nanowires, and carbon nanotubes.

Nanoscience and Nanotechnology have a clear distinction. The first is related to the study of objects of size varying from 1 nm to 100 nm, at least in one dimension. However, Nanotechnology uses these tiny substances to fabricate products with some practical applications. Manipulation, regulation and integration of atoms and molecules are essential in developing materials, components, structures and devices at the nanoscale level. The inclusion of nanomaterials can improvise the materials used by industrial sectors. On the ground, the factor which will decide the utility of nanotechnologies for the industrial sector is cost versus added benefit [18 - 20].

Requirement of Nanotechnology in Electronics

Today microelectronics is used and resolves the utmost of problems. However, the exceptional disadvantages of microelectronics are:

a. We are optimizing electronic gadgets' displays. This helps make the screens lighter and thinner while also consuming less power.
b. Memory chip densities are being raised. For example, one terabyte of memory is expected to fit onto each square inch of a memory chip that scientists are currently working on designing.
c. In integrated circuits, transistors are shrunk. As a result, according to studies, the power of all today's computers might be placed in the palm of a hand.
d. Lower power usage.

Impact of Nanotechnology on Electronics

In this section, we will discuss the impact of nanotechnology on electronics:

a. Many communications, computers, and electronic applications make use of nanotechnology. These applications enable quicker, more compact, and more portable systems that can handle and store ever-increasing volumes of data. The purpose of nanoelectronics is to transmit, process and store information using characteristics of materials different from their macroscopic properties.

b. Nanotechnology is also utilized for printed electronics such as smart cards, RFID and smart packaging, video games and flexible displays for the readers of the e-book. In addition, nanotechnology has made high-speed and energy-efficient nanoscale transistors possible.

c. Organic Light Emitting Diodes (OLEDs) are nanostructured polymer films utilized in many present T.V.s, laptop computers, digital cameras, and cell phones. OLED screens consume less power and have a longer lifespan than traditional LCD screens.

d. Nanotechnology facilitates Magnetic Random-Access Memory (MRAM). This can quickly and effectively save the data, including encrypted data, in case of a shutdown or crash of a system.

e. Introducing nanomaterials, printable and flexible electronics, quantum computing, data storage, magnetic nanoparticles for data storage and nanotechnology in electronics enables faster, more accessible, smaller, and better handheld devices.

f. Several electronic devices, processes and applications, including nano diodes, nano transistors, plasma displays, OLEDs and quantum computers, can be transformed with nanotechnology.

g. The advent of nanotechnology in electronics has enhanced the functionality of devices due to improvements in the density of memory chips and decreasing the transistor size used in integrated circuits. As a result, electronic devices have reduced weight and consume less power.

h. Electronic devices with improved display screens using less power, lesser weight, and thinner screens are available.

Advantages of using Nanotechnology in Electronics

Electronics using nanotechnology are more portable, smaller, and faster. Nanoelectronics improves the density of memory chips, expands the functionality of electronic devices, and decreases power consumption and transistor size in integrated circuits. Nanotechnology is essential to communication engineering, has several uses, and can have a variety of effects on the telecommunications sector [21 - 23].

There are the following advantages which are listed below:

a. The density of the memory chip increased.
b. The thickness and weight of screens decreased.
c. Nanolithography is used for the fabrication.
d. Reducing the size of the transistor in integrated circuits.
e. Improvement in the display screen of electronic devices.
f. Reduction in power consumption.

NANOELECTRONICS DEVICES

The nanoelectronics devices and systems are producing emerging nanoelectronics devices and systems that will power the next generation of electronics. It includes vertically integrated activities ranging from incorporating functional materials into cutting-edge nanoelectronics devices to investigating fascinating applications in numerous fields. Such coordinated efforts may open the door to the Hyper-Scaling era of scaling.

Beyond the silicon CMOS device scaling roadmap: logic devices, the project's scope includes innovative nanoelectronics materials, such as graphene and carbon nanotube, novel nanoelectromechanical (NEM) relays concepts, device physics, modelling, and circuit design and device fabrication. Fig. (**6**) presents different types of Nanoelectronics devices.

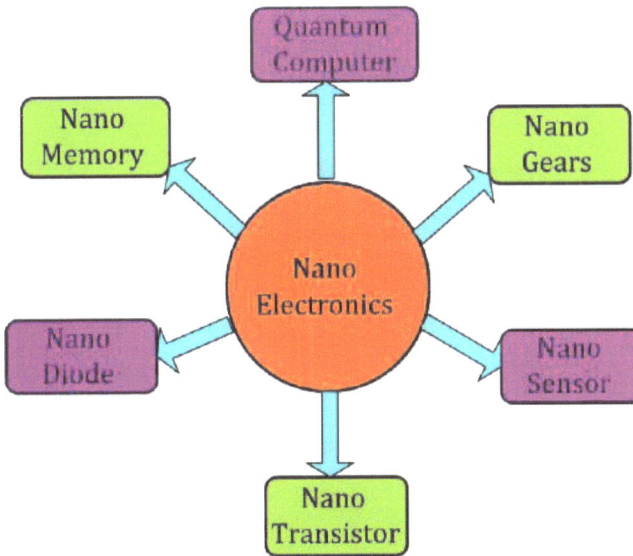

Fig. (6). Different types of nano electronics devices.

Nanoelectronics Transistors

While nanotechnology has already been subtly integrated into current high-technology production processes employing nanoelectronic transistors, these processes are still dependent on conventional top-down approaches. Computer processors could potentially become more potent with nanoelectronics than is currently conceivable with traditional semiconductor production methods. Numerous strategies are being investigated right now, such as novel types of nanolithography and the substitution of conventional CMOS components with nanomaterials like nanowires or tiny molecules [24].

Both heterostructured nanowires and semiconducting carbon nanotubes have been used to create field-effect transistors. The MOSFET transistor with an 18 nm diameter was tested in 1999 by a CMOS transistor designed at the Grenoble; a France-based Laboratory for Electronics and Information Technology. Approximately 70 atoms were placed side by side in this design. The smallest industrial transistor in 2003 was almost ten times smaller than this (130 nm in 2003, 90 nm in 2004, 65 nm in 2005 and 45 nm in 2007). Theoretically, to fit seven billion connections on a coin of 1 Euro was made feasible. In 1999, the CMOS transistor was invented. It demonstrated the function of CMOS technology when operated on a molecular scale. Executing the same on an industrial scale and controlling the coordinated assembly of these transistors on a circuit have been challenging. Different device architecture is presented in the IRDS more Moore roadmap, as shown in Fig. (**7**) [3].

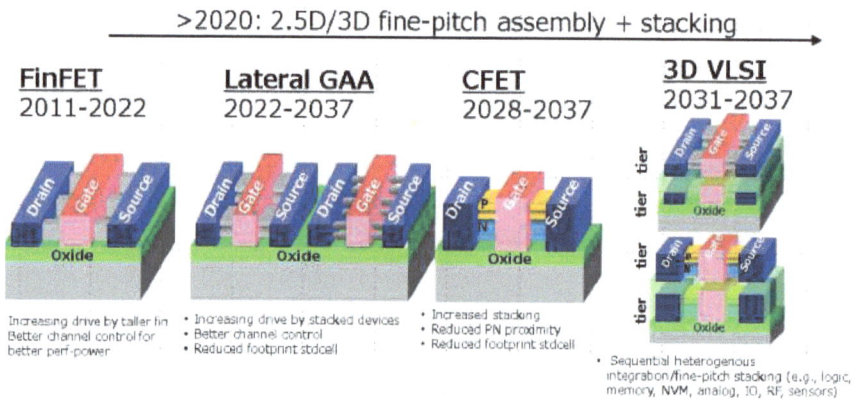

Fig. (7). Evolution of device architectures in the IRDS More Moore roadmap [3].

Nano Memory (Spintronic)

The study of and application of magnetic moment and electric charge with electron spin in solid-state devices is known as spintronics.

Spintronics are electronic components that execute logic operations based on a carrier's spin and electrical charge. For instance, the spin-up or spin-down states of electrons could transfer or store information. The injection, transport, and detection of spin-polarized carriers are among the problems in this recent field of study. Current research includes the influence of the ferromagnetic-semiconductor interface's electrical structure and nanoscale structure on the spin injection process, the controlled creation of new ferromagnetic semiconductors and the potential application of nanostructured features to spin manipulation [25].

By the IRDS 2022 stacking structure of 3D DRAM cell, the stacking design of 3D Flash technology is used. In the 2022 report is stated that 96 staked layers are already in volume production, and 128 layers are achievable.

FeRAM is used for radio frequency identification (RFID), Smart cards, ID cards, and other embedded applications. Recently HfO_2 based ferroelectric field effect transistor (FET) memory element has been used for low power and fast switching [3]. Spin transfer torque magnetic RAM (STT-MRAM) is used for low-power IoT applications.

Nano-Sensors

Numerous products, exciting disciplines and uses for photonic sensors are made possible by nanotechnology. Existing applications can be improved, such as digital cameras view accommodation of more pixels on a sensor compared to what is currently possible. Additionally, sensors can be created at the nanoscale, improving their quality and possibly eliminating defects. Ultimately, this would lead to larger, more precise pictures. A communication network will use photonic sensors to transform optical data into electricity (*i.e.*, photons to electrons transformation). Photonic sensors made at the nanoscale will be more effective and benefit from the same advantages as other nanoscale materials [26, 27].

PROGRESS IN THE FIELD OF NANOELECTRONICS

Significant developments are happening in the field of nanoelectronics each day. Details of a few projects/ developments are as follows:

1. The University of Michigan has developed a method for growing hexagonal boron nitride in single layers atop graphene. This technique may provide extremely thin graphene wafer-level sheets, insulated by very thin boron nitride, since the latter can be utilized as an insulator.

2. Harvard researchers used nanofabrication techniques to create a laser for lithium niobate photonic circuits.

3. In comparison to other designs of a similar size, NIST researchers have built a Light Emitting Diode (LED) with fin-like zinc oxide nanostructures. The scientists also found that the above structure emits laser light to increase the current.

4. CMOS integrated circuits with the integration of silicon nanophotonics components. A higher rate of speed for data transmission between integrated circuits can be achieved by utilizing this optical technique.

5. Scholars at UC Berkeley presented a low-power method to use nanomagnets as switches like transistors in electrical circuits. It is feasible that this power consumption in electrical circuits can be lowered than in transistor-based circuits.

6. Scholars at Georgia Tech, the University of Tokyo and Microsoft Research have devised a methodology wherein standard inkjet printers print prototype circuit boards. The conductive lines of the circuit boards are made with Silver nanoparticle ink.

7. The Caltech researchers have developed a nanopatterned silicon surface used in a laser. It produces light with significantly better frequency control than existing ones. Using this, information transmission across fibre optics would happen at considerably faster data speeds.

8. They are creating carbon nanotube transistors to enable a few nanometers minimum transistor size and developing methods to produce integrated circuits containing nanotube transistors.

9. Stanford University researchers have designed a technique for producing functional integrated circuits utilizing carbon nanotubes.

10. Lead-free solder has been developed using copper nanoparticles for space missions and other high-stress conditions.

11. Production of more slim and flexible flat panel displays than existing ones by utilizing electrodes composed of nanowires.

12. Using semiconductor nanowires, transistors and integrated circuits are designed.

13. An innovative technique for creating graphene P.N. junctions, a crucial part of transistors, has been devised by researchers. First, the p and n areas of the substrate were patterned. Then, electrons were either supplied or removed from the graphene, depending on the mechanism in which the substrate was doped when the graphene layer was deposited. According to the researchers, the disturbance of the graphene lattice that can happen with other approaches is reduced by this method.

14. Nanoparticle Organic Memory Field-Effect Transistor (NOMFET) transistors are produced by combining an organic compound with nanoparticles of gold.

15. It is creating a thin, millimetre-thick "nano emissive" display panel that is lightweight by using carbon nanotubes. It directs the electrons toward illuminating pixels.

16. Construct integrated circuits with nanometer-scale characteristics, such as the method used to create integrated circuits with transistor gates 22 nm wide.

17. Magnetoresistive Random-Access Memory is created using magnetic rings with a nanoscale (MRAM).

18. Researchers developed a magnetoelectric random access memory technique that uses nanoscale magnets to produce lower power and higher density (MeRAM).

19. They are making nanoscale integrated circuits with self-aligned nanostructures.

20. Construction of transistors without p-n junctions using nanowire.

21. With the help of bucket balls, dense and low-power memory systems are constructed.

22. Spintronic semiconductor devices were developed by using magnetic quantum dots. Contrary to current semiconductor devices, which measure electronics groups, spintronic devices evaluate the spin of electronics to decide a 1 or 0. Therefore, these are considered to have higher density and lesser power consumption.

23. Construction of compact memory devices by using nanowires made of iron-nickel alloyed. Magnetized sections are developed along the wire's length by applying a current. As magnetic portions move across the wire length, the data is read by sensors. This technique is termed Racetrack Memory.

24. IMEC and Nantero are designing a memory chip made of carbon nanotubes. This is termed Nanotube-based Nonvolatile Random-Access Memory (NRAM) and is aimed to replace compact Flash memory chips.

25. Researchers have developed organic nano glue. This creates a nanometer-thick coating between a computer chip and a heat sink. The research findings show that applying this nano glue increases the thermal conductivity between the components above. This helps maintain the temperature of other features, such as chips.

APPLICATION OF NANOTECHNOLOGY

The applications of Nanotechnology are vast and can be grouped as shown in Fig. (**8**).

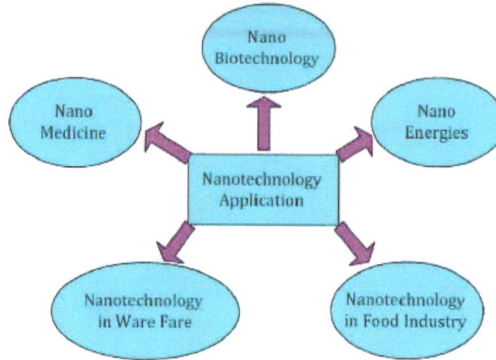

Fig. (8). Application of nanotechnology.

Nano Medicines

The use of nanotechnology in medicine is called Nanomedicine. Though nanomedicine is in its starting phase, revolutionary changes in the healthcare sector are anticipated utilizing it. Public investment has significantly assisted the advancement of nanomedical research. The emerging developments in this field can provide advantages such as enhanced efficacy, bioavailability, dose-response, targeting capability, personalization, and safety over existing medications. One of the most remarkable ideas in nanomedical research is the construction of multifunctional nanoparticle (N.P.) complexes; these may simultaneously carry diagnostic and therapeutic chemicals to desired areas. The main objective of nanotechnology in this field is to observe and improvise the human biological systems working from the molecular level. Details of a few nanotechnology-based drugs that are commercially available or in human clinical trials are tabulated in Table **3** [28 - 30].

Table 3. Nanotechnology-based Drug [31].

Ser	Product	Approved in Year	Indication
(a)	Adagn	1990	Severe combined immunodeficiency disease
(b)	Oncaspar	1994	Acute lymphoblastic leukaemia
(c)	Abelcet	1995	Invasive fungal infections
(d)	Doxil/Caelyx	1995	Various cancers
(e)	Daunoxome	1996	HIV-related Kaposi's sarcoma
(f)	Copaxone	1996	Multiple sclerosis
(g)	AMBiSome	1997	Fungal and protozoal infections
(h)	Inflexal	1997	Influenza
(i)	Depocyt	1999	Lymphomatous meningitis

(Table 3) cont.....

Ser	Product	Approved in Year	Indication
(j)	Visudyne	2000	Wet age-related macular degeneration
(k)	Renagel	2000	Chronic kidney disease
(l)	PegIntrom	2001	Hepatitis c
(m)	Eligard	2002	Advanced prostate cancer
(n)	Neulasta	2002	Chemotherapy-induced neutropenia
(o)	Pegasys	2002	Hepatitis c
(p)	Somavert	2003	Acromegaly
(q)	DepoDor	2004	Relief of postsurgical pain
(r)	Macugen	2004	Neovascular age-related macular degeneration
(s)	Abraxane	2005	Breast cancer
(t)	Mircera	2007	Symptomatic anaemia associated with chronic kidney disease
(u)	Cimzia	2008	Crohn's disease

Nanotechnology Applications in Nanomedicine

Fig. (**9**) shows applications of Nanotechnology in Nanomedicine.

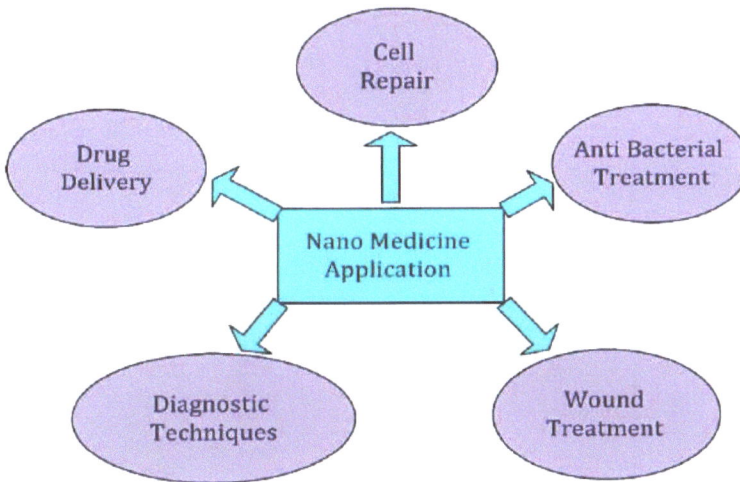

Fig. (9). Application of nano medicine.

Benefits

Nanomedicine has various benefits such as reduced side effects, better efficiency, simple and fast detection of diseases and easy cure without surgery.

Nano Biotechnology

The interdisciplinary field of nanotechnology covers advancements in Engine-ering, Chemistry, Biology and Physics. The intersection of nanotechnology and biology is termed nanobiotechnology. It is crucial for developing molecular electronics, food processing and medicine. Therapeutic and diagnostic applications are the two categories of nanotechnology applications in nanobiotechnology [32, 33].

Nano-biotechnology, which is valuable for numerous applications, especially in healthcare and pharmaceuticals, places a lot of emphasis on creating objects and biological systems at the atomic level. The development of nanotechnology depends on freshly made nanomaterials, which dominate all fields due to their distinctive features. In industries such as drug delivery, diagnostics, cosmetics, tissue engineering, and agriculture, including biological components with nanoparticles is crucial. Many nanomaterials are influencing the nano-biotechnology industries of agriculture, tissue engineering, and medicine, as shown in Figs. (**10** and **11**).

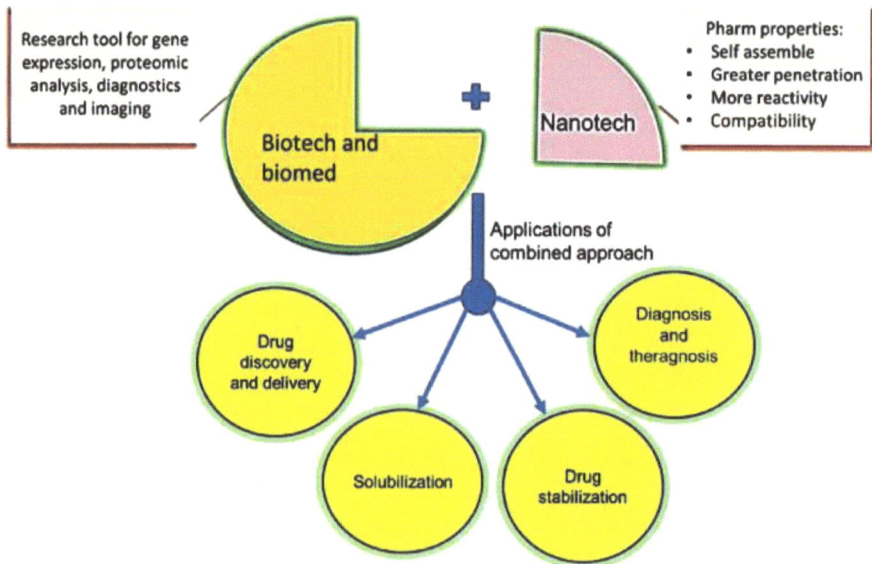

Fig. (10). Combined application of nanotech with biotech [32].

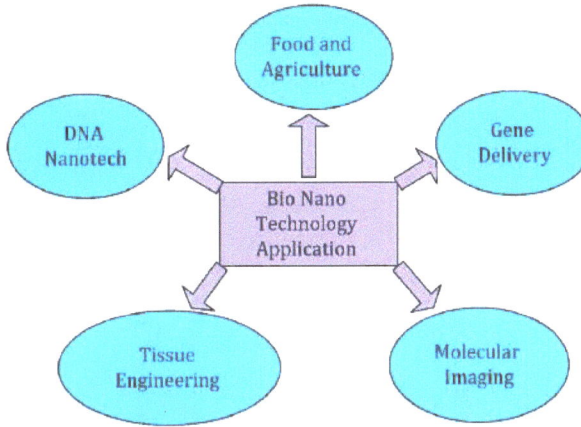

Fig. (11). Application of nano biotechnology.

Also, there are following challenges occur when biotechnology combines with nanotechnology:

1. Monitored exposure to Nano Materials.

2. Applicable Method Development.

3. Predict the effects of nanomaterial.

4. Health impact of nanomaterial.

5. Measure the risk of health.

Applications of Nanotechnology in Warfare

Nanotechnology's implementation in warfare is a specific subfield or branch of nanoscience. This field deals with developing molecular systems constructed and built to be nanoscale compatible. It has facilitated the design of various nano-weapons across wide-ranging categories such as miniature robotic robots, hyper-reactive explosives and super electromagnetic materials. In addition, chemical warfare agents such as choking agents, nerve and blood agents and incapacitating agents have also been developed.

The recent military nanotechnological weapon research focuses on developing defensive military equipment to improve extant designs of light, resilient and flexible materials. By using sensors and manipulating electromechanical qualities, these state-of-the-art systems have elements that work for the betterment of offensive tactics as well. In the past 20 years, nations like China, U.K., Russia and the U.S. have swiftly supported this technology for military use during the last 20 years [34, 35].

The application of Nanotechnology in the field of the military is covered in the following paragraphs:

Nanotechnologies can enhance efficiency across all industry sectors through innovative technology solutions and improved production technologies. As a result, it will influence the production of renewable energy economically. In Fig. (**12**), the energy industry's value chain is impacted by developments in nanotechnology [36, 37].

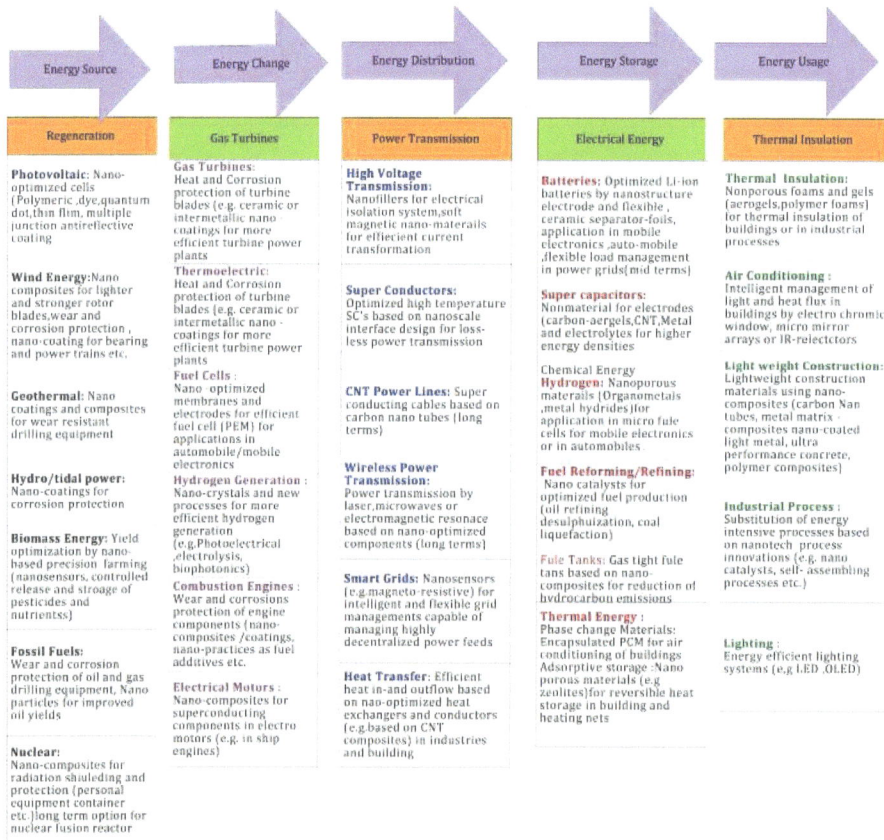

Energy Source	Energy Change	Energy Distribution	Energy Storage	Energy Usage
Regeneration	**Gas Turbines**	**Power Transmission**	**Electrical Energy**	**Thermal Insulation**
Photovoltaic: Nano-optimized cells (Polymeric ,dye,quantum dot,thin film, multiple junction antireflective coating	**Gas Turbines:** Heat and Corrosion protection of turbine blades (e.g. ceramic or intermetallic nano coatings for more efficient turbine power plants	**High Voltage Transmission:** Nanofillers for electrical isolation system,soft magnetic nano-materails for effiecient current transformation	**Batteries:** Optimized Li-ion batteries by nanostructure electrode and flexible , ceramic separator-foils, application in mobile electronics ,auto-mobile ,flexible load management in power grids(mid terms)	**Thermal Insulation:** Nonporous foams and gels (aerogels,polymer foams) for thermal insulation of buildings or in industrial processes
Wind Energy: Nano composites for lighter and stronger rotor blades,wear and corrosion protection , nano-coating for bearing and power trains etc.	**Thermoelectric:** Heat and Corrosion protection of turbine blades (e.g. ceramic or intermetallic nano - coatings for more efficient turbine power plants	**Super Conductors:** Optimized high temperature SC's based on nanoscale interface design for loss-less power transmission	**Super capacitors:** Nonmaterial for electrodes (carbon-aergels,CNT,Metal and electrolytes for higher energy densities	**Air Conditioning :** Intelligent management of light and heat flux in buildings by electro chromic window, micro mirror arrays or IR-relectctors
Geothermal: Nano coatings and composites for wear resistant drilling equipment	**Fuel Cells :** Nano -optimized membranes and electrodes for efficient fuel cell (PEM) for applications in automobile/mobile electronics	**CNT Power Lines:** Super conducting cables based on carbon nano tubes (long terms)	Chemical Energy **Hydrogen:** Nanoporous materails (Organometals ,metal hydrides)for application in micro fule cells for mobile electronics or in automobiles	**Light weight Construction:** Lightweight construction materials using nano-composites (carbon Nan tubes, metal matrix - composites nano-coated light metal, ultra performance concrete, polymer composites)
Hydro/tidal power: Nano-coatings for corrosion protection	**Hydrogen Generation :** Nano-crystals and new processes for more efficient hydrogen generation (e.g.Photoelectrical ,electrolysis, biophotonics)	**Wireless Power Transmission:** Power transmission by laser,microwaves or electromagnetic resonace based on nano-optimized components (long terms)	**Fuel Reforming/Refining:** Nano catalysts for optimized fuel production (oil refining desulphuization, coal liquefaction)	**Industrial Process :** Substitution of energy intensive processes based on nanotech process innovations (e.g. nano catalysts, self- assembling processes etc.)
Biomass Energy: Yield optimization by nano-based precision farming (nanosensors, controlled release and strooge of pesticides and nutrientss)	**Combustion Engines :** Wear and corrosions protection of engine components (nano-composites /coatings, nano-practices as fuel additives etc.	**Smart Grids:** Nanosensors (e.g.magneto-resistive) for intelligent and flexible grid managements capable of managing highly decentralized power feeds	**Fule Tanks:** Gas tight fule tans based on nano-composites for reduction of hydrocarbon emissions	
Fossil Fuels: Wear and corrosion protection of oil and gas drilling equipment, Nano particles for improved oil yields	**Electrical Motors :** Nano-composites for superconducting components in electro motors (e.g. in ship engines)	**Heat Transfer:** Efficient heat in-and outflow based on nao-optimized heat exchangers and conductors (e.g.based on CNT composites) in industries and building	**Thermal Energy :** Phase change Materials: Encapsulated PCM for air conditiuning of buildings Adsorptive storage :Nano porous materials (e.g zeulites)for reversible heat storage in building and heating nets	**Lighting :** Energy efficient lighting systems (e.g LED ,OLED)
Nuclear: Nano-composites for radiation shiuleding and protection (personal equipment container etc.)long term option for nuclear fusion reactor				

Fig. (12). Implémentation of Nanotechnology in the Energy Industry [36].

Contribution of Nanotechnology to the Food Sector

The food industry has changed because of nanotechnology, which has gained popularity over the past few decades. Researchers are driven to investigate ways to improve food quality while having the least possible impact on the product's nutritional value by the growing customer concerns about food quality and health advantages. Since many nanoparticle-based products include vital nutrients and are non-toxic, their demand has surged in the food business. Nanotechnology offers total food solutions, from production to processing and packaging. Nanomaterials significantly improve food quality and safety as well as the health benefits that food offers. Fig. (**13**) presents the application of Nanotechnology in the food industry [38 - 43].

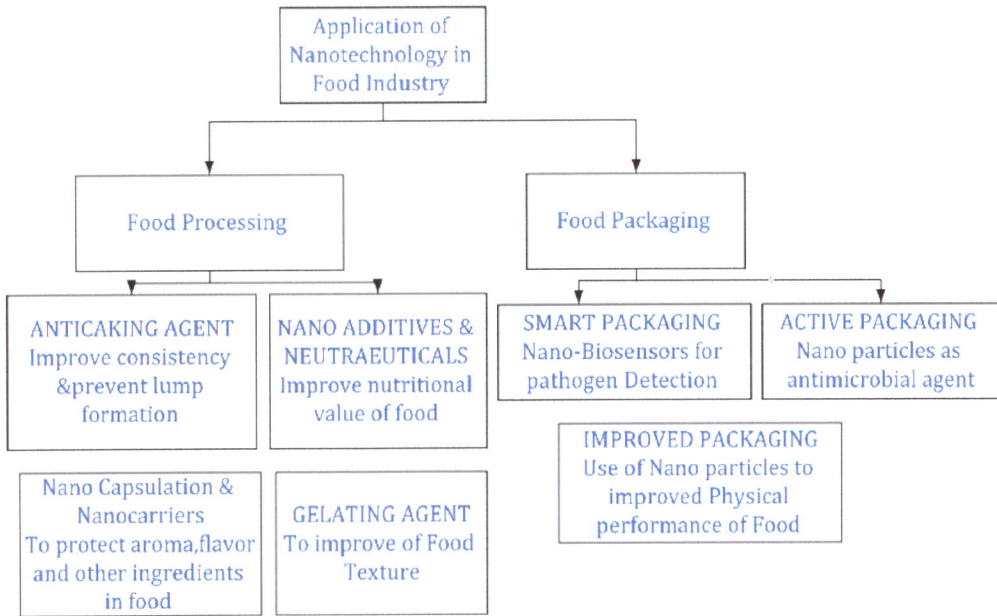

Fig. (13). Role of nanotechnology in the food industry [39].

Environmental Restoration

One of the most demanding jobs is employing nanotechnology to minimize environmental degradation, identify solutions to current environmental issues, protect our environment from further deterioration, and contribute to sustainable development. For example, soil, groundwater, air, and surface contamination can be treated, remedied, monitored in real-time, and detected using nanotechnology [44, 45].

In the following section, we have discussed different applications:

1. Energy Efficiency can be improved with the utilization of Nanotechnology. It can assist in the detection of pollutants and their elimination.

2. Researchers have designed a molybdenum disulphide (MoS_2) membrane, which filters water 2-5 times more than conventional filters. This is a tool for energy-effective desalination. Further, the chemical methods that can make these pollutants harmless are underway to reduce the effect of industrial pollutants on the groundwater. This would be cheaper than treating polluted water after taking it from the earth's surface. Thus, clean drinking water can be made accessible by treating water impurities using fast and cheaper screening methods.

3. For use in cleanup applications, the researchers created a "paper towel" using microscopic wires of potassium manganese oxide as nano fabric. This paper towel can absorb a good amount of oil, approximately twenty times its weight. In addition, magnetic and water-repellent nanoparticles are also used to remove fat from water mechanically in oil spill areas.

4. Many filter cartridges fixed in aircraft cabins and other air purification systems are also based on nanotechnology. These systems enable "mechanical filtration" in which the fibre material's nanoscale pores catch particles more significantly than the pores. Layers of charcoal that absorb odours could also be present in the filters.

5. With more sensitivity than ever, sensors and solutions made possible by nanotechnology can detect and indicate the chemical and biological material present in the air and soil. For different types of toxic site remediation, scientists are examining particles of Self-Assembled Monolayers on Mesoporous Supports, carbon nanotubes and dendrimers to ascertain the employability of their distinctive physical and chemical features. In addition, NASA has created another sensor for firefighters to utilize as a smartphone add-on to monitor the air quality near flames.

FUTURE OF NANOTECHNOLOGY

Currently, the possibilities in the field of nanotechnology can be seen in a favourable position. This sector is undoubtedly expected to grow worldwide due to technological advancements, government backing, investment from the private sector and increasing demand for smaller / compact devices/ equipment. Nevertheless, the environmental, safety and health risks and apprehensions related to its commercialization may limit the market expansion for nanotechnology.

The countries leading the nanotechnology industry in 2024 include the U.S., Brazil, Germany, and Asian countries like China, India, Japan, South Korea, Malaysia and Taiwan. The leading sectors in the future would be Electronics and Energy, followed by the cosmetics sector, and nanotechnology would contribute majorly to the former's advancement, as shown in Table **4**.

Table 4. Projected Power and Performance scaling of SoC [3].

Year of Production	2022	2025	2028	2031	2034	2037
-	G48M24	G45M20	G42M16	G40M16/T2	G38M16/T4	G38M16/T6
Logic industry "Node Range" Labeling	"3nm"	"2nm"	"1.5nm"	"1.0nm eq"	"0.7nm eq"	"0.5nm eq"
Fine-pitch 3D integration scheme	Stacking	Stacking	Stacking	3DVLSI	3DVLSI	3DVLSI
Logic device structure options	finFET LGAA	LGAA	LGAA CFET-SRAM	LGAA-30 CFET-SRAM	LGAA-3D CFET-SRAM	LGAA-3D CFET-SRAM
Platform device for logic	finFET	LGAA	LGAA CFET-SRAM	LGAA-3D CFET-SRAM-30	LGAA-30 CFET-SRAM-30	LGAA-3D CFET-SRAM-3D
Logic Technology Integration Capacity						
Number of stacked tiers (1)	1	1	1	2	4	6
NAND2-eq gate count (Mgates/mm2)	21	29	39	81	171	284
L3 cache SRAM density (Mbits/mm2)	27	27	41	87	191	286
CPU thruput sceling target node-to-node	1.70	1.70	1.70	1.70	1.70	1.70
#GPU cores in SoC based on integration capacity	36	49	66	136	286	477
#CPU cores in SoC based on thruput target, #CPUxfmax	12	20	33	55	92	155
#MAC units in SoC based on integration capacity	8192	11038	14980	30966	65191	108652
Analog 10 scaling	1.00	0.85	0.72	0.61	0.52	0.44

(Table 4) cont.....

SoC footprint scaling	1.00	0.85	0.61	0.33	0.19	0.14
POWER AND PERFORMANCE SCALING FACTORS					
HP frequency improvement	1.00	1.03	1.06	1.08	1.09	1.10
HP block power at iso frequency	1.00	0.83	0.78	0.59	0.50	0.48
HD block power at iso frequency	1.00	0.81	0.72	0.56	0.50	0.49
HP power at fmax	1.00	0.80	0.74	0.55	0.46	0.44
HP power at fmax	1.00	0.80	0.74	0.55	0.46	0.44
Power density at fmax	1.00	1.03	1.20	2.29	4.85	7.99
CPU clock frequency (GHz)	3.18	3.28	3.36	3.42	3.47	3.50
CPU clock frequency at constant power density (GH1)	3.18	3.17	2.79	1.49	0.71	0.44
CPU throughput at fmax (TFLOPS(sec)	0.31	0.52	0.88	1.50	2.55	4.33
CPU throughput at constant power density (TFLOPS/sec)	0.31	0.50	0.73	0.65	0.53	0.54

Upcoming Advancements in Nanoelectronics

1. Power consumption and loss of heat/energy produced by integrated circuits is the main issue facing high-performance data centres and personal computers.

2. By incorporating photosensitive organic molecules into small semiconductor material particles known as quantum dots, researchers are investigating novel methods for designing and fabricating logic circuits.

3. The TUT Optoelectronics Research Centre (ORC) is designing a quantum dot-based logic circuit technology framework.

4. A multidisciplinary project called "Photonic QCA" has been started by researchers at the Tampere University of Technology. It integrates the knowledge of semiconductor growth, organic chemistry, and nanofabrication.

5. Design a new logic circuit that consumes no power due to the motion of a single electron within quantum dots.

6. The Department of Chemistry and Bioengineering's researchers are attempting to figure out how to close the gap between the microscale and nanoscale. The project benefits from the knowledge of new designs by the Department of Electronics researchers.

7. To design solar panels which produce more energy with the help of nanoengineering.

8. To design nanotech batteries with extended batteries, lighter in weight and more efficient.

CONCLUSION

The study of microscopic particles is known as nanotechnology.

Nanotechnologies' main prospect is that it provides adequate, economical methods for utilizing renewable energy sources and maintaining a clean environment. Many scientists and engineers are developing innovative strategies to employ nanotechnology to better the globe. Electronics, chemical engineering, biology and robotic electronics are a few nanotechnology areas.

However, with the aid of nanotechnology, clinicians can identify diseases in their early stages and treat conditions like cancer, diabetes and heart disease with safer and more effective medications. Researchers also envision new technological solutions for conventional and chemical weapons defence for citizens and military personnel.

LEARNING OBJECTIVES

• Using nanotechnology, many fields, including chemistry, medicine, materials science, and engineering, have been predicted to experience a revolution soon.

• Using nanotechnology in the electronics field, we designed systems/circuits with low power, higher performance and lesser area.

• Using nanotechnology, the cost of design is lesser.

MULTIPLE CHOICE QUESTIONS

1. Which one of the following is an example of semiconducting nanowires?

  a. Nickel
  b. Platinum
  c. Silicon
  d. All of the above

2. Who discovered nanotubes?

  a. Gerd Binning
  b. Alex Zettl
  c. PM Ajayan
  d. Sumio Iijima

3. Who is the co-discover of the buck minister fullerenes?

  a. Gerd Binning
  b. Hary Kroto
  c. PM Ajayan
  d. Sumio Iijima

4. Who built the first molecular motor based on CNT?

  a. Gerd Binning
  b. Harry Kroto
  c. PM Ajayan
  d. Alex Zettl

5. Which one of the following is used in solar cells?

  a. Carbon Nanotube
  b. Nanorods
  c. Nanobots
  d. None of the above

6. What is the standard form of SEM?

a. Scanning Electron Microscope
b. Scanning Electrode Microscope
c. Scanning Electrical Microscope
d. None of the above

7. Which one of the following is an example of insulating nanowires?

a. SiO_2
b. InP
c. Si
d. All of the above

8. The size of polymeric nanoparticle nanosystem is around _____.

a. 1-300cm
b. 1-500cm
c. 10-1000nm
d. None of the above

9. The nanostructures are categorized into _____ types according to their dimensions?

a. One
b. Two
c. Three
d. Four

10. Who defined the term nanotechnology?

a. Gerd Binning
b. Alex Zettl
c. PM Ajayan
d. Norio Taniguchi

11. Which one of the following is a one-dimensional nanomaterial?

a. Whisker and bells
b. Nanodiscs

c. Nanolayers

d. Both b and c

12. Which one of the following is an example of a two-dimensional nanomaterial?

a. Colloids

b. Nanowires

c. Thin films

d. All of the above

13. Which one of the following is a two-dimensional nanomaterial?

a. Embedded Clusters

b. Nanodiscs

c. Nanolayers

d. Both b and c

14. Which one of the following is used in drug delivery?

a. Carbon Nanotubes

b. Nanorods

c. Nanobots

d. All of the above

15. What are the advantages of using nano-carbon tubes?

a. Consume less energy

b. High Speed

c. High capacity memory

d. All of the above

ANSWER KEY

1. (c)

2. (c)

3. (b)

4. (d)

5. (a)

6. (a)

7. (a)

8. (c)

9. (d)

10. (d)

11. (a)

12. (b)

13. (d)

14. (c)

15. (d)

REFERENCES

[1]   W. Arden, M. Brillouët, P. Cogez, M. Graef, B. Huizing, and R. Mahnkopf, "More-than-Moore white paper", *Version 2,* vol. 14, 2010.

[2]   Salah, K. (2017, December). "More than moore and beyond CMOS: new interconnects schemes and new circuits architectures", In *2017 IEEE 19th Electronics Packaging Technology Conference (EPTC)*, pp. 1-6. IEEE.
[http://dx.doi.org/10.1109/EPTC.2017.8277427]

[3]   "IEEE International roadmap for devices and systems", *IRDS,* 2022.

[4]   W.-T. J. Chan, A. Kahng, S. Nath, and I. Yamamoto, "The ITRS MPU and SoC system drivers: calibration and implications for design-based equivalent scaling in the roadmap," *Proc. IEEE Int. Computer Design (ICCD)*, pp. 153-160, 2014.
[http://dx.doi.org/10.1109/ICCD.2014.6974675]

[5]   M. Kandil, "The role of nanotechnology in electronic properties of materials",

[6]   R.F. Pierret, and G.W. Neudeck, *Advanced semiconductor fundamentals*. Vol. 6. Addison-Wesley: Reading, MA, 1987.

[7]   W. Bian, "The united states of america redesigned national nanotechnology initiative", *Science Focus,* vol. 17, no. 2, pp. 47-53, 2022.

[8]   Jr. Poole, P. Charles, and Owens. Frank J, "Introduction to nanotechnology",

[9]   B. Bhushan, "Introduction to nanotechnology", In: *Springer handbook of nanotechnology.* Springer: Berlin, Heidelberg. pp. 1-19, 2017.
[http://dx.doi.org/10.1007/978-3-662-54357-3_1]

[10] H. Bruus, *Introduction to nanotechnology.* Department of Micro and Nanotechnology, Technical University of Denmark, 2004.

[11] N. A. Melosh, A. Boukai, F. Diana, B. Gerardot, A. Badolato, P. M. Petroff, and J. R. Heath, *Ultrahigh-density nanowire lattices and circuits. Science, 300,* vol. 5616, no. 3, pp. 112-115, 2003.

[12]　S. Das, A.J. Gates, H.A. Abdu, G.S. Rose, C.A. Picconatto, and J.C. Ellenbogen, "Designs for ultra-tiny, special-purpose nanoelectronic circuits", *IEEE Trans. Circuits Syst. I Regul. Pap.,* vol. 54, no. 11, pp. 2528-2540, 2007.
[http://dx.doi.org/10.1109/TCSI.2007.907864]

[13]　B. D. Malhotra, and M. A. Ali, "Nanomaterials in biosensors: Fundamentals and applications", *Nanomaterials for Biosensors.,* pp. 1-74, 2018.

[14]　K. Habiba, V. I. Makarov, B. R. Weiner, and G. Morell, "Fabrication of nanomaterials by pulsed laser synthesis", *Manufacturing nanostructures,* vol. 10, pp. 263-292, 2014.

[15]　I.M. Chung, I. Park, K. Seung-Hyun, M. Thiruvengadam, and G. Rajakumar, "Plant-mediated synthesis of silver nanoparticles: Their characteristic properties and therapeutic applications", *Nanoscale Res. Lett.,* vol. 11, no. 1, p. 40, 2016.
[http://dx.doi.org/10.1186/s11671-016-1257-4] [PMID: 26821160]

[16]　D. Goldhaber-Gordon, M.S. Montemerlo, J.C. Love, G.J. Opiteck, and J.C. Ellenbogen, "Overview of nanoelectronic devices", *Proc. IEEE,* vol. 85, no. 4, pp. 521-540, 1997.
[http://dx.doi.org/10.1109/5.573739]

[17]　W. Zhu, P.J.M. Bartos, and A. Porro, "Application of nanotechnology in construction", *Mater. Struct.,* vol. 37, no. 9, pp. 649-658, 2004.
[http://dx.doi.org/10.1007/BF02483294]

[18]　W. Hannah, and P.B. Thompson, "Nanotechnology, risk and the environment: A review", *J. Environ. Monit.,* vol. 10, no. 3, pp. 291-300, 2008.
[http://dx.doi.org/10.1039/b718127m] [PMID: 18392270]

[19]　M.T. Bohr, "Nanotechnology goals and challenges for electronic applications", *IEEE Trans. Nanotechnol.,* vol. 1, no. 1, pp. 56-62, 2002.
[http://dx.doi.org/10.1109/TNANO.2002.1005426]

[20]　N. Dwivedi, S. Kumar, J.D. Carey, and C. Dhand, "Functional nanomaterials for electronics, optoelectronics, and bioelectronics", *J. Nanomater.,* vol. 2015, pp. 1-2, 2015.
[http://dx.doi.org/10.1155/2015/136465]

[21]　D. Grace, "Special feature: Emerging technologies", *Medical Product Manufacturing News.,* vol. 12, pp. 22-23, 2008.

[22]　R.W. Whatmore, "Nanotechnology—what is it? should we be worried?", *Occup. Med.,* vol. 56, no. 5, pp. 295-299, 2006.
[http://dx.doi.org/10.1093/occmed/kql050] [PMID: 16868126]

[23]　M.H. Fulekar, *Nanotechnology: Importance and applications.* I.K. International Pvt Ltd., 2010.

[24]　T.T. Tran, and A. Mulchandani, "Carbon nanotubes and graphene nano field-effect transistor-based biosensors", *Trends Analyt. Chem.,* vol. 79, pp. 222-232, 2016.
[http://dx.doi.org/10.1016/j.trac.2015.12.002]

[25]　J. Grollier, D. Querlioz, and M.D. Stiles, "Spintronic nanodevices for bioinspired computing", *Proc. IEEE,* vol. 104, no. 10, pp. 2024-2039, 2016.
[http://dx.doi.org/10.1109/JPROC.2016.2597152] [PMID: 27881881]

[26]　B.A. Parviz, D. Ryan, and G.M. Whitesides, "Using self-assembly for the fabrication of nano-scale electronic and photonic devices", *IEEE Trans. Adv. Packag.,* vol. 26, no. 3, pp. 233-241, 2003.
[http://dx.doi.org/10.1109/TADVP.2003.817971]

[27]　Y. Zhou, M. Zhang, Z. Guo, L. Miao, S.T. Han, Z. Wang, X. Zhang, H. Zhang, and Z. Peng, "Recent advances in black phosphorus-based photonics, electronics, sensors and energy devices", *Mater. Horiz.,* vol. 4, no. 6, pp. 997-1019, 2017.
[http://dx.doi.org/10.1039/C7MH00543A]

[28]　D.F. Emerich, and C.G. Thanos, "Nanotechnology and medicine", *Expert Opin. Biol. Ther.,* vol. 3, no.

4, pp. 655-663, 2003.
[http://dx.doi.org/10.1517/14712598.3.4.655] [PMID: 12831370]

[29]   M. Ferrari, "Cancer nanotechnology: Opportunities and challenges", *Nat. Rev. Cancer,* vol. 5, no. 3, pp. 161-171, 2005.
[http://dx.doi.org/10.1038/nrc1566] [PMID: 15738981]

[30]   G.A. Silva, "Introduction to nanotechnology and its applications to medicine", *Surg. Neurol.,* vol. 61, no. 3, pp. 216-220, 2004.
[http://dx.doi.org/10.1016/j.surneu.2003.09.036] [PMID: 14984987]

[31]   S.K. Sahoo, S. Parveen, and J.J. Panda, "The present and future of nanotechnology in human health care", *Nanomedicine,* vol. 3, no. 1, pp. 20-31, 2007.
[http://dx.doi.org/10.1016/j.nano.2006.11.008] [PMID: 17379166]

[32]   N. Durán, and P.D. Marcato, "Nanobiotechnology perspectives. role of nanotechnology in the food industry: A review", *Int. J. Food Sci. Technol.,* vol. 48, no. 6, pp. 1127-1134, 2013.
[http://dx.doi.org/10.1111/ijfs.12027]

[33]   R. Amin, S. Hwang, and S.H. Park, "Nanobiotechnology: An interface between nanotechnology and biotechnology", *Nano,* vol. 6, no. 2, pp. 101-111, 2011.
[http://dx.doi.org/10.1142/S1793292011002548]

[34]   D. Simić, M. Marjanović, M. Vitorović-Todorović, S. Bauk, D. Lazić, A. Samolov, and N. Ristović, "Nanotechnology for military applications: A survey of recent research in military technical institute", *Scientific Technical Review,* vol. 68, no. 1, pp. 59-72, 2018.
[http://dx.doi.org/10.5937/str1801059S]

[35]   J.J. Ramsden, "Nanotechnology for military applications", *Collegium,* vol. 30, p. 99, 2012.

[36]   E. Serrano, G. Rus, and J. García-Martínez, "Nanotechnology for sustainable energy", *Renew. Sustain. Energy Rev.,* vol. 13, no. 9, pp. 2373-2384, 2009.
[http://dx.doi.org/10.1016/j.rser.2009.06.003]

[37]   K.W. Guo, "Green nanotechnology of trends in future energy: A review", *Int. J. Energy Res.,* vol. 36, no. 1, pp. 1-17, 2012.
[http://dx.doi.org/10.1002/er.1928]

[38]   T. Singh, S. Shukla, P. Kumar, V. Wahla, V.K. Bajpai, and I.A. Rather, "Application of nanotechnology in food science: Perception and overview", *Front. Microbiol.,* vol. 8, p. 1501, 2017.
[http://dx.doi.org/10.3389/fmicb.2017.01501] [PMID: 28824605]

[39]   B.S. Sekhon, "Food nanotechnology: An overview", *Nanotechnol. Sci. Appl.,* vol. 3, pp. 1-15, 2010.
[PMID: 24198465]

[40]   C.F. Chau, S.H. Wu, and G.C. Yen, "The development of regulations for food nanotechnology", *Trends Food Sci. Technol.,* vol. 18, no. 5, pp. 269-280, 2007.
[http://dx.doi.org/10.1016/j.tifs.2007.01.007]

[41]   N. Sozer, and J.L. Kokini, "Nanotechnology and its applications in the food sector", *Trends Biotechnol.,* vol. 27, no. 2, pp. 82-89, 2009.
[http://dx.doi.org/10.1016/j.tibtech.2008.10.010] [PMID: 19135747]

[42]   L. Rashidi, and K. Khosravi-Darani, "The applications of nanotechnology in food industry", *Crit. Rev. Food Sci. Nutr.,* vol. 51, no. 8, pp. 723-730, 2011.
[http://dx.doi.org/10.1080/10408391003785417] [PMID: 21838555]

[43]   B. K. Choudhary, K. Majumdar, and S. Deb, "An overview of application of nanotechnology in environmental restoration", 2016.

[44]   B. Karn, T. Kuiken, and M. Otto, "Nanotechnology and in situ remediation: A review of the benefits and potential risks", *Environ. Health Perspect.,* vol. 117, no. 12, pp. 1813-1831, 2009.
[http://dx.doi.org/10.1289/ehp.0900793] [PMID: 20049198]

CHAPTER 2

Self-Assembled Monolayer-Based Molecular Electronic Devices

Jaismon Francis[1], **Aswin Ramesh**[1] and **C. S. Suchand Sangeeth**[1,*]

[1] Department of Physics, National Institute of Technology Calicut, Calicut-673601, Kerala, India

Abstract: This chapter focuses on molecular tunnel junctions (MTJ), the basic building block of molecular electronics (ME), which consist of either a single molecule or an ensemble of molecules in the form of a self-assembled monolayer (SAM) sandwiched between two electrodes. MTJs based on SAMs find practical applications such as diode rectifiers, switches, and molecular memory devices. The predominant charge transport mechanism in two-terminal junctions is tunneling; therefore, perturbances in the bond length scale will translate into nonlinear electrical responses, allowing MTJ to induce and control electronic activity on nanoscopic length scales with various inputs. For this reason, the subject is now progressing to devices based on finite ensembles of molecules, and many studies are underway to develop devices that can augment and complement traditional semiconductor-based electronics. SAM-based tunnel junctions are like single molecular junctions, demonstrating effects like quantized conductance, tunneling, hopping, and rectification; they also possess a unique set of properties. In addition, several new problems that need to be addressed arise from the unique characteristics of SAM-based junctions. General aspects of the two terminal molecular junctions, roles of the electrode, molecule, and molecule electrode interfaces, and how to differentiate the components of a molecular junction using impedance spectroscopy are discussed in this chapter. Different testbeds to measure the charge transport in SAM-based tunnel junctions are discussed, and a comparison of the reported charge transport data on alkanethiolate SAMs is presented. Finally, the molecular rectifiers are briefly discussed.

Keywords: Charge transport, Impedance spectroscopy, Molecular tunnel junctions, Molecular rectifier, Self-assembled monolayers.

INTRODUCTION

The history of electronics is well connected to the progress in the miniaturization of electronic components. For instance, the first computers were still cupboard-sized, whereas small-sized laptops available these days surpass those computers in

---

* Corresponding author C. S. Suchand Sangeeth: Department of Physics, National Institute of Technology Calicut, Calicut-673601, Kerala, India; E-mail: sangeeth@nitc.ac.in

Gopal Rawat & Aniruddh Bahadur Yadav (Eds.)

memory, computational power, and all other performance properties by several orders of magnitude. Having more transistors per chip, more sophisticated integrated circuits are realized, which eventually transformed society in an unimaginative way. Due to fundamental scaling limitations, efforts to design various new nanoelectronic devices and materials as substitutes for complementary metal oxide semiconductor (CMOS) transistors-based electronic circuits have been intensive [1]. As the dimensions of conventional semiconductor devices are pushed below 100 nanometers, it creates many fundamental and technical challenges. For example, traditional scaling methods influence vital parameters like the device's threshold voltage and on/off currents. The increased financial cost of associated equipment required to produce semiconductor devices is holding back efforts to improve the capabilities of these devices [2].

In the devices mentioned above, electronic components based on molecules are crucial. When molecules are used as the main component of electronic devices, it offers numerous degrees of freedom intrinsic to molecular structure. Tuning the electrical characteristics of the molecular devices could be easily achieved by altering the structure of the molecules. Therefore, they are a potential candidate for next-generation electrical components [1]. ME studies electron transport across one molecule or an ensemble of molecules sandwiched between two electrodes [3]. Nowadays, ME is rapidly evolving. Molecular-mechanical switches operated by single electron transfer, three-input mechanical Boolean sorter, electrically gated single molecule switches, *etc.*, are some of the milestones in this field so far [4, 5].

Aviram and Ratner's paper in 1974 and subsequent studies have shaped ME into what it is today. They suggested a single molecular rectifier having a Donor-Bridge-Acceptor construct [6]. The ultimate goal of ME is to conquer the fundamental scaling limits of conducting polymer or traditional silicon-based inorganic devices by building nanometer-sized active or passive electronic components [7]. The exciting aspect of ME is that the electronic functions exhibited by inorganic semiconductors can be mimicked by functionalizing a single molecule sandwiched between a two-terminal junction [8]. Experimental development of such a concept using a variety of methods like scanning probe microscopes, break junctions, soft metallic contact, nano transfer printing (nTP), lift-off-float-on electrodes, *etc.*, have been realized through the years [9, 10]. These solutions achieved single MTJ, exhibiting effects like quantum interference effect, quantized conductance, tunneling, hopping and rectification [11].

Reproducibility, robustness, integration, and upscaling limit the incorporation of single-molecule into practical applications. The problem lies in controlling the geometry, structure, and orientation of the individual molecules with the interface.

Due to *n* number of geometries and orientations that a molecule can have at the interface of an electrode and various defects in the junction, the measured value of current varies to a large extent. Thus, each device fabricated using the exact method and material tend to be distinguishable, posing the issue of reproducibility and making the device irrelevant. It should also be noted that there is an inherent difficulty for the molecules to be stable when subjected to an external bias. The contact resistance at the molecule/metal interface was one of the most challenging problems due to the variation in the performance of the junctions. In 1971, Mann and Kuhn performed charge transport measurements in SAMs of cadmium salts fatty acids between two metal electrodes [12]. They observed an exponential decay of the conductivity independent of temperature as the chain length of the monolayer increased, which they ascribed to incoherent tunneling through the monolayer. Molecular electronic junctions with an ensemble of molecules are statistically more favorable than a junction consisting of a single molecule due to the inconsistency in results produced by the latter. The concept of a 'finite ensemble of molecules' is realized through SAM. SAMs are ordered arrays of molecules that spontaneously form by adsorption of constituent molecules from solutions or the vapour phase [13], and Kuhn performed charge transport measurements in SAMs of alkane(di)thiol in a two-terminal configuration [12]. This was a significant breakthrough as it was later established that SAMs are a practical alternative to the electronic functionality that a single molecule offers.

ME is a broad field, and excellent reviews are available that discuss the progress of two-terminal MTJ [14]. These reviews mainly focus on materials and practical challenges in the design and fabrication of an MTJ. Our attempt is not to replicate this; instead, this chapter focuses on the methods to analyze the obtained data. We also discuss the usefulness of impedance spectroscopy in identifying the circuit elements in the junction and determining the charge transport bottlenecks in the MTJ. The device details primarily addressed in this chapter are two-terminal MTJ (molecules or SAMs of n-alkanethiols, where n is 2 to 16 sandwiched between two stationary electrodes). Although two-terminal devices are widely studied, many conflicting results can be found in the literature [13 - 15]. These contradic-tions arise since different groups use different electrodes or testbeds for the same molecules resulting in MTJs with different characteristics. The top and bottom electrode and the kind of interface it makes with the monolayer and the bulk of the monolayer influence the device's electrical characteristics. Recently Mukhopadhyay *et al.* conducted a cross-laboratory study to understand charge transport in protein-based junctions. In their research, they tried to answer how results from various platforms and labs can be compared and how to discern the distinction between platform-induced and true protein electrical transport characteristics [16]. Results from the study show that the effective contact area is determined by junction geometry, and the charge transport mechanism is not dependent on the particular testbed. Their study also suggests that transport effectiveness across interfaces is governed by electrode-molecule coupling, and solid-state electron transport across protein-based junctions is predominated by tunneling. The observed current density variations are ascribed to variations in the actual contacts (proper electrical contact area) in comparison to the geometric contact area (which is challenging to identify) and electronic contact-protein coupling [16].

MOLECULAR ELECTRONIC JUNCTIONS

This section is divided into two sub-sections. General aspects of two-terminal MTJ are discussed in the first section. Details of a molecular junction and the roles of the electrode, molecule, and molecule electrode interfaces are provided in this section. Here we separate junctions by their elements; the bottom electrodes to which the SAMs are anchored, the top electrode that forms a single junction by defining its geometry; the molecular layer whose length, composition, orientation, and packing determine the functionality of the device; and the molecule–electrode interface. The last section concerns different test beds used to study charge transport.

SAM-BASED MOLECULAR TUNNEL JUNCTIONS

The Electrodes

The active layer in an MTJ is either one molecule or a few numbers of molecules Fig. (**1**); therefore, conventional methods of depositing or implanting electrodes directly on the active layer for studying the electrical characteristics fail desperately. We often end up with a short circuit, and if one gets lucky enough to make a good device, the chances of reproducing the results are very low. Unlike ordinary two-terminal devices, the number of molecules available for contact with electrodes is much less; thus, averaging out the local effects due to orientation and geometries does not occur. Due to these reasons, designing electrodes for ME devices should be dealt with utmost care. The molecule(s) or the SAM is grown on the "bottom electrode," which determines the nature of the active layer. Therefore, the bottom electrode is chosen depending on the SAM layer intended for the study. Planar and highly curved nanostructures are used as substrates. As the bottom electrode, metal-thin films of gold (Au), silver (Ag), copper (Cu), *etc.*, and different alloys are coated over various substrates such as glass, quartz, and mica using different synthesis techniques available, which include physical vapour deposition (PVD), electron beam epitaxy, sputtering, *etc.* The roughness and grain size of the films can be varied during the deposition and is discussed in the literature [13].

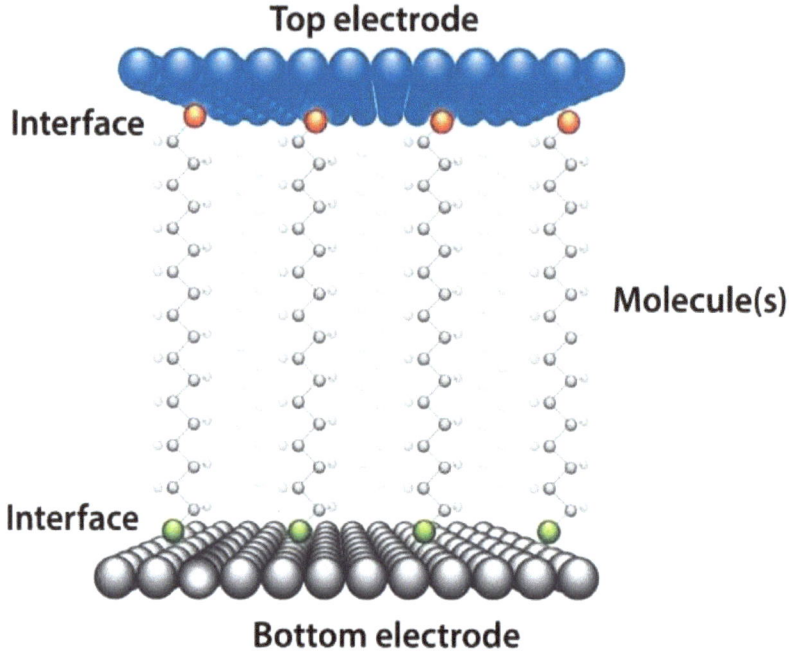

Fig. (1). Schematic of an ideal (defect free) SAM-based tunnel junction. Molecules are sandwiched between the electrodes.

The type of thin metal films, their roughness, and grain size are found to affect the direction of growth of the SAM and its packing density. The SAM grows well-defined when we have fewer defects on the electrode. The defects include phase domain boundaries, grain boundaries, and step edges where the packing of SAMs is not uniform. This causes the distance between the two electrodes to vary, creating defects in the SAM grown [17]. Thus, a larger-sized grain and narrow grain boundaries of the bottom electrode help to improve the reproducibility of the device by forming an improved SAM layer with decreased defects. Once a good SAM is ready on the bottom electrode, the crucial part is to deploy a top electrode without damaging or penetrating the monolayer. In the subsequent section, we will discuss some of the standard methods successfully used.

Molecules

Due to their ability to self-assemble onto highly ordered structures of Ag/Au surfaces, SAMs consisting of saturated alkanethiol molecules have received much attention in the early years [13]. There is substantial information in the literature on charge transport through alkanethiol-based SAM, particularly on *J-V* behavior and length dependence [18]. Since the structure and configuration of SAMs made of alkanethiolates are characterized and agreed upon by the scientific community, thiolates are regarded as the standard for any future study [19]. The charge transport across an MTJ is mediated by its electrodes, the molecular layer in between, and the interface thus formed [13]. As this area of research focuses on technological advancement using tunable chemo-electronic properties of molecules, many researchers have developed various molecules that are suitable for ME devices. In this section, we will mention the molecules that are known in this regard and the physical nature that enable the molecule to self-assemble.

Molecular self-assembly is the ordered arrangement of molecules onto a specific substrate. The structure of such molecules has three distinct regions, as shown in Fig. (**2**). The first region is the tail functional group which shows the affinity to adsorb (chemisorption/physisorption) on the bottom electrode. The functionality of the end group is specific to the type of bottom electrode. The middle portion is the 'backbone' of the molecule, which constitutes the bulk of the device. In contrast, the 'head' is to interact with the top electrode and is crucial for device operation [20]. Strong bonding of one of the anchoring groups with the electrode is inevitable for SAM formation. The anchor groups used are substrate(electrode) specific, and the most common names that appear in ME are discussed here. Molecules compatible with most metal electrodes and dielectrics discussed in the literature are thiolates and silanes. Thiols (-SH) are known for their strong affinity towards Au and other coinage metals, which chemisorbs the substrate [21]. While the silanes ($SiCl_3$) and carboxylic acid groups (-COOH) are used as anchoring groups for oxide surfaces (SiO_2, ITO, Al_2O_3, *etc.*) [22]. Other anchoring groups like NR_2 and OH are also used [20].

In an MTJ, the backbone of the molecules determines the nature of charge transport. Different kinds of aliphatic and aromatic molecules, σ-π and fully conjugated molecules, are implemented as the backbone of the SAM. Aliphatic molecules are good candidates for studying the interface properties of the ME device. Alkanethiol molecules, due to their intrinsic behavior to form SAM on a wide variety of electrodes, led to their widespread electrical study. The significant amount of available literature and data specifically on electrical characteristics make it a benchmark molecule for further analysis [13]. But due to the large conduction gap between HOMO-LUMO levels, the charge transport across the device is limited. Here σ-π and fully conjugated molecules can help in achieving better device conductivity. Molecules like DNA assemblies, protein complexes, and homopeptides are an interesting and active area of research as far as charge transport study is concerned [23].

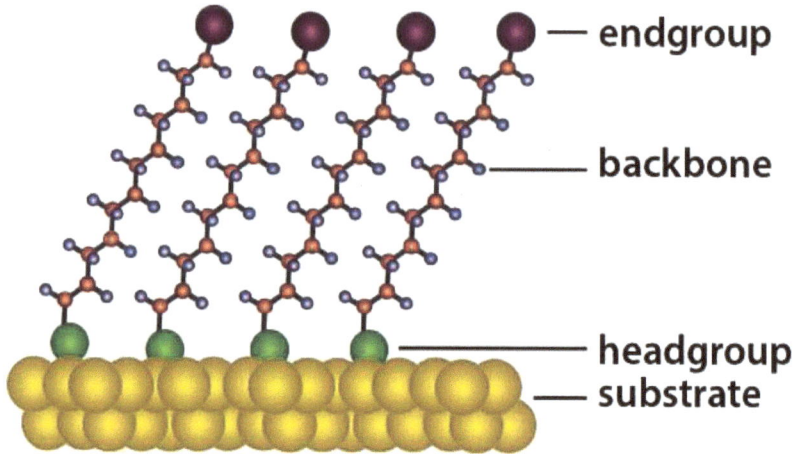

Fig. (2). Schematic of alkane thiolate SAM on a gold surface, highlighting the anatomy of SAM.

Molecule Electrode Interface

A deep understanding of the interface between the molecule and the electrode is necessary to comprehend the electrical characteristics of ME devices. The electronic coupling of molecular levels to the electrode and the eventual modification of the Fermi level defines the barrier for the charges to flow through. Sometimes, the intrinsic property of the active layer (monolayer) gets masked by the molecule electrode interface [24]. There have been instances where rectification, which is non-molecular in origin, was reported [25]. Hence, it is reliable if the contacts are Ohmic, ensuring that the active layer is in command of the electrical properties that the device exhibits [26]. The adsorption of anchoring groups onto the electrode alters the electronic distribution at the interface modifying the Fermi energy level. This will eventually affect the charge transport across the device. Chemisorption of functional groups like thiols onto the metal electrode like Au is favored by a covalent bond which causes strong coupling compared to physisorbed anchoring groups favored by Van der Waals interactions [27]. Coupling affects the transmission probability of electrons across the junction, affecting its conductance [28]. Depending on the interaction at the interface, the device's conductance will vary by a few orders. Because, as per the Landauer formula, the conductance is dependent on the transmission coefficient through the interface and the molecules equally. Therefore, it is difficult to distinguish between molecular and interface effects [28]. Because of the numerous bottom electrode/SAM/top electrode combinations available, it is essential to assess the effectiveness of each contact. The value of β describes the scale of attenuation of the tunneling current with the length of the molecule. A low value of β from the expected value implies the deviation from the ideal tunneling nature of charge transport. When coupled with the value of β with respect to different electrodes, this helps us to understand the interface thus formed. Parallelly, the range of value of J_0 observed for different electrodes implies the effect of contact resistance with respect to various electrodes. Therefore, a deviation from the consensus value reported in the literature will enable us to assess the effectiveness of the interface formed by the contacts. For example, the intercalation of SAM into the electrodes or the penetration of the electrodes into the active layer will increase J_0. Thus, A measure of J_0 and $β$ acts as an effective parameter for assessing the interface [29].

Molecular Electronics Test-Beds

Once a good SAM is ready on the bottom electrode, the crucial part is to deploy a top electrode without damaging or penetrating the monolayer. Traditional methods like thermal evaporation cause the metal atoms to penetrate the SAM layer resulting in a short circuit. We will discuss some of the standard methods successfully used so far.

Single Molecular Junctions

Scanning Tunneling Microscopy (STM): The current through the molecule(s) is measured using an STM tip in this technique, as depicted in Fig. (**3a**). An external voltage is supplied across the bottom electrode, the STM/AFM tip, and the tunneling current is monitored [30]. The tip of STM is sharp on an atomic scale, due to which we can take localized measurements in size range of a molecule. Usually, a tunneling gap between the STM tip and the molecule reduces the measured current drastically, making it unreliable for measuring currents at larger molecular lengths [31]. The size of the tip, however, holds the advantage of measuring localized current across single molecules.

Conducting probe atomic force microscope (CP-AFM): It can perform charge transport measurements on well-defined points inside the SAM, as shown in Fig. (**3b**). The main drawback of this method is the influence of force [51] put in by the CP-AFM tip and ambiguity in the number of molecules [52] involved in the measurement. AFM tip is considerably bigger and can scan over an ensemble of molecules. The tip of AFM can be modified by coating different metal thin films like gold and silver, which causes chemisorption of molecules like alkanedithiol, increasing the measured current, which in turn allows us to study the interface effects due to chemisorption [32].

Break Junction: Generalized into two types, mechanically controllable see Fig. (**3c**) and electromigration [33]. Its key benefit is the mechanical break junction's (MCBJ) capacity to control contacts at a nanometer scale and measure for single molecules. In an electromigration break junction, a high-magnitude electrical current is employed to produce the gap between the two electrodes. Electromigration junctions cannot carry out a substantial repetitive collection of measurements, in contrast to MCBJ. Therefore, many devices need to be created to perform statistical analysis [34].

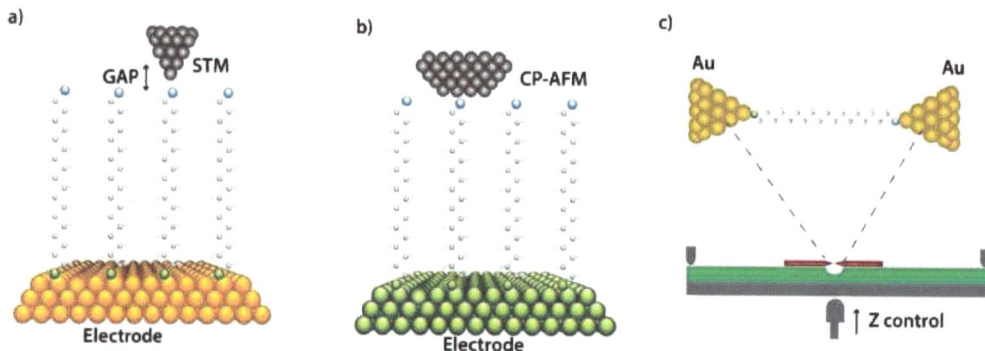

Fig. (3). Schematic representation of molecular tunnel junctions **a**) STM-based single-molecule junction. The tunneling current across the STM tip and bottom electrode determines the molecule's electrical characteristics. **b**) CP-AFM molecular junction. Transport through the molecules is measured by applying a voltage between the AFM tip and the bottom electrode. **c**) A mechanically controllable break junction is depicted schematically.

Ensemble Molecular Junctions

Metal top electrode: In the early stages of the studies on ME devices, the top electrode of the MTJ is fabricated by direct thermal evaporation of the metal, which results in a low yield [35]. Even with careful electrode preparation, it is impossible to conclude the charge transport mechanism of the tunnel junction unambiguously in this method. For these reasons, a variety of test beds are developed.

Crossed wire junctions: Two metal wires are held at right angles while the SAM is grown on one of the wires [36]. The whole system is placed under a magnetic field. A current is passed through the wire to manipulate the Lorentz force to alter the contact between the monolayers and the bare wire [36]. Different metals can be used to make out an asymmetric device, and the device area can be altered by the precise deflection of the wire in contact. But due to the curvature of the contact, an exact number of molecules per device area is difficult to estimate.

Nanopores: Nanopores are formed on an insulating material (*e.g.*, Si_3N_4, SiO_2) by different lithography techniques [37], and on one end of the nanopore, a metal electrode is placed on which the SAM is grown see Fig. (**4F**). On the other end, the electrode is coated by sputtering. This approach can produce more devices in an array in the same insulating material from which the data can be statistically analyzed [37]. Calculating the number of molecules per device area is straightforward when the packing density of the molecule is known. The main drawback is that the yield of the device depends on the packing density of the SAM in the nanopore. Thus, when the spacing between the molecule increases, the chances of the growth of metal filament through the monolayer also increases.

Conducting protective layer (PL): Many research groups have employed fabrication methods that sandwich a protection layer (PL) which is conductive between the top metal electrode and active layer (SAM) [38 - 40]. The PL blocks the thermally evaporated metal atoms, which acts as a buffer so that the atoms do not penetrate the SAM forming an alloy with the bottom electrode [19]. This helps produce a robust and reproducible top contact. Apart from this, it significantly improves the junction yield [38]. On the other hand, incorporating PL, an additional resistive component, and sometimes capacitance complicates the comprehension of the electrical properties of the MTJ. This issue is noticeable in DC measurements since it determines the total current impeded by all the junction components. Examples of such systems include carbon paint (CP) [39] and multi-layered graphene (MLG) [40].

Large area MTJ using conducting polymers: A photoresist is spin-coated over the bottom electrode, and holes of desired device area are etched out using lithography, allowing us to control the growth of SAMs into a specific location. This is then immersed into the solvent to grow SAM in the designated area. Then a conducting polymer; poly(3,4-ethylene-dioxythiophene): poly(4-styrene-sulphonic acid) (PEDOT:PSS), as shown in Fig. (**4E**), is spin coated over the SAM to act as a protective layer, after which the metal electrode is thermally evaporated to serve as the top electrode. The protective layer blocks the thermally evaporated atoms from penetrating and forming an alloy with the bottom electrode. The large device area obtained here has a statistical advantage over the other techniques. Also, the conductivity of PL is very high compared to the alkanethiolates which can help in ruling out the role of the PL used [34].

Carbon electrodes: Allotropes of carbon have found application as electrodes due to their ability to form different forms and shapes, chemical stability, surface chemistry properties, and strong covalent bonds present not only between both carbons but also with other modifiers, usually on the surface. Single and multilayer graphene, graphene oxides, and carbon paint have produced a high yield of ~90% as bottom and top electrodes in ME devices see Figs. (**4B and C**) [40]. The transfer of single-layer graphene (SGL) grown by chemical vapor deposition onto a conducting substrate was reported by Xuelei Liang *et al.* [41]. This principle was used to produce SGL electrodes over the gold substrate, and

SAMs of arylalkanes were grown on top of SGL by Seock-Hyeon Hong *et al.* [42]. The covalent bond formation is incidentally due to the molecule's deionization of sp2 hybridized carbon. As mentioned above, the device prepared by Seock-Hyeon Hong *et al.* had both top and bottom electrodes as SGL, with its bottom electrode being covalently bonded with the aryl group and the top electrode physically adsorbed *via* Van der Waals interaction. These devices reported a high yield (>80%), demonstrating controlled molecular charge transport by arylalkanes.

Soft Liquid-Metal Electrodes

Mercury as the top electrode: Several groups have successfully used soft liquid-metal electrodes such as Hg and EGaIn [43] to circumvent the difficulties encountered with thermal evaporation of metal contacts. SAM is grown on a mercury drop extended from a syringe or capillary tube. Then the SAM so formed is made to contact the SAM produced on a bottom electrode. The bottom electrode could be any different metal or Hg itself, forming a metal (Hg)-SAM//SAM-Hg structure. This method is easy to carry out and reproducible if the SAMs grown on the bottom electrode are relatively regular in height (which implies SAMs have fewer defects). If the device area is measured accurately and the tilt angle and roughness can be controlled, a statistically significant amount of data can be acquired easily. A short-chain SAM requires a PL between the molecule and top electrode in systems with Hg top electrode see Fig. (**4D**). The charge transport characteristics of relatively larger junctions are possible (0.1-1.0 mm^2) with this test bed. Still, the main drawback is low yield (~25%), and also, the Hg forms an amalgam with most of the metal bottom electrode [44].

A top electrode made of Eutectic of Indium and Gallium (EGaIn): The group of Whitesides showed the use of a liquid metal alloy (based on a eutectic of gallium and indium (EGaIn)) as a soft top electrode to establish electrical contact with SAMs. Their approach uses metal films as a bottom electrode and EGaIn as a top electrode [29]. The SAM grown on the metal film acts as the insulator through which charge tunnels; see Fig. (**4A**). The outer layer of EGaIn oxidizes spontaneously to become GaO_x as soon as it contacts air. This layer of GaO_x prevents Ga and In from penetrating the monolayers and providing us with reliable electrical contact. EGaIn was successfully used as electrodes by forming SAM/GaO_x/EGaIn junction [45]. This is achieved by deploying a cone-shaped EGaIn suspended from a syringe, or the contact is given through microchannels.

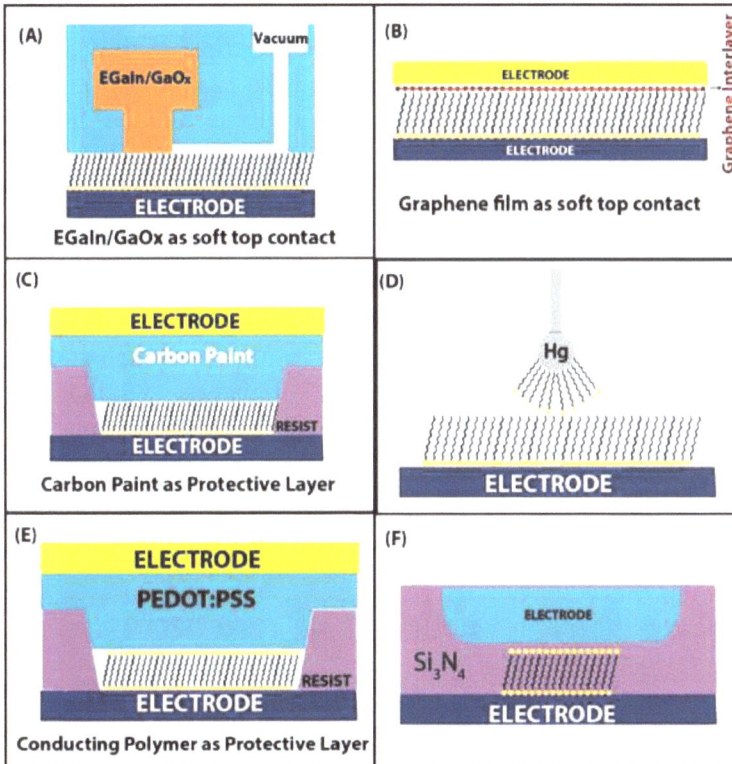

Fig. (4). Schematic representation of molecular tunnel junctions mentioned in this chapter). **A**) microfluidic device composed of PDMS (polydimethylsiloxane) is used to stabilize the EGaIn/GaOx top electrode so that connections to SCn SAMs can be made [28]. **B**) Multilayered graphene as a bridging interfacial layer between the active molecules and a top electrode [40]. **C**) Junctions composed of SAMs of n-alkanethiolates supported by Au and a layer of carbon paint (CP) as PL [39]. **D**) A SAM-coated controlled drop of Hg is brought into contact with a metal surface that also has another SAM to form a Hg-SAM-SAM-metal junction [46]. **E**) In this junction, PEDOT:PSS conductive polymer layer acts as the top electrode [38]. **F**) Nanopores are formed on an insulating material by different lithography techniques, and on one end of the nanopore, a metal electrode is placed on which the SAM is grown [37].

Polydimethylsiloxane (PDMS) is a commonly used insulator that makes microchannels [19]. This method of transferring a thin metal film layer into a designated surface through microchannels of elastomeric or rigid insulators is called nano-transfer printing. The metals intended to form the electrode are filled or evaporated through the microchannel, and the metal layer makes a chemical bond with the SAM. The EGaIn/GaO$_x$ top electrode has proven more effective by producing high yields. The possibility of collecting a significant amount of data makes it a promising candidate as the top electrode.

In general, the electrical properties of molecular junctions have historically had poor repeatability. Developing techniques for fabricating MTJs that generate data with high reproducibility in terms of precision (distribution widths) and replicability (closeness to a reference value) is critical if they are to be used in applications. One of the advantages of MTJs with EGaIn-based top electrodes is that they can generate highly reproducible J(V) data in terms of both precision and accuracy (for detailed information, refer to [47]) see Fig. (**5b**). Apart from this, these MTJs are electrically stable with good yields and β close to the consensus value.

Sangeeth *et al.* characterized (both DC and AC) a set of systematically synthesized conjugated oligophenyleneimine wires of molecular length varying between 1.5 and 7.5 nm using the EGaIn top contact approach and CP-AFM. The EGaIn junction allows access to AC and DC measurements and high-precision variable temperature measurements, which are essential for comprehending transport processes. The DC bias J-V experiment findings and the AC measurements showed a definite crossover from tunneling to hopping transport at about 4 nm [48]. It was also concluded that the EGaIn and CP-AFM methods have an excellent agreement in intensive observables [48].

CHARGE TRANSPORT IN MOLECULAR JUNCTIONS

Methods to Understand the Charge Transport Across Molecular Junctions

DC Measurements

Typically, a measure of current density J (A/cm^2) as a function of DC voltage V (V) will give us an idea about the charge transport in MTJ. The total current flowing through the molecular layer is determined in DC measurements [49]. The kind of charge transport mechanism is determined from the J-V traces along with the "Simmons tunneling equation" (equation (**1**)) [29].

$$J = J_0\, e^{-\beta d_{SAM}} \tag{1}$$

The pre-exponential term J_0 represents the current across the junction in the absence of SAM (*i.e.*, $d_{SAM} = 0\ n_c$), β represents the tunneling decay constant (in n_c^{-1}), and d_{SAM} is the thickness of the monolayer (in n_c). The constants β and J_0 are obtained from a fit of equation (**1**) to the J-V curve. $J(V)$ measurements conducted on various testbeds using a variety of SAMs have generated an abundance of data with sometimes not accurate and has uncharacterized reliability. Reasons inherent in the methods used for measurement, such as poor experimental design, often make data noisy and complicate data analysis. Studies sometimes lacked a robust statistical approach in earlier periods and were based on selected data instead. J is often plotted on a log scale at a specific voltage as a function of counts. Histograms of $\log_{10}|J|$ data taken from Ref [50]. for n- alkanethiol SAMs, $S(CH_2)_{n-1}CH_3$, for n = 9 and 18 are shown in Fig. (**5a**). Fig. (**5**) conveys the magnitude of the difficulty faced by any analysis of charge transport in MTJs. Despite being at different ends of the series, two histograms for n-alkanethiols are superimposing. It indicates that measured current density values are noisy and contain artifacts that are difficult to separate from actual data without rigorous statistical analysis. Whitesides *et al.* proposed statistical tools for analyzing charge transport measurements in their paper [50]. They used a statistical model to describe how values of J emerge from a population of Ag^{TS}- S/CH_3-//Ga_2O_3/EGaIn junctions. Mainly four approaches to analyzing charge transport are covered in their work, which can be found elsewhere [50].

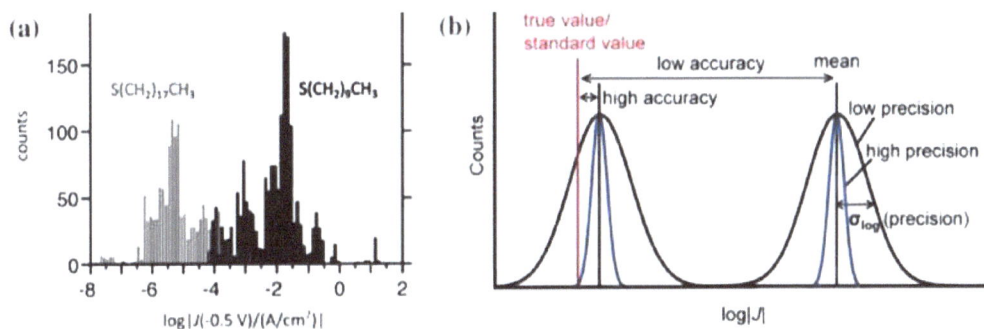

Fig. (5). (a) Histograms of $\log_{10}|J|$ data taken from Ref [50]. for n-alkanethiol SAMs, $S(CH_2)_{n-1}CH_3$, for n = 9 and 18. Reproduced with permission from ref [50]. Copyright, American Chemical Society. (b) Diagram showing the concept of the accuracy and precision (σ_{log}) of the electrical measurements for MTJs based on SAMs. Reproduced with permission from ref [47]. Copyright © 1999-2022 John Wiley & Sons.

One may determine the contact resistance from the DC measurement by extrapolating the resistance against chain length data (d_{SAM}=0). But this value turns out to be underestimated owing to the long extrapolation [49].

Impedance Measurements

AC impedance spectroscopy is a swift and non-destructive approach that lets us differentiate between the bulk and the interface. A wide range of materials, as well as organic semiconductor devices (such as organic solar cells, organic field-effect transistors (OFET), light-emitting diodes (LED), *etc.*), can be electrically characterized using this technique. In impedance spectroscopy (IS), an applied ac voltage serves as the input, and the current is measured. The impedance is determined by the relationship between the current and voltage. Since the measurements also involve phase values, impedance (Z) is a complex quantity. The Z is expressed in polar form as shown below (equation (**2**)) [51].

$$Z = |Z|e^{j\phi} \qquad (2)$$

where $|Z|$ is the magnitude and ϕ is the phase difference by which current (I) lags the voltage (V) [51, 52]. Transforming into cartesian form, Z is represented as:

$$Z = Z' + jZ'' \qquad (3)$$

where Z' and Z" are the real and imaginary parts of complex impedance Z. In impedance measurements, the capacitance of each component of MTJ is reflected as the alternating current is used for measurement. Especially at high frequencies, the capacitance term dominates the impedance. Capacitive reactance X_C is related to applied AC frequency by the following relation.

$$X_C = \frac{1}{C\omega} \qquad (4)$$

For an ideal circuit element with a resistor of resistance R and a capacitor of capacitance C connected in parallel, the corresponding Z is:

$$\frac{1}{Z} = \frac{1}{R} + j\omega C \qquad (5)$$

The Bode and the Nyquist plot help in understanding the impedance spectra. Bode's plot consists of Z ($|Z|$) and the phase angle (ϕ) plotted against the frequency, whereas Nyquist plots are created by plotting the imaginary portion of impedance Z" against the real part of the impedance Z'. A measure of capacitive and resistive elements of the sample is obtained from the Nyquist plot. The physical properties of the molecular layer and the interface is visualized by simulating an equivalent circuit to fit the impedance data. Usually, a complex nonlinear least-squares fitting is carried out to get a good fit, and the residual plots and χ^2 values are evaluated to measure the accuracy of the fit. The bias voltage, the AC perturbation voltage employed, and the frequency range analyzed are crucial parameters in IS [52]. For these reasons, when compared to DC measurements, AC measurements provide a better understanding of the charge transport characteristics of a molecular thin film. AC techniques allow us to compare the capacitance (dielectric constant) in an equivalent circuit with the capacitance found by other experiments, which verifies how accurate the equivalent circuit is.

Impedance spectroscopy in SAM-based MTJ: Since AC impedance measurements record the phase shift between the applied bias and the response current, through equivalent circuit modeling, one can determine R_{SAM}, R_C, capacitance, and hence dielectric constant ε_r of SAMs. Thus, AC measurements complement the DC measurements by granting sufficient information to separate the individual components of the junction. Fig. (**6b**) shows an MTJ based on alkanethiol SAM. The equivalent circuit in Figs. (**6a and 6c**) can be used to fit the impedance data of many of the reported MTJs. This equivalent circuit is well established in the case of alkanethiol-based large-area tunnel junctions with the EGaIn electrode. The following equation determines the impedance of this junction.

$$Z = R_C + \left(\frac{R_{SAM}}{1 + \omega^2 R_{SAM}^2 C_{SAM}^2} \right) + j \left(\frac{\omega C_{SAM} R_{SAM}^2}{1 + \omega^2 R_{SAM}^2 C_{SAM}^2} \right) \tag{6}$$

Where the varying frequency ($\omega = 2\pi f$) is expressed in rad/s.

In recent years, potentiodynamic and temperature-dependent impedance measurements have been used to characterize MTJs [26]. In potentiodynamic IS, sinusoidal perturbation is superposed on a DC bias. Thus this approach allows us to measure impedance data over a range of applied DC bias. The measurements and equivalent circuit modeling help determine the nature of the electrical contact with the SAMs [26]. In temperature-dependent IS, measurements are performed at various temperatures to determine whether there is any charge trapping [26].

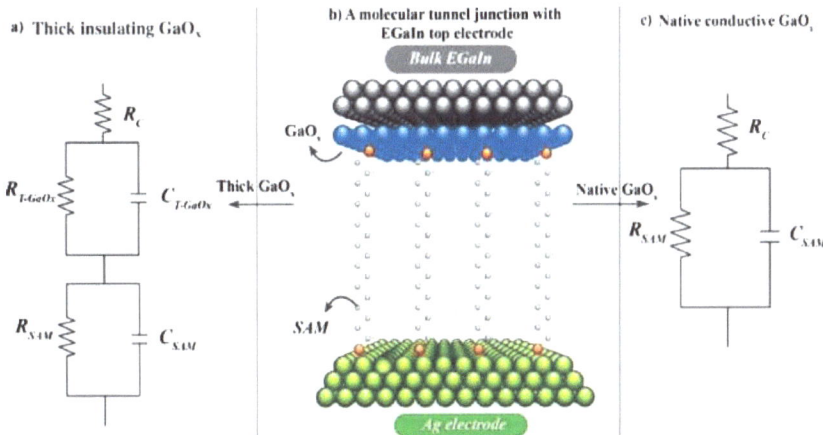

Fig. (6). (**a**) A typical circuit representation of an MTJ when PL is thick, and the layer's capacitance is no longer ignorable (**b**) a pictorial representation of a typical MTJ. (**c**) An equivalent circuit is determined, assuming the PL is a resistor. R_{SAM} stands for SAMs resistance, C_{SAM} for its capacitance, and R_C for contact resistance.

Landauer's Formalism

The single-level Landauer model has been widely utilized [53, 54] to model the charge transport in MTJ. This theoretical model would help to simulate *I-V* curves of single-molecule [55] and SAM-based junction [56]. In this model, a SAM-based junction current is given by equation **(7)** [56].

$$I = \frac{Nq}{h} \iint_{-\infty}^{\infty} dEdE' D_E'(E) G_{\delta E_{ME}}(E') \frac{\gamma_L \gamma_R}{\gamma_L + \gamma_R} [f_L(E) - f_R(E)] \tag{7}$$

where γ_L and γ_R represents the tunneling rates at the left and right interface of the device and D_E' is the electronic density of states at the molecular energy level given by [56].

$$D_E' = \frac{\frac{\gamma}{2\pi}}{\left(E - \left(E' + \left(\eta - \frac{1}{2}\right) \times V\right)\right)^2 + \left(\frac{\gamma}{2}\right)^2} \tag{8}$$

It's a Lorentzian centred at energy $E' + \left(\eta - \frac{1}{2}\right) \times V$ and with a level width $\gamma = \gamma_L + \gamma_R$. Here, $\eta = \frac{V_R}{(V_L + V_R)}$ is the voltage division parameter accounting for the capacitive coupling with the left and right electrode. The $f_{L,R}(E)$ are two distinct fermi distributions for each electrode and are given by:

$$f_{L,R}(E) = \frac{1}{1 + \exp\left[\dfrac{E \pm \dfrac{V}{2}}{k_B T}\right]} \tag{9}$$

Further a Gaussian is attached to take account of the inherent dispersion (σ) of the molecular energy (δE_{ME}) in an ensemble of molecule.

$$G_{\delta E_{ME}}(E') = A \, exp\left(\frac{(E' - \delta E_{ME})^2}{2\sigma^2}\right) \tag{10}$$

The Landauer approach connects conductance (G) of a two-terminal MTJ to transmission probability per mode (T) and the number of conduction channels (M) (the transverse modes) through equation **(11)** [57].

$$G_{\text{junction}} = \frac{2e^2 M}{h} + \frac{2e^2 M}{h} \frac{T}{1-T} = G_C + G_{\text{SAM}} \qquad (11)$$

Where h is the Planck's constant, e represents the elementary charge, G_C denotes the conductance correlated with the interfaces of a low-dimensional conductor with M sub-bands and two large contacts with many sub-bands (it does not depend on T). GSAM gives the conductance across the active layer (SAM). Since from a two-terminal architecture, one can measure the 'total' resistance, not the 'device' resistance, the frequently used terms such as "molecular effects" and "interface effects" in the SAM-based junction context are not well defined in this theoretical perspective. In impedance measurement, it is possible to separate the capacitive and resistive contributions in the transport through the molecular junction, so Landauer formalism is relevant in this context [19].

Charge Transport Through Alkanethiol-based MTJ and Comparison of Transport Data Over Different Testbeds

DC Measurements

Much information is available in the literature regarding charge transport through alkanethiol and alkane dithiol-based SAM, particularly on *J-V* behaviour and length dependence. This section compares reported *J-V* data from the literature for different molecular test beds. Fig. (**7a**) presents plots of $log_{10}|J|$ through SAMs of n-alkanethiolates against the total number of carbon atoms in the alkyl chain (n_c) for a variety of testbeds using the *J(V)* data that was previously reported. In all test beds, $log_{10}|J|$ values are calculated at $V = -0.5$ V. In each of the seven test beds, the $log_{10}|J|$ values, which represent the tunneling current density *J*, drop linearly as n_c rises [see Fig. (**7a**)]. From *J(V)* data, the resistance of molecular tunnel junction ($R_{Junction}$) is obtained. The $R_{Junction}$ is the total resistance of the junction, including the contact resistance (including the interface resistance) (R_0), the resistance due to the tunneling barrier posed by the SAM (R_{SAM}), and the resistance offered by the protective layer (R_{PL}) if present. This can be mathematically expressed as [58].

$$R_{Junction} = R_{SAM} + R_0 + R_{PL} \qquad (12)$$

It should be noted that the conductivity of the protective layer should be much larger than that of the SAM (in other words, $R_{PL} << R_{SAM} + R_0$) so that the protective layer does not mask the electrical properties of the tunneling junction. To compare the molecular resistance of alkanethiol junctions across different testbeds, one needs to estimate the single molecular resistance. Certain assumptions are made to evaluate single molecule resistance (r) from the total junction resistance (R_{Junction}). First, we assume a maximum grafting density of alkanethiol on Au (4.6×10^{18} m^{-2}) when the number of molecules in the MTJ is not stated. Secondly, if the junction's area varies in size and is not explicitly stated, the number of molecules in the MTJ is estimated using an average junction area [49]. The estimated values won't be significantly affected by these presumptions. In addition, throughout the chapter, rather than focusing on the actual lengths of molecules at MTJs, which generally include bond lengths to the electrodes, we have utilized the n_c.

A SAM-based MTJ consists of parallel molecules between a top and bottom electrode. To determine the overall resistance of the two terminal tunnel junctions (see equation **(13)**), equal resistance is attributed to each molecule in the monolayer, and oblique charge transfer between molecules is disregarded [19].

$$\frac{1}{R_{SAM}} = \frac{1}{r} + \frac{1}{r} + \frac{1}{r} + \cdots = \frac{n_{mol}}{r} = \frac{\Gamma_{SAM}}{r} \tag{13}$$

Here, n_{mol} represents the number of molecules in the MTJ, and SAMs surface coverage determines this. Single-molecule resistance is directly derived from $R_{Junction}$ based on the previously mentioned assumptions. By fitting equation **(14)** to r vs. n_c, we can determine the single-molecule contact resistance (r_0) and decay constant per carbon atom (β_N). Here β_N is the slope of this plot. Fig. (**7b**) shows the semi-log plot of r vs. n_c measured with different electrode combinations (testbeds). An exponentially increasing single molecular resistance (r) with the molecular length (measured in n_c) is the characteristic feature of coherent tunneling. It is possible to rewrite Simmons equation (equation **(1)**) as follows [59].

$$R = R_0 \, e^{\beta_N n_c} \tag{14}$$

In Fig. (**7b**) red line indicates the extrapolation performed to determine r_0. Table **1** lists the calculated decay factor (β_N) for various testbeds. For all the MTJ other than the Hg drop and PEDOT:PSS top electrode, the β_N is nearly unity. The reason for a considerable variation of β_N, compared to the values obtained by most of the experiments, is not evident. In general, no direct correlation of this type is observed in the literature between the different values of β_N obtained from comparable MTJs in various test beds.

Table 1. Comparison of the decay constant (β_N) and single-molecule resistance (r) of alkane-based MTJs discussed in this chapter. It should be noted that both the DC technique and impedance spectroscopy have been used to determine single molecule resistance.

Number of Carbons (N)	Contacts	Technique	Number of Molecules [a]	β_N(Per Carbon) [b]	r_0(MΩ) Extrapolation	r_0(MΩ) Impedance Spectroscopy	Refs.
2,4,6,8,10,12,14,16,18	Ag-SC$_n$//GaO$_x$/EGaIn	EGaIn	4.32×10^9	1.05	4.52×10^5	2.49×10^6	[28]
4,6,8,10,12,14	Au-SC$_n$//CP//Au	Carbon Paint (Large area junction)	1.41×10^9	1.19	2.25×10^2	3.50×10^4	[39]
8,10,12,14	Au-SC$_n$S//PEDOT//Au	PEDOT:PSS(Large area junction)	1.77×10^{10}	0.67	3.97×10^3	1.7×10^5	[38]
20, 24, 28	Ag-SC$_n$//C$_n$S-Hg	Hanging Hg drop junction	2.25×10^{11}	0.7325	1.25×10^5	-	[46]
6, 8, 10, 12	Au-SC$_n$//Au	CP-AFM	100-1000	0.97	19.76	-	[59]
8,12,16	Au-SC$_n$//MLG//Au	Graphene multilayer	5.67×10^7	1.06	0.9	-	[40]
8,10,12	Au-SC$_n$//RGO//Au	Reduced Graphene Oxide	4.5×10^8	1.02	5.73×10^3	-	[60]
8,12,16	Au-SC$_n$//GO-Ti-Au	Graphene Oxide	1.4×10^7	0.89	8	-	[61]

[a] The number of molecules in MTJs is estimated using the maximal grafting density of alkanethiol on Au (111), which is 4.6×10^{14}cm^{-2}. [b] Take into account that stated β_N is obtained from Fig. (**7b**).

Based on the single molecular resistance, these testbeds can be roughly divided into three categories; low resistance group (graphene oxide, graphene multilayer, and CP-AFM), medium resistance group (PEDOT:PSS, carbon paint, and reduced graphene oxide) and high resistance group (hanging Hg drop junction and EGaIn). The observed variation in single molecule resistance measured using various testbeds is ascribed to the type of contact (chemisorption/physisorption) formed at the interface. When utilizing EGaIn, the top electrode's roughness lowers the effective electrical area, which may be fixed by multiplying by an area multiplicative factor of roughly 10^4 [19]. Since Hg–SAM–SAM–Ag systems have a PL between the SAM and Hg see Fig. (**4D**), it has a higher resistance compared to other junctions (*i.e.*, n_c is comparatively large due to the symmetric bilayer of SAMs). Contact resistance significantly affects electrical transport in SAM-based MTJ. Contact resistance indicates the existence of a charge transport barrier at the metal–molecule interface. The value of contact resistance is the primary concern, and it depends on the type of contact the molecule makes with the electrodes. For instance, it has been shown that in the case of molecular junctions with chemisorbed contact at both molecule electrode interfaces, the contact resistance is much lower than (one to 2 orders) the contact resistance of the molecular junctions with one physisorbed molecule–electrode contact [19].

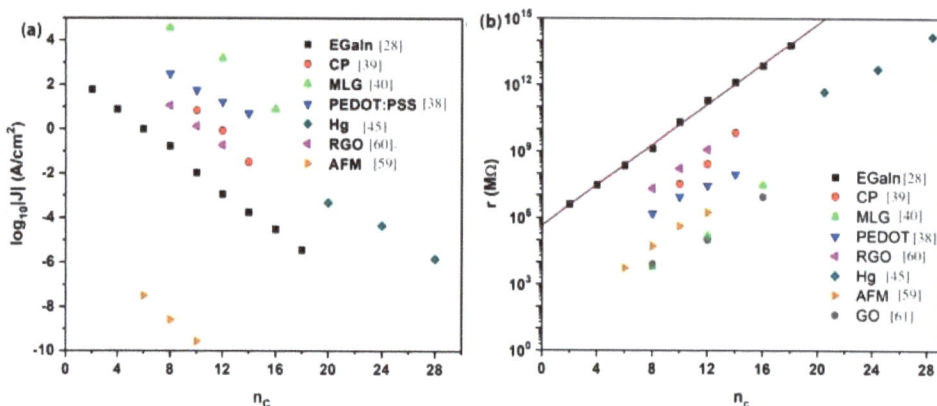

Fig. (7). (**a**) The plot depicts the log10|J| variation with the length of the alkane chain in nc on different testbeds. The input voltage is kept at 0.5V in all testbeds. The linear decrease of log10|J| with nc implies an exponential reduction of J with length, indicating tunneling of charge carriers. (**b**) A semi-log plot of the single molecule resistance (r) *vs.* nc for various test benches. The decay factor βN is estimated from the slope of this plot. In Fig. 7 (**b**) red line indicates the extrapolation performed to estimate *r*0. Reproduced with permission from ref [49]. Copyright, IOP Publishing.

Applications of AC Measurements

This section provides insight into how AC measurements make it possible to directly quantify the contact resistance, identify the role of PL, and differentiate the role of defects within MTJs. It also helps in estimating the dielectric constant of the SAM. Even though numerous MTJs were reported, equivalent circuits have only been established to date for EGaIn junctions with a thin layer of ~ 0.7 nm GaO_x [19], MTJ with carbon paint top electrode [39], and junctions with ~90 nm thick protective layer of PEDOT:PSS [34].

Determination of Contact Resistance

In this section, a comparison is made on reported values of contact resistance calculated from the DC data as well as AC data for three different tunnel junctions (Ag-SC_n//GaO_x/EGaIn, Au-SC_n//CP//Au, and Au–SC_nS//PEDOT//Au) [28, 38, 39]. Fig. (**8**) shows the single molecule resistance *r* (M-Ohm) determined from AC and DC measurements. The comparison shows that, in all testbeds, the magnitude of *r* estimated from ac measurement is higher than that calculated from dc measurements. The linear fit (marked by the red line) of EGaIn *J-V* data is extrapolated to estimate r_0, and the black line just shows the averaged value of r_0

estimated from AC data [49]. In some junctions, the exact contact resistance and the extrapolated value differ by orders of magnitude. This shows that impedance measurements are more accurate at delineating contact resistance from the intrinsic charge transfer resistance of the SAM molecules. When dealing with PEDOT:PSS, the equivalent circuit model, as stated in [3], is used to derive single-molecule resistance at n_c = 14 and contact resistance using provided impedance data. Because the impedance tests were conducted not at higher bias but zero bias, the r value for carbon paint and PEDOT:PSS is estimated at 0.05 V.

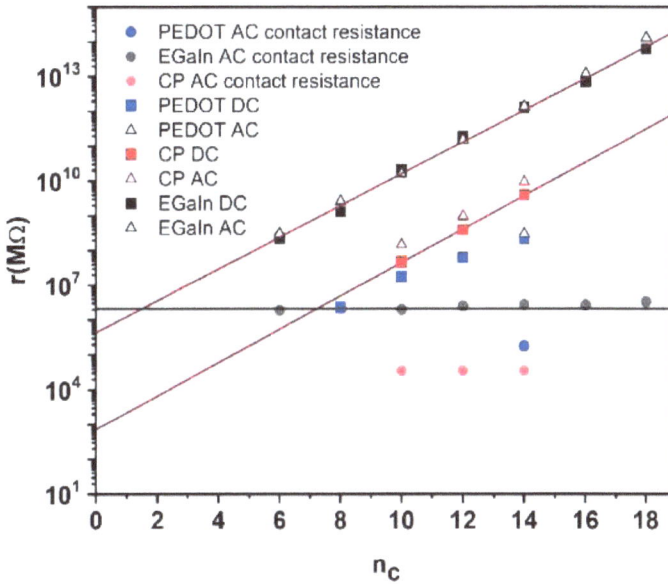

Fig. (8). A comparison of the single-molecule resistance measured by the AC and DC methods. To estimate r_0, the linear fit of the EGaIn and carbon paint J-V data is extrapolated (red line). The black line represents the average value of r_0 determined from the impedance data. Reproduced with permission from ref [49]. Copyright, IOP Publishing.

To Identify the Role of PL in Charge Transport

In this section, we present how IS could be effectively utilized to understand the effect of PL on the electron transport characteristics of Ag-SC$_n$//GaO$_x$/EGaIn MTJ. Using impedance spectroscopy, Nijhuis *et al.* evaluated the charge transport in SAM-based MTJs with EGaIn top electrode [19]. The experimental parameters are given in Table **2**. The first EGaIn junction has a thin native GaO$_x$ layer of thickness of about 0.7 nm between the EGaIn top electrode (Ag-SC$_n$//GaO$_{Native}$/EGaIn) and SAM. In the second junction, a thick layer of GaO$_x$ of around (5.2 ±0.2 nm) is formed *via* electrochemical oxidation on EGaIn top electrode (Ag-SC$_n$//GaO$_{Thick}$/EGaIn).

Table 2. Junction parameters extracted from impedance measurements.

SAM	d_{GaOx} (nm)	R_{SAM} ($\Omega.cm^2$)	C_{SAM} ($\mu F/cm^2$)	R_{PL} ($\Omega.cm^2$)	C_{PL} ($\mu F/cm^2$)
SC_9CH_3	0.7	7.5±0.4	2.16 ± 0.10	-	-
SC_9CH_3	5.2	7.8±0.5	2.31±0.12	626 ± 33	0.61±0.07

The corresponding Nyquist plots of SC_9CH_3 native and thick junctions are shown in Fig. (**9**). The Nyquist plot for the EGaIn junction with the native thin GaOx layer consists of only one semicircle (which means the presence of one capacitance). In contrast, the Nyquist plot for the EGaIn junction with thick GaO_x layer shows two semicircles. It could also be possible that the contact capacitance value exceeds the measurement limit of the impedance analyzer used. The data is fitted with equivalent circuits in Fig. (**6**) and obtained junction parameters are shown in Table **2**.

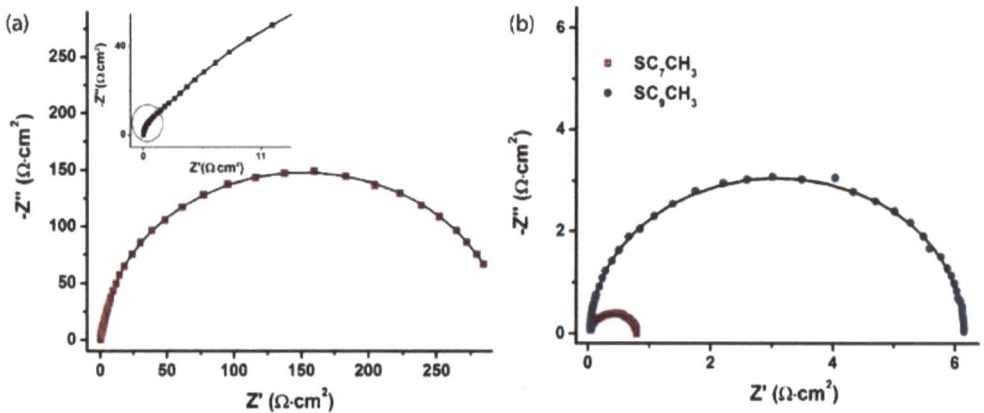

Fig. (9). Nyquist plots of Ag-SCn//GaOThick/EGaIn and Ag-SCn//GaONative/EGaIn junctions. Reproduced with permission from ref [32]. Copyright, American Chemical Society.

The main findings from this work are as follows, in the case of the junction with an electrochemically grown protective layer, the R_{SAM} obtained from impedance spectroscopy is accurate; however, R_{SAM} derived from the DC measurements is overestimated by order of magnitude. Also, the dielectric constant (ε_r) was calculated (3.4 ± 0.4) from the AC data, which agrees well with the previously reported value in alkanethiol (2.7 ± 0.3) SAMs [19].

To Identify the Nature of the Contact

The potentiodynamic impedance measurements and equivalent circuit modeling are crucial in finding the nature of the contacts in ME devices. Some junctions show rectification resulting from the Schottky contact, and potentiodynamic impedance measurements are necessary to eliminate the origin of such possibilities. Nijhuis *et al.* performed potentiodynamic impedance measurements on Ag-SC$_n$//GaO$_x$/EGaIn junction (where n = 10, 12, 14, 16 or 18) [26]. Fig. (**10**) shows Nyquist plots of the SC$_{10}$ junction. In this junction, impedance decreases with V. A tunneling transport process is indicated from the impedance data as R_{SAM} falls off exponentially with external bias in the high-bias domain. A constant R_C value across the applied range of biases suggests that the SAM//top-electrode interface (SC$_n$-GaO$_x$) exhibits an ohmic behaviour. As a result, molecular effects dominate in the junction's electrical characteristics. In all these experiments, the stability of potentiodynamic impedance measurements was ensured by comparing the values obtained from impedance measurements with the J (V) characteristics obtained from the DC measurements.

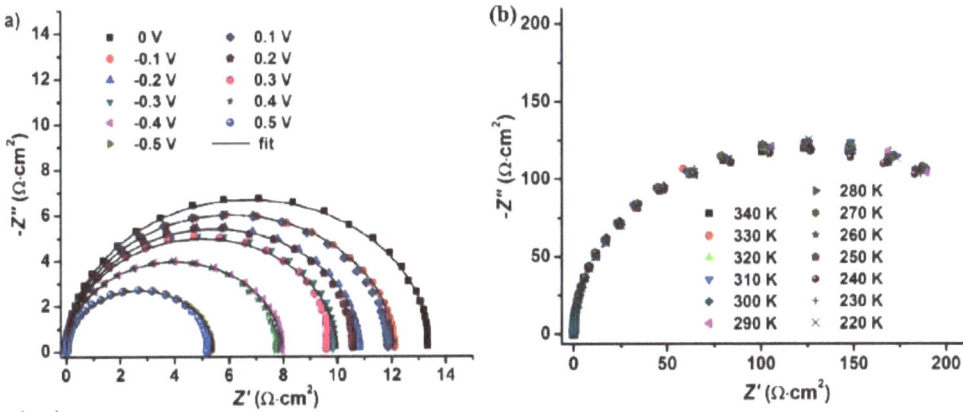

Fig. (10).   The Nyquist plots of Ag-SCn//GaOThick/EGaIn junctions (**a**) potentiodynamic and (**b**) temperature-dependent impedance spectroscopy. Reproduced with permission from ref [26] Copyright, Royal Society of Chemistry.

How the impedance depends on temperature can reveal the nature of the molecule electrode interface of a SAM-based junction. In these measurements, the SAMs resistance is temperature independent, as shown in Fig. (**11a**) and falls in an exponential manner as bias voltage increases. It demonstrates that coherent tunneling is the charge transfer mechanism across the junctions. In addition, temperature-independent R_C also means molecule electrode contacts are Ohmic. Here, C_{SAM} does not change with temperature, indicating that this junction does not contain appreciable numbers of charge traps see Fig. (**11b**).

Dielectric Measurements

Understanding and controlling the dielectric response of molecular-scale materials is essential for various applications, including OFET and supercapacitors [62, 63].

The dielectric characteristics of materials at a molecular scale can be studied using MTJ. Since the separation between electrodes (nm) is much less than the area of the electrodes in a large area MTJ, it can be modeled as a parallel plate capacitor.

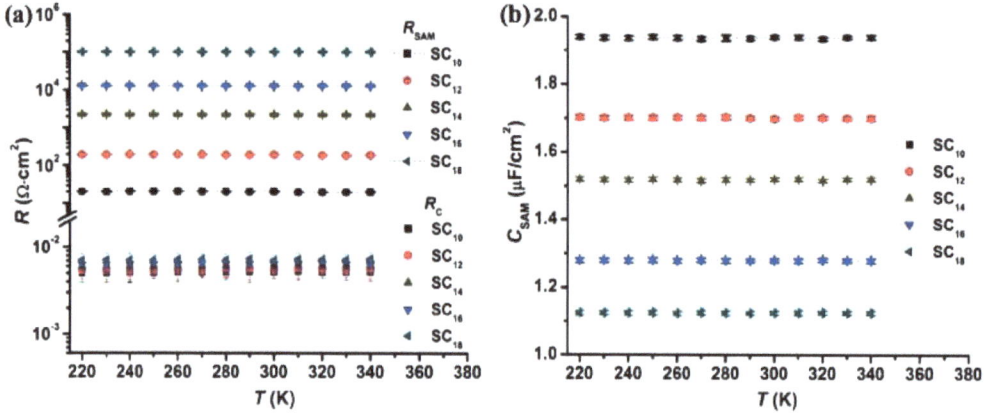

Fig. (11). The impedance data as a function of temperature at zero bias. **a)** The values of R_{SAM} and R_C vs. T. **b)** The values of C_{SAM} vs. T. Reproduced with permission from ref [26], Royal Society of Chemistry.

$$C = \varepsilon_r \varepsilon_0 \frac{A}{d} \tag{15}$$

The dielectric behaviour of SAMs depends on the molecular backbone, functional groups, packing density, tilt angle, and the molecular chain length [64]. Impedance measurements were well used previously to estimate the dielectric constant of SAMs. Alkanethiolate SAM showed its potential in replacing the conventional dielectrics used in organic FETs due to nanometer thickness and excellent insulating behaviour. Ratner *et al.* theoretically proposed that by tuning the SAMs' molecular properties, their dielectric constant can be enhanced. Nijhuis *et al.* realized it experimentally [63].

They characterized $Ag^{A\text{-}TS}$–$SC_{11}X$//GaOx/EGaIn with SAMs of $S(CH_2)_{11}X$ ($SC_{11}X$ with X = H, F, Cl, Br, or I) tunneling junctions see Fig. (**12**), with impedance spectroscopy and estimated contact resistance (R_C), the resistance of the SAM (R_{SAM}), and capacitance (C_{SAM}) are independent of one another. With the increasing polarizability of X, it has been discovered that the tunneling rate rises by three orders of magnitude, and the relative dielectric constant ε_r increases by a factor of 4. Also, according to the impedance measurements, ε_r rises by a factor of 4 from $\varepsilon_r = 2.0 \pm 0.1$ for X = F to $\varepsilon_r = 7.9 \pm 0.7$ for X =I [65].

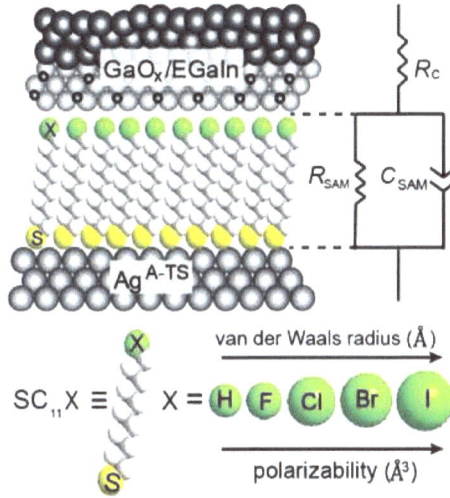

Fig. (12).   Schematic depiction of the $Ag^{A\text{-}TS}$–$SC_{11}X$//GaOx/EGaIn junction with X = H, F, Cl, Br, or I. Reproduced with permission from ref [65]. Copyright © 1999-2022 John Wiley & Sons.

Role of Defects

Defects will inevitably occur in MTJs due to impurities and roughness-induced defects in electrode materials [66]. Many reports on the effects of defects in MTJs are available in the literature, especially from the Nijhuis group [43]. Figs. (**13a and b**) illustrate an ideal molecular junction and a defective molecular junction, respectively. These defects impact the electrical properties of the MTJs. Therefore, understanding the effect of defects on the charge transport of MTJs is of paramount importance [67].

The *J(V)* measurements show that defects caused by the bottom electrode change the electrical properties of MTJs based on SAM. According to Nijhuis *et al.*, the β value in SAM-based MTJs significantly reduces as the surface roughness of the bottom substrate increases [68]. Sangeeth *et al.* utilized the potentiodynamic impedance spectroscopy to compare the charge transport properties of junction $Ag^{A\text{-}TS}$–$SC_{11}X$//GaOx/EGaIn and junction $Ag^{A\text{-}TS}$–$SC_{11}X$//GaOx/EGaIn. Fig. (**14**) shows AFM images of bottom electrodes (Template stripped (TS) and as-deposited (DE)) used for this study. Potentiodynamic impedance measurements and equivalent circuit modeling show that leakage currents exclusively affect the SAMs' resistance (R_{SAM}) in this junction see Fig. (**15**). The bottom electrode's surface roughness has no effect on contact resistance (R_C) or capacitance of the junction (C_{SAM}), but alters the value of R_{SAM} significantly (by two orders). In other words, defective junctions act as leaky capacitors [43].

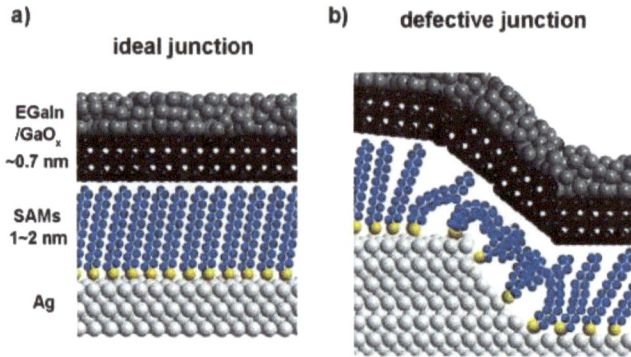

Fig. (13). Schematic representation of (**a**) an ideal molecular junction and (**b**) a defective molecular junction with a disordered SAM formed on a defect in the bottom electrode. Reproduced with permission from ref [43]. Copyright, Royal Society of Chemistry.

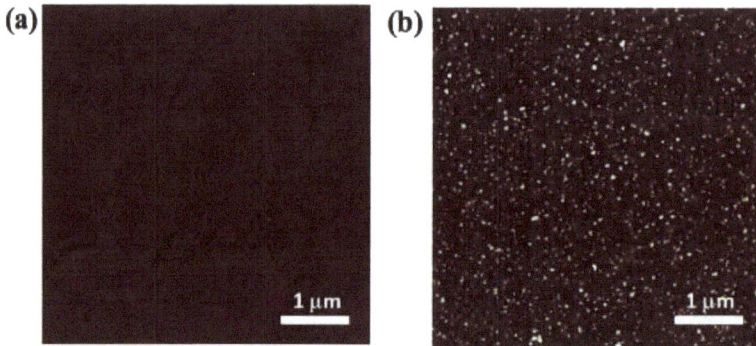

Fig. (14). AFM images of the (**a**) Ag^{A-TS} bottom electrodes made with template stripping (TS) having an rms surface roughness of 0.5 nm, (**b**) $Ag^{DE,1}$ while as-deposited (DE) electrodes made by thermal deposition with an rms surface roughness of 4.8 nm (both measured over an area of 5×5 μm^2) [43]. Reproduced with permission from ref [43]. Copyright, Royal Society of Chemistry.

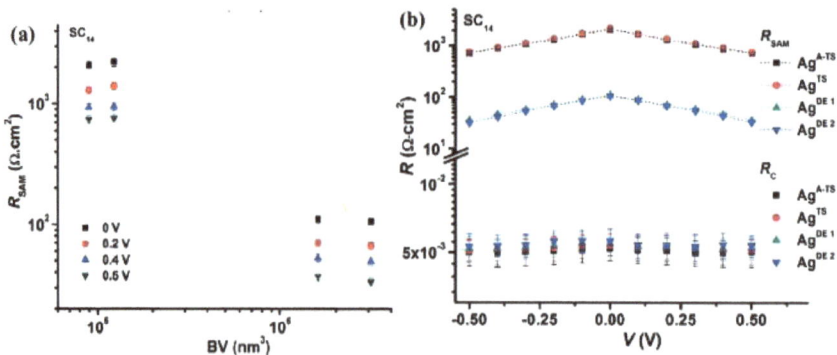

Fig. (15). (**a**) The R_{SAM} values for the junction with SC_{14} *vs* bearing volume BV. (**b**) The SAM resistance (R_{SAM}) and contact resistance (R_C) *vs.* DC bias voltages for the junction with SC_{14} SAMs [43]. Reproduced with permission from ref [43]. Copyright, Royal Society of Chemistry.

These examples demonstrate how defects may be crucial in MTJs. Using only the DC method, it is difficult to pinpoint which component of the junction is impacted by defects. A detailed study based on the IS measurements is necessary to identify the interface's and SAM's role in charge transport in defective MTJ [49].

MOLECULAR RECTIFIERS

Aviram and Ratner postulated a donor−insulator−acceptor (D-σ-A) molecular diode in 1974, which is a crucial component in molecular electronics [6]. As we know, the inorganic diode rectifier contains a p-n junction. They wanted to extend this idea to molecular diode rectifiers using donor and acceptor molecules. A molecule having a donor and acceptor moiety separated by an insulating bridge makes up the Aviram-Ratner diode [6]. The charge transfer from cathode to anode takes place in three steps. Electrons hop from the cathode to the acceptor moiety of the molecule, and then from the acceptor group, the carrier tunnels through the sigma bridge and reach the donor group. Finally, the electrons are transported from the donor part of the molecule to the anode. It was explained that at one particular bias (forward), the frontier molecular orbitals (HOMO or LUMO) of the Donor/Acceptor system align with the electrode Fermi levels.

In contrast, the alignment doesn't take place at reverse bias, resulting in asymmetric I-V characteristics [6]. One must be careful when comparing the D–σ–A unimolecular rectifier with an inorganic pn junction. The A moiety (electron-poor) is similar to the p region, whereas the D moiety (electron-rich) resembles the n-doped area of the PN junction. From Fig. (**16**), it can be clearly understood that electron flows from the first electrode to A, then to D, and eventually to the second metallic electrode under the forward bias condition. In the case of an inorganic pn junction rectifier, the preferred direction of flow for electrons is from the n-region to the p-region.

Moreover, the inorganic pn junction always consists of electrons and holes with different carrier concentrations. It is challenging to implement the Aviram-Ratner-type experimentally. However, molecular rectifiers and diodes have been realized in Langmuir−Blodgett (LB) films with SAMs and uni-molecular devices [69]. Most of the molecular rectifiers demonstrated were primarily of single-molecule junctions and were used for studying the charge transport characteristics rather than for practical applications. Whitesides *et al.* showed practical SAM-based rectifier junctions which contain ferrocene [44, 70], and bipyridyl groups [71], and the rectification appears to be due to a combination of hopping and tunneling transport. Rectification can also be observed in a purely tunneling system, provided the energy barrier is sufficiently asymmetric [72].

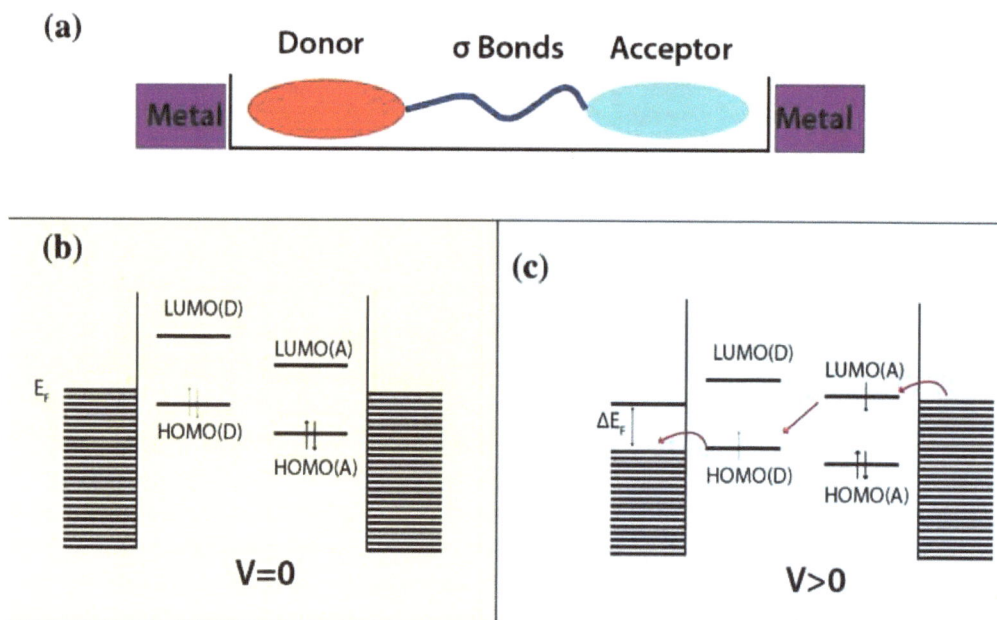

Fig. (16). (**a**) A uni-molecular rectifier that Aviram and Ratner proposed. (**b**) Energy level diagrams in schematic form.

Metzger *et al.* developed a D-π-A rectifier system, basically an LB monolayer of g-hexadecylquinolinium tricyanoquinodimethanide [73]. In their work, they proposed that the rectification could be attributed to the position of the HOMO or LUMO levels with respect to the Fermi levels (E_F) of the electrodes. Another reason could be an asymmetric reduction in the electrostatic potential over the molecule due to the presence of a substantially large dipole [74]. The theoretical work by Ratner's group suggested that the asymmetry in I-V curves in molecular junctions is possibly due to the disruption in the potential profile of the molecular junction. Under forward bias, the molecular orbital gets nearer to the E_F of an electrode, thereby reducing the tunneling barrier. On the other hand, reverse bias voltage shifts the molecular energy level in the opposite way and increases the tunneling barrier. In the other mechanism, they explained that rectification could be due to molecular asymmetry. The molecule electrode asymmetric couplings can lead to an asymmetric electrostatic potential profile across the molecule and lead to rectification at the molecular level. Apart from the Aviram-Ratner model, there were a couple of models, the Kornilovitch−Bratkovsky−Williams [75] and Datta−Paulsson model [76], widely used to explain the rectification in the molecular diodes. Usually, these approaches result in molecular diodes with small rectification ratios (R <10).

Recently molecular junctions containing redox-active molecules have attracted much attention since these molecules provide accessible energy levels enabling preferential electronic functionality at the molecular level [77]. The frontier orbitals nearer to the electrode E_F can readily take part in the transport mechanism. In the case of redox-active molecules, the carriers probably interact strongly with the molecules [77]. The charge transport in these molecular systems depends on the redox properties of the molecules and the strength of molecule electrode coupling. In case of a weak coupling limit, the carriers interact strongly with molecules leading to typical redox-type reactions in the molecular junctions, and the transport mechanism is hopping in such cases. Whereas in the case of a strong coupling limit, there is only weak interaction between carriers and the molecules. Hence redox reactions cannot occur and lead to a coherent tunneling mechanism [77]. Landauer's theory is used to explain coherent tunneling transport, while Marcus' theory is used to describe incoherent tunneling. However, in reality, the electron transport mechanism in molecular junctions is not very straightforward but complex, and a simple version of both theories does not hold well. Landauer and Marcus' theories with extensions and a combination of both approaches have been proposed to fit the practical molecular junction. There is a more extensive description of the charge transport mechanism in such systems elsewhere [77, 78].

The rectification ratio in the molecular diode is given by

$$R = \left| \frac{J_{forward}}{J_{reverse}} \right| \tag{16}$$

where $J_{forward}$ and $J_{reverse}$ are the current density in the forward and reverse bias conditions, respectively. Comparing various molecular diode rectifiers reported, it has been noted that redox-active molecule-based molecular diodes show the highest rectification ratio [77]. They show a change in transport mechanism (hopping under the forward bias condition and tunneling in the reverse bias) when we change bias polarity, as shown in Fig. (**17**). The rectification in these redox-active molecular diodes is explained considering the asymmetrical potential drop across the molecule which is essential for observing the diode action. Large areas of two terminal SAM-based molecular junctions showing the rectification phenomenon of molecules have been investigated extensively by research groups of Whitesides and Nijhuis [44, 70, 79]. They measured the *J-V* characteristic of a molecular junction consisting of SAMs of 11-(ferrocenyl)-1-undecanethiol (SC$_{11}$Fc) on the surface of ultra-smooth Ag^{TS} bottom electrodes. A soft liquid metal EGaIn is used as the top electrode in their molecular junction measurements. The EGaIn was either taken in a capillary tube or a microfluidic channel sealed using PDMS. They report a high yield, usually 70-90% for such junctions, and a rectification ratio R > 100 [70].

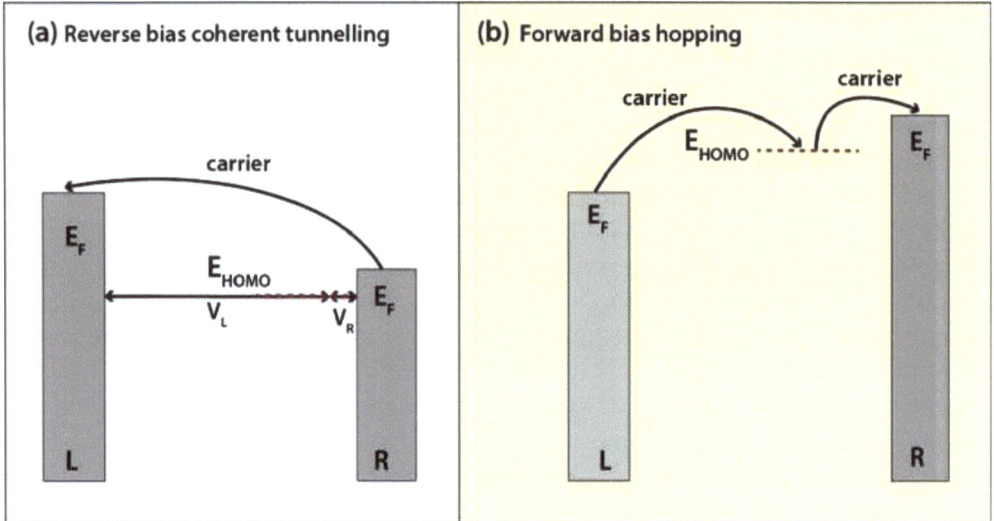

Fig. (17). Schematic illustration of the energy level diagrams for MTJs for various applied biases (**a**) Reverse bias when tunneling dominates (**b**) forward bias when hopping dominates.

A representative graph of *J vs. V* for the junction in a semi-log scale is shown in Fig. (**18b**). The histogram of the rectification ratio R with a Gaussian fit is shown in the inset for a set of five devices for 300-800 J-V curves [70]. They explained that the HOMO of the ferrocene group (Fc) couples more strongly to the GaOx/EGaIn top electrode compared bottom Ag electrode since it is situated close to the EGaIn electrode and separated from the Ag by the long SC_{11} group. The rectification can be explained as shown in Fig. (**16**). At a negative bias, the Fermi level of the top electrode GaOx/EGaIn increases, which causes the HOMO level of the Fc to rise into the window between the Fermi levels of the two electrodes [70, 79]. The HOMO level can now participate in conduction as a resonant tunneling state. On the other hand, at a positive bias, the Fermi level of the GaOx/EGaIn electrode shift down. The HOMO level of the Fc also decreases, and it remains below the Fermi levels of the electrodes and, therefore, will not take part in the electron transport [70, 79]. In this case, the only possibility of charge conduction is through a tunneling mechanism [70, 79].

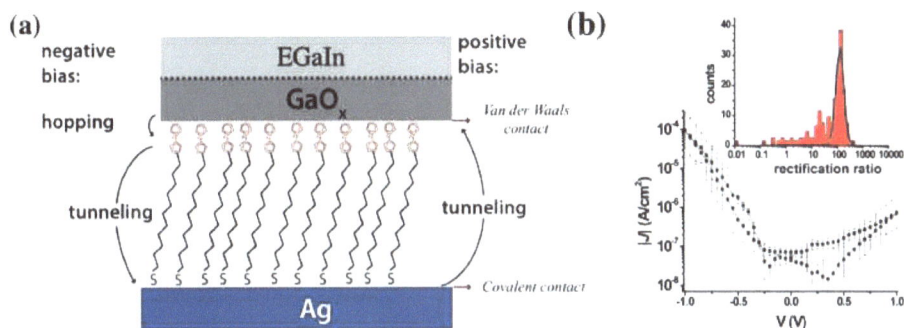

Fig. (18). (**a**) A schematic representation of the molecular rectifier consists of Ag^{TS}-$SC_{11}Fc$//Ga_2O_3/EGaIn. (**b**) A semi-log plot of the junctions' voltage vs. the current density's average absolute value. A Gaussian fit to the histogram of the rectification ratios is shown in the inset. Reproduced with permission from ref [70]. Copyright 2010, American Chemical Society.

The research group of Nijhuis has explored the Fc-based monolayer system to understand the factors affecting the performance of the diode. For instance, the role of the odd-even effect in determining the rectification ratio and how precursors' purity and topography of the bottom electrode affect the performance of the diode were studied in depth [80]. They also considered how rational molecular design might improve the rectification ratio [81]. Nijhuis *et al.* recently reported a very high rectification ratio of R= 6.3×10^5 in molecular diode, which is considered equivalent to traditional diodes based on inorganic semiconductors. An asymmetrically coupled Fc-C≡C-Fc redox unit was the foundation for this molecular junction. The mechanism was explained in a similar way to the Fc-based junction rectification was stated. This large rectification ratio is due to the Fc units' charging, which boosts the molecule electrode coupling strength in just one bias direction. It, in turn, raises the R-value by two to three orders of magnitude as compared to the previously reported Fc-based diodes [82].

Although the molecular junctions showing rectification were deeply investigated experimentally in the case of single-molecule junctions and large-area molecular junctions, the practical applications of these junctions were rarely explored. Whitesides *et al.* demonstrated a half-wave rectifier that used SAMs with Fc termini-based molecular diode [83]. Molecular junctions with SAMs of 11-(ferrocenyl)-1-undecanethiol were grown on Ag^{TS} bottom electrodes and contacted by EGaIn [84]. The rectifier circuit was examined at 50 Hz and showed an excellent response, as presented in Fig. (**19**) [83].

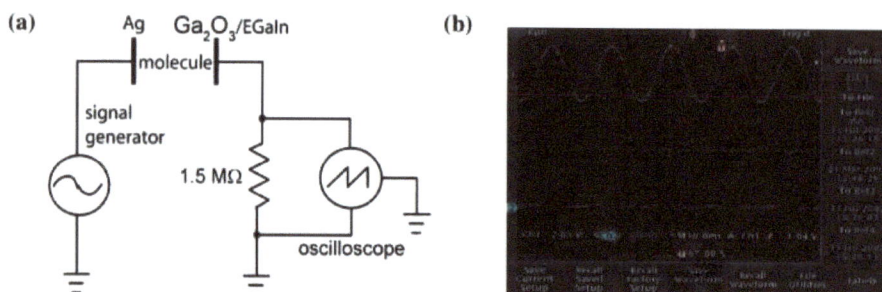

Fig. (19). (**a**) A circuit comprising an AC signal generator and a resistor in series with an MTJ acting as the diode (**b**) An oscilloscope screen showing an EGaIn top electrode junction functioning as a half-wave rectifier. Reproduced with permission from ref [83]. Copyright 2011, American Chemical Society.

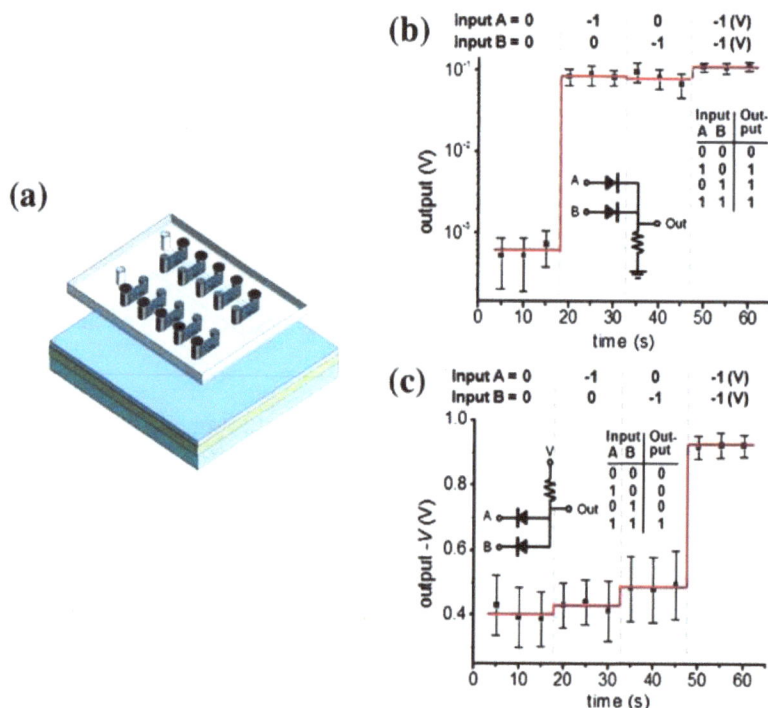

Fig. (20). (**a**) Schematic of a molecular junction utilized to fabricate molecular diode-based logic gates. (**b**) Schematic representation of the electronic circuits and the experimentally obtained truth tables for (**b**) OR and (**c**) AND logic gates, which are made up of arrays of MTJs with molecular diodes incorporated in the device. Reproduced with permission from Ref [84], Copyright, Royal Society of Chemistry.

Group of Nijhuis developed a method to fabricate a microfluidic top-electrode consisting of GaOx/EGaIn mechanically stabilized in polydimethylsiloxane (PDMS) microchannel that may be used to create arrays of SAM-based molecular junctions [84]. The top-electrodes can be simply placed on the SAM supported by Ag^{TS} or Au^{TS} bottom-electrodes. This method doesn't require lithography for patterning the bottom-electrode, which is necessary in the case of the usual method to form multiple junctions. Also, they showed that the EGaIn stabilized in a microfluidic PDMS channel as the top electrode results in molecular junction with high reproducibility (yield of ~80%) and stability. They demonstrated the working of logic gates (AND and OR gates) using the arrays of molecular diodes, as shown in Fig. (**20**). The possible architecture for real-world molecular electronic devices is two-terminal molecular junctions. Diodes based on molecular junctions with redox-active molecules provide exciting possibilities, and it offers to control the charge transport mechanism at the nanometer length scales and induce electronic function [78]. The performance of the device is determined by molecule electrode interfaces, the molecular and supramolecular structure of the junctions, and changes in the charge transport mechanism, which make rationally designing molecular devices a challenge [77, 85].

CONCLUSION

In this chapter, we covered charge transport in SAM-based molecular junctions. Various testbeds used to make molecular junctions, and their use in practical applications such as molecular diodes are discussed. We presented the recent progress in different fabrication methods which give rise to molecular junctions with high reproducibility and their potential for use in electrical devices in the future. In the case of SAM-based molecular tunnel junctions, the molecule electrode interfaces are crucial in determining the charge transport, and we discussed the significance of impedance measurements in differentiating the role of each electrical component of a molecular device. The use of redox-active molecules for constructing molecular diodes with a high rectification factor is also presented towards the end of the chapter. A more profound comprehension of electron transport across the molecules and molecular electrode coupling is mandatory for developing efficient molecular electronic devices. In addition to reducing lateral size, production costs, reliability, performance, and cost of operation are critical for emerging technology in the case of electronic device applications. Although organic molecule-based electronics/ molecular electronics have made significant progress, there is a slight possibility of it growing to replace silicon-based technology in all electronic devices. In particular, molecular electronics have not been able to compete with silicon-based technology in terms of operating speed and stability. If molecular electronics cannot meet these requirements, its future will be low-end applications with low processing costs. SAMs will eventually find use in electronic devices because they are easier to process than single-molecule junctions while retaining molecular functionality.

LEARNING OBJECTIVES

• Understanding the general aspects of self-assembled monolayer (SAM)-based two-terminal molecular junctions

• Knowing different testbeds to measure the charge transport in SAM-based tunnel junctions is discussed in detail, and the single molecular resistance of alkanethiolate SAMs is compared across these testbeds

• To understand the importance of impedance measurements in differentiating the role of each electrical component in a molecular device is explained.

• The use of redox-active molecules for constructing molecular diodes with a high rectification factor is briefly presented.

MULTIPLE CHOICE QUESTIONS

1. Which of the following is not a large area molecular electronic junction?

  a. EGaIn-based SAM junction
  b. Hg drop SAM junction
  c. Graphene-based molecular junction
  d. STM molecular junction

2. Which of the following molecular junction can have the lowest contact resistance?

  a. AFM-based molecular junction
  b. Hg drop SAM junction
  c. PEDOT: PSS-based molecular junction
  d. EGaIn-based SAM junction

3. In a molecular junction with conductive molecule(s), the electron transport can be:

  a. Tunneling
  b. Hopping
  c. Ballistic transport
  d. None of these

4. Considering the Landauer model, the minimum conductance is:

a. 0
b. 7.75×10^{-5} S
c. It cannot be predicted
d. 15.5×10^{-5} S

5. An ac voltage is applied to a pure resistor. The current through it is:

a. 90^0 out of phase with the voltage
b. 180^0 out of phase with the voltage
c. Always in phase with the voltage
d. Phase cannot be determined

6. Assuming the maximal grafting density of alkanethiol on Au (111), estimate the number of alkanethiol molecules in the cross-section of a nanopore with a diameter of 50 nm.

a. $\sim1.6\times10^3$
b. $\sim2.6\times10^4$
c. $\sim3.6\times10^4$
d. $\sim4.6\times10^5$

7. A voltage of 200 mV is applied to an atomic contact with a resistance of 5 kΩ at room temperature. The power dissipated is:

a. 0.3 μW
b. 1.6 nW
c. 2 mW
d. 10 mW

8. Consider the statement: 'DC measurement overestimates the value of resistance of the SAM layer. A plausible reason would be:

a. In DC measurements, the conductivity of the protective layer should be much larger than that of the SAM.
b. In DC measurements, one measures the total junction resistance, including the resistance of the interfaces.
c. In DC, resistances caused by the wires, probes, *etc.*, are considerable.
d. None of these.

9. An exponential increase of single molecular resistance (r) with the molecular length is the characteristic feature of.

a. Coherent tunneling
b. In-coherent tunneling
c. Hopping
d. Both b and c

10. Total junction resistance is converted into single molecule resistance (r) while comparing different test beds because:

a. It enables us to quantify relative changes due to different end groups
b. It enables us to quantify relative changes due to different electrodes
c. Both a and b
d. None of the above

11. Dielectric relaxation in the active layer contributes to …………………….. element in the equivalent circuit.

a. Resistive
b. Inductive
c. Capacitive
d. Both b and c

12. Temperature-dependent conductivity of the active layer indicates:

a. Charge transport through hopping
b. Absence of tunneling transport
c. Charge transport through tunneling
d. Ohmic conduction

13. Which of the following molecular junction cannot be used as a rectifier?

a. Junctions with redox-active SAMs
b. Junctions with donor-acceptor molecules
c. Junctions with alkanephosphate SAMs
d. None of these

14. In an experiment, an unknown polymer is used as a protective layer for the top electrode. It is found that the resistance of the monolayer varies with temperature. One may infer that:

a. The inherent transport mechanism is tunneling, negating the effect of the polymer layer.
b. The inherent transport mechanism is not tunneling, and the electrical characteristics of the polymer override that of the SAM.
c. The inherent transport mechanism is tunneling, and the electrical characteristics of the polymer override that of the SAM.
d. The inherent transport mechanism is not tunneling negating the effect of the polymer layer.

15. Temperature-independent contact resistance indicates molecule electrode contacts are:

a. Ohmic
b. Schottky
c. Both
d. None

ANSWER KEY

1. (d)

2. (a)

3. (c)

4. (b)

5. (c)

6. (c)

7. (b)

8. (b)

9. (a)

10. (c)

11. (c)

12. (b)

13. (a)

14. (c)

15. (a)

ACKNOWLEDGMENTS

C. S. Suchand Sangeeth acknowledges the Faculty Research Grant (FRG2019) NITC for financial support. Aswin Ramesh, thank UGC for the financial assistance.

REFERENCES

[1]   D. Xiang, X. Wang, C. Jia, T. Lee, and X. Guo, "Molecular-scale electronics: From concept to function", *Chem. Rev.,* vol. 116, no. 7, pp. 4318-4440, 2016.
[http://dx.doi.org/10.1021/acs.chemrev.5b00680] [PMID: 26979510]

[2]   W. LU, and C. M. LIEBER, "Nanoelectronics from the bottom up", In: *Nanoscience and Technology.* Co-Published with Macmillan Publishers Ltd: UK. pp. 137-146, 2009.
[http://dx.doi.org/10.1142/9789814287005_0014]

[3]   P.A. Van Hal, E.C.P. Smits, T.C.T. Geuns, H.B. Akkerman, B.C. De Brito, S. Perissinotto, G. Lanzani, A.J. Kronemeijer, V. Geskin, J. Cornil, P.W.M. Blom, B. De Boer, and D.M. De Leeuw, "Upscaling, integration and electrical characterization of molecular junctions", *Nat. Nanotechnol.,* vol. 3, no. 12, pp. 749-754, 2008.
[http://dx.doi.org/10.1038/nnano.2008.305] [PMID: 19057596]

[4]   A. J. Heinrich, C. P. Lutz, J. A. Gupta, and D. M. Eigler, "Molecule cascades", *Science.,* vol. 298, no. 5597, pp. 1381-1387, 2002.
[http://dx.doi.org/10.1126/science.1076768]

[5]   W. Liang, M.P. Shores, M. Bockrath, J.R. Long, and H. Park, "Kondo resonance in a single :Molecule transistor", *Nature,* vol. 417, no. 6890, pp. 725-729, 2002.
[http://dx.doi.org/10.1038/nature00790] [PMID: 12066180]

[6]   A. Aviram, and M.A. Ratner, "Molecular rectifiers", *Chem. Phys. Lett.,* vol. 29, no. 2, pp. 277-283, 1974.
[http://dx.doi.org/10.1016/0009-2614(74)85031-1]

[7]   A. Salomon, D. Cahen, S. Lindsay, J. Tomfohr, V.B. Engelkes, and C.D. Frisbie, "Comparison of electronic transport measurements on organic molecules", *Adv. Mater.,* vol. 15, no. 22, pp. 1881-1890, 2003.
[http://dx.doi.org/10.1002/adma.200306091]

[8]   H. Jeong, D. Kim, D. Xiang, and T. Lee, "High:Yield functional molecular electronic devices", *ACS Nano,* vol. 11, no. 7, pp. 6511-6548, 2017.
[http://dx.doi.org/10.1021/acsnano.7b02967] [PMID: 28578582]

[9]   J.M. Beebe, V.B. Engelkes, L.L. Miller, and C.D. Frisbie, "Contact resistance in metal-molecule-metal junctions based on aliphatic SAMs: Effects of surface linker and metal work function", *J. Am. Chem. Soc.,* vol. 124, no. 38, pp. 11268-11269, 2002.
[http://dx.doi.org/10.1021/ja0268332] [PMID: 12236731]

[10] E. Moons, M. Bruening, A. Shanzer, J. Beier, and D. Cahen, "Electron transfer in hybrid molecular solid-state devices", *Synth. Met.,* vol. 76, no. 1-3, pp. 245-248, 1996.
[http://dx.doi.org/10.1016/0379-6779(95)03463-T]

[11] W. Wang, T. Lee, and M.A. Reed, "Elastic and inelastic electron tunneling in alkane self-assembled monolayers", *J. Phys. Chem. B,* vol. 108, no. 48, pp. 18398-18407, 2004.
[http://dx.doi.org/10.1021/jp048904k]

[12] B. Mann, and H. Kuhn, "Tunneling through fatty acid salt monolayers", *J. Appl. Phys.,* vol. 42, no. 11, pp. 4398-4405, 1971.
[http://dx.doi.org/10.1063/1.1659785]

[13] J.C. Love, L.A. Estroff, J.K. Kriebel, R.G. Nuzzo, and G.M. Whitesides, "Self-assembled monolayers of thiolates on metals as a form of nanotechnology", *Chem. Rev.,* vol. 105, no. 4, pp. 1103-1170, 2005.
[http://dx.doi.org/10.1021/cr0300789] [PMID: 15826011]

[14] L. Newton, T. Slater, N. Clark, and A. Vijayaraghavan, "Self assembled monolayers (SAMs) on metallic surfaces (gold and graphene) for electronic applications", *J. Mater. Chem. C Mater. Opt. Electron. Devices,* vol. 1, no. 3, pp. 376-393, 2013.
[http://dx.doi.org/10.1039/C2TC00146B]

[15] R.L. McCreery, and A.J. Bergren, "Progress with molecular electronic junctions: meeting experimental challenges in design and fabrication", *Adv. Mater.,* vol. 21, no. 43, pp. 4303-4322, 2009.
[http://dx.doi.org/10.1002/adma.200802850] [PMID: 26042937]

[16] S. Mukhopadhyay, S.K. Karuppannan, C. Guo, J.A. Fereiro, A. Bergren, V. Mukundan, X. Qiu, O.E. Castañeda Ocampo, X. Chen, R.C. Chiechi, R. McCreery, I. Pecht, M. Sheves, R.R. Pasula, S. Lim, C.A. Nijhuis, A. Vilan, and D. Cahen, "Solid-state protein junctions: Cross-laboratory study shows preservation of mechanism at varying electronic coupling", *Science.* vol. 23, no. 5, p. 101099, 2020.
[http://dx.doi.org/10.1016/j.isci.2020.101099] [PMID: 32438319]

[17] L. Yuan, L. Jiang, D. Thompson, and C.A. Nijhuis, "On the remarkable role of surface topography of the bottom electrodes in blocking leakage currents in molecular diodes", *J. Am. Chem. Soc.,* vol. 136, no. 18, pp. 6554-6557, 2014.
[http://dx.doi.org/10.1021/ja5007417] [PMID: 24738478]

[18] M.M. Thuo, W.F. Reus, C.A. Nijhuis, J.R. Barber, C. Kim, M.D. Schulz, and G.M. Whitesides, "Odd-even effects in charge transport across self-assembled monolayers", *J. Am. Chem. Soc.,* vol. 133, no. 9, pp. 2962-2975, 2011.
[http://dx.doi.org/10.1021/ja1090436] [PMID: 21323319]

[19] C.S.S. Sangeeth, A. Wan, and C.A. Nijhuis, "Equivalent circuits of a self-assembled monolayer-based tunnel junction determined by impedance spectroscopy", *J. Am. Chem. Soc.,* vol. 136, no. 31, pp. 11134-11144, 2014.
[http://dx.doi.org/10.1021/ja505420c] [PMID: 25036915]

[20] M. Halik, and A. Hirsch, "The potential of molecular self-assembled monolayers in organic electronic devices", *Adv. Mater.,* vol. 23, no. 22-23, pp. 2689-2695, 2011.
[http://dx.doi.org/10.1002/adma.201100337] [PMID: 21823250]

[21] F. Schreiber, "Structure and growth of self-assembling monolayers", *Prog. Surf. Sci.,* vol. 65, no. 5-8, pp. 151-257, 2000.
[http://dx.doi.org/10.1016/S0079-6816(00)00024-1]

[22] S.A. DiBenedetto, A. Facchetti, M.A. Ratner, and T.J. Marks, "Molecular self-assembled monolayers and multilayers for organic and unconventional inorganic thin-film transistor applications", *Adv. Mater.,* vol. 21, no. 14-15, pp. 1407-1433, 2009.
[http://dx.doi.org/10.1002/adma.200803267]

[23] N. Amdursky, D. Marchak, L. Sepunaru, I. Pecht, M. Sheves, and D. Cahen, "Electronic transport via proteins", *Adv. Mater.,* vol. 26, no. 42, pp. 7142-7161, 2014.

[http://dx.doi.org/10.1002/adma.201402304] [PMID: 25256438]

[24]   S.M. Lindsay, and M.A. Ratner, "Molecular transport junctions: Clearing mists", *Adv. Mater.,* vol. 19, no. 1, pp. 23-31, 2007.
[http://dx.doi.org/10.1002/adma.200601140]

[25]   K.S. Wimbush, R.M. Fratila, D. Wang, D. Qi, C. Liang, L. Yuan, N. Yakovlev, K.P. Loh, D.N. Reinhoudt, A.H. Velders, and C.A. Nijhuis, "Bias induced transition from an ohmic to a non-ohmic interface in supramolecular tunneling junctions with Ga 2 O 3 /EGaIn top electrodes", *Nanoscale,* vol. 6, no. 19, pp. 11246-11258, 2014.
[http://dx.doi.org/10.1039/C4NR02933J] [PMID: 25132523]

[26]   C.S. Suchand Sangeeth, A. Wan, and C.A. Nijhuis, "Probing the nature and resistance of the molecule–electrode contact in SAM-based junctions", *Nanoscale,* vol. 7, no. 28, pp. 12061-12067, 2015.
[http://dx.doi.org/10.1039/C5NR02570B] [PMID: 26119496]

[27]   Y. Liu, X. Qiu, S. Soni, and R.C. Chiechi, "Charge transport through molecular ensembles: Recent progress in molecular electronics", *Chemical Physics Reviews.* vol. 2, no. 2, p. 021303, 2021.
[http://dx.doi.org/10.1063/5.0050667]

[28]   L. Jiang, C.S.S. Sangeeth, and C.A. Nijhuis, "The origin of the odd–even effect in the tunneling rates across egain junctions with self-assembled monolayers (SAMs) of n -alkanethiolates", *J. Am. Chem. Soc.,* vol. 137, no. 33, pp. 10659-10667, 2015.
[http://dx.doi.org/10.1021/jacs.5b05761] [PMID: 26230732]

[29]   C.A. Nijhuis, W.F. Reus, J.R. Barber, and G.M. Whitesides, "Comparison of SAM-based junctions with Ga 2 O 3 /EGain top electrodes to other large-area tunneling junctions", *J. Phys. Chem. C,* vol. 116, no. 26, pp. 14139-14150, 2012.
[http://dx.doi.org/10.1021/jp303072a]

[30]   G. Binnig, H. Rohrer, C. Gerber, and E. Weibel, "Tunneling through a controllable vacuum gap", *Appl. Phys. Lett.,* vol. 40, no. 2, pp. 178-180, 1982.
[http://dx.doi.org/10.1063/1.92999]

[31]   S.V. Aradhya, and L. Venkataraman, "Single-molecule junctions beyond electronic transport", *Nat. Nanotechnol.,* vol. 8, no. 6, pp. 399-410, 2013.
[http://dx.doi.org/10.1038/nnano.2013.91] [PMID: 23736215]

[32]   D.J. Wold, and C.D. Frisbie, "Formation of metal–molecule–metal tunnel junctions: Microcontacts to alkanethiol monolayers with a conducting AFM Tip", *J. Am. Chem. Soc.,* vol. 122, no. 12, pp. 2970-2971, 2000.
[http://dx.doi.org/10.1021/ja994468h]

[33]   J.R. Heath, "Molecular electronics", *Annu. Rev. Mater. Res.,* vol. 39, no. 1, pp. 1-23, 2009.
[http://dx.doi.org/10.1146/annurev-matsci-082908-145401]

[34]   H.B. Akkerman, P.W.M. Blom, D.M. de Leeuw, and B. de Boer, "Towards molecular electronics with large-area molecular junctions", *Nature,* vol. 441, no. 7089, pp. 69-72, 2006.
[http://dx.doi.org/10.1038/nature04699] [PMID: 16672966]

[35]   T.W. Kim, G. Wang, H. Lee, and T. Lee, "Statistical analysis of electronic properties of alkanethiols in metal–molecule–metal junctions", *Nanotechnology.* vol. 18, no. 31, p. 021303, 2007.
[http://dx.doi.org/10.1088/0957-4484/18/31/315204]

[36]   J.G. Kushmerick, J. Naciri, J.C. Yang, and R. Shashidhar, "Conductance scaling of molecular wires in parallel", *Nano Lett.,* vol. 3, no. 7, pp. 897-900, 2003.
[http://dx.doi.org/10.1021/nl034201n]

[37]   J. Chen, M. A. Reed, A. M. Rawlett, and J. M. Tour, "Large on-off ratios and negative differential resistance in a molecular electronic device", *Science.,* vol. 286, no. 5444, pp. 1550-1552, 1999.
[http://dx.doi.org/10.1126/science.286.5444.1550]

[38] H.B. Akkerman, R.C.G. Naber, B. Jongbloed, P.A. van Hal, P.W.M. Blom, D.M. de Leeuw, and B. de Boer, "Electron tunneling through alkanedithiol self-assembled monolayers in large-area molecular junctions", *Proc. Natl. Acad. Sci.,* vol. 104, no. 27, pp. 11161-11166, 2007. [http://dx.doi.org/10.1073/pnas.0701472104] [PMID: 17592120]

[39] S.K. Karuppannan, E.H.L. Neoh, A. Vilan, and C.A. Nijhuis, "Protective layers based on carbon paint to yield high-quality large-area molecular junctions with low contact resistance", *J. Am. Chem. Soc.,* vol. 142, no. 7, pp. 3513-3524, 2020. [http://dx.doi.org/10.1021/jacs.9b12424] [PMID: 31951129]

[40] G. Wang, Y. Kim, M. Choe, T.W. Kim, and T. Lee, "A new approach for molecular electronic junctions with a multilayer graphene electrode", *Adv. Mater.,* vol. 23, no. 6, pp. 755-760, 2011. [http://dx.doi.org/10.1002/adma.201003178] [PMID: 21287637]

[41] X. Liang, B.A. Sperling, I. Calizo, G. Cheng, C.A. Hacker, Q. Zhang, Y. Obeng, K. Yan, H. Peng, Q. Li, X. Zhu, H. Yuan, A.R. Hight Walker, Z. Liu, L. Peng, and C.A. Richter, "Toward clean and crackless transfer of graphene", *ACS Nano,* vol. 5, no. 11, pp. 9144-9153, 2011. [http://dx.doi.org/10.1021/nn203377t] [PMID: 21999646]

[42] S.H. Hong, D.H. Seo, and H. Song, "Demonstration of molecular tunneling junctions based on vertically stacked graphene heterostructures", *Crystals,* vol. 12, no. 6, p. 787, 2022. [http://dx.doi.org/10.3390/cryst12060787]

[43] C.S.S. Sangeeth, L. Jiang, and C.A. Nijhuis, "Bottom-electrode induced defects in self-assembled monolayer (SAM)-based tunnel junctions affect only the SAM resistance, not the contact resistance or SAM capacitance", *RSC Advances,* vol. 8, no. 36, pp. 19939-19949, 2018. [http://dx.doi.org/10.1039/C8RA01513A] [PMID: 35541643]

[44] C.A. Nijhuis, W.F. Reus, and G.M. Whitesides, "Molecular rectification in metal-SAM-metal oxide-metal junctions", *J. Am. Chem. Soc.,* vol. 131, no. 49, pp. 17814-17827, 2009. [http://dx.doi.org/10.1021/ja9048898] [PMID: 19928851]

[45] C.A. Nijhuis, W.F. Reus, J.R. Barber, and G.M. Whitesides, "comparison of sam-based junctions with Ga 2 O 3 /EGaIn top electrodes to other large-area tunneling junctions", *J. Phys. Chem. C,* vol. 116, no. 26, pp. 14139-14150, 2012. [http://dx.doi.org/10.1021/jp303072a]

[46] E.A. Weiss, R.C. Chiechi, G.K. Kaufman, J.K. Kriebel, Z. Li, M. Duati, M.A. Rampi, and G.M. Whitesides, "Influence of defects on the electrical characteristics of mercury-drop junctions: Self-assembled monolayers of n-alkanethiolates on rough and smooth silver", *J. Am. Chem. Soc.,* vol. 129, no. 14, pp. 4336-4349, 2007. [http://dx.doi.org/10.1021/ja0677261] [PMID: 17358061]

[47] A. Wan, L. Jiang, C.S.S. Sangeeth, and C.A. Nijhuis, "Reversible soft top-contacts to yield molecular junctions with precise and reproducible electrical characteristics", *Adv. Funct. Mater.,* vol. 24, no. 28, pp. 4442-4456, 2014. [http://dx.doi.org/10.1002/adfm.201304237]

[48] C.S.S. Sangeeth, A.T. Demissie, L. Yuan, T. Wang, C.D. Frisbie, and C.A. Nijhuis, "Comparison of DC and AC Transport in 1.5–7.5 nm Oligophenylene Imine Molecular Wires across Two Junction Platforms: Eutectic Ga–In *versus* Conducting Probe Atomic Force Microscope Junctions", *J. Am. Chem. Soc.,* vol. 138, no. 23, pp. 7305-7314, 2016. [http://dx.doi.org/10.1021/jacs.6b02039] [PMID: 27172452]

[49] J. Francis, S.A. Bassam, and C.S.S. Sangeeth, "Importance of impedance spectroscopy in self-assembled monolayer-based large-area tunnel junctions", *J. Phys. D Appl. Phys..* vol. 55, no. 7, p. 075301, 2022. [http://dx.doi.org/10.1088/1361-6463/ac30fc]

[50] W.F. Reus, C.A. Nijhuis, J.R. Barber, M.M. Thuo, S. Tricard, and G.M. Whitesides, "Statistical tools for analyzing measurements of charge transport", *J. Phys. Chem. C,* vol. 116, no. 11, pp. 6714-6733,

2012.
[http://dx.doi.org/10.1021/jp210445y]

[51]   J. R. Macdonald, and W. B. Johnson, "Fundamentals of impedance spectroscopy", In: *Impedance Spectroscopy: Theory, Experiment, and Applications.* Wiley, 2005.
[http://dx.doi.org/10.1002/0471716243.ch1]

[52]   B. Boukamp, "A nonlinear least squares fit procedure for analysis of immittance data of electrochemical systems", *Solid State Ion.,* vol. 20, no. 1, pp. 31-44, 1986.
[http://dx.doi.org/10.1016/0167-2738(86)90031-7]

[53]   A.R. Garrigues, L. Yuan, L. Wang, E.R. Mucciolo, D. Thompon, E. del Barco, and C.A. Nijhuis, "A single-level tunnel model to account for electrical transport through single molecule- and self-assembled monolayer-based junctions", *Sci. Rep.,* vol. 6, no. 1, p. 26517, 2016.
[http://dx.doi.org/10.1038/srep26517] [PMID: 27216489]

[54]   A. Vilan, D. Aswal, and D. Cahen, "Large-area, ensemble molecular electronics: motivation and challenges", *Chem. Rev.,* vol. 117, no. 5, pp. 4248-4286, 2017.
[http://dx.doi.org/10.1021/acs.chemrev.6b00595] [PMID: 28177226]

[55]   L.A. Zotti, T. Kirchner, J.C. Cuevas, F. Pauly, T. Huhn, E. Scheer, and A. Erbe, "Revealing the role of anchoring groups in the electrical conduction through single-molecule junctions", *Small,* vol. 6, no. 14, pp. 1529-1535, 2010.
[http://dx.doi.org/10.1002/smll.200902227] [PMID: 20578111]

[56]   X. Chen, B. Kretz, F. Adoah, C. Nickle, X. Chi, X. Yu, E. del Barco, D. Thompson, D.A. Egger, and C.A. Nijhuis, "A single atom change turns insulating saturated wires into molecular conductors", *Nat. Commun.,* vol. 12, no. 1, p. 3432, 2021.
[http://dx.doi.org/10.1038/s41467-021-23528-8] [PMID: 34103489]

[57]   S. Datta, "Electronic transport in mesoscopic systems",
[http://dx.doi.org/10.1017/CBO9780511805776]

[58]   F.C. Simeone, H.J. Yoon, M.M. Thuo, J.R. Barber, B. Smith, and G.M. Whitesides, "Defining the value of injection current and effective electrical contact area for EGaIn-based molecular tunneling junctions", *J. Am. Chem. Soc.,* vol. 135, no. 48, pp. 18131-18144, 2013.
[http://dx.doi.org/10.1021/ja408652h] [PMID: 24187999]

[59]   V.B. Engelkes, J.M. Beebe, and C.D. Frisbie, "Length-dependent transport in molecular junctions based on SAMs of alkanethiols and alkanedithiols: effect of metal work function and applied bias on tunneling efficiency and contact resistance", *J. Am. Chem. Soc.,* vol. 126, no. 43, pp. 14287-14296, 2004.
[http://dx.doi.org/10.1021/ja046274u] [PMID: 15506797]

[60]   S. Seo, M. Min, J. Lee, T. Lee, S.Y. Choi, and H. Lee, "Solution-processed reduced graphene oxide films as electronic contacts for molecular monolayer junctions", *Angew. Chem. Int. Ed.,* vol. 51, no. 1, pp. 108-112, 2012.
[http://dx.doi.org/10.1002/anie.201105895] [PMID: 22076709]

[61]   M. Kühnel, S.V. Petersen, R. Hviid, M.H. Overgaard, B.W. Laursen, and K. Nørgaard, "Monolayered graphene oxide as a low contact resistance protection layer in alkanethiol solid-state devices", *J. Phys. Chem. C,* vol. 122, no. 18, pp. 9731-9737, 2018.
[http://dx.doi.org/10.1021/acs.jpcc.7b12606]

[62]   H.M. Heitzer, T.J. Marks, and M.A. Ratner, "Maximizing the dielectric response of molecular thin films *via* quantum chemical design", *ACS Nano,* vol. 8, no. 12, pp. 12587-12600, 2014.
[http://dx.doi.org/10.1021/nn505431p] [PMID: 25415650]

[63]   H.M. Heitzer, T.J. Marks, and M.A. Ratner, "Computation of dielectric response in molecular solids for high capacitance organic dielectrics", *Acc. Chem. Res.,* vol. 49, no. 9, pp. 1614-1623, 2016.
[http://dx.doi.org/10.1021/acs.accounts.6b00173] [PMID: 27576058]

[64] C. Van Dyck, T.J. Marks, and M.A. Ratner, "Chain length dependence of the dielectric constant and polarizability in conjugated organic thin films", *ACS Nano,* vol. 11, no. 6, pp. 5970-5981, 2017. [http://dx.doi.org/10.1021/acsnano.7b01807] [PMID: 28575578]

[65] D. Wang, D. Fracasso, A. Nurbawono, H.V. Annadata, C.S.S. Sangeeth, L. Yuan, and C.A. Nijhuis, "Tuning the tunneling rate and dielectric response of SAM-Based Junctions via a single polarizable atom", *Adv. Mater.,* vol. 27, no. 42, pp. 6689-6695, 2015. [http://dx.doi.org/10.1002/adma.201502968] [PMID: 26414779]

[66] L. Jiang, C.S.S. Sangeeth, L. Yuan, D. Thompson, and C.A. Nijhuis, "One-nanometer thin monolayers remove the deleterious effect of substrate defects in molecular tunnel junctions", *Nano Lett.,* vol. 15, no. 10, pp. 6643-6649, 2015. [http://dx.doi.org/10.1021/acs.nanolett.5b02481] [PMID: 26340232]

[67] G.D. Kong, J. Jin, M. Thuo, H. Song, J.F. Joung, S. Park, and H.J. Yoon, "Elucidating the role of molecule–electrode interfacial defects in charge tunneling characteristics of large-area junctions", *J. Am. Chem. Soc.,* vol. 140, no. 38, pp. 12303-12307, 2018. [http://dx.doi.org/10.1021/jacs.8b08146] [PMID: 30183277]

[68] L. Yuan, L. Jiang, B. Zhang, and C.A. Nijhuis, "Dependency of the tunneling decay coefficient in molecular tunneling junctions on the topography of the bottom electrodes", *Angew. Chem. Int. Ed.,* vol. 53, no. 13, pp. 3377-3381, 2014. [http://dx.doi.org/10.1002/anie.201309506] [PMID: 24615875]

[69] J. Hihath, C. Bruot, H. Nakamura, Y. Asai, I. Díez-Pérez, Y. Lee, L. Yu, and N. Tao, "Inelastic transport and low-bias rectification in a single-molecule diode", *ACS Nano,* vol. 5, no. 10, pp. 8331-8339, 2011. [http://dx.doi.org/10.1021/nn2030644] [PMID: 21932824]

[70] C.A. Nijhuis, W.F. Reus, J.R. Barber, M.D. Dickey, and G.M. Whitesides, "Charge transport and rectification in arrays of SAM-based tunneling junctions", *Nano Lett.,* vol. 10, no. 9, pp. 3611-3619, 2010. [http://dx.doi.org/10.1021/nl101918m] [PMID: 20718403]

[71] H.J. Yoon, K.C. Liao, M.R. Lockett, S.W. Kwok, M. Baghbanzadeh, and G.M. Whitesides, "Rectification in tunneling junctions: 2,2′-bipyridyl-terminated n-alkanethiolates", *J. Am. Chem. Soc.,* vol. 136, no. 49, pp. 17155-17162, 2014. [http://dx.doi.org/10.1021/ja509110a] [PMID: 25389953]

[72] J. Hihath, C. Bruot, H. Nakamura, Y. Asai, I. Díez-Pérez, Y. Lee, L. Yu, and N. Tao, "Inelastic transport and low-bias rectification in a single-molecule diode", *ACS Nano,* vol. 5, no. 10, pp. 8331-8339, 2011. [http://dx.doi.org/10.1021/nn2030644] [PMID: 21932824]

[73] D. Vuillaume, B. Chen, and R.M. Metzger, "Electron transfer through a monolayer of hexadecylquinolinium tricyanoquinodimethanide", *Langmuir,* vol. 15, no. 11, pp. 4011-4017, 1999. [http://dx.doi.org/10.1021/la990099r]

[74] C. Krzeminski, C. Delerue, G. Allan, D. Vuillaume, and R.M. Metzger, "Theory of electrical rectification in a molecular monolayer", *Phys. Rev. B Condens. Matter.* vol. 64, no. 8, p. 085405, 2001. [http://dx.doi.org/10.1103/PhysRevB.64.085405]

[75] P.E. Kornilovitch, A.M. Bratkovsky, and R. Stanley Williams, "Current rectification by molecules with asymmetric tunneling barriers", *Phys. Rev. B Condens. Matter.* vol. 66, no. 16, p. 165436, 2002. [http://dx.doi.org/10.1103/PhysRevB.66.165436]

[76] F. Zahid, A.W. Ghosh, M. Paulsson, E. Polizzi, and S. Datta, "Charging-induced asymmetry in molecular conductors", *Phys. Rev. B Condens. Matter Mater. Phys..* vol. 70, no. 24, p. 245317, 2004. [http://dx.doi.org/10.1103/PhysRevB.70.245317]

[77]   Y. Han, and C.A. Nijhuis, "Functional redox-active molecular tunnel junctions", *Chem. Asian J.,* vol. 15, no. 22, pp. 3752-3770, 2020.
[http://dx.doi.org/10.1002/asia.202000932] [PMID: 33015998]

[78]   M. Galperin, M.A. Ratner, and A. Nitzan, "Molecular transport junctions: Vibrational effects", *J. Phys. Condens. Matter,* vol. 19, no. 10, p. 103201, 2007.
[http://dx.doi.org/10.1088/0953-8984/19/10/103201]

[79]   C.A. Nijhuis, W.F. Reus, and G.M. Whitesides, "Mechanism of rectification in tunneling junctions based on molecules with asymmetric potential drops", *J. Am. Chem. Soc.,* vol. 132, no. 51, pp. 18386-18401, 2010.
[http://dx.doi.org/10.1021/ja108311j] [PMID: 21126089]

[80]   N. Nerngchamnong, L. Yuan, D.C. Qi, J. Li, D. Thompson, and C.A. Nijhuis, "The role of van der Waals forces in the performance of molecular diodes", *Nat. Nanotechnol.,* vol. 8, no. 2, pp. 113-118, 2013.
[http://dx.doi.org/10.1038/nnano.2012.238] [PMID: 23292010]

[81]   L. Yuan, R. Breuer, L. Jiang, M. Schmittel, and C.A. Nijhuis, "A molecular diode with a statistically robust rectification ratio of three orders of magnitude", *Nano Lett.,* vol. 15, no. 8, pp. 5506-5512, 2015.
[http://dx.doi.org/10.1021/acs.nanolett.5b02014] [PMID: 26196854]

[82]   X. Chen, M. Roemer, L. Yuan, W. Du, D. Thompson, E. del Barco, and C.A. Nijhuis, "Molecular diodes with rectification ratios exceeding 105 driven by electrostatic interactions", *Nat. Nanotechnol.,* vol. 12, no. 8, pp. 797-803, 2017.
[http://dx.doi.org/10.1038/nnano.2017.110] [PMID: 28674457]

[83]   C.A. Nijhuis, W.F. Reus, A.C. Siegel, and G.M. Whitesides, "A molecular half-wave rectifier", *J. Am. Chem. Soc.,* vol. 133, no. 39, pp. 15397-15411, 2011.
[http://dx.doi.org/10.1021/ja201223n] [PMID: 21842878]

[84]   A. Wan, C.S. Suchand Sangeeth, L. Wang, L. Yuan, L. Jiang, and C.A. Nijhuis, "Arrays of high quality SAM-based junctions and their application in molecular diode based logic", *Nanoscale,* vol. 7, no. 46, pp. 19547-19556, 2015.
[http://dx.doi.org/10.1039/C5NR05533D] [PMID: 26537895]

[85]   H. Jeong, D. Kim, D. Xiang, and T. Lee, "High-yield functional molecular electronic devices", *ACS Nano,* vol. 11, no. 7, pp. 6511-6548, 2017.
[http://dx.doi.org/10.1021/acsnano.7b02967] [PMID: 28578582]

CHAPTER 3

Performance Analysis of Rectangular Core-Shell Double Gate Junctionless Transistor (RCS-DGJLT)

Vishal Narula[1,*], Shekhar Verma[1], Amit Saini[2] and Mohit Agarwal[3]

[1] *Lovely Professional University, Jalandhar, Punjab, India*

[2] *Cadre Design Systems, Ghaziabad, India*

[3] *Thapar Institute of Engineering and technology, Patiala, Punjab 147004, India*

Abstract: The shrinking of the device parameters' dimensions could be a solution for improving the performance and high transistor density of traditional MOSFETs. However, the short-channel effects could create a problem in the performance of the device. This chapter examines and performs comprehensive simulations of the standard junctionless double-gate transistor. In this research, silicon thickness and work function engineering are used to better understand the junctionless transistor's operation. As silicon thickness increases, the junctionless double-gate FET's performance begins to decline. Additionally, the typical double-gate junctionless FET is modelled, and the change in silicon thickness, work function, gate dielectric, and doping concentration is studied. The findings of the analysis and simulation are found to be quite similar. As a result, the device is referred to as a rectangular core-shell double-gate junctionless transistor because of the core being sandwiched between the two shells of the device (RCS-DGJLT). While the core-shell is doped with acceptor impurities in an n-type RCS-DGJLT, donor impurities are used in the shells. The device performance parameters have been improved such that I_{OFF} of order $\sim10^{-14}$A, $I_{ON} \sim10^{-5}$A, $I_{ON}/I_{OFF} \sim10^{9}$, SS nearly 68.9mV/decade, DIBL nearly 52.6mV/V are obtained at a total silicon thickness of 12nm and channel length of 20nm. The effect of channel length variation on RCS-DGJLT is also studied. RCS-DGJLT is found to have better performance than conventional DGJLT.

Keywords: Core-Shell, DIBL, Junctionless, MOSFET, Multigate, Oxide Material, Subthreshold-Slope, Transconductance, Threshold Voltage, VLSI.

INTRODUCTION

The development of the electronics sector is greatly stepped up by increasing consumer spending around the globe. With the increase in economies, consumer

---

* **Corresponding author Vishal Narula:** Lovely Professional University, Jalandhar, Punjab, India;
E-mail: vishal.narula2302@gmail.com

Gopal Rawat & Aniruddh Bahadur Yadav (Eds.)

demand also increases. It was in the 20th century that the electronic industry was born. The most profitable sector within electronics is the semiconductor industry which has grown to become more than $400 billion globally as of 2018.

When Brattain, Shockley, and Bardeen built the first transistor in 1947, this revolution in the semiconductor industry began. This is followed by an introduction to the "workhorse of the electronics industry," which is the MOSFET. For the last 30 years, microelectronics which is a subfield of electronics that relates to the study of small electronic components and design, has benefited marvelously from MOSFET miniaturization. Complementary metal-oxide-semiconductor (CMOS) has revolutionized the human lifestyle. The larger packaging density and improved performance showed the supremacy of reduced physical size. According to Moore's law, the transistor density on a chip will double every 18 months, and improvements in many properties would accompany that increase.

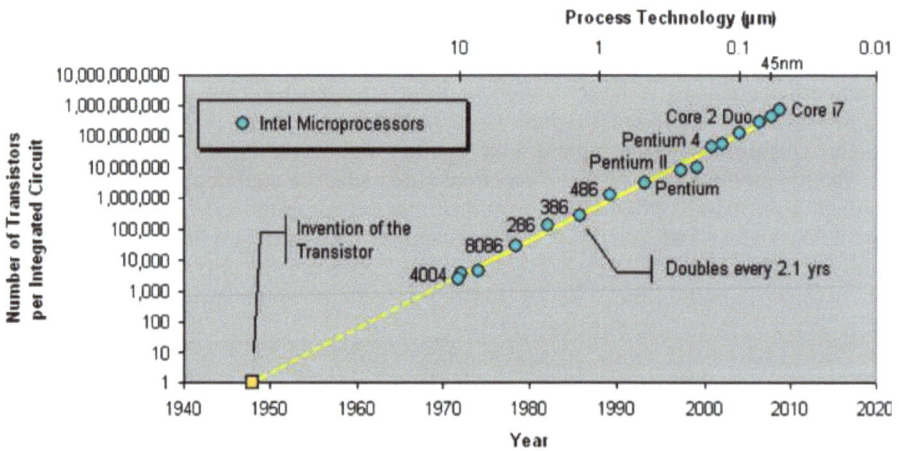

Fig. (1). Transistor density along with process technology node opted by intel microprocessors from the year of the invention of the transistor.

Fig. (**1**) shows an illustration of the number of transistors per integrated circuit and process technology node opted by intel microprocessor from the invention year of the transistor, which follows Moore's law. The latest development of an artificial intelligence (AI) chip called "Cerebras'" with 1.2 trillion transistors is now commercially available. Further, the miniaturization of dimensions of the MOSFET is becoming difficult due to short channel effects (SCE's) and higher gate leakage currents that limit the device performance, such as higher microprocessor speed, along with less power dissipation. There are various challenges during the scaling of the transistors, like gate leakage current, better gate control, output current, power/performance ratio and reliability of the device.

The challenges under gate control lie in different short channel effects, which include gate-induced drain leakage (GIDL), subthreshold slope degradation, gate oxide tunneling of carriers, source/drain direct tunneling, DIBL, which degrades the performance of the device and thus affects the subthreshold characteristics.

For further downscaling of the dimension of the MOSFET, the multi-gate-based MOSFET's (MugFET) comes into the picture such that the control on the carriers of the device can be enhanced [1]. MugFET improves the SCE's, reduces the device variability and increases the carrier mobility without the need for high doping concentration. The SCE's are highly reduced in MugFET's not by scaling the oxide dielectric but by scaling the thickness of the channel, which brings down the gate tunneling current. Therefore, due to these advantages, according to International Roadmap for Semiconductors (ITRS), MugFET's have been predicted as a successor to planar transistors since 2001 [1]. Fig. (**2**) shows different multigate device architectures. The double gate [2], triple gate [3], quadruple gate [4], cylindrical gate [5] and bulk FinFET [6] architectures are thus shown below.

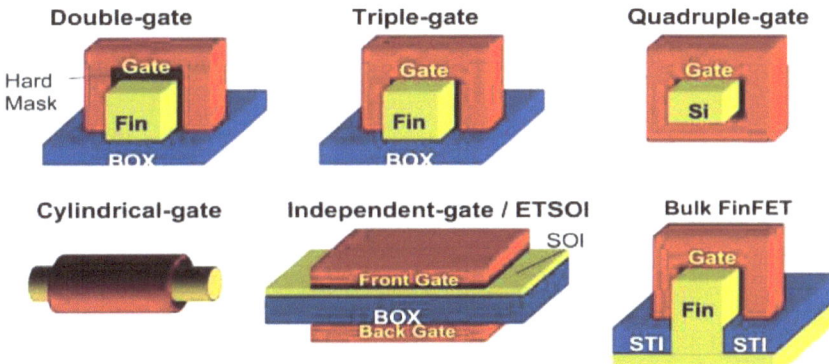

Fig. (2). Different device architectures of multigate FET.

The reduction in SCE's gives a great motivation towards achieving the best performance of the miniaturized device. Unfortunately, the presence of source and drain region does not nullify the SCE's completely in MugFET's and brings challenges in doping profile techniques. The requirement of abrupt source and drain junctions in ultra-short channel devices invites multiple complications, even in MugFET's [7]. The drain and source regions of n-type MOSFETs are extensively populated. Acceptor and donor concentrations of 10^{15} cm^{-3} and 10^{20} cm^{-3}, respectively, are used to dope the channel region. To realize such abrupt doping concentration in short-channel devices is a tedious task. The statistical nature of the distribution of dopant atoms and the law of diffusion restricts to realization of such abrupt junctions [7]. The minimization of diffusion of the atoms is done by the flash annealing technique. Even if diffusion is completely absent, other techniques like ion implantation and many more could not achieve such perfectly abrupt junctions [8 - 12].

The junctionless technology is already in use by some of the semiconductor industries. Imec is one of the nanoelectronics industries that developed junctionless nanowire FET's and explored their potential by proposing a novel SRAM cell design with two vertically stacked junctionless vertical NWFET's reducing the SRAM area per bit by 39%. According to Imec, the JLT could be a viable option for analog/RF applications as well. JL-FETs are thought to be a promising research area, and many businesses and academic groups are actively investigating their potential advantages and difficulties.

DEVICE STRUCTURE & WORKING PRINCIPLE OF JUNCTIONLESS TRANSISTOR

Junctionless Transistor with Double Gate

In the world of devices, junctionless transistors, with no p-n junctions at the source/channel and channel/drain interfaces, are becoming increasingly prevalent. Due to tiny cross-sections and high-density integration, the device has become more commonly used [8, 10, 13 - 15]. Source, channel, and drain dopants in a junctionless transistor have zero concentration gradients. Reduced short-channel effects are attributed to the absence of p-n junction. Junctionless devices have been pursued by researchers because of the ease and simplicity of their manufacturing. Contrary to traditional MOSFETs, junctionless devices have more bulk conduction than MOSFETs with surface conduction. The ability to turn off junctionless devices is contingent on the disparity in work function between the gate and the semiconductor [9]. Channel thickness is kept minimum in order to reduce leakage current [16] and deplete the carriers. Process parameters must be carefully controlled for fabricating devices with a small thickness, though [17]. Conventional JLT design factors such as total silicon thickness and gate work function must be studied before using a novel suggested RCS-DGJLT structure, which has been presented. Scaling of the dimensions of a conventional MOSFET introduces several short-channel effects.

It is feasible to lessen short-channel effects if the effective channel length of the device is decreased in junctionless-based transistors. The dopant concentration is the same from the drain to the channel to the source in a junctionless device. It is demonstrated in Figs. (**3a & b**) that n and p-type conventional double gate junctionless transistors have the same construction.

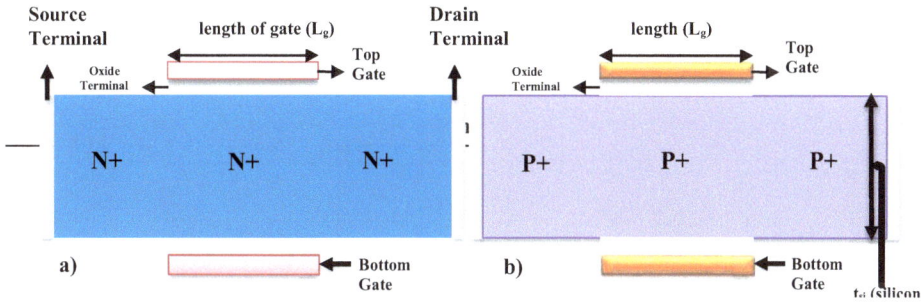

Fig. (3a & b). Schematic of traditional double-gate JL transistor n and p-type, respectively.

To understand the operating principle, there is a need to know the selection of Φ (work function) of gate material in junctionless transistors through the band diagram of the metal oxide semiconductor given in Fig. (**4**).

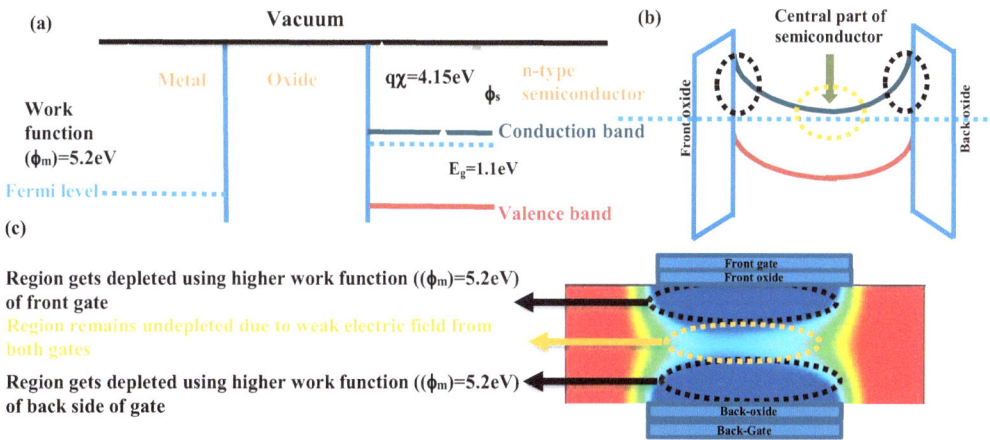

Fig. (4). a) Position of Fermi levels of Metal and n-type semiconductor without the contact between them **b)** Band diagram of double gate junctionless transistor **c)** Pictorial representation of conventional DGJLT.

Fig. (**4a**) shows the position of the fermi level of metal and n-type semiconductors without contact between them. Upon the formation of contact, electrons begin to flow from a higher level to a lower level region, which leads to the band bending shown in Fig. (**4b**) [18 - 20]. At equilibrium, the fermi level becomes constant. The band bending of the valence band and conduction band is shown for double gate junctionless MOSFET. It can be observed that for a higher value of Φ (work-function) of 5.2 eV (for n-type), a band bending representing the depletion at the oxide/semiconductor interface, whereas at the central part of the n-type semiconductor, the $E_{conduction}$ is closer to the fermi level representing the non-depletion in the middle part of the device which can be observed from pictorial representation given in Fig. (**4c**) When the lower work-function of gate material will be selected, the band bending at the oxide/semiconductor interface will be less, exhibiting the non-depletion of the carriers. Hence, a higher work-function is needed for junctionless MOSFET. However, the realization of metal with Φ greater than 5.2eV is challenging thus, it limits the device thickness.

Rectangular Core-shell Double Gate JL Transistor

An RCS-DGJLT has been proposed, which is represented in Fig. (**5**) and comprises a rectangular core sandwiched between two rectangular shells. The kind of RCS-DGJLT is determined by the doping type in the rectangular shell. When it comes to n and p-type RCS-DGJLT, rectangular shells are doped with donor and acceptor impurities, respectively. With the core in place, the device's central section will have less charge carrier density than it would in a normal double-gate junctionless transistor (DGJLT).This could also reduce the restriction of keeping the silicon thickness lesser for junctionless devices. The placement of the oppositely doped core can deplete more carriers in the device and hence improves the device's performance.

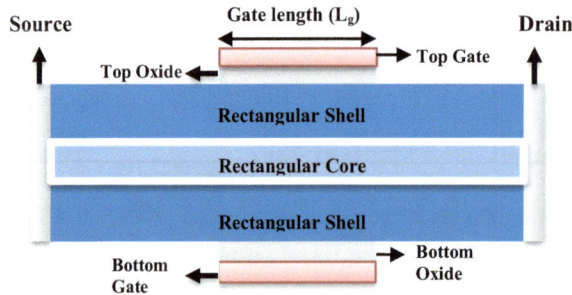

Fig. (5). Schematic of RCS-DGJLT.

RESULTS AND DISCUSSION

Silicon Thickness and Work Function Impact on Conventional DGJLT

Thickness effects on n-DGJLT performance are examined in this part using extensive TCAD simulation. Subthreshold-slope, I_{on}/I_{off} current ratio, I_{off} current, threshold voltage, and I_{on} current are just a few of the performance metrics that are measured and discussed. Fig. (**6**) (**left**) depicts a typical n-type DGJLT. The n-type DGJLT is doped with donor impurities in all source/channel/drain regions. The top gate and bottom gate are connected together. It has been utilized to develop and simulate the device architectures employing oxide as SiO_2 with k=3.9 and 5.2 eV work-function gate material [21]. Recombination, Boltzmann carrier statistics, and shortening of the band gap are all employed in the simulation. Lombardi's mobility model, which incorporates the effects of surface roughness scattering and doping-dependent bulk mobility, is utilized during the simulation of the device. The width of the device is set to 1 um by default. During the simulation of a standard DGJLT, the silicon thickness (t_{si}) was adjusted from 6nm to 12nm, while all other parameters remained fixed Fig. (**6**) (**right**).

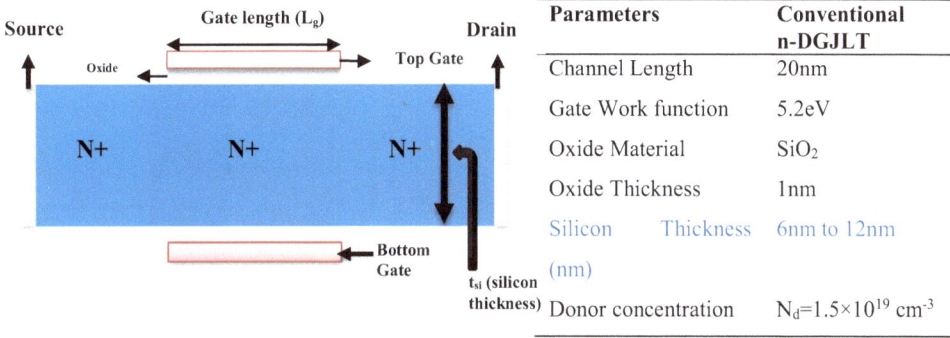

Parameters	Conventional n-DGJLT
Channel Length	20nm
Gate Work function	5.2eV
Oxide Material	SiO_2
Oxide Thickness	1nm
Silicon Thickness (nm)	6nm to 12nm
Donor concentration	$N_d=1.5\times10^{19}$ cm^{-3}

Fig. (6). (Left) Schematic of conventional n-type DGJLT **(Right)** Design parameters used during simulation.

The I_d vs. V_{gs} characteristic of n-type DGJLT for different t_{si}, which changes from 6nm to 12nm with work-function kept at 5.2eV in both log scale and linear scale, are presented in Figs. (**7a & b**) respectively. The figure exhibits the increase in the leakage current on increasing the silicon thickness, and carriers could not be depleted for larger silicon thickness due to the weak strength of the gate. Further, the OFF current was found to be increasing with $t_{silicon}$ in Fig. (**7c**), which leads to the decrease in I_{on}/I_{off} current ratio. Off current rises seven orders of magnitude with increasing silicon thickness from 6 to 12 nm. When using 6nm silicon thickness, the I_{on}/I_{off} current ratio is 10^9, however when using 12nm silicon, that ratio has decreased by a factor of five. Fig. (**7d**) shows the rise in barrier height at the source/channel interface as t_{si} reduces from 12nm to 6nm, as seen in the energy band diagram. The charge carriers in the channel are completely depleted for 6nm silicon thickness [22]. Performance characteristics, including threshold voltage, ON current, subthreshold slope, I_{off} current and I_{on}/I_{off} current ratio, are shown in Table **1**.

Another important performance parameter that explains the transition of the device is called the subthreshold slope (SS). The SS is calculated as (SS= $dV_{gs}/d(log(I_D))$ [10]. The subthreshold slope was found to be increasing on increasing the silicon thickness. The SS is calculated as 66.2mV/dec for 6nm silicon thickness and 75.1mV/dec for 10nm silicon thickness. The threshold voltage for silicon thickness 12nm is negative for n-type DGJLT, and thus subthreshold slope has not been calculated. Also, the contour plots of conventional DGJLT at different silicon thickness in OFF state (V_{gs}=0V) is given in Fig. (**7e**). The contour plots give a clear picture of the non-depletion of the carriers in OFF state for larger silicon-thickness. It can be observed that the gate work function of 5.2eV can effectively deplete the charge carriers till t_{si}~4nm on either side of the gate, and thus for t_{si}=8nm, the OFF current is found to be ~10^{-12}A.

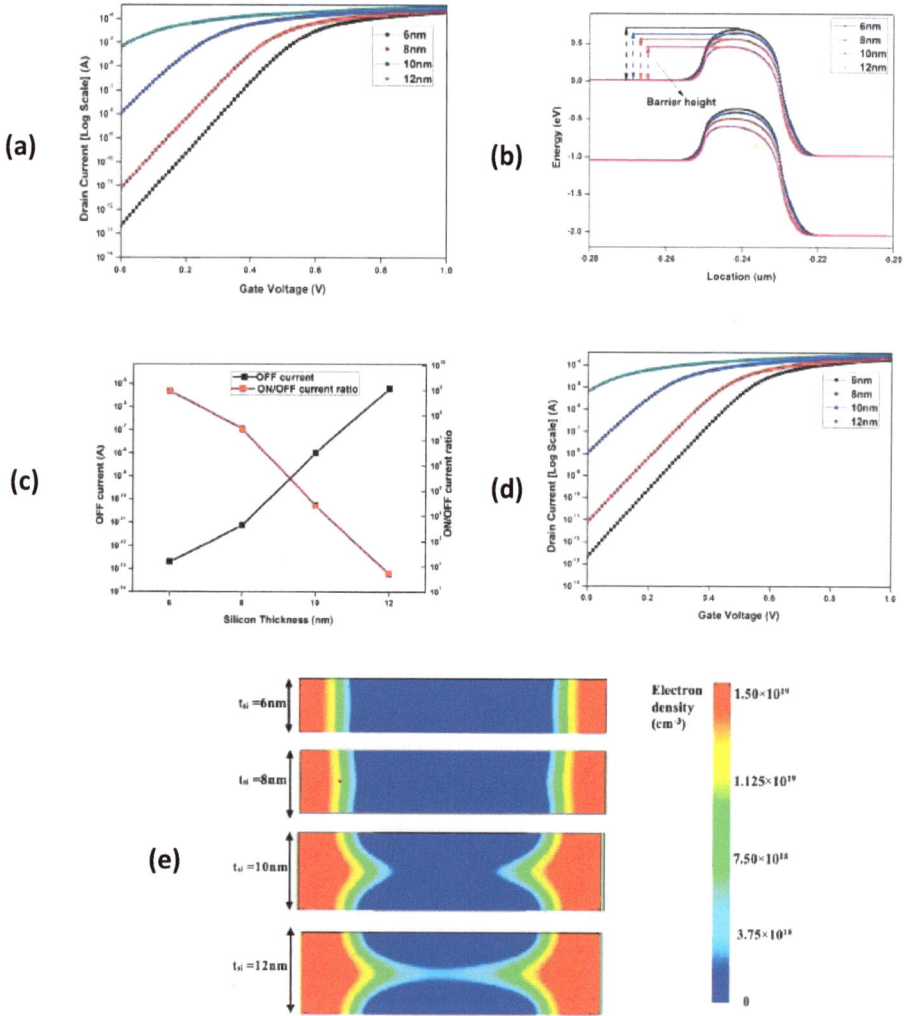

Fig. (7). a & b) Drain Current *vs* Gate-source characteristics of DGJLT at different silicon thickness in log and linear scale respectively at V_{ds}=0.05V **c)** I_{on}/I_{off} Current ratio and I_{off} current (V_{gs}=0V) plotted at different t_{si} **d)** Energy Band diagram at V_{ds}=1V and V_{gs}=0V **e)** Contour plots showing electron density at different silicon thickness in OFF state (V_{gs}=0V). L_g=20nm, W.F=5.2eV, V_{ds}=0.05V, N_d=1.5×10^{19} cm^{-3}.

Table 1. Different performance characteristics of n-type DGJLT for different silicon-thickness with W.F =5.2eV.

Silicon Thickness (nm)	Performance Parameters				
	SS (mV/decade)	I_{on} Current (A)	I_{on}/I_{off} current ratio	I_{off} current (A)	V_{th} (V)
6	66.2	1.90×10^{-4}	9.4×10^{8}	2.02×10^{-13}	0.38
8	68.9	2.17×10^{-4}	2.98×10^{7}	7.27×10^{-12}	0.29
10	75.1	2.72×10^{-4}	3.08×10^{4}	8.81×10^{-9}	0.12
12	-	3.2×10^{-4}	5.22×10^{1}	6.12×10^{-6}	-0.1

Fig. (8) shows the I_{off} current and I_{on}/I_{off} current-ratio with varying work functions for 12nm silicon thickness. I_{off} current drops by four orders of magnitude while I_{on}/I_{off} current ratio increases by five orders of magnitude when Φ rises from 5.2eV to 5.6eV. Therefore, according to this study, it can be concluded that even if silicon-thickness is larger, the carriers volume depletion in the OFF-state can be acquired by increasing the work-function of the gate-material. However, the realization of such practical work-functions is very difficult and expensive [17]. This leaves a scope to work for novel architecture, which helps to obtain good performance parameters at higher silicon thicknesses along with available practical work function of the gate material.

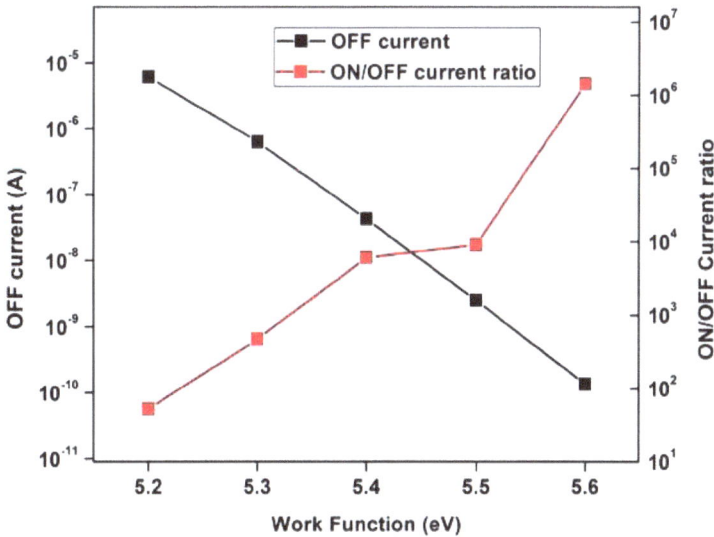

Fig. (8). I_{off} current and I_{on}/I_{off} current-ratio of DGJLT at t_{si}=12nm on varying the gate work function. L_g =20nm, W.F=5.2eV, V_{ds}=0.05V, N_d=1.5×10^{19}cm^{-3}, t_{si}=12nm.

Basic Modelling of Double Gate Junctionless Transistor

The electrostatic potential profile is given using 2-D poisson's equation in **(1)**, where $\varphi_A(x,y)$ represents an electrostatic potential of region A(*i.e.*, channel region), q is electronic charge, N_d represents donor concentration, n_i intrinsic concentration of Si, V_{qf} is a quasi-fermi voltage of electron, KT/q represents thermal-voltage. x and y represent the orientation through the channel and from the top gate to the bottom gate, respectively.

$$\frac{\partial^2 \varphi_A(x,y)}{\partial x^2} + \frac{\partial^2 \varphi_A(x,y)}{\partial y^2} = \frac{-qN_d}{\varepsilon_{silicon}} + \frac{qn_i e^{q(\varphi - V_{qf})/KT}}{\varepsilon_{silicon}} \quad 0 \le x \le L_g, \ 0 \le y \le t_{silicon} \quad (1)$$

Since, we are interested in subthreshold region, therefore we can neglect the contribution of mobile charges and therefore, **(1)** is reduced to:

$$\frac{\partial^2 \varphi_A(x,y)}{\partial x^2} + \frac{\partial^2 \varphi_A(x,y)}{\partial y^2} = \frac{-qN_d}{\varepsilon_{silicon}} \tag{2}$$

Therefore, the approximate solution of the above equation is given as follows, where the potential is assumed to be parabolic shape along the y-axis:

$$\varphi_A(x,y) = \eta_1(x) + \eta_2(x).y + \eta_3(x).y^2 \tag{3}$$

Where, $\eta_1(x)$, $\eta_2(x)$ and $\eta_3(x)$, are coefficients. The values of the coefficients can be found using boundary conditions given as:

1) The potential at the surface at y=0 (top gate) and $t_{silicon}$ (bottom gate) is $\varphi_{surface}(x)$.

$$\varphi_A(x,0) = \varphi_A(x, t_{silicon}) = \varphi_{surface}(x) \tag{4}$$

2) Electric field at y=0 (top gate) and $t_{silicon}$ (bottom gate) is given by gate voltage (V_{gs}) and flat band voltage (V_{FB}).

$$\frac{\partial \varphi_A(x,y)}{\partial y} = \frac{\varepsilon_{ox}(\varphi_{surface}(x) - (V_{gs} - V_{FB}))}{t_{ox}.\varepsilon_{silicon}} \quad \text{at } y = 0 \tag{5}$$

$$\frac{\partial \varphi_A(x,y)}{\partial y} = -\frac{\varepsilon_{ox}(\varphi_{surface}(x) - (V_{gs} - V_{FB}))}{t_{ox}.\varepsilon_{silicon}} \quad \text{at } y = t_{silicon} \tag{6}$$

Now, finding coefficients $\eta_1(x)$, $\eta_2(x)$ and $\eta_3(x)$ using above boundary conditions.

On solving Equations **(3 & 4)**, $\eta_1(x)$ the can be expressed as:

$$\eta_1(x) = \varphi_{surface}(x) \tag{7}$$

$\eta_2(x)$ can be found using Equations **(3 & 5)**,

$$\frac{\partial \varphi_A(x, y)}{\partial y} = 0 + \eta_2(x) + 2.y.\eta_3(x)$$

Placing y=0 and using (5), we get:

$$\eta_2(x) = \frac{\frac{\partial \varphi_A(x,y)}{\partial y}}{\frac{\varepsilon_{ox}(\varphi_{surface}(x)-(V_{gs} - V_{FB}))}{t_{ox}.\varepsilon_{silicon}}} = \tag{8}$$

Similarly, using boundary condition (6) in (3), we get:

$$\frac{\partial \varphi_A(x, y)}{\partial y} = 0 + \eta_2(x) + 2.t_{silicon}.\eta_3(x)$$

Using boundary condition (6) and putting the value of $\eta_2(x)$ from (8) to get $\eta_3(x)$ as:

$$-\frac{\varepsilon_{ox}(\varphi_{surface}(x)-(V_{gs} - V_{FB}))}{t_{ox}.\varepsilon_{silicon}} = \frac{\varepsilon_{ox}(\varphi_{surface}(x)-(V_{gs} - V_{FB}))}{t_{ox}.\varepsilon_{silicon}} + 2.t_{silicon}.\eta_3(x) \tag{9}$$

$$\eta_3(x) = -\frac{\varepsilon_{ox}(\varphi_{surface}(x)-(V_{gs} - V_{FB}))}{t_{ox}.\varepsilon_{silicon}.t_{silicon}}$$

Now, substitute all coefficients from (7-9) in (3) to get $\varphi_A(x,y)$ as:

$$\varphi_A(x, y) = \varphi_{surface}(x) + \frac{\varepsilon_{ox}(\varphi_{surface}(x)-(V_{gs} - V_{FB}))}{t_{ox}.\varepsilon_{silicon}}.y - \frac{\varepsilon_{ox}(\varphi_{surface}(x)-(V_{gs} - V_{FB}))}{t_{ox}.\varepsilon_{silicon}.t_{silicon}}.y^2 \tag{10}$$

For the symmetric mode of operation in conventional DGJLMOS, the central potential $\varphi_{central}(x)$, at $y = \varphi_{silicon}/2$ can be calculated after finding the relationship between $(\varphi_{surface}(x))$ and $(\varphi_{central}(x))$.

Put $y = t_{silicon}/2$ in equation **(10)**, we get

$$\varphi_A\left(x, \frac{t_{silicon}}{2}\right) = \varphi_{central}(x) =$$

$$\varphi_{surface}(x) + \frac{\varepsilon_{ox}(\varphi_{surface}(x) - (V_{gs} - V_{FB}))}{t_{ox}.\varepsilon_{silicon}}.\frac{t_{silicon}}{2}$$
$$- \frac{\varepsilon_{ox}(\varphi_{surface}(x) - (V_{gs} - V_{FB}))}{t_{ox}.\varepsilon_{silicon}.t_{silicon}}.\left(\frac{t_{silicon}}{2}\right)^2$$

$$= \varphi_{surface}(x) + \frac{\varepsilon_{ox}(\varphi_{surface}(x)}{t_{ox}.\varepsilon_{silicon}}.\frac{t_{silicon}}{2} - \frac{(V_{gs} - V_{FB}).\varepsilon_{ox}.t_{silicon}}{t_{ox}.\varepsilon_{silicon}}.\frac{t_{silicon}}{2} - \frac{\varepsilon_{ox}(\varphi_{surface}(x))}{t_{ox}.\varepsilon_{silicon}}.\frac{t_{silicon}}{4}$$
$$\frac{(V_{gs} - V_{FB}).\varepsilon_{ox}.t_{silicon}}{t_{ox}.\varepsilon_{silicon}}.\frac{t_{silicon}}{4} \qquad\qquad +$$

$$\varphi_{central}(x) = \varphi_{surface}(x)\left[1 + \frac{\varepsilon_{ox}.t_{silicon}}{2.t_{ox}.\varepsilon_{silicon}} - \frac{\varepsilon_{ox}.t_{silicon}}{4.t_{ox}.\varepsilon_{silicon}}\right] + \frac{(V_{gs} - V_{FB}).\varepsilon_{ox}.t_{silicon}}{t_{ox}.\varepsilon_{silicon}}\left[\frac{1}{4} - \frac{1}{2}\right] \qquad \textbf{(11)}$$

$$\varphi_{central}(x) + \frac{(V_{gs} - V_{FB}).\varepsilon_{ox}.t_{silicon}}{4.t_{ox}.\varepsilon_{silicon}} = \varphi_{surface}(x)\left[1 + \frac{\varepsilon_{ox}.t_{silicon}}{2.t_{ox}.\varepsilon_{silicon}} - \frac{\varepsilon_{ox}.t_{silicon}}{4.t_{ox}.\varepsilon_{silicon}}\right]$$

$$\varphi_{surface}(x)$$

$$= \frac{4.\varepsilon_{silicon}.\varphi_{central}(x) + \dfrac{\varepsilon_{ox}.(V_{gs} - V_{FB}).t_{silicon}}{t_{ox}}}{4.\varepsilon_{silicon} + \dfrac{\varepsilon_{ox}.t_{silicon}}{t_{ox}}}$$

In finding the relationship between $\varphi_{surface}(x)$ and $\varphi_{central}(x)$ in **(11)**, the 1-D poisson equation governing $\varphi_{central}(x)$ can be obtained at $y = t_{silicon}/2$ as,

Put the value of **(10)** in **(2)**, and then using **(11)**, we get,

$$\frac{\partial^2 \varphi_{(central)}(x)}{\partial x^2} - \frac{2.\varepsilon_{ox}.\varphi_{surface}(x)}{\varepsilon_{silicon}.t_{ox}.t_{silicon}} - \frac{2.\varepsilon_{ox}.(V_{gs} - V_{FB})}{\varepsilon_{silicon}.t_{ox}.t_{silicon}} = \frac{-qN_d}{\varepsilon_{silicon}}$$

$$\frac{\partial^2 \varphi_{(central)}(x)}{\partial x^2} - \frac{2.\varepsilon_{ox}.\left(4.\varepsilon_{silicon}.\varphi_{central}(x) + \dfrac{\varepsilon_{ox}.(V_{gs} - V_{FB}).t_{silicon}}{t_{ox}}\right)}{\left[4.\varepsilon_{silicon} + \dfrac{\varepsilon_{ox}.t_{silicon}}{t_{ox}}\right].(\varepsilon_{silicon}.t_{ox}.t_{silicon})} - \frac{2.\varepsilon_{ox}.(V_{gs} - V_{FB})}{\varepsilon_{silicon}.t_{ox}.t_{silicon}} \qquad \textbf{(12)}$$
$$= \frac{-qN_d}{\varepsilon_{silicon}}$$

Equation **(12)** can be re-written as:

$$\frac{\partial^2 \varphi_{(central)}(x)}{\partial x^2} - \frac{\varphi_{(central)}(x)}{\lambda^2_N} + \frac{\xi}{\lambda^2_N} = 0 \qquad \textbf{(13)}$$

Where λ_N represents characteristics length which depicts the short channel effects

$$\lambda^2{}_N = \frac{\varepsilon_{silicon}.t_{silicon}.t_{ox}}{2.\varepsilon_{ox}} + \frac{t_{silicon}{}^2}{8} \text{ and } \xi = \frac{q.N_d}{\varepsilon_{silicon}}.\lambda^2{}_N + (V_{gs} - V_{FB})$$

Now, the general solution of **(13)** can be represented as:

$$\varphi_{(central)}(x) = Me^{\frac{x}{\lambda_N}} + Ne^{-\frac{x}{\lambda_N}} + \xi \qquad (14)$$

The constants M and N of **(14)** can be found using boundary conditions at the gate edges:

$$\varphi_{(central)}(0) = 0 \text{ and } \varphi_{(central)}(L_g) = V_{ds} \qquad (15)$$

Using boundary conditions given in **(15)** M and N are as follows:

$$\boldsymbol{M} = \frac{-V_{ds} - \xi(e^{-\frac{L_g}{\lambda_N}} - 1)}{e^{-\frac{L_g}{\lambda_N}} - e^{\frac{L_g}{\lambda_N}}} \quad \mathbf{N} = \frac{-V_{ds} - \xi(e^{\frac{L_g}{\lambda_N}} - 1)}{e^{\frac{L_g}{\lambda_N}} - e^{\frac{L_g}{\lambda_N}}}$$

Therefore, the central potential of conventional DGJLMOS can be written as:

$$\boldsymbol{\varphi_{(central)}}(\boldsymbol{x}) = \left(\frac{-V_{ds} - \xi(e^{-\frac{L_g}{\lambda_N}} - 1)}{e^{-\frac{L_g}{\lambda_N}} - e^{\frac{L_g}{\lambda_N}}} \right) e^{\frac{x}{\lambda_N}} + \left(\frac{-V_{ds} - \xi(e^{\frac{L_g}{\lambda_N}} - 1)}{e^{\frac{L_g}{\lambda_N}} - e^{\frac{L_g}{\lambda_N}}} \right) e^{-\frac{x}{\lambda_N}}$$
$$+ \left(\frac{q.N_d}{\varepsilon_{silicon}}.\lambda^2{}_N + (V_{gs} - V_{FB}) \right)$$

Fig. **(9)** shows the central potential profile of n-type DGJLMOS at different silicon thicknesses (t_{si}). It is noted that the minimum potential shifts upwards on increasing the t_{si}. This indicates the lesser depletion of the charge carriers at the OFF state, *i.e.,* V_{gs}=0V, due to the weakening of the control of the gate. As the gate work-function is responsible to deplete the carriers at 0V therefore, on increasing the t_{si} the gate work function needs to be tuned to get a lower central potential of the device. The results of the analysis are found to be in good agreement with those of the simulation.

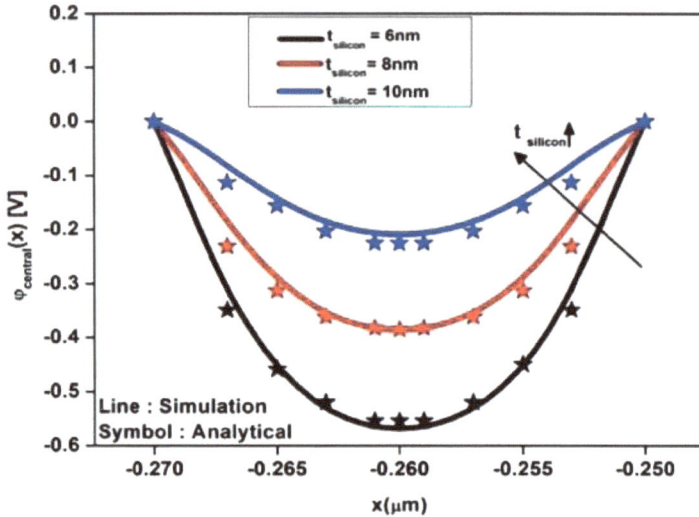

Fig. (9). Body central potential distribution along lateral direction from source to drain at different silicon thickness. **Line: Simulation results, Symbol: Analytical results.**

Fig. (**10**) shows the central potential profile of n-type DGJLMOS at different Φ_m of the gate. It is observed that when the work function is increased to 5.4eV from 5.2eV, the resultant carriers at the weakest gate coupling region, *i.e.,* the central part of the silicon film in the device depletes more effectively and thus, the potential shift downwards, indicating better performance in terms of leakage current [30]. Analytical and simulated findings are found to be quite well aligned. Fig. (**11**) shows the variation of central potential distribution along the source-to drain region of n-type DGJLMOS at a different dielectric constant of the gate. When the ε (dielectric constant) of the gate varies from 3.9 to 22, the central potential graph shifts downwards, indicating the larger depletion of the carriers as barrier potential for the carriers in case of k=22 increases and thus, lesser leakage current is observed [32].

Fig. (**12**) shows the central potential curve at different donor doping levels of n-type DGJLMOS. It is quite obvious that if carrier concentration is less, the depletion of the carriers is more and thus, a steep valley of the potential curve can be obtained, indicating the lesser potential for $N_d = 1 \times 10^{17} \text{cm}^{-3}$.

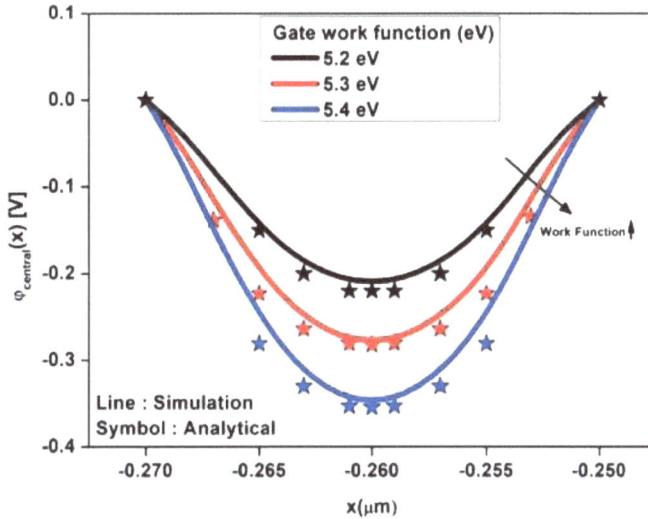

Fig. (10). Body central potential distribution along the lateral direction from source to drain at a different work function. Line: Simulation results, Symbol: Analytical results.

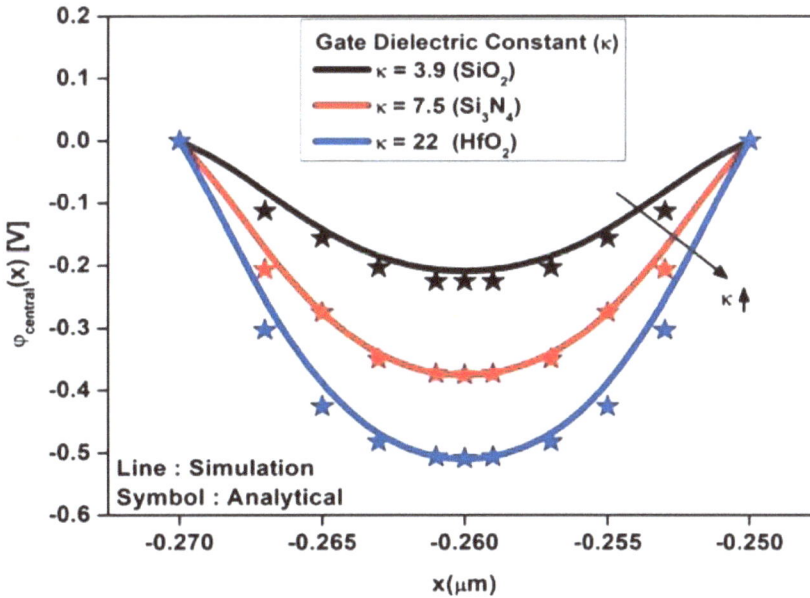

Fig. (11). Body central potential distribution along a lateral direction from source to drain at different gate dielectric constant(k). Line: Simulation results, Symbol: Analytical results.

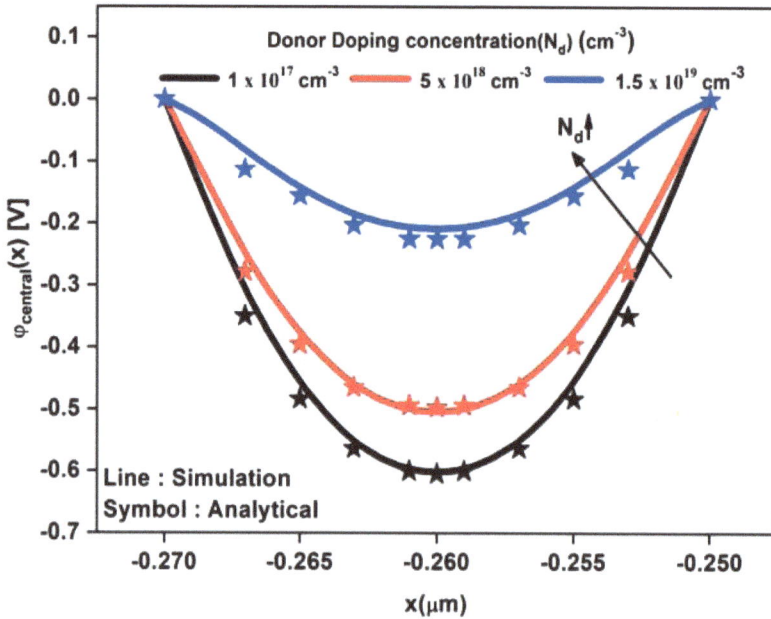

Fig. (12). Body central potential distribution along a lateral direction from source to drain at different donor doping concentrations (Nd). Line: Simulation results, Symbol: Analytical results.

Impact of RCS-DGJLT Device Parameters on the Device's Performance

The poor performance of conventional DGJLT for larger silicon thickness at a suitable practical work function is improved by the proposed structure called RCS-DGJLT. As we have observed that the gate field is weakest in the middle region of the conventional DGJLT, a lightly doped or oppositely doped film is inserted in the DGJLT such that RCS-DGJLT can be obtained. Incorporating an oppositely doped film in the device's center has greatly enhanced performance even for thicker layers of silicon. Depleting carriers in the RCS structure allows the device's performance to be reworked by utilizing a new work function and a core that is either light-doped or oppositely-doped [23 - 27].

As seen in Fig. (**13**), the suggested structure has a rectangular core sandwiched between two rectangular shells, which we refer to as the RCS-DGJLT. The RCS-DGJLT type is determined by the doping type in the rectangular shell [28, 29]. Donor and acceptor impurities are added to rectangular shells for RCS-DGJLT in order to make the device work in both the n-and p-type configurations. The idea behind placing the core is to reduce the charge carrier density in the middle region of the device, which is left undepleted in the conventional DGJLT. The values of device parameters used during simulation Table **2**.

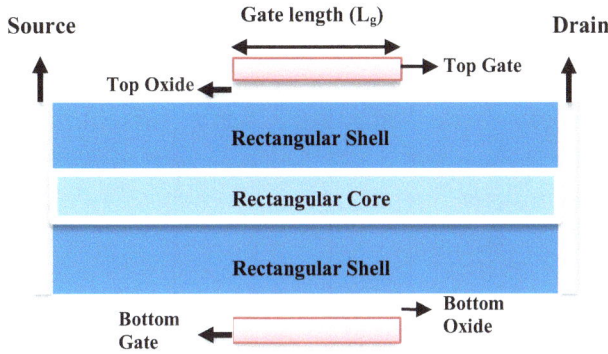

Fig. (13). Schematic of RCS-DGJLT.

Table 2. Parameters that are used to model the devices' behavior.

Parameters	RCS-DGJLT
Channel Length (L_g)	20nm
Core thickness (t_{core})	4nm
Total silicon thickness (t_{si})	12nm
Source/drain extension length (L_{ext})	10nm
Oxide thickness	1nm
Work function	5.2eV
Each Shell thickness (t_{shell})	4nm

Impact of Channel Length on RCS-DGJLT

It is also critical to investigate the impact of channel length on n-type RCS-DGJLT performance characteristics. I_{on} current, I_{off} current, On/OFF current ratio, SS, and V_{th} are all examined in detail with varying channel lengths in RCS-DGJLT.

Figs. (**14a & b**) shows transfer characteristics of RCS-DGJLT at different channel lengths in linear and log scale, respectively. It is observed that as channel length reduces from 50nm to 10nm, the OFF current increases. The OFF current at L_g=50nm and 10nm is of the order $\sim 10^{-15}$A and $\sim 10^{-10}$A, respectively. The increase in OFF current of ~5 orders of magnitude is due to the closer proximity of drain and source. The volume depletion of the carriers at the OFF state becomes difficult due to the lesser channel length, which implies a lesser distance between the source and channel. Also, it can be observed that ON current at V_{gs}=0.8V increases on decreasing the channel length. The higher ON current for L_g=10nm is due to the lesser distance for the carriers to travel from the drain to the source region of the device. Further, it can also be observed that I_{on}/I_{off} current ratio increases on increasing the L_g. The I_{on}/I_{off} current ratio at $L_g=50$nm is found to be 10^{11} which is ~5 orders greater than I_{on}/I_{off} current ratio at $L_g=10$nm given in Fig. **(14c)**. The increase in I_{on}/I_{off} current ratio is due to a decrease in OFF current. Also, the V_{th} keeps on increasing with the channel length shown in Fig. **(14c)**.

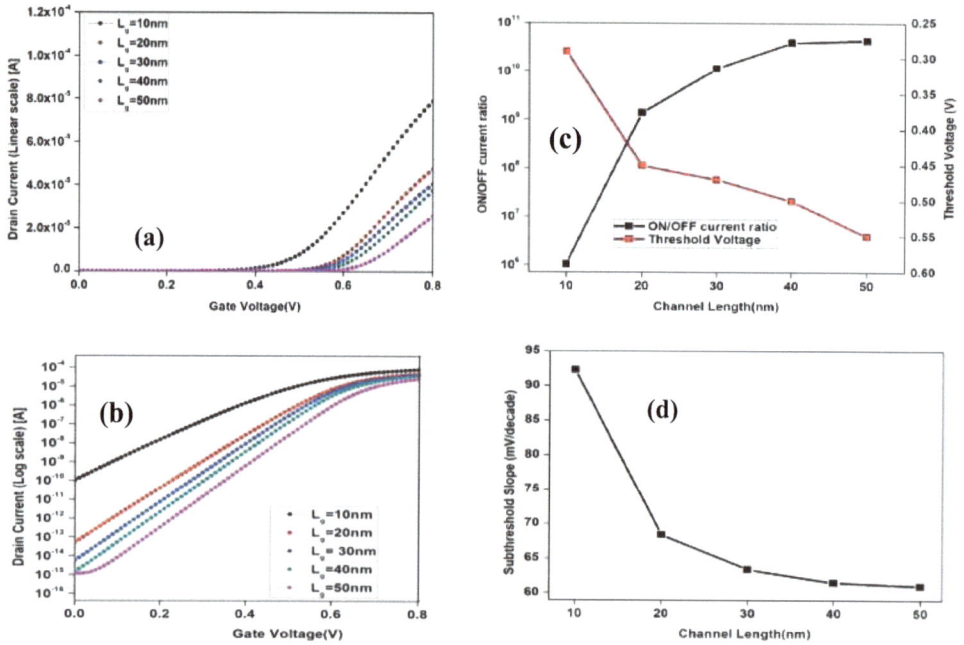

Fig. (14). (a & b) Transfer characteristics of RCS-DGJLT at different channel length in linear and log scale respectively (c) I_{on}/I_{off} current ratio and V_{th} at different channel length (d) SS at different channel length. W.F=5.2eV, V_{ds}=0.05V, N_d (shell)=1.5×10^{19}cm^{-3}, N_a (Core) = 1×10^{19}cm^{-3}, t_{top_shell}+ t_{bottom_shell} = 8nm, t_{core}=4nm.

Another reason for the decrease in OFF current at larger channel lengths is due to the surge in the V_{th}. Due to this, a lesser number of charge carriers are available at V_{gs}=0.8V, exhibiting the lower ON current. Another performance parameter to analyze the performance of RCS DGJLT on varying the channel length is subthreshold slope (SS). SS is a physical quantity that measures the conversion speed, when the device changes its state from OFF to ON. With an increase in the channel length, the subthreshold slope was found to be decreased, as given in Fig. **(14d)** [30]. The SS reaches the theoretical value of 60mV/decade at a channel length 50nm.

CONCLUSION

The high performance and low-power devices are in huge demand for today's microelectronics industry. The junctionless transistor has proved to be a promising device in terms of better performance and ease of fabrication in this device province. Double gate junctionless transistors and a newly suggested device, rectangular core-shell double gate junctionless transistors, are studied in detail in this chapter (RCS-DGJLT). It is examined how silicon-thickness and work-function engineering affect the functioning mechanism of a junctionless transistor. By increasing the silicon-thickness, the performance parameters of the device start degrading. Further, by increasing the work function difference between the semiconductor film and gate material, the performance starts improving. Also, the analytical model of conventional DGJLT is presented. The variation of the gate dielectric, silicon thickness, work function, and donor dopant concentration was studied. The findings of this study show that the simulation results closely match the analytical results. The performance parameters have been improved for RCS-DGJLT such that I_{OFF} of order $\sim 10^{-14}$A, $I_{ON} \sim 10^{-5}$A, I_{ON}/I_{OFF} $\sim 10^{9}$, SS nearly 68.9mV/decade, DIBL nearly 52.6mV/V are obtained at total silicon thickness of 12nm. The effect of channel length variation and oxide thickness on RCS-DGJLT is also studied. The RCS-DGJLT performance in terms of parameters is more magnificent than DGJLT. The SS reaches the theoretical value of 60mV/decade at channel length 50nm.

LEARNING OBJECTIVES

• To understand the working mechanism of Double Gate Junctionless Transistor.

• To study the detailed analysis of the performance of Rectangular Core-Shell Double Gate Junctionless Transistors.

• To study the analytical modelling of Conventional Double Gate Junctionless Transistors.

MULTIPLE CHOICE QUESTIONS

1. What is the full form of CMOS?

 a. Complementary metal-oxide semiconductor
 b. Compliment Metal-Oxide semiconductor
 c. Complementary metal of semiconductor
 d. Complementary metal of oxide semiconductor.

2. The majority charge carriers in p-Type FET are:

  a. Holes
  b. Electrons
  c. Positron
  d. None of these

3. Speed and Power factor are always acts as a tradeoff while designing a high-performance based VLSI circuits:

  a. Yes
  b. No

4. Chanel Length of any device is defined as a distance between the drain and source region:

  a. Yes
  b. No

5. RCS-DGJLT stands for:

  a. Rectangular Core Shell-Double gate Junction less transistor.
  b. Right Core Shell-Double Gate gate Junction less transistor.
  c. Rectangular Compliment Shell-Double gate Junction less transistor.
  d. Rectangular Core Shell-Depletion gate Junction less transistor.

6. The ideal value of the subthreshold swing for MOSFET device will be:

  a. 60mV/decade
  b. 70mV/decade
  c. 80mV/decade
  d. 90mV/decade

7. Junctionless transistor consist of same doping in all three regions.

  a. Yes
  b. NO

8. The majority charge carriers in n-Type FET are:

a. Holes
b. Electrons
c. Both a & b
d. None of these

9. In MOSFET the region between the source and drain is known as:

a. Channel region
b. Doping region
c. Junction region
d. Acceptor region

10. What is the full form of SCE's & GIDL?

a. Short channel effects & Gate Induced Drain Leakage.
b. Short current effects & Gate Indicator Drain Leakage.
c. Short channel effects & Gate Induction Drain Leakage.
d. Short current efficiency & Gate Induction Drain Leakage.

11. How many transistors used for the development of artificial intelligence (AI) based chip called as "Cerebras":

a. 1.2 trillion transistors.
b. 1.3 trillion transistors.
c. 1.5 trillion transistors.
d. 2.0 trillion transistors.

12. The units used for defining the work function is:

a. Volts
b. Electron Volt
c. Emitter per Volt
d. Emitter Volt

13. ITRS stands for International Roadmap for Semiconductors:

a. Yes
b. No

14. Expression N_a defines the concentration of the Acceptor impurities:

a. Yes
b. No

15. Expression N_d define the concentration of the Donor impurities:

a. Yes
b. No

16. In n-Type RCS-DGJLT, by increasing the channel length of device, the threshold voltage will:

a. Increase
b. Decrease
c. Does not change
d. None of these

17. As per the Moore's Law transistor density will double in every:

a. 18 months
b. 16 months
c. 14 months
d. 12 months

18. Which of the following is the fastest switching device?

a. JFET
b. BJT
c. MOSFET
d. Triode

19. The I_d-V_{gs} characteristics of a MOSFET in a saturation region is:

a. exponential
b. quadratic
c. logarithmic
d. hyperbolic

20. Power dissipation is negligibly small in:

a. SCR
b. BJT
c. MOSFET
d. CMOS

21. The three terminals of the MOSFET are:

a. Drain, source, gate.
b. Anode, source, gate.
c. Cathode, anode, gate.
d. Two main terminals and a gate terminal.

22. Why is N-channel MOSFET preferred over a P-channel MOSFET?

a. It allows fast switching.
b. It is TTL compatible.
c. It has low input impedance.
d. Low noise

23. The regions of operation of a MOSFET to work as a linear resistor and linear amplifier are:

a. Cut off and saturation respectively.
b. Triode and cut off respectively.
c. Triode and saturation respectively.
d. Saturation and triode respectively.

24. Which of the following is not correctly matched?

a. MOSFET- Unipolar device.
b. BJT- Switching power loss.
c. MOSFET- high ON state conduction power loss.
d. IGBT- voltage-controlled device.

25. Which of the following parameters are affected due to short channel MOSFET geometry:
i. Mobility of carriers
ii. Threshold voltage

iii. Drain current

  a. Only i
  b. Only ii
  c. Both i and ii
  d. i, ii, iii

ANSWER KEY

1. (a)

2. (a)

3. (a)

4. (a)

5. (a)

6. (a)

7. (a)

8. (b)

9. (a)

10. (a)

11. (a)

12. (b)

13. (a)

14. (a)

15. (a)

16. (a)

17. (a)

18. (c)

19. (b)

20. (d)

21. (a)

22. (a)

23. (c)

24. (b)

25. (d)

ACKNOWLEDGEMENTS

The author would like to thank Lovely Professional University, Cadre Design Systems, for giving me the opportunity to work in the labs.

REFERENCES

[1]   J-P. Colinge, "The soi mosfet: From single gate to multigate", In: *Finfets and other multi-gate transistors.* Springer, pp. 1-48, 2008.
       [http://dx.doi.org/10.1007/978-0-387-71752-4_1]

[2]   A. Amara, and O. Rozeau, *Planar double-gate transistor: From technology to circuit.* Springer Science & Business Media, 2009.
       [http://dx.doi.org/10.1007/978-1-4020-9341-8]

[3]   M. Prasad, and U.B. Mahadevaswamy, "Performance analysis for tri-gate junction-less fet by employing trioxide and rectangular core shell (rcs) architecture", *Wirel. Pers. Commun.,* vol. 118, no. 1, pp. 619-630, 2021.
       [http://dx.doi.org/10.1007/s11277-020-08035-1]

[4]   Jong-Tae Park, J.P. Colinge, and C.H. Diaz, "Pi-gate soi mosfet", *IEEE Electron Device Lett.,* vol. 22, no. 8, pp. 405-406, 2001.
       [http://dx.doi.org/10.1109/55.936358]

[5]   Sang-Hyun Oh, D. Monroe, and J.M. Hergenrother, "Analytic description of short-channel effects in fully-depleted double-gate and cylindrical, surrounding-gate mosfets", *IEEE Electron Device Lett.,* vol. 21, no. 9, pp. 445-447, 2000.
       [http://dx.doi.org/10.1109/55.863106]

[6]   V. Subramanian, B. Parvais, J. Borremans, A. Mercha, D. Linten, P. Wambacq, J. Loo, M. Dehan, C. Gustin, N. Collaert, S. Kubicek, R. Lander, J. Hooker, F. Cubaynes, S. Donnay, M. Jurczak, G. Groeseneken, W. Sansen, and S. Decoutere, "Planar bulk mosfets versus finfets: an analog/rf perspective", *IEEE Trans. Electron Dev.,* vol. 53, no. 12, pp. 3071-3079, 2006.
       [http://dx.doi.org/10.1109/TED.2006.885649]

[7]   J.P. Colinge, A. Kranti, R. Yan, C.W. Lee, I. Ferain, R. Yu, N. Dehdashti Akhavan, and P. Razavi, "Junctionless nanowire transistor (JNT): Properties and design guidelines", *Solid-State Electron.,* vol. 65-66, no. 1, pp. 33-37, 2011.
       [http://dx.doi.org/10.1016/j.sse.2011.06.004]

[8]   Nazarov A, Colinge JP, Balestra F, Raskin JP, Gamiz F, Lysenko VS, "Junctionless transistors: physics and properties," In *Semiconductor-On-Insulator Materials for Nanoelectronics Applications*, pp. 187–200, 2011.
       [http://dx.doi.org/10.1007/978-3-642-15868-1_10]

[9]   J.P. Colinge, C.W. Lee, A. Afzalian, N.D. Akhavan, R. Yan, I. Ferain, P. Razavi, B. O'Neill, A. Blake, M. White, A.M. Kelleher, B. McCarthy, and R. Murphy, "Nanowire transistors without junctions", *Nat. Nanotechnol.,* vol. 5, no. 3, pp. 225-229, 2010.
[http://dx.doi.org/10.1038/nnano.2010.15] [PMID: 20173755]

[10]   C.H. Park, M-D. Ko, K-H. Kim, R-H. Baek, C-W. Sohn, C.K. Baek, S. Park, M.J. Deen, Y-H. Jeong, and J-S. Lee, "Electrical characteristics of 20-nm junctionless Si nanowire transistors", *Solid-State Electron.,* vol. 73, pp. 7-10, 2012.
[http://dx.doi.org/10.1016/j.sse.2011.11.032]

[11]   E. Gnani, A. Gnudi, S. Reggiani, and G. Baccarani, "Theory of the junctionless nanowire FET", *IEEE Trans. Electron Dev.,* vol. 58, no. 9, pp. 2903-2910, 2011.
[http://dx.doi.org/10.1109/TED.2011.2159608]

[12]   S. Kaundal, and A.K. Rana, "A review of junctionless transistor technology and its challenges", *Journal of Nanoelectronics and Optoelectronics,* vol. 14, no. 3, pp. 310-320, 2019.
[http://dx.doi.org/10.1166/jno.2019.2508]

[13]   V. Narula, C. Narula, and J. Singh, "Investigating short channel effects and performance parameters of double gate junctionless transistor at various technology nodes," *2015 2nd Int. Conf. Recent Adv. Eng. Comput. Sci. RAECS 2015,* no. December, pp. 3–7, 2016.
[http://dx.doi.org/10.1109/RAECS.2015.7453429]

[14]   V. Narula, C. Narula, and J. Singh, "Simulation and characterization of junction less CMOS inverter at various technology nodes", *Indian J. Sci. Technol.,* vol. 9, no. 47, pp. 1-7, 2016.
[http://dx.doi.org/10.17485/ijst/2015/v8i1/106897]

[15]   V. Narula and M. Agarwal, "Effect of gate oxide thickness on the performance of rectangular core-shell based junctionless field effect transistor," In *AIP Conference Proceedings*, vol. 2220, no. 1, p. 40004, 2020.
[http://dx.doi.org/10.1063/5.0001657]

[16]   X. Qian, Y. Yang, Z. Zhu, S.-L. Zhang, and D. Wu, "Evaluation of DC and AC performance of junctionless MOSFETs in the presence of variability," In *2011 IEEE International Conference on IC Design & Technology*, pp. 1–4, 2011.
[http://dx.doi.org/10.1109/ICICDT.2011.5783243]

[17]   S. Gundapaneni, S. Ganguly, and A. Kottantharayil, "Bulk planar junctionless transistor (BPJLT): An attractive device alternative for scaling", *IEEE Electron Device Lett.,* vol. 32, no. 3, pp. 261-263, 2011.
[http://dx.doi.org/10.1109/LED.2010.2099204]

[18]   S. M. Sze, VLSI technology. McGraw-hill, 1988.

[19]   B. G. Streetman and S. Banerjee, Solid state electronic devices. Prentice-Hall of India, 2001.

[20]   D. A. Neamen, Semiconductor physics and devices: basic principles. New York, NY: McGraw-Hill, 2012.

[21]   V. TCAD, "VisualTCAD," vol. Version 1., 2008.

[22]   S. Gundapaneni, M. Bajaj, R.K. Pandey, K.V.R.M. Murali, S. Ganguly, and A. Kottantharayil, "Effect of band-to-band tunneling on junctionless transistors", *IEEE Trans. Electron Dev.,* vol. 59, no. 4, pp. 1023-1029, 2012.
[http://dx.doi.org/10.1109/TED.2012.2185800]

[23]   V. Narula, and M. Agarwal, "Enhanced performance of double gate junctionless field effect transistor by employing rectangular core–shell architecture", Semicond. Sci. Technol., vol. 34, no. 10, p. 105014, 2019.
[http://dx.doi.org/10.1088/1361-6641/ab3cac]

[24]   V. Narula, and M. Agarwal, "Doping engineering to enhance the performance of rectangular core shell double gate junctionless field effect transistor", *Semicond. Sci. Technol.,* vol. 35, p. 075003, 2020.

[25]   V. Narula, and M. Agarwal, "Impact of core thickness and gate misalignment on rectangular core–shell based double gate junctionless field effect transistor", *Semicond. Sci. Technol.*. 35, no. 3, p. 035010, 2020.
[http://dx.doi.org/10.1088/1361-6641/ab6bb2]

[26]   V. Narula, and M. Agarwal, "Study of analog performance of common source amplifier using rectangular core–shell based double gate junctionless transistor", *Semicond. Sci. Technol.*. vol. 35, no. 10, p. 105022, 2020.
[http://dx.doi.org/10.1088/1361-6641/abaaed]

[27]   V. Narula, and M. Agarwal, "Performance enhancement of core-shell JLFET by gate/dielectric engineering", *Int. J. Electron,* vol. 107, no. 6, pp. 966-984, 2019.

[28]   V. Narula and M. Agarwal, "Correlation betweenWork Function and Silicon Thickness of Double Gate Junctionless Field Effect Transistor," In *2019 Women Institute of Technology Conference on Electrical and Computer Engineering (WITCON ECE)*, 2019, pp. 227–230.
[http://dx.doi.org/10.1109/WITCONECE48374.2019.9092902]

[29]   V. Narula, A. Saini, and M. Agarwal, "Correlation of core thickness and core doping with gate & spacer dielectric in rectangular core shell double gate junctionless transistor", *J. Inst. Electron. Telecommun. Eng.,* pp. 1-12, 2021.
[http://dx.doi.org/10.1080/03772063.2021.1946437]

[30]   B.C. Paz, F. Ávila-Herrera, A. Cerdeira, and M.A. Pavanello, "Double-gate junctionless transistor model including short-channel effects", *Semicond. Sci. Technol.*. vol. 30, no. 5, p. 055011, 2015.
[http://dx.doi.org/10.1088/0268-1242/30/5/055011]

CHAPTER 4

Performance Analysis of Electrical Characteristics of Hetero-junction LTFET at Different Temperatures for IoT Applications

Sweta Chander[1,*] and **Sanjeet Kumar Sinha**[1]

[1] School of Electronics and Electrical Engineering, Lovely Professional University, Phagwara, Punjab, India

Abstract: Scaling down the metal-oxide- semiconductor (MOS) technology in the nanometer regime has been performed to achieve high device performance, but reliability and power consumption are the main concern for the semiconductor industry. In the past few years, area-scaled tunneling field-effect transistors (TFETs) have been researched aggressively to enhance the tunneling cross-sectional area of devices. Although the area-scaled Tfet increases the device footprint for the same channel length when compared to the conventional TFET structure. This problem can be resolved by considering a nonplanar device structure. The LTFET structure enhances the on-state current and reduces the device footprint area. In the present study, a detailed analysis of the electrical characteristics of L-shaped TFET (LTFET) through 2-D TCAD simulations is presented. The proposed hetero-junction LTFET with 20 nm gate length exhibits a high I_{ON} of 1.08×10^{-4} A/µm, low I_{OFF} of 1.57×10^{-14} A/µm, high I_{ON}/I_{OFF} of 10^{10}, and steep sub-threshold slope (SS) of 25 mV/dec at room temperature. The analysis has been carried out to encounter the effect of Gaussian traps at the channel–gate oxide interface at a wide range of temperatures from 250 K to 350 K. An extensive study on the influence of temperature variations on various DC analysis, AC analysis, linearity analysis, and electrical noise analysis has been carried out. The study reveals that the electrical parameters like I_{ON}, IOFF, and SS, on which all figures of merit (FOMs) of the device depend, show a small variation with increasing temperature. The drain current noise spectral density (S_{ID}) changes from 2.12×10^{-26} A^2/Hz to 2.42×10^{-20} A^2/Hz, and voltage noise spectral density (S_{VG}) changes from 1.79×10^{-11} V^2/Hz to 1.97×10^{-5} V^2/Hz on increasing temperature from 250 K to 350 K. The change in temperature does not impact the on-current of the device, while a small variation in the off-current occurs. The various FOMs of the device also show small variations in the results with increasing temperature. The only unfavorable factor where the evident change in the results has been observed is the electrical noise characteristics of the device. The reliability analysis clarifies that the proposed LTFET device performs well at a wide range of temperatures and can be well-suited for low-power applications.

---
* Corresponding author Sweta Chander: School of Electronic and Electrical Engineering, Lovely Professional University, Phagwara, Punjab, India; E-mail: sweta.chander@gmail.com

Gopal Rawat & Aniruddh Bahadur Yadav (Eds.)

Keywords: Noise analysis, Interface Traps, Gaussian Traps, Reliability Analysis, Band-To-Band Tunneling (BTBT), Temperature.

INTRODUCTION

The nanoelectronic industry has emerged rapidly and continuously as it merges nanotechnology in electronic components. The technological improvement in the design, use of different materials, and suitability of the device in applications motivate the growth of the nano-electronic industry [1]. The size of nanoelectronic devices ranges between 1 nm to 100 nm. The primary engine of the progress of the industry is miniaturization [2]. Scaling down the metal-oxide semiconductor (MOS) technology in the nanometer regime has been performed to achieve high device performance, but reliability and power consumption are the main concern for the semiconductor industry [3, 4]. Moreover, under the minimum fundamental limit of subthreshold slope (SS), the scaling of V_{DD} gets restricted. These issues get resolved using an outstanding device, *i.e.*, tunnel field effect transistors (TFETs), that come into the scene to attain low SS [5 - 7]. The band-to--to-band tunneling (BTBT) mechanism of TFET offers steep SS and low off-state current [8]. In the past few years, area-scaled TFETs have been researched aggressively to increase the device tunneling cross-sectional area. The area-scaled Tfet increases the device footprint compared to conventional TFET of the same channel length [9, 10]. This problem can be resolved by considering a nonplanar device structure. In this study, to enhance the on-state current and reduce the device footprint area, LTFET is proposed [11 - 14].

Despite the development of different TFET structures, the effect of noise on the device performance is still not explored extensively [15 - 19]. Nonetheless, the impact of introducing traps on the effect of device performance has been reported in a few studies [20 - 24]. Ghosh and Bhowmick reported the low-frequency noise analysis in the presence of uniform and Gaussian interface traps. The improved current ratio and SS were obtained in the case of uniform traps than the Gaussian trap [25]. Wangkheirakpam *et al*. presented the effect of uniform and Gaussian traps in D-MOS TFET and discussed the effect of temperature variation on device characteristics [26]. Another simulation study of the effect of trap on oxide/semiconductor interface for different charge distribution was reported by Talukdar and Mummaneni. From the study, it has been found that the Gaussian trap degrades the device performance compared to the uniform trap in terms of current ratio and SS [27].

This chapter deals with the effect of variation on transfer, output characteristics and electrical noise behavior, linearity, and reliability of a hetero-junction LTFET device with a 20 nm gate length. The effect of the Gaussian trap on the noise behavior of the device also has been performed. The simulation study has been carried out using the Sentaurus TCAD tool. The chapter is organized into four sections. The chapter starts with the introduction section, followed by the proposed device structure and simulation methodology. Moreover, the fabrication process flow of the proposed LTFET has been discussed in detail. Moreover, the obtained results and discussion consists of DC analysis, AC analysis, linearity analysis, electrical noise analysis, and reliability analysis. Finally, the study has been concluded by discussing the important findings and the future scope of the proposed device.

DEVICE STRUCTURE

Fig. (**1a**) depicts the schematic of the proposed TFET along with the device dimen- sions used for the simulations. In n-type TFET, the P++ region is defined as the source region of Ge doped with a boron concentration of 1×10^{19}cm^{-3}. While the N++ region is defined as the drain region of Si with a phosphorous concentration of 1×10^{19}cm^{-3}. An offset region with a boron concentration of 1×10^{15}cm^{-3} has been created to enhance the TFET performance. The channel is doped with a boron concentration of 1×10^{15}cm^{-3}. The offset region is a channel region, as both are doped with the same concentration. The height of the offset region is optimized as it is the tunneling junction, and the maximum tunneling of charge carriers occurs in the offset region. A 2 nm thick HfO$_2$ with a dielectric constant of 22 is used as the gate oxide, while Al, with a work function of 4.3, is used as the gate metal. The calibration of simulated data is done with the experimental data to verify the accuracy of the simulation set-up [28]. Fig. (**1b**) depicts that the obtained data is in good agreement with the experimental data, hence it validates the selected models.

PROPOSED SIMULATION FRAMEWORK

All simulations have been carried out in Sentaurus TCAD [29]. The TCAD tool works with an algorithm to determine the nonlocal BTB tunneling rate without prior knowledge of tunneling locations [30]. The TCAD models used for the simulations are summarized in Fig. (**2**).

(a)

Parameters	Value
L_S	40 nm
L_G	20 nm
L_D	40 nm
H_S	60 nm
H_{OFF}	5 nm
H_D	20 nm

(b)

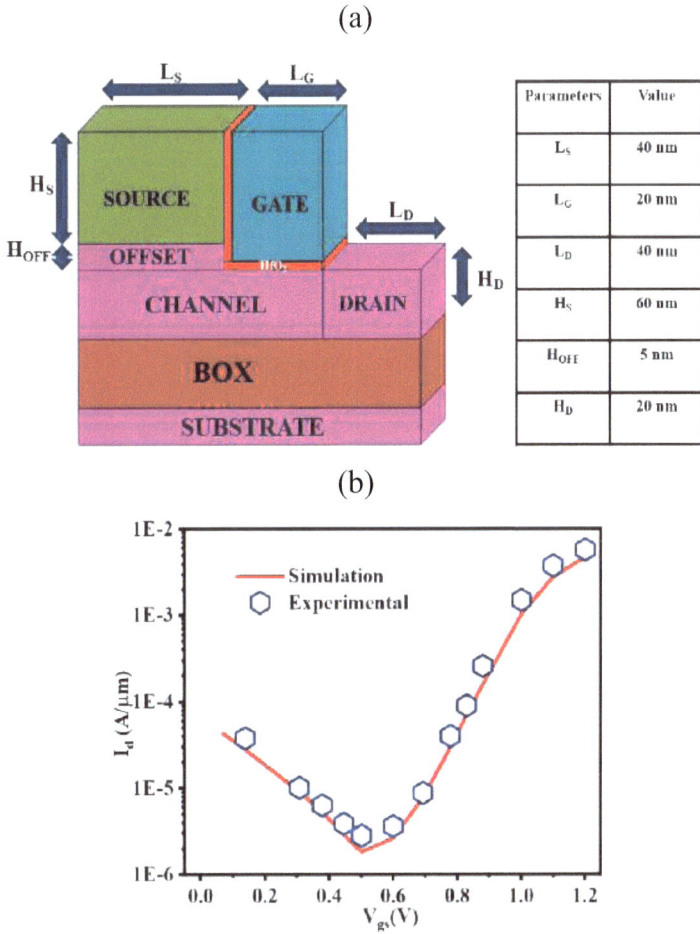

Fig. (1). (**a**) Schematic of proposed LTFET device along with dimension (**b**) Calibration plot [28].

Model Name	Function
Non-local BTBT	Inter-band tunneling
Bandgap narrowing	High doping concentration
Mobility	Carrier mobility
Fermi	Carrier concentration
Trap	Gaussian trap

Fig. (2). Sentaurus TCAD models.

Fig. (**3**) shows the required steps for the fabrication of the proposed LTFET.

(a). Deposition of intrinsic silicon layer over Silicon on an insulator (SOI).

(b). Oxide deposition over the channel region can be done using the oxidation method with dope n-type drain region.

(c). Passivation of the drain region, the silicon layer was deposited and doped to form offset.

(d). Ge layer was formed to create the source region using the LPCVD method, and p+ doping can be done using the diffusion method.

(e). HfO$_2$ deposition using atomic layer deposition method.

(f). Metal gate formation using the sputtering method.

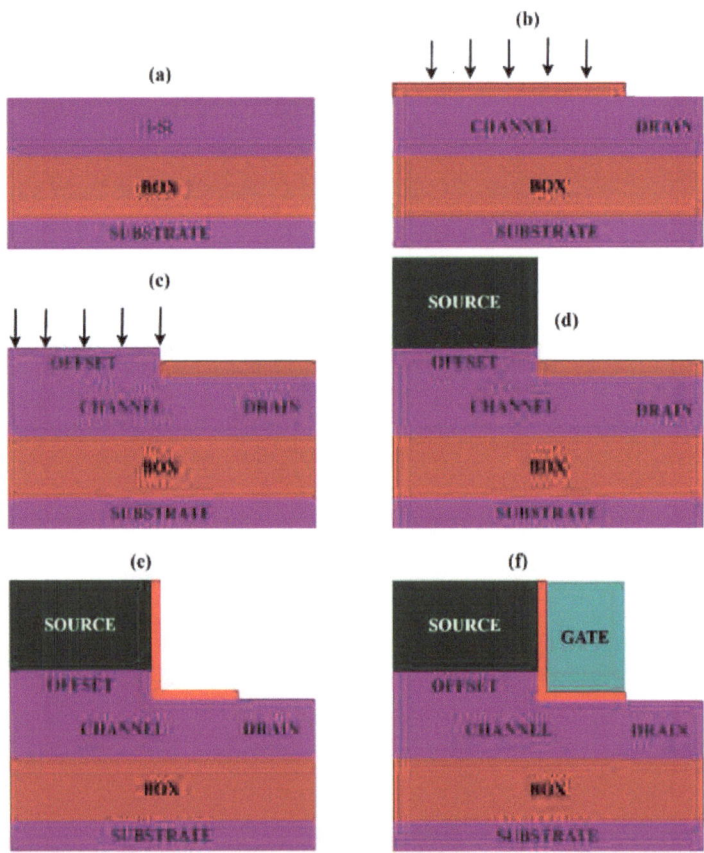

Fig. (3). Fabrication process flow.

RESULTS AND DISCUSSION

D.C. Analysis

The transfer characteristics (I_d-V_{gs}) of the hetero-junction LTFET device are depicted in Fig. (**4a**). The gate voltage was varied from -0.3 V to 1.5 V by keeping the drain-source voltage constant at 0.5 V and the back gate voltage at ground potential. In TCAD, for the calculation of the BTB, the generation rate (G) using Kane's model can be expressed by equation (**1**) [31]:

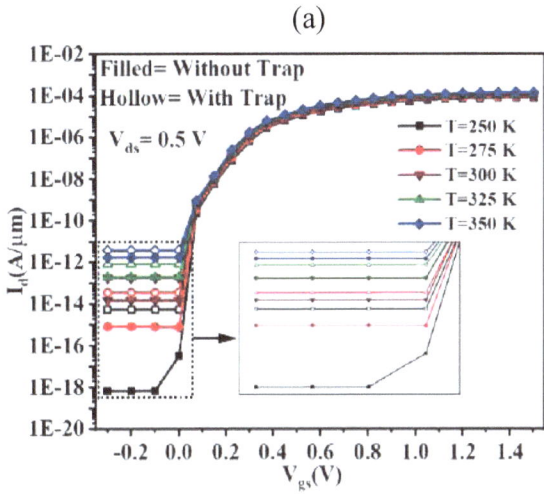

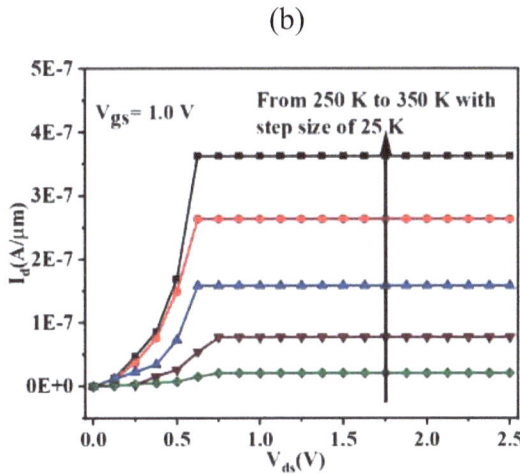

Fig. (4). Proposed LTFET device (**a**) Id-Vgs, (**b**) I_d-V_{ds}.

$$G = A \left(\frac{F}{F_o} \right)^P \exp \left(-\frac{B}{F} \right) \tag{1}$$

Here A and B are Kane's parameters, F is the electric field, $F_o = 1\,\text{V/m}$, and P= 2.5.

The results of drain current are obtained without including the effects of traps under different temperature ranges from 250 K to 350 K with a step size of 25 K. A slight variation in the current at low temperatures has been observed, while as temperature increases, the variation in the leakage current reduces. However, a small variation is observed in the saturation region of id. The current dependence on temperature can be expressed by equation (2) [32]:

$$I \propto \exp \left(-\frac{E_a}{k_B T} \right) \tag{2}$$

Where T is the absolute temperature, and E_a is the activation energy. The activation energy does not remain constant; it varies with V_{gs} [33]. When the V_{gs} is at the turn-off point, E_a is roughly half of the bandgap, which means it results mainly from SRH recombination [34]. Once the device is turned ON and id saturated E_a changes slightly as V_{gs} is varied. Since I depends upon E_a and E_a depends upon temperature. So, in the state, large variations in E_a occur, so It changes as the temperature is changed. But in the saturation region, a small variation in E_a causes a small variation in on-current with varying temperatures. Fig. (4b) shows the output characteristics plotted at V_{gs} = 1.0 V at different temperatures. With increasing V_{ds}, the drain current first increases rapidly because of the high band-to--to-band tunneling rate of charge carriers. Although at a certain voltage, the drain current saturates as the maximum band-to--to-band tunneling rate has been achieved. Moreover, as the temperature increases from 250 K to 350 K, the drain current increases because at high temperatures, the tunneling band gap decreases and the drain current increases.

Figs. (5a & b) shows the contour plot of G and the energy band diagram of the proposed LTFET. With increasing temperature, G increases, which further increases the drain current. By making the variation in temperature from 250 K to 350 K, the energy band diagram is plotted. The results clearly illustrate that the rise in temperature causes a fall in the energy band. Since it shows dependence on the energy band gap, it further depends upon the temperature.

(a)

(b)

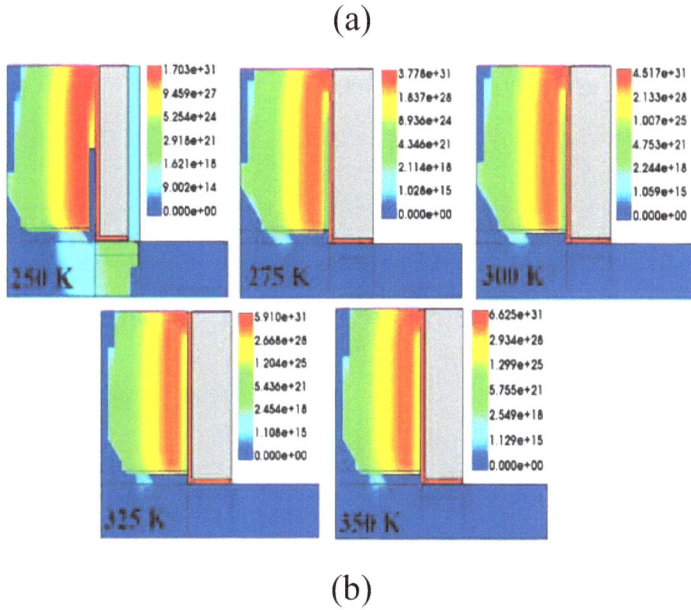

Fig. (5). LTFET (**a**) Contour plot of G (**b**) Energy band diagram at different temperatures.

AC Analysis

For AC analysis, the investigation of the device's capacitance is required. Figs. (**6a-d**) shows the change in capacitance as a function of V_{gs} at different temperatures without traps and with Gaussian traps at both low frequency (LF at 1 MHz) and high frequency (HF at 1GHz). For the case of trap analysis, the rise in temperature causes a small variation in gate-drain capacitance (C_{gd}) and negligible variation in gate-source capacitance (C_{gs}) both at LF and HF. However, on including the Gaussian traps, noteworthy changes in the curves are observed on rising temperature. The C_{gd} is large because of the tunneling barrier at the source side of the TFET device. So, the total gate capacitance is dominated by C_{gd}. It is very evident from the results that on changing the temperature, C_{gd} and C_{gs} are not much affected [35].

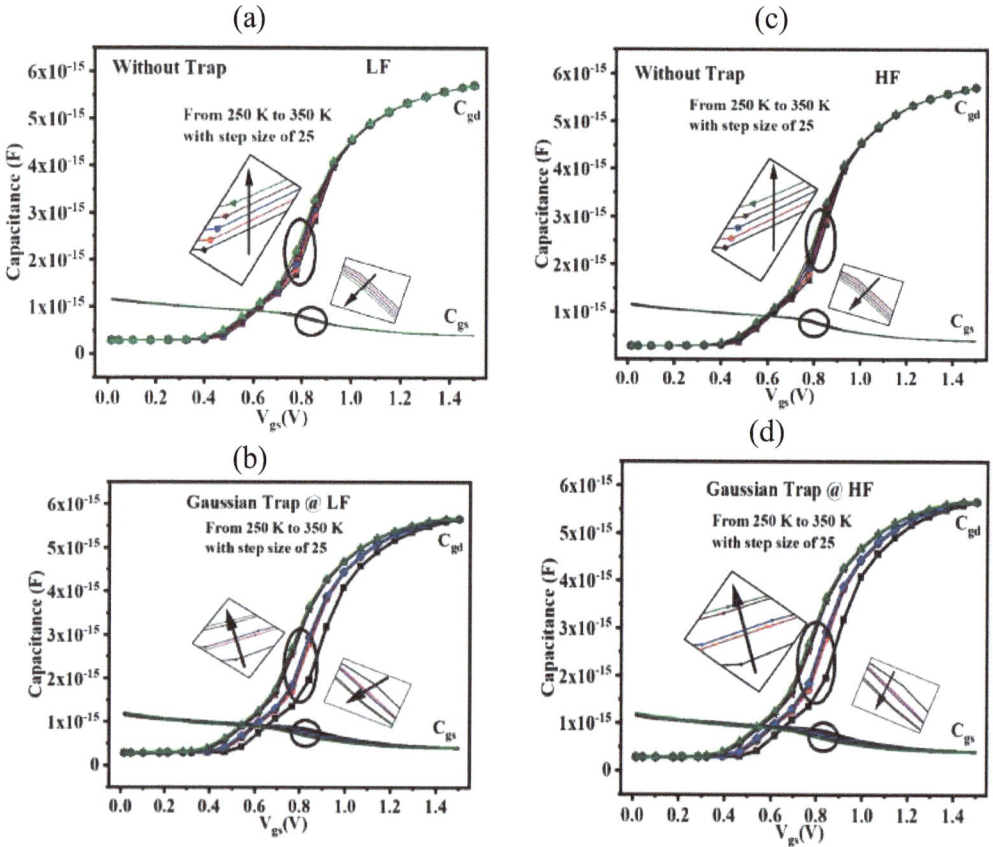

Fig. (6). Proposed LTFET capacitance plot without a trap and with Gaussian Trap at (**a**) LF, (**b**) HF, (**c**) LF, (**d**) HF.

For analog applications, transconductance (g_{m1}) is another important parameter of the TFET device. The transconductance can be expressed by equation (**3**) [36]:

$$g_{m1} = \left.\frac{\partial I_d}{\partial V_{gs}}\right|_{Vds} \tag{3}$$

Where Vgs and V_{ds} are the gate-source voltage and drain-source voltage.

The inverse of the output resistance is known as g_d and is expressed by Equation **(4)** [36]:

$$g_d = \frac{\partial I_d}{\partial V_{ds}}\bigg|_{Vgs} \tag{4}$$

The obtained results for g_{m1} and g_d are given in Figs. (**7a & b**). The rise in gate voltage results in the injection of a large number of carriers from the source to the channel region, which further increases g_{m1}. Since g_{m1} depends upon I_d, the variation in g_{m1} is observed as the temperature is increased.

The ratio of g_{m1}/g_d is known as intrinsic gain or maximum voltage gain (A_v) [36]. Fig. (**8a**) shows the intrinsic gain at different temperatures. At high V_{ds}, Gd decreases due to the decrease in tunneling rate. The peak increases due to an increase in temperature as the carrier tunneling increases at high temperatures.

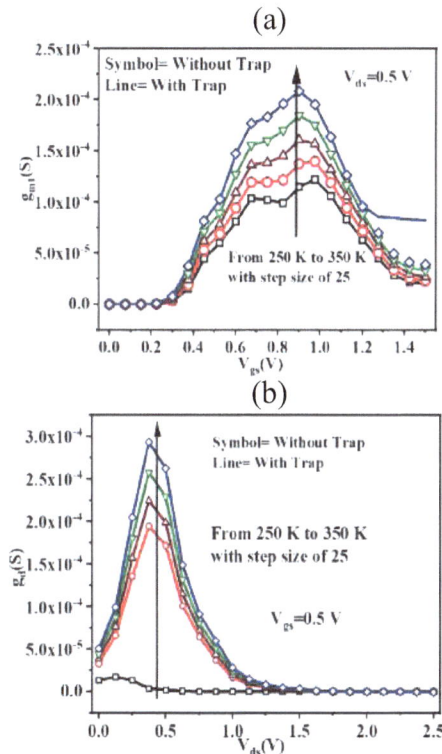

Fig. (7). Proposed LTFET (**a**) Transconductance at V_{ds}= 0.5 V, (**b**) Output conductance at V_{gs}= 1.0 V.

(a)

(b)

(c)

(d)

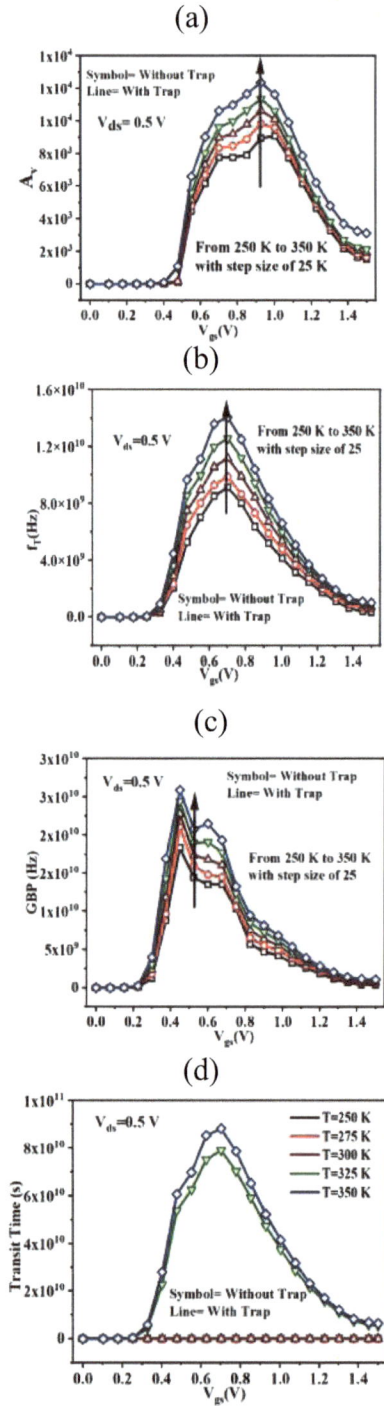

Fig. (8). Proposed LTFET (**a**) A_v (**b**) f_T (**c**) GBP (**d**) Transit time.

The RF analysis of the device requires the investigation of two parameters, *i.e.*, cut-off frequency (f_T) and gain bandwidth (GBP). A high value of f_T and GBP is required for the devices used in high-frequency circuits. Equation **(5)** states that f_T directly depends on g_{m1} while inversely depending upon the total gate capacitance. While GBP is calculated at a fixed static gain and expressed by equation **(6)** [37]. The variation in f_T and GBP at different temperatures are plotted in Figs. (**8a & b**). It is evident from the results that because of the rapid rise in g_m, f_T increases and attains a maximum value around $V_{gs} = 0.7$ V. Afterward, the steep fall in g_{m1} and a rise in Cgg reduces the f_T. Similarly, the maximum value of GBP is attained at $V_{gs} = 0.4$ V, and then a fall in GBP occurs because of the reduction in g_{m1}. Since f_T and GBP depend upon g_{m1}, that further depends upon temperature. Thus, as the temperature increases, f_T and GBP increase.

While another crucial parameter for RF analysis is, transit time (τ). It defines the time required by the charge carriers to reach from the source to drain [37] and is expressed by Equation **(7)**. The change in transit time by varying temperatures is plotted in Fig. (**8d**).

$$f_T = \frac{g_{m1}}{2\pi(C_{gd} + C_{gs})} \tag{5}$$

Transconductance frequency product (TFP) is the product of the efficiency of the device and is given by Equation **(8)**, and the obtained results are plotted in Fig. (**9a**). It depicts a trade-off between power and bandwidth and is used for moderate to high-speed designs. Transconductance generation efficiency (TGF) is given by Equation **(9)**, and the simulated results are plotted in Fig. (**9b**). TGF shows the efficiency of current-to-transconductance conversion. Although the higher values of TGF can ensure the amplification per unit drain current. However, the device with high TGF may exhibit linearity issues and degrade the linearity of circuits [36]. Both TGF and TFP increase with increasing temperature.

$$GBP = \frac{g_{m1}}{2\pi C_{gd}} \tag{6}$$

$$\tau = \frac{1}{2\pi f_T} \tag{7}$$

$$TFP = \frac{g_{m1} f_T}{I_d} \tag{8}$$

$$TGF = \frac{g_{m1}}{I_d} \tag{9}$$

(a)

(b)

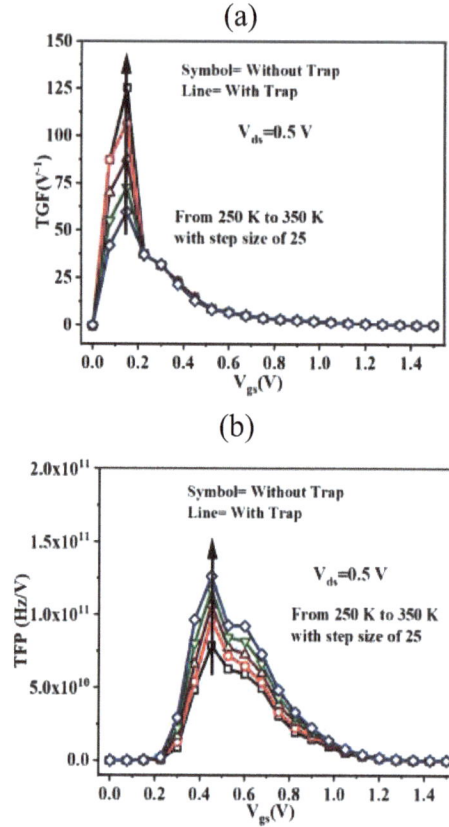

Fig. (9). Proposed LTFET (**a**) TGF (**b**) TFP.

Linearity Analysis

A highly linear device is required for wireless applications. To analyse the linearity and distortion performance of any FET device, g_{m1} and its higher-order derivatives g_{m2} and g_{m3} are evaluated. g_{m2} and g_{m3} are the second and third-order transconductance and can be given by Equations (**10 & 11**) [38]. The variations in g_{m2} and g_{m3} with temperature variation are plotted in Figs. (**10 & 11**).

$$g_{m2} = \frac{\partial g_{m1}}{\partial V_{gs}} \tag{10}$$

$$g_{m3} = \frac{\partial g_{m2}}{\partial V_{gs}} \tag{11}$$

(a)

(b)

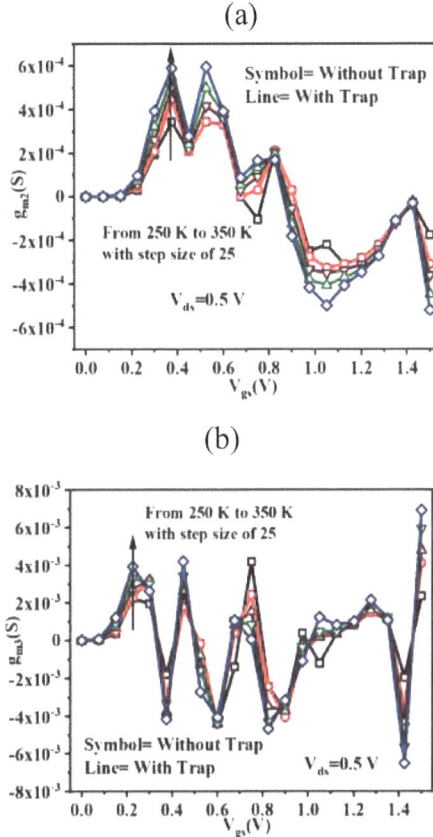

Fig. (10). Proposed LTFET (**a**) g_{m2} *vs.* Vgs, (**b**) g_{m3} *vs.* V_{gs} at V_{ds} = 0.5 V.

The linearity parameters HD2, HD3, and IIP3 are used to evaluate the linearity and distortions of the system. The variations in harmonic distortion at different temperatures are shown in Figs. (**11a & b**); the harmonic distortion HD2 is more at low Vgs, *i.e.*, when the device is turned ON. The increase in V_{gs} decreases the distortion. The harmonic distortion HD3 has a low effect at low and high V_{gs}, but for the middle range, more distortions occur. The harmonic distortions HD2 and HD3 can be expressed by equations (**12 & 13**) [39]:

$$HD2 = \frac{1}{2} \times V_i \times \frac{g_{m2}}{2g_{m1}} \, dBm \qquad (12)$$

$$HD3 = \frac{1}{4} \times V_i^2 \times \frac{g_{m3}}{6g_{m1}} \, dBm \qquad (13)$$

Here, V_i is the amplitude of the input signal.

IIP3 is the extrapolated input power at which the first and the third harmonic become equal. This parameter is used to compare the linearity of different circuits and devices and can be expressed by equation **(14)** [40]. The obtained results are plotted in Fig. **(11c)**.

$$IIP3 = \frac{2}{3} \times \frac{g_{m1}}{g_{m3} \times R_s} \, dBm \qquad \qquad \textbf{(14)}$$

(a)

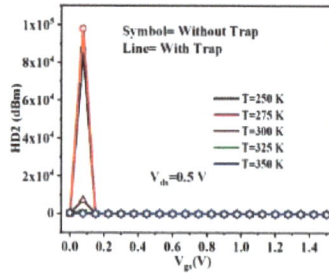

(b)

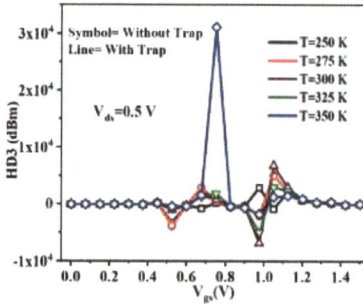

(c)

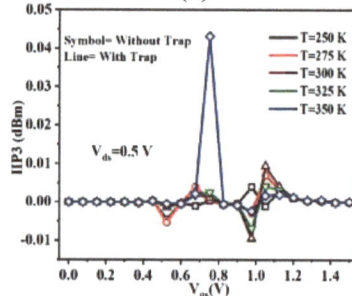

Fig. (11). Proposed LTFET (**a**) HD2 *vs.* V_{gs} (**b**) HD3 *vs.* V_{gs} (**c**) IIP3 *vs.* V_{gs}.

Fig. (**12a**) shows the influence of temperature variation on the gain frequency product (GFP) of the proposed LTFET. The GFP can be expressed by equation (**15**). Both A_v and f_T increase with increasing temperature, thus, the GFP value also increases with increasing temperature [41].

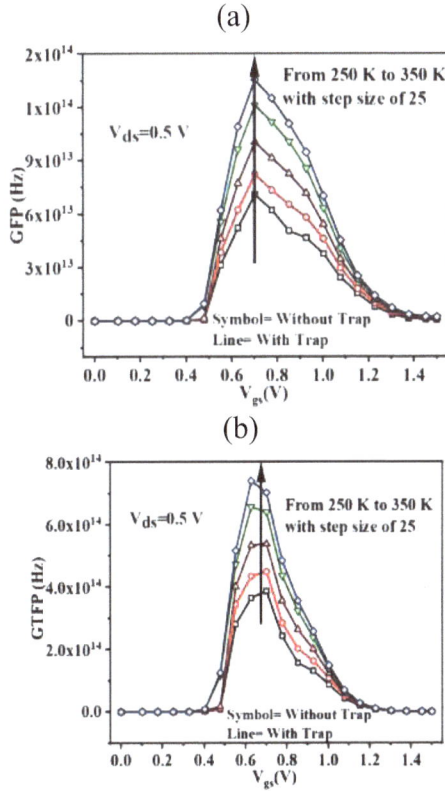

Fig. (12). Proposed LTFET (**a**) GFP *vs.* Vgs, (**b**) GTFP *vs.* V_{gs}.

$$GFP = \frac{g_{m1}}{g_d} * f_T \qquad (15)$$

Fig. (**12b**) depicts the gain, transconductance, and frequency product (GTFP) variation with varying temperatures. GFTP can be expressed by equation (**16**) [41].

$$GFTP = \frac{g_{m1}}{g_d} * \frac{g_{m1}}{I_d} * f_T \qquad (16)$$

It offers a trade-off between av, TGF, fT, and GTFP increases with temperature.

Electrical Noise Analysis without Traps

A thorough investigation of electrical noise analysis parameters like drain current noise spectral density (S_{ID}) and noise voltage spectral density (S_{vg}) at LF and HF, at different temperature ranges, is performed. The analysis has been carried out by including the traps and not including the traps. Both S_{ID} and S_{VG} are investigated as a function of the gate voltage (V_{gs}) at various temperature ranges from 250 to 350 K with a step size of 25 K. The general LF noise of a semiconductor device can be represented by equation **(17)** [42]:

$$S_{ID} = \frac{q^2 kT \lambda N_t}{C_{ox}^{\ 2} WL_{eff} f^{\gamma}} \frac{I_d^{\ 2}}{\left(V_{gs} - V_t\right)^2} \tag{17}$$

Where, T, R, q, N_t, k, L_{eff}, W, C_{ox}, λ, and f are the temperature, resistance, charge, interface trap density, Boltzmann constant, effective channel length, width of the device, oxide capacitance, tunneling parameter, and frequency of the device respectively.

$$g_m = \frac{I_d}{\left(V_{gs} - V_t\right)} \tag{18}$$

$$I_d = I_{tunneling} = AF^2 \exp\left(\frac{-B}{F}\right) \tag{19}$$

Where, F, A and B are the electric field and empirical, parameters, respectively.

The input voltage noise PSD (S_{VG}) can be expressed by equation **(20)** [42]:

$$S_{VG} = \frac{q^2 kT \lambda N_t}{C_{ox}^{\ 2} WL_{eff} f^{\gamma}} \tag{20}$$

In the proposed LTFET, the effect of Gaussian trap charges in Si/HfO_2 and Ge/HfO_2 with a trap charge density of $N_f = \pm 1 \times 10^{13}$ cm^{-2} is considered. By introducing Gaussian traps, the characteristics of the device get degraded and Gaussian distribution can be defined by equation **(21)** [27]:

$$D_{Gaussian} = N_o \exp\left(-\frac{|E - E_O|^2}{2E_S^2}\right) \tag{21}$$

N_o is the maximum trap concentration, E_o and E_s are energy mid and energy sig in Gaussian distribution. From Equation **(21)**, it is clear that the Gaussian distribution of traps is related to the concentration of trap charges and the exponential function of energy. In Figs. (**13a-d**), the simulated results of S_{ID} and S_{VG} were plotted at different temperatures without the trap and with the trap. As the temperature increases, the effect of noise increases, and in HF, the noise is less pronounced than in LF.

From the analysis, it has been found that as the temperature increases, both S_{ID} and S_{VG} are more pronounced at high temperatures. Almost the same value of S_{ID} is achieved for the entire range of Vgs, while S_{VG} has a small variation in the value at low V_{gs} and high V_{gs}. This variation is observed because of the strong SRH dependent on the temperature at lower V_{gs} and less at higher V_{gs} due to BTBT. It is quite evident from equations **(18-20)** that both S_{ID} and S_{VG} depend upon temperature. As the temperature increases, the effect of noise increases. The drain current noise has inverse dependence on frequency; the higher the frequency, the less the impact of noise, and *vice versa* [35].

Reliability Analysis

The reliability concern of the proposed device over a wide range of temperatures from 250 K to 350 K also has been well performed. The results in Section 3.1 clearly illustrate that the impact of changing the temperature on on-current is very low. However, a small variation in the leakage current can be seen, but still, the SS of the device remains high ($\sim 10^8$). With increasing the temperature from 250 K to 350 K, a small change in I_{ON}/I_{OFF} and SS can be seen. The device shows slight ambipolarity at 250 K, while on changing the temperature, the device performance has not been affected by ambipolar conduction. In TFET, on-current, off-current, and SS are crucial parameters that decide the figure of merit (FOMs) of the device. For a clear illustration, Figs. (**14a-d**) shows the impact of change in the temperature on I_{ON}, I_{OFF}, I_{ON}/I_{OFF}, and SS without considering the traps. A small variation in the on-current and a slight change in the off-current at low temperatures is observed. The change in I_{ON}/I_{OFF} ratio is evident at low temperatures, while at high temperatures, the ratio of 10^8 and SS of 27 mV/dec can still be achieved. Even different linearity parameters are not much affected by the variation in the temperature. The proposed hetero-junction LTFET device shows good performance and is reliable over a wide range of temperatures.

(a)

(b)

(c)

(d)

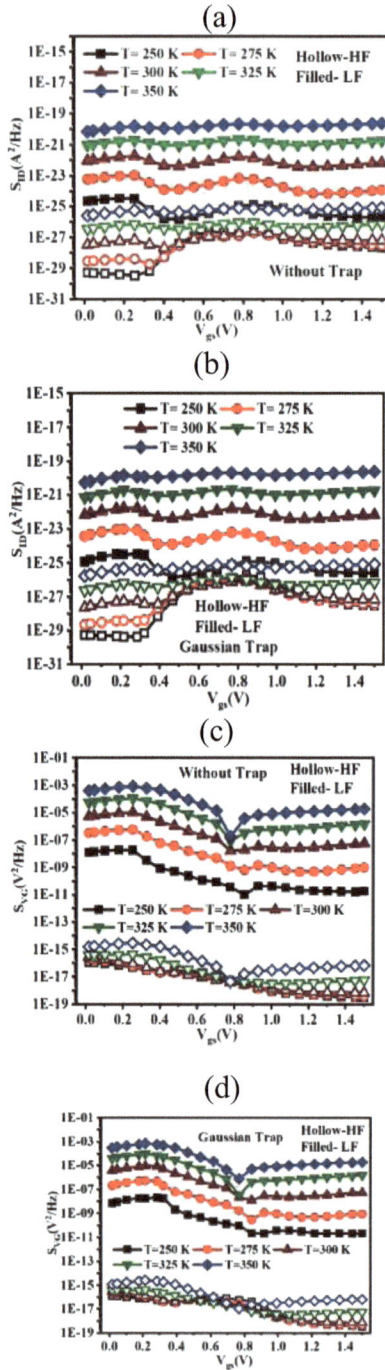

Fig. (13). Proposed LTFET with trap (**a**) S_{ID} *vs.* V_{gs} without trap (**b**) S_{ID} *vs.* V_{gs} with trap (**c**) S_{VG} *vs.* V_{gs} without trap (**d**) S_{VG} *vs.* V_{gs} with the trap.

(a)

(b)

(c)

(d)

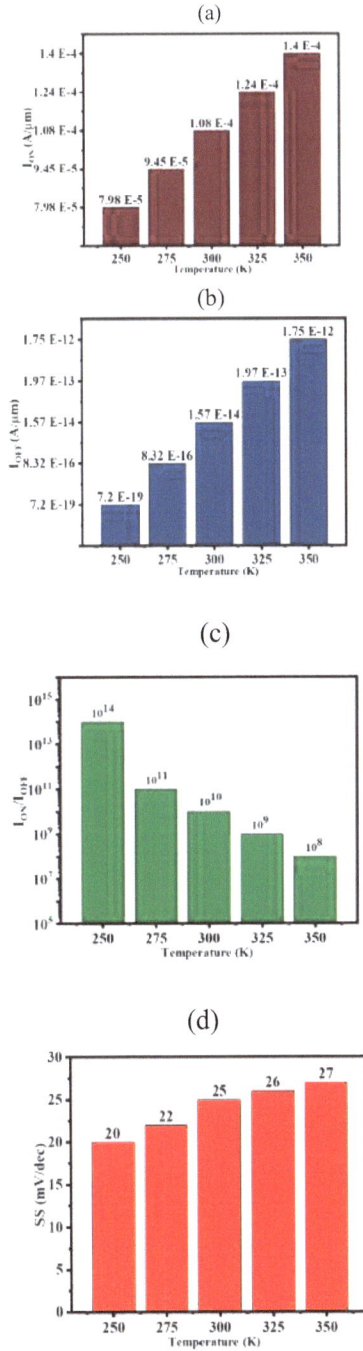

Fig. (14). Impact of temperature variation on (**a**) I_{ON}, (**b**) I_{OFF}, (**c**) I_{ON}/I_{OFF}, (**d**) SS.

CONCLUSION

A detailed study of LTFET, like AC analysis, analog/RF, linearity analysis, and electrical noise analysis with traps and without traps, has been reported in the present work. As the temperature increases from 250 K to 350 K, the off-current increases. From the analysis, it has been found that the temperature dependency is high in the area where current flows because of SRH than in band-to-band tunneling-dominated areas. The change in temperature does not impact the on-current of the device, while a small variation in the off-current occurs. The various FOMs of the device also show small variations in the results with rising temperature. The only unfavorable factor where the evident change in the results has been observed is the electrical noise characteristics of the device. The reliability analysis clarifies that the proposed LTFET device performs well at a wide range of temperatures.

The proposed LTFET shows good performance because of high on-current, low off-current, and steep SS. The less impact of noise and reliable performance over a wide range of temperatures makes it suitable for low-power Internet-of-things (IoT) applications.

LEARNING OBJECTIVES

• A thorough study of the electrical characteristics of the nanoelectronic devices, i.e., Tunnel field-effect transistors (TFET), has been presented using the SYNOPSYS TCAD tool.

• The detailed investigation consists of DC analysis, AC analysis, linearity analysis, electrical noise analysis, and reliability analysis.

• The proposed device shows good performance in terms of different electrical characteristics and is highly reliable over a wide range of temperatures.

• The design of a nanoelectronic device using germanium as source material shows superior characteristics and can be well-suited for low-power IoT applications.

MULTIPLE CHOICE QUESTIONS

1. Nanoelectronic device TFET stands for:

  a. Tunnel field-effect transistor.
  b. Transistor field-effect tunneling.
  c. Tunneling field-effect transistor.

d. None of the above.

2. The doping concentration in n-type TFET is:

a. N-I-P
b. P-I-N
c. P-I-N-P
d. N-I-P-N

3. With increasing the temperature, the leakage current of TFET_____.

a. Remains same
b. Decreases
c. Increases
d. Slight variation is seen

4. The activation energy of TFET does not remain constant, it varies with___.

a. Drain voltage
b. Gate voltage
c. Back voltage
d. All of above

5. The temperature dependence equation of drain current shows that Itdepends upon___and___ depends upon temperature.

a. Activation energy, Activation energy.
b. Drain voltage, Activation energy.
c. Activation energy, Gate voltage.
d. Drain voltage, Gate voltage.

6. Traps are any location within a solid (generally a semiconductor or an insulator) that restricts the movement of _____.

a. Electrons
b. Holes
c. Electrons and holes
d. None of the above

7. The tunneling barrier at the source side is responsible for the large _____of TFET device. Thus, total _____is dominated by_____.

a. C_{gs}, C_{gg}, C_{gd}
b. C_{gd}, C_{gs}, C_{gd}
c. C_{gd}, C_{gg}, C_{gs}
d. C_{gd}, C_{gg}, C_{gd}

8. The temperature variations may have a negligible impact on _____ at LF.

a. C_{gs}
b. C_{gd}
c. Both C_{gd}, C_{gs}
d. None of the above

9. Transconductance is the ratio of:

a. Rate of change of drain current with changing gate-source voltage.
b. Rate of change of gate current with changing gate-source voltage.
c. Rate of change of gate current with changing drain-source voltage.
d. Rate of change of drain current with changing drain-source voltage.

10. The ratio of_____ is known as intrinsic gain or maximum voltage gain (A_v).

a. g_d/g_m
b. g_m/g_d
c. C_{gs}/C_{gd}
d. C_{gg}/C_{gd}

11. The cut-off frequency (f_T) of TFET depends directly on _____ and inversely on_____.

a. g_m, C_{gg}
b. g_d, C_{gg}
c. g_m, C_{gd}
d. g_m, C_{gs}

12. In RF analysis, TFP stands for:

a. Transconductance frequency efficiency
b. Transconductance frequency product
c. Transfer characteristics frequency product
d. None of the above

13. The linearity parameters _____ are used to evaluate the linearity and distortions of the system.

a. g_{m1}, g_{m2}, g_{m3}
b. C_{gd}, C_{gs}, C_{gg}
c. g_m, g_d
d. HD2, HD3, and IIP3

14. With increasing the frequency, the electrical noise in TFET____.

a. Decreases
b. Increases
c. Remains same
d. Changes negligibly

15. Both TGF and TFP_____ with increasing temperature.

a. Decreases
b. Increases
c. Remains same
d. Changes negligibly

ANSWER KEY

1. (a)

2. (b)

3. (c)

4. (b)

5. (a)

6. (c)

7. (d)

8. (c)

9. (a)

10. (b)

11. (a)

12. (b)

13. (d)

14. (a)

15. (b)

ACKNOWLEDGEMENT

This work is supported by DST-SERB, CRG grant, Govt. of India, CRG/2020/006229, dated: 05/04/2021.

REFERENCES

[1]   C. Martin, "Towards a new scale", *Nat. Nanotechnol.,* vol. 11, no. 2, pp. 112-112, 2016.
[http://dx.doi.org/10.1038/nnano.2016.8] [PMID: 26839254]

[2]   S.P. Beaumont, "III–V nanoelectronics", *Microelectron. Eng.,* vol. 32, no. 1-4, pp. 283-295, 1996.
[http://dx.doi.org/10.1016/0167-9317(95)00367-3]

[3]   R. Asra, M. Shrivastava, K.V.R.M. Murali, R.K. Pandey, H. Gossner, and V.R. Rao, "A tunnel FET for VDD scaling below 0.6 V with a CMOS-comparable performance", *IEEE Trans. Electron Dev.,* vol. 58, no. 7, pp. 1855-1863, 2011.
[http://dx.doi.org/10.1109/TED.2011.2140322]

[4]   A.M. Ionescu, and H. Riel, "Tunnel field-effect transistors as energy-efficient electronic switches", *Nature,* vol. 479, no. 7373, pp. 329-337, 2011.
[http://dx.doi.org/10.1038/nature10679] [PMID: 22094693]

[5]   U.E. Avci, D.H. Morris, and I.A. Young, "Tunnel field-effect transistors: Prospects and challenges", *IEEE J. Electron Devices Soc.,* vol. 3, no. 3, pp. 88-95, 2015.
[http://dx.doi.org/10.1109/JEDS.2015.2390591]

[6]   R. Pandey, S. Mookerjea, and S. Datta, "Opportunities and challenges of tunnel FETs", *IEEE Trans. Circuits Syst. I Regul. Pap.,* vol. 63, no. 12, pp. 2128-2138, 2016.
[http://dx.doi.org/10.1109/TCSI.2016.2614698]

[7]   S. Chander, S.K. Sinha, and R. Chaudhary, "Comprehensive review on electrical noise analysis of TFET structures", *Superlattices Microstruct..* p. 107101, 2021.

[8]   S. Datta, H. Liu, and V. Narayanan, "Tunnel FET technology: A reliability perspective", *Microelectron. Reliab.,* vol. 54, no. 5, pp. 861-874, 2014.
[http://dx.doi.org/10.1016/j.microrel.2014.02.002]

[9] K. Hemanjaneyulu, and M. Shrivastava, "Fin enabled area scaled tunnel FET", *IEEE Trans. Electron Dev.,* vol. 62, no. 10, pp. 3184-3191, 2015.
[http://dx.doi.org/10.1109/TED.2015.2469678]

[10] R. Molaei Imenabadi, M. Saremi, and W.G. Vandenberghe, "A novel PNPN-like Z-shaped tunnel field-effect transistor with improved ambipolar behavior and RF performance", *IEEE Trans. Electron Dev.,* vol. 64, no. 11, pp. 4752-4758, 2017.
[http://dx.doi.org/10.1109/TED.2017.2755507]

[11] Z. Yang, "Tunnel field-effect transistor with an L-shaped gate", *IEEE Electron Device Lett.,* vol. 37, no. 7, pp. 839-842, 2016.
[http://dx.doi.org/10.1109/LED.2016.2574821]

[12] S. Chander, S.K. Sinha, R. Chaudhary, and R. Goswami, "Effect of noise components on L-shaped and T-shaped heterojunction tunnel field effect transistor IOP Science", *Semicond. Sci. Technol.,* vol. 37, no. 7, pp. 1-11, 2022.
[http://dx.doi.org/10.1088/1361-6641/ac696e]

[13] Z. Yang, "Tunnel field-effect transistor with an L-shaped gate", *IEEE Electron Device Lett.,* vol. 37, no. 7, pp. 839-842, 2016.
[http://dx.doi.org/10.1109/LED.2016.2574821]

[14] F. Najam, and Y. Yu, "Optimization of Line-Tunneling Type L-Shaped Tunnel Field-Effect-Transistor for Steep Subthreshold Slope", *Electronics,* vol. 7, no. 11, p. 275, 2018.
[http://dx.doi.org/10.3390/electronics7110275]

[15] A. Singh, S.K. Sinha, and S. Chander, "Impact of Fe Material thickness on performance of raised source overlapped negative capacitance tunnel field effect transistor (NCTFET)", *Silicon,* vol. 14, no. 14, pp. 9083-9090, 2022.
[http://dx.doi.org/10.1007/s12633-022-01696-6]

[16] K. Baruah, R.G. Debnath, and S. Baishya, "Study of noise behavior of heterojunction double-gate PNPN TFET for different parameter variations", In: *Micro and Nanoelectronics Devices, Circuits and Systems.* Springer, 2022, pp. 91-97.
[http://dx.doi.org/10.1007/978-981-16-3767-4_8]

[17] S. Chander, and S.K. Sinha, "Effect of raised buried oxide on characteristics of tunnel field effect transistor", *Silicon,* vol. 14, no. 14, pp. 8805-8813, 2022.
[http://dx.doi.org/10.1007/s12633-022-01681-z]

[18] K.S. Singh, S. Kumar, and K. Nigam, "Impact of interface trap charges on analog/RF and linearity performances of dual-material gate-oxide-stack double-gate TFET", *IEEE Trans. Device Mater. Reliab.,* vol. 20, no. 2, pp. 404-412, 2020.
[http://dx.doi.org/10.1109/TDMR.2020.2984669]

[19] B.V. Chandan, K. Nigam, D. Sharma, and S. Pandey, "Impact of interface trap charges on dopingless tunnel FET for enhancement of linearity characteristics", *Appl. Phys., A Mater. Sci. Process.,* vol. 124, no. 7, p. 503, 2018.
[http://dx.doi.org/10.1007/s00339-018-1923-8]

[20] S. Chander, S. K. Sinha, and R. Chaudhary, "Ge-source based l-shaped tunnel field effect transistor for low power switching application", *Silicon, Oct.,* vol. 14, pp. 4735-7448, 2021.

[21] R. Goswami, and B. Bhowmick, "Circular gate tunnel FET: Optimization and noise analysis", *Procedia Comput. Sci.,* vol. 93, pp. 125-131, 2016.
[http://dx.doi.org/10.1016/j.procs.2016.07.191]

[22] R. Pandey, B. Rajamohanan, H. Liu, V. Narayanan, and S. Datta, "Electrical noise in heterojunction interband tunnel FETs", *IEEE Trans. Electron Dev.,* vol. 61, no. 2, pp. 552-560, 2014.
[http://dx.doi.org/10.1109/TED.2013.2293497]

[23] R. Goswami, B. Bhowmick, and S. Baishya, "Electrical noise in circular gate tunnel FET in presence

of interface traps", *Superlattices Microstruct.,* vol. 86, pp. 342-354, 2015.
[http://dx.doi.org/10.1016/j.spmi.2015.07.064]

[24]   S. K. Sinha, S. Chander, and R. Chaudhary, "Investigation of noise characteristics in gate-source overlap tunnel field-effect transistor", *Silicon,* vol. 14, pp. 10661-10668, 2022.
[http://dx.doi.org/10.1007/s12633-022-01806-4]

[25]   P. Ghosh, and B. Bhowmick, "Low-frequency noise analysis of heterojunction SELBOX TFET", *Appl. Phys., A Mater. Sci. Process.,* vol. 124, no. 12, p. 838, 2018.
[http://dx.doi.org/10.1007/s00339-018-2264-3]

[26]   V. D. Wangkheirakpam, B. Bhowmick, and P. D. Pukhrambam, "Investigation of temperature variation and interface trap charges in dual MOSCAP TFET", *Silicon,* vol. 13, pp. 2971-2978, 2020.

[27]   J. Talukdar, and K. Mummaneni, "A non-uniform silicon TFET design with dual-material source and compressed drain", *Appl. Phys., A Mater. Sci. Process.,* vol. 126, no. 1, p. 81, 2020.
[http://dx.doi.org/10.1007/s00339-019-3266-5]

[28]   S. W. Kim, W. Y. Choi, M. C. Sun, H. W. Naj, and B. G. Park, "Design guideline of Si-based L-shaped tunneling field-effect transistors", *Jpn. J. Appl. Phys.,* vol. 51, p. 06FE09, 2012.

[29]   "Sentaurus Device User Guide, Synopsys Inc.,Version D-2021.0. 2021",

[30]   S. Dash, and G.P. Mishra, "A new analytical threshold voltage model of cylindrical gate tunnel FET (CG-TFET)", *Superlattices Microstruct.,* vol. 86, pp. 211-220, 2015.
[http://dx.doi.org/10.1016/j.spmi.2015.07.049]

[31]   K.H. Kao, A.S. Verhulst, W.G. Vandenberghe, B. Soree, G. Groeseneken, and K. De Meyer, "Direct and indirect band-to-band tunneling in germanium-based TFETs", *IEEE Trans. Electron Dev.,* vol. 59, no. 2, pp. 292-301, 2012.
[http://dx.doi.org/10.1109/TED.2011.2175228]

[32]   J. Lee, D. W. Kwon, H. W. Kim, J. H. Kim, E. Park, T. Park, and B. G. Park, "Analysis on temperature dependent current mechanism of tunnel field-effect transistors", *Jpn. J. Appl. Phys.,* vol. 55, p. 06GG03, 2016.
[http://dx.doi.org/10.7567/JJAP.55.06GG03]

[33]   A. Vandooren, A.M. Walke, A.S. Verhulst, R. Rooyackers, N. Collaert, and A.V.Y. Thean, "Investigation of the subthreshold swing in vertical tunnel-FETs using H_2 and D_2 anneals", *IEEE Trans. Electron Dev.,* vol. 61, no. 2, pp. 359-364, 2014.
[http://dx.doi.org/10.1109/TED.2013.2294535]

[34]   Yue Yang, Xin Tong, Li-Tao Yang, Peng-Fei Guo, Lu Fan, and Yee-Chia Yeo, "Tunneling field-effect transistor: Capacitance components and modeling", *IEEE Electron Device Lett.,* vol. 31, no. 7, pp. 752-754, 2010.
[http://dx.doi.org/10.1109/LED.2010.2047240]

[35]   S. Kumar, and D.S. Yadav, "Temperature analysis on electrostatics performance parameters of dual metal gate step channel TFET", *Appl. Phys., A Mater. Sci. Process.,* vol. 127, no. 5, p. 324, 2021.
[http://dx.doi.org/10.1007/s00339-021-04457-1]

[36]   S. Chen, H. Liu, S. Wang, W. Li, X. Wang, and L. Zhao, "Analog/RF performance of t-shape gate dual-source tunnel field-effect transistor", *Nanoscale Res. Lett.,* vol. 13, no. 1, p. 321, 2018.
[http://dx.doi.org/10.1186/s11671-018-2723-y] [PMID: 30315380]

[37]   B.R. Raad, K. Nigam, D. Sharma, and P.N. Kondekar, "Performance investigation of bandgap, gate material work function and gate dielectric engineered TFET with device reliability improvement", *Superlattices Microstruct.,* vol. 94, pp. 138-146, 2016.
[http://dx.doi.org/10.1016/j.spmi.2016.04.016]

[38]   S. Ghosh, K. Koley, and C.K. Sarkar, "Deep insight into linearity and NQS parameters of tunnel FET with emphasis on lateral straggle", *Micro & Nano Lett.,* vol. 13, no. 1, pp. 35-40, 2018.
[http://dx.doi.org/10.1049/mnl.2017.0326]

[39]   A. K. Gupta, A. Raman, and N. Kumar, "Performance tuning and reliability analysis of the electrostatically configured nanotube tunnel fet with impact of interface trap charges", *Silicon,* vol. 13, pp. 4553-4564, 2020.

[40]   R. Saha, "Simulation study on ferroelectric layer thickness dependence RF/Analog and linearity parameters in ferroelectric tunnel junction TFET", *Microelectronics.* vol. 113, p. 105081, 2021. [http://dx.doi.org/10.1016/j.mejo.2021.105081]

[41]   M. von Haartman, and M. Östling, "Low-frequency noise in advanced MOS devices", [http://dx.doi.org/10.1007/978-1-4020-5910-0]

[42]   W. Shin, G. Jung, S. Hong, Y. Jeong, J. Park, D. Jang, B.G. Park, and J.H. Lee, "Low frequency noise characteristics of resistor- and Si MOSFET-type gas sensors fabricated on the same Si wafer with In2O3 sensing layer", *Sens. Actuators B Chem..* 318, p. 128087, 2020. [http://dx.doi.org/10.1016/j.snb.2020.128087]

CHAPTER 5

Device Structure Modifications in Conventional Tunnel Field Effect Transistor (TFET) for Low-power Applications

Amandeep Singh[1], Sanjeet Kumar Sinha[1] and **Sweta Chander[1,*]**

[1] *School of Electronics and Electrical Engineering, Lovely Professional University, Phagwara, Punjab, India*

Abstract: With the rapid scaling of transistors in the nanometer regime, various short-channel effects emerge in short-channel devices; researchers are looking for an alternative device to replace complementary MOSFET (CMOS) in circuit applications. TFETs are considered to be a good replacement for the conventional MOSFET in the upcoming technologies. The methods used for making I_{ON} higher also impacts the I_{OFF} current. So, the overall current ratio remains unaltered. To overcome this problem, a technique has been developed and adopted in this work that not only improves the current ratio but also makes the subthreshold swing steeper. The major improvements are the reduction of short channel effects, enhancing current ratio reducing dynamic power consumption. Negative capacitance being a new phenomenon, helps in providing improvised results. The device optimized in this work has given values of Subthreshold swing as 53.75 mV/decade, I_{ON} and I_{OFF} as $4.295*10^{-5}$ A/µm, $6.01*10^{-15}$ A/µm, respectively. DIBL calculated for conventional NCTFET is 61.2 mV/V, and for proposed NCTFET is 31.92 mV/V. So DIBL improvement of 52.2% has been achieved.

Keywords: Complementary MOSFET, Current ratio, Drain Induced Barrier Lowering (DIBL), I_{ON}, I_{OFF}, Negative Capacitance, Sub-threshold swing, Tunnel Field Effect Transistor.

INTRODUCTION

With the continuous scaling of transistors, the channel length of semiconductor devices so formed is causing serious problems with regard to current ratio, subthreshold swing, and leakage current. In the new age devices, the channel is becoming shorter, making the channel length the same as the depletion width of the drain and source junction. Other short-channel effects such as mobility degra-

* **Corresponding author Sweta Chander:** School of Electronics and Electrical Engineering, Lovely Professional University, Phagwara, Punjab, India; E-mail: sweta.chander@gmail.com

Gopal Rawat & Aniruddh Bahadur Yadav (Eds.)

dation, drain-induced barrier lowering (DIBL), and drain punch-through have also been introduced in devices with short channels. The researchers are now adopting various alternate device structures to replace the MOSFET device with an improved and better device incorporating the latest device modifications and structures [1]. The new-age devices presented in the literature are successful in achieving better device performance and overcoming the short channel effects, which in turn, provide promising results for low-power applications [2].

According to Moore's law, the density of transistors (*i.e.*, the number of transistors) roughly doubles every two years. It gives a historic trend and provides an estimate of future requirements. The next important parameter that demands attention is Power consumption. The most common solution to higher power consumption is to choose a lesser supply voltage but there is an upper limit to this. After a specific value of threshold voltage (V_{TH}), it restricts the further downfall of supply voltage [3]. V_{TH} voltage depicts the lowest value below which the transistor will not operate. Device structure modifications also have a direct impact on power consumption and device characteristics. Conventional MOSFET has been replaced by new-age FET devices, such as Nano Wire Field Effect Transistor, Carbon Nanotube Field Effect Transistors (CNTFET), Tunnel Field Effect Transistors (TFET), L-Shaped TFET, U-Shaped TFET, *etc.* to overcome the problem of leakage current and less stepper subthreshold swing.

TFET STRUCTURE AND ITS WORKING PRINCIPLE

Tunnel field effect transistor (TFET) consists of PIN structure (*i.e.* p-type, intrinsic, n-type). It is similar to conventional MOSFET but source and drain terminals are of opposite biasing and the electrostatic potential of the intrinsic area is controlled by the gate voltage. When the gate voltage is applied to the intrinsic region of the TFET, the electron buildup takes place. When this applied voltage crosses the threshold, band-to-band tunneling occurs as the conduction band of the intrinsic region comes in line with the valence band of the P-region as depicted in Fig. (**1**). This in turn is responsible for the current flow in the device. As soon as the gate bias reduces, the bands disalign, making the device turn off [4]. Due to its ability to provide a higher current ratio and stepper subthreshold swing, TFET can act as a good option to be utilized in low-power applications in place of conventional MOSFET. Moreover, as TFET is dependent on a band-t--band tunneling mechanism, it provides i) lesser I_{OFF}; ii) Cuts high fermi energy tail; and iii) I_{ON} current is controlled by tunneling area and width [5]. With the advantage of lowering the SS slope below 60 mV/dec and achieving a better current ratio at a lesser supply voltage, TFET can effectively be used in practical circuit applications.

(a)

(b)

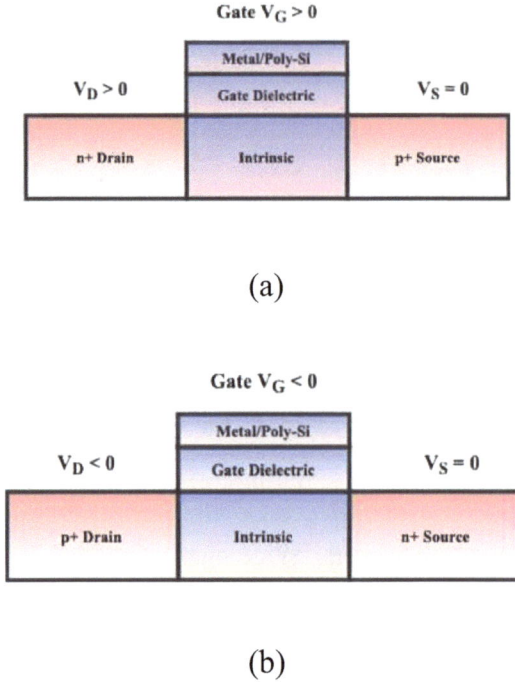

Fig. (1). Supply voltages in case of (**a**) n-TFET (**b**) p-TFET.

By incorporating a germanium source, a much lesser I_{OFF} current is achieved as compared to a conventional MOSFET device, which has a direct impact on power consumption (dynamic as well as static). The major shortcoming of using TFET is the ambipolarity and lesser I_{ON} current. These limitations can be overcome by using new-age device architectures and choosing the materials wisely [6]. Fig. (**2**) shows the schematic diagram of Conventional TFET.

Concept of Negative Capacitance: Applied to TFETs

Due to the continued demand for high-performance devices, the power dissipation of devices is also increasing. This is leading to serious problems, such as reduced reliability, higher weight, increased operating cost, *etc.* Nowadays, researchers are looking for alternate low-power techniques to overcome the listed problems. As reported by Giovanni in 2007, the Negative capacitance phenomenon intervened by using the Fe layer is the main factor associated with lowering subthreshold slope in NCFET transistors. Another advantage of negative capacitance is better voltage amplification and a higher current ratio [7]. Salahuddin and Datta [1] presented in their work that the dynamic power of the device can be reduced by replacing the insulating material with ferroelectric material in conventional devices. With the increasing performance requirements, the continuous scaling of transistors leads to more and more leakage current and increase in the operating temperature of the device. To overcome these problems, researchers are looking for:

a. Strained Si (for increasing mobility)
b. Multicore processing units
c. High-k Dielectric Materials (for reducing leakage current)
d. FiFET structures (for reducing short channel effects)

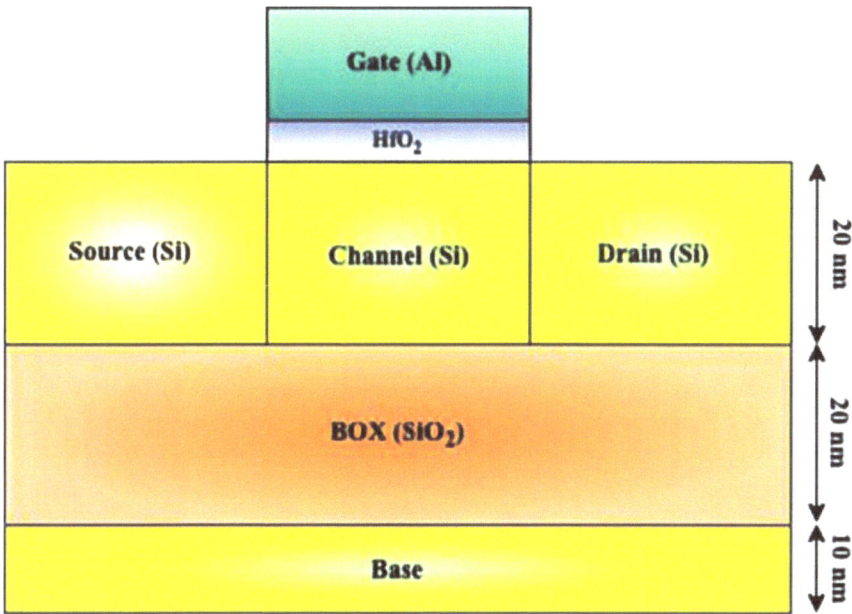

Fig. (2). Device Structure of Conventional TFET.

NCFETs fabricated using HfO_2 material achieve a much lesser subthreshold swing and overcomes the physical limitation of 60 mV/decade. Negative capacitance phenomenon is implemented with the use of a thin Fe material to make a gate stack as shown in Fig. (**3**). The ferroelectric material is added over the conventional SiO_2 layer and the gate stack acts as a series combination of C_i (positive) and C_d (negative).

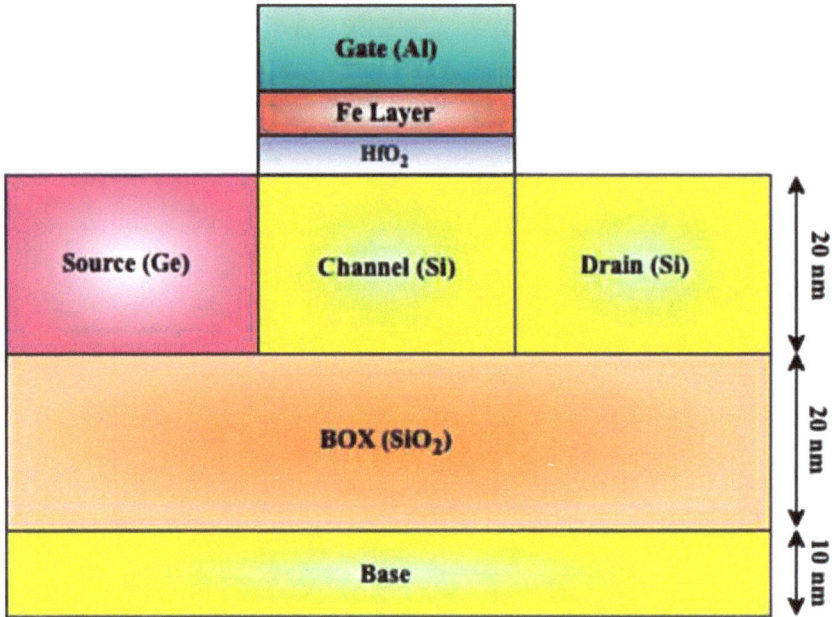

Fig. (3). Device Structure of Negative Capacitance TFET.

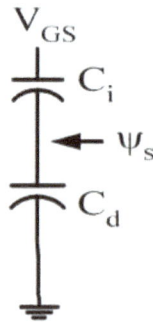

Fig. (4). Capacitance Model.

As per Fig. **(4)**, the overall capacitance of two capacitances in series, can be written as follows:

$$C_{eq} = \frac{|Ci|Cd}{|Ci|-Cd} \qquad (1)$$

It can be observed that if Ci is greater than Cd, the overall capacitance will be more than both the individual capacitances. So, for the same supply voltage, more charge can be accumulated or it can be said that a lesser applied voltage can achieve the same performance and store the same amount of charge in the channel.

$$SS = 60 \text{ x } [1 + \frac{CMOS}{CFE}] \tag{2}$$

C_{FE} denotes the capacitance of ferroelectric material and C_{MOS} is the overall capacitance of oxide capacitance (C_{ox}) and channel-to-source capacitance (C_S).

As per equation (2), if C_{FE} becomes negative and is greater than $C_{MOS,}$ it would be possible to achieve SS lesser than 60 mV/decade. Also, the I_{ON} of the optimized device becomes higher as the transport dynamics of the carriers alter. The effect generated with the stack so formed is helpful in attaining a steeper SS slope and better I_{ON} current. The same can be observed in Fig. (5). This unique property of NCTFETs can be used in low-power applications.

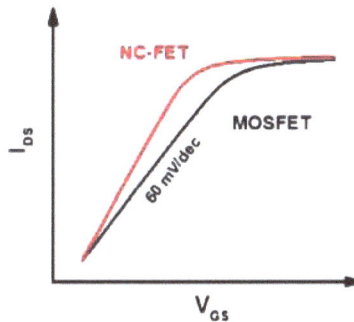

Fig. (5).   Subthreshold swing in case of Conventional MOSFET and NCFET.

By including the Ferroelectric (Fe) layer, the Total gate voltage becomes:

$$V_g = V_{FB} + V_{fe} + V_{ox} + \psi_s \tag{3}$$

Where voltage across the oxide is given by: $V_{ox} = Q/C_{ox}$; C_{ox} denotes oxide capacitance and V_{FB} is flat band voltage. For finding voltage across the ferroelectric layer (V_{fe}), Landau-Khalatnikov model can be used [8]. This model defines the relation between polarization and electric field as:

$$\delta \frac{dP}{dt} = -\frac{\delta G}{\delta P} \tag{4}$$

$$G = \alpha P^2 + \beta P^4 + \gamma P^6 - EP \tag{5}$$

Where P is the polarization of Fe material used, G defines the Gibb's energy, and E is Electric Field, α, β, γ are Landau coefficients associated with Ferroelectric materials as summarized in Table **1**.

Table 1. Landau coefficients for various ferroelectric materials

Landau Coeff. Material	α (cm/F)	β (cm^5/F/Coul^2)	γ (cm^9/F/Coul^4)
PZT	-2.25e9	1.3e18	9.8333e25
BaTiO3	-5e8	-2.225e18	7.5e27
P(VDF-TrFE)	-1.8e11	5.8e22	0

Above equations yields:

$$E = \frac{V_{fe}}{T_{fe}} = 2\alpha P + 4\beta P^3 + 6\gamma P^5 + \delta \frac{dP}{dt} \tag{6}$$

Where, T_{fe} is thickness of ferroelectric layer, V_{fe} is Voltage across ferroelectric material, and denotes the polarization damping factor.

DEVICE ARCHITECTURE

Various device modifications have been done for improvising the device performance. A raised source gate overlapped NCTFET Device has made I_{ON}/I_{OFF}, DIBL and subthreshold swing better as shown in Fig. (**6**).

In raised source NCTFET device, Ge material is used as source material to create a heterojunction at the source-channel junction and a 3nm thick BaTiO$_3$ Fe material layer is stacked over the HfO$_2$ layer to form a gate stack of NCTFET [9]. It has been seen that the reported device is capable of achieving a subthreshold slope of 53.7 mV/decade and I_{ON}/I_{OFF} in order of ~1E10.

To validate the application of the proposed NCTFET device in power-efficient circuits, dynamic random access (DRAM) memory cell has been designed and compared with conventional devices in terms of power consumed. It has been observed from the result that the proposed NCTFET inverter can be put to use in practical applications to replace conventional MOSFET [10].

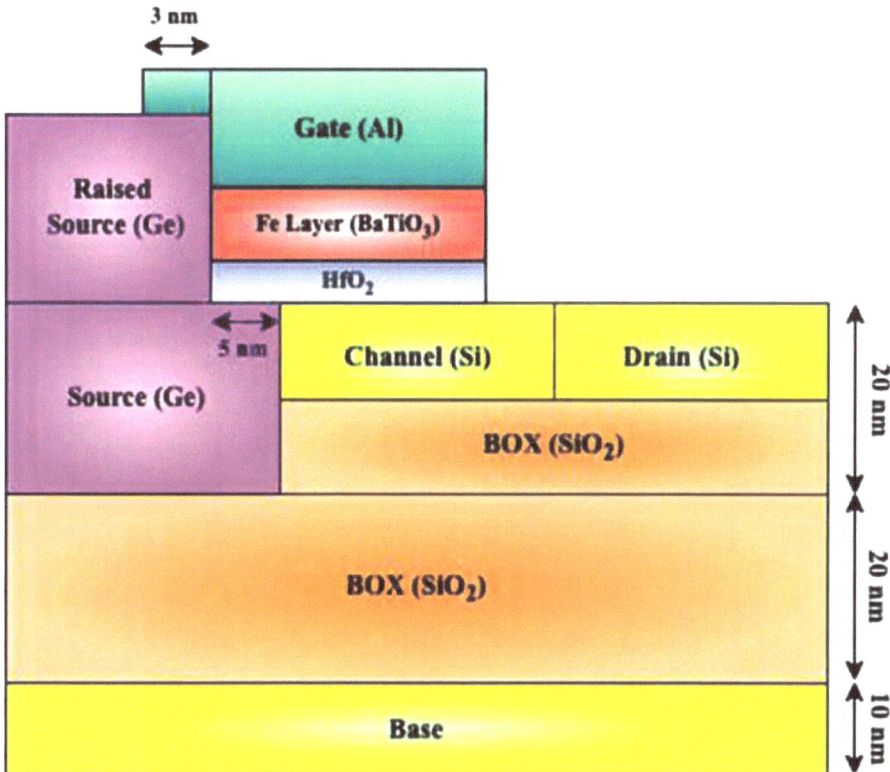

Fig. (6). Schematic of Raised Source NCTFET.

Effect of Ferroelectric Materials and their Thickness on Device Performance

Ferroelectric material used in gate stack directly influences the working of the device. Having different dielectric properties the Fe materials have a varying band-to-band generation, electric potential generation, *etc.* Electron band-to-band generation in the case of different ferro materials can be seen in Figs. (**7a-c**). Landau-Khalatnikov (LK) equation (**8**) states the relation between voltage and charge for Ferroelectric materials and can be defined as:

$$V_{FE} = T_{FE} \left(\rho \frac{dP}{dt} + 2\alpha P + 4\beta P^3 + 6\gamma P^5 \right) (1) \, dt \qquad (7)$$

As seen in Fig. (**7c**), the highest band-to-band generation is achieved for BaTiO$_3$ material.

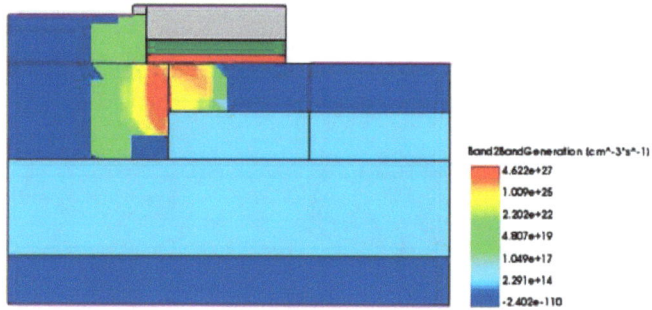

(a)

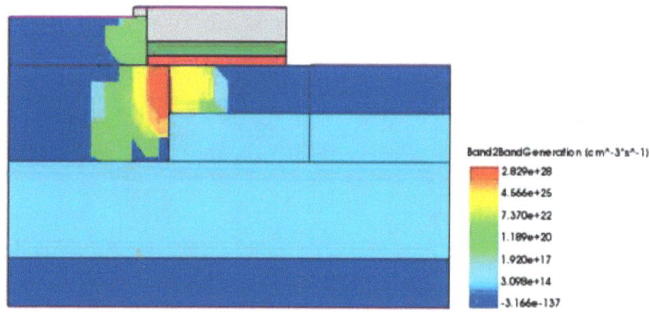

(b)

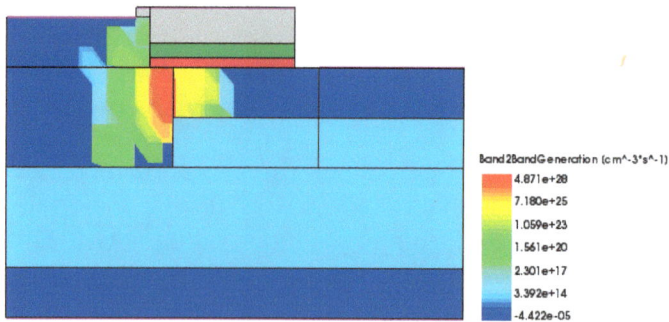

(c)

Fig. (7). e-Band to Band Generation in case of (**a**) Barium Titanate (**b**) P(VDF-trifluoroethylene) (**c**) Lead Zirconate Titanate Fe Materials.

Various performance plots (such as $I_d V_{gs}$, $I_d V_d$, electrostatic potential generation and energy band diagram) can be seen in Fig. (**8**) with various Fe materials. It can be observed from the plots that $BaTiO_3$ material is giving the best results in terms of electrostatic potential, transfer and output characteristics [11].

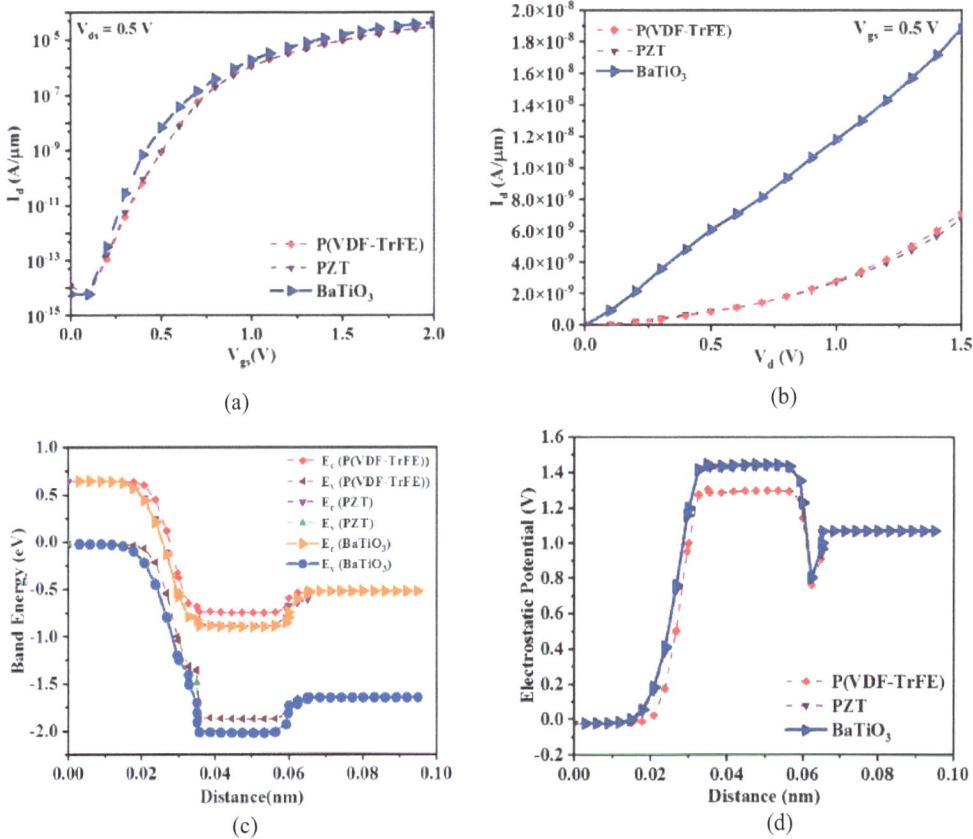

Fig. (8). (**a**) I_D V_{GS} characteristics of NCTFET (**b**) I_D V_D characteristics of NCTFET (**c**) Energy Band Diagram (EBD) of NCTFET (**d**) Electrostatic Potential of NCTFET.

For suppressing the leakage current and enhancing the current ratio, a negative capacitance transistor is designed with a ferroelectric layer in the gate stack that provides a stronger BTBT rate as seen in Fig. (**9**).

The optimization of Fe material thickness is equally important as choosing an appropriate Fe material.

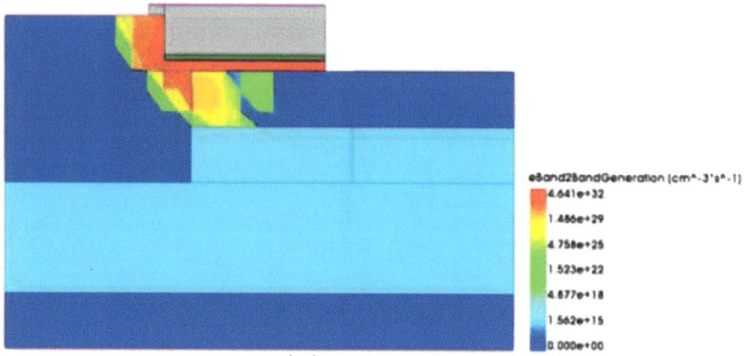

(a)

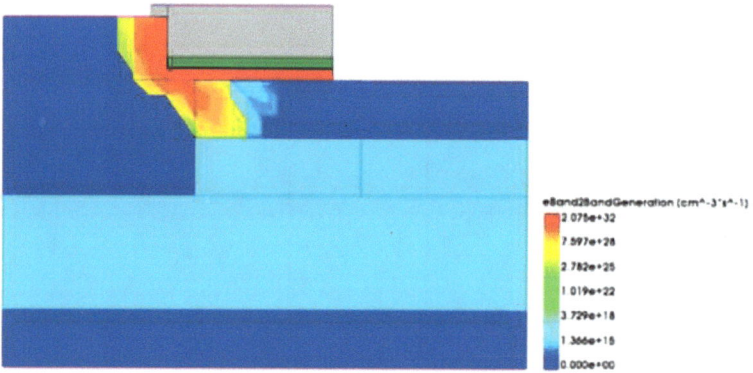

(b)

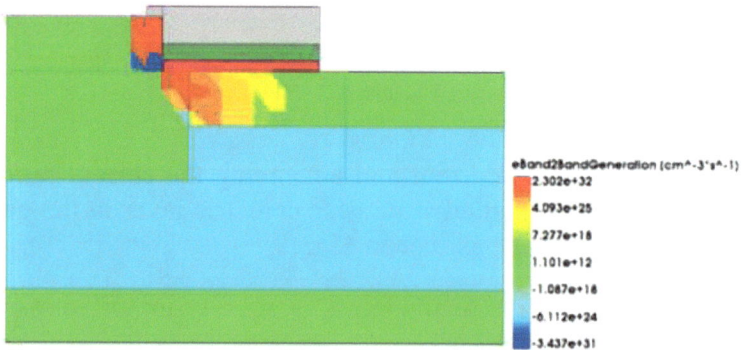

(c)

(Fig. 9) contd.....

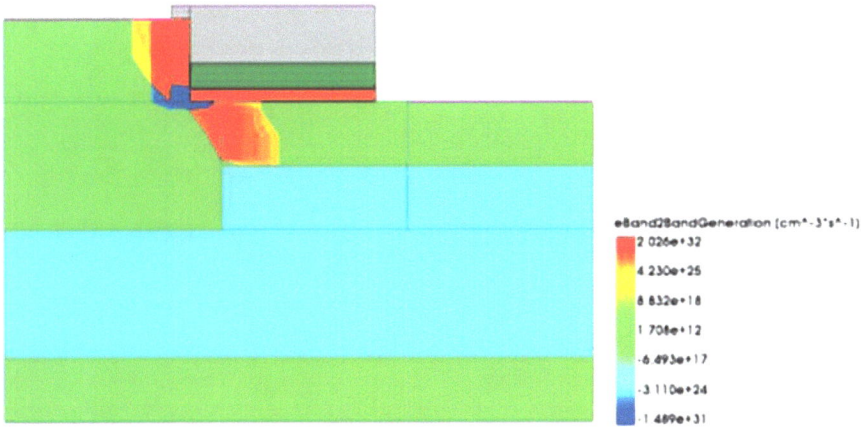

(d)

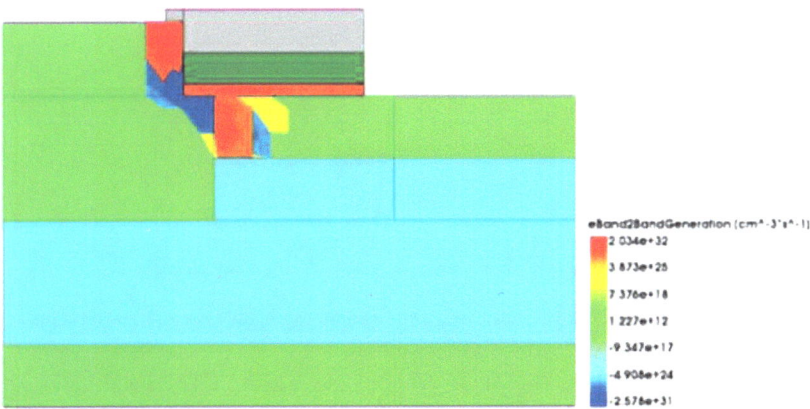

(e)

Fig. (9). e-Band to Band generation rate for various thicknesses of ferroelectric layer (**a**) 1nm (**b**) 2 nm (**c**) 3 nm (**d**) 4 nm (**e**) 5nm.

Transfer characteristics and g_m plot of optimized NCTFET device are plotted in Fig. (**10a**) and Figs. (**10a & b**).

Fig. **10 (a)** depicts that for Fe thickness of 3nm, the transfer characteristics have shown a stepper curve and better I_{ON}/I_{OFF} ratio. Moreover, g_m plot depicts that there is no downfall in the graph at 3nm thickness [12].

(a)

(b)

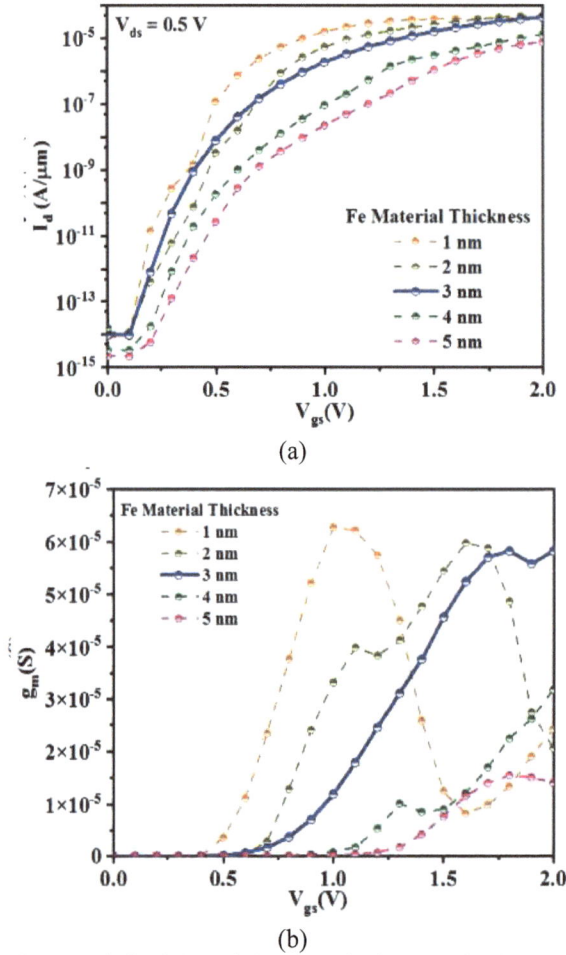

Fig. (10). **(a)** Transfer Characteristics ($I_d V_{gs}$ Plot) **(b)** g_m plot in case of NCTFET with various Ferroelectric material thickness.

Table 2. Device parameters at various Fe Thickness.

Parameter	1 nm	2 nm	3 nm	4 nm	5 nm
$V_{T(Transconductance)}$	0.74	1.04	1.22	1.54	1.50
$V_{T(Common\ Current)}$	0.48	0.66	0.65	1.01	1.19
Transconductance (g_m)	6.27	5.97	5.82	3.18	1.54
I_{ON}/I_{OFF} **Ratio**	2.9E+9	4.3E+9	**7.1E+9**	1.7E+9	1.7E+9
Subthreshold Swing (SS)	58	78	**53.7**	63	79

It is observed that a steeper slope is achieved at smaller I_D and for larger values of drain current, further improvisation of the gate stack is required so as to enhance the effect of negative capacitance. Table **2** summarizes the device parameters obtained at different Fe thicknesses (ranging from 1 nm to 5 nm).

Device Parameters and Material Used

There are different materials reported and used for the source, with each combination of source and channel providing different results in terms of current ratio I_{ON}/I_{OFF} and ambipolarity. Table **3** summarizes the device parameters for NCTFET used in this work.

Table 3. NCTFET Device Parameters.

Quantities	Symbol	Value
Thickness of Oxide Layer	T_{OX}	**5 nm**
Thickness of Fe Layer	T_{FE}	**3 nm**
Channel Length	L_{CH}	**30 nm**
Doping Concentration (Source)	N_A	**1e19 cm^{-3}**
Doping Concentration (Channel)	N_I	**1e16 cm^{-3}**
Doping Concentration (Drain)	N_D	**1e19 cm^{-3}**

Various Ferroelectric materials have been tried for incorporating NC effect in the device (like PZT, BaTiO$_3$, P(VDF-TrFE). It can be observed from the results obtained that 3nm thick BaTiO$_3$ has provided a higher current ratio and better device characteristics [13 - 16].

LOW-POWER CIRCUIT IMPLEMENTATION

With the advancement and day-to-day increasing requirements of adding more number of bits per unit area in Si die, the memory manufacturing companies are looking for high-speed and energy efficient memory elements. The basic memory cell can be made using a capacitance or transistor. The capacitance-based memories are difficult to scale in comparison to the traditional transistors so optimization of 1T/1C DRAM memory cell is more popular and provides an efficient solution for achieving better memory density without compromising the speed and memory capacity. The NCTFET being optimized in this work is based on the negative capacitance phenomenon and incorporates the band-to-band tunneling mechanism. The extra hole carriers present in the body are capable of maintaining on state current in memory cells [17, 18]. 1T DRAM cell so formed is compact and operates at a much low supply voltage and offers a high data rate. Thus, memory density and speed are maintained while power consumption is reduced.

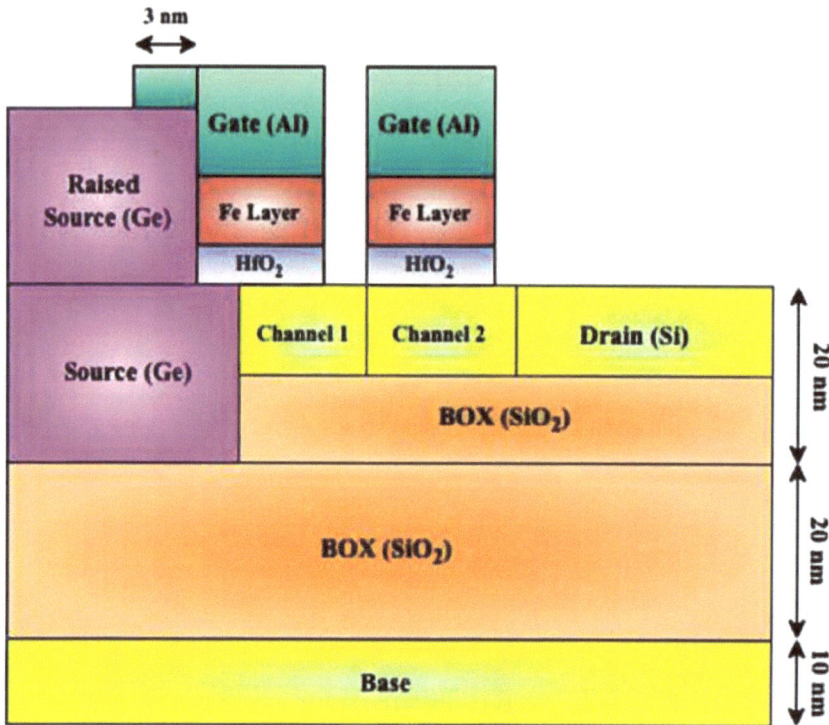

Fig. (11). 1-T DRAM cell using raised source negative capacitance field effect transistor.

The device structure of 1T DRAM cell is shown in Fig. (**11**). The L-Shaped GATE1 is made to cope with raised source section and provides a wider cross-section area to enhance the BTBT tunneling. The potential well so formed stores the extra charge carriers generated when back gate voltage is accumulated [19 - 21]. Two different GATE sections are being made for read and write operations respectively. GATE1 is used for performing read operations and GATE2 incorporates the hold and write operation. Write '1' and Write '0' operations are represented by hole concentration in the storage (GATE2) area.

Table **4** depicts the device parameters used to make the 1T DRAM cell. Channel areas under GATE1 and GATE2 have been kept at different doping concentrations so as to efficiently shift and hold the charge carriers as and when required [22].

BTBT mechanism is the main factor behind the working of DRAM cells and various new age structures have been engineered for attaining better tunneling [23].

Table 4. IT DRAM Device Parameters.

Parameter	1T NCTFET DRAM
Silicon Thickness	10 nm
Gate Oxide Thickness	2 nm
Fe Material Thickness	3 nm
Channel 1 Doping (cm^{-3})	1 x 10^{16}
Channel 2 Doping (cm^{-3})	1 x 10^{18}
S/D Doping	1 x 10^{20}
Gate Work Function	4.45
Gate 1 Length	14 nm
Gate 2 Length	12 nm
Source Height	10 nm

Write '1' Operation

As depicted in Table **5**, for the write '1' operation V_{G2S} is kept at a negative bias of -0.75V and V_{G1S} at zero bias. EBD (Energy Band Diagram) for write '1' operation is plotted in Fig. (**12**). With mentioned biasing at both gate terminals, as the barrier width decreases the electrons tunnel from the valence band of the channel to the conduction band of drain, thus hole concentration in the GATE2 region increases [23 - 25].

Table 5. Biasing Conditions for 1T DRAM.

Operations	V_{DS} (V)	V_{G1S} (V)	V_{G2S} (V)
Write '1'	0.5	0.0	-0.75
Write '0'	0.0	0.0	0.75
Hold	0.0	0.25	-0.25
Read	0.5	0.75	0.0

Write '0' Operation

In write '0' operation, V_{G2S} is kept at the positive bias of 0.75V. In comparison to write '1', the number of holes in GATE2 (storage) area reduces [26] and the supply voltage applied at GATE2 evacuates the carriers to the drain region from the storage region.

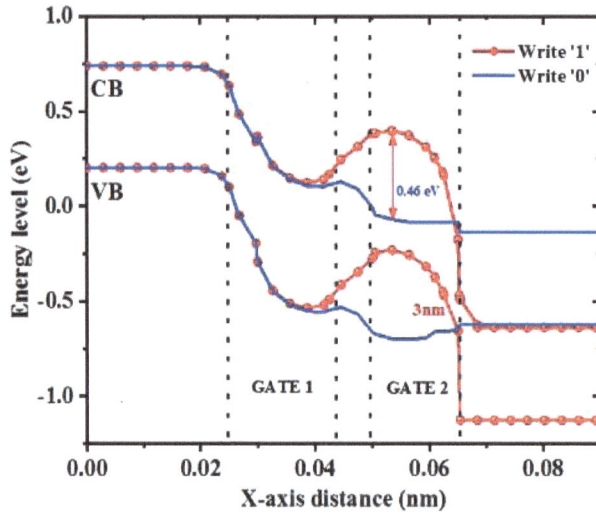

Fig. (12). Energy band diagram for Write operations.

Hold Operation

In this operation, V_{G2S} is kept at a negative bias of -0.25V. As depicted in Fig. (**13**), a potential barrier of 0.22eVis formed beneath G2 in comparison with the hold '1' operation [27]. Negative bias applied across the GATE2 helps to keep the hole carriers intact in the storage area.

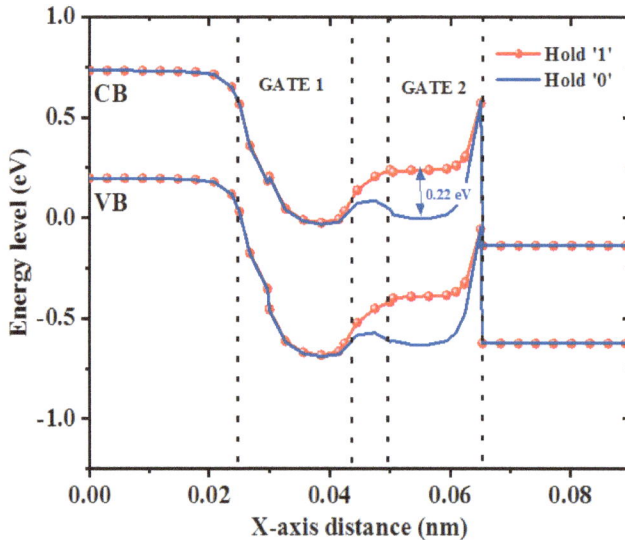

Fig. (13). Energy band diagram for Hold operations.

Read Operation

Unlike Hold and Write, Read operation depends on the values of GATE1 voltage and drain voltage. V_{GIS} and V_{DS} are biased with 0.75V and 0.5V respectively. As GATE1 is kept at a positive potential, the tunneling width near source-channel junction narrows down thus the electrons tunnel from the valence band of the source to the conduction band of the channel [28].

Fig. (14) clearly depicts that during both read operations there occurs a difference of approx 0.43 eVs in potentials maintained in the case of the read '0' operation and read '1' operation. Depending upon the value to be written the data line is pulled up or pulled down thus the read operation has to be followed by a write-back so that the data to be read must not return vague or corrupted bit [29, 30].

For data = 1, the data line is to be pre-charged to $V_{DD}/2$. If the stored data is '1', due to charge sharing, the voltage of the data line rises slowly. Whenever the sense amplifier detects a small change in the voltage of the data line and returns '1' as a stored value. The same is followed for the read '0' operation [31 - 33]. The charge/discharge takes a longer time as the data line is having large capacitance, so sense amplifiers are required to efficiently detect the logic '1' or logic '0' values.

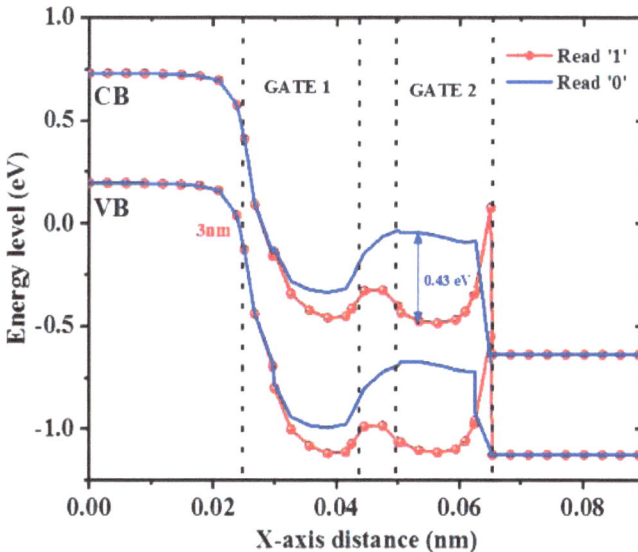

Fig. (14). Energy band diagram for Read operations.

It can be observed that the supply voltages are much lesser in comparison to the conventional TFET device. Thus the optimized 1T DRAM memory cell being presented in this work can be utilized in ultra-low power circuit applications [34, 35].

CONCLUDING REMARKS

With day to day scaling of VLSI circuits, several shortcomings in short channel devices occur, which is affecting the device performance and overall power consumption of VLSI devices and circuits. As the chip size is reducing and density is increasing, more power-efficient and compact devices are required to meet the need of the hour. The biggest challenge being faced in this regard is that as the feature size is reducing the threshold voltage is getting scaled proportionately but V_{DD} is not getting scaled. So the researchers are now looking forward to replacing conventionally used devices with newly engineered devices so as to cope with today's requirements.

TFET emerges to be a good option in replacing the conventional MOSFET as it incorporates band to band tunneling mechanism. Various new age techniques have been reported in the literature which subsequently solve the problems/shortcomings of TFETs as well so as to give an optimized device structure that can prove to be the best replacement without compromising on the speed and device performance. The parameters such as current ratio, subthreshold swing and DIBL play an important role in defining the device performance as compared to the conventional devices, so the conventional TFET structure is modified by introducing a ferroelectric layer (phenomenon of negative capacitance), and then certain device structure changes have been made to get an improved device performance in terms of listed parameters. The results obtained show that the choice of ferroelectric material and its thickness plays an important role in improvising the current ratio and making the subthreshold swing steeper. Further improvements are been done in the transfer characteristics and DIBL by raising the source and making use of different source materials to form a heterojunction. Once the device is optimized, it is utilized in the circuit level implementations, where a 1T DRAM memory cell has been made and results reveal that it operates at a much lower supply voltage and is suitable for making low-power memory cells.

LEARNING OBJECTIVES

• Study of device characteristics and performance of tunnel field effect transistor in Synopsys TCAD tool.

• Application of Negative Capacitance phenomenon to TFET for improvisation of device characteristics.

• Study of device structure modifications for increasing current ratio (ION/IOFF) and making subthreshold swing steeper.

• Optimization of device structure by raising source and replacing source material with germanium.

• Application of optimized device in making 1T DRAM memory cell.

MULTIPLE CHOICE QUESTIONS

1. Which of the following component of TFET allows charge carriers to exit from TFET channel?

a. Gate
b. Source
c. Drain
d. Channel

2. Which of the following component of FET modulates conductivity of channel?

a. Gate
b. Source
c. Drain
d. Channel

3. What is the P of TFET transistor?

a. Ambipolar Behaviour
b. Lesser I_{ON} current
c. Both a. and b.
d. None of these

4. What is the fundamental limit on subthreshold swing?

a. 60 mV/dec
b. More than 60mV/dec
c. Less than 60 mV/dec
d. None of these

5. Which out of the following accounts for short channel effects in FETs?

a. DIBL
b. Surface Scattering
c. Punch Through
d. All of the above

6. Which model is used to define the negative capacitance across ferroelectric material?

a. Fermi Dirac Model
b. Landau-Khalatnikov Model
c. Gaussian Trap Model
d. Recombination Model

7. The voltage across ferroelectric material is dependent on:

a. Thickness of Fe Material
b. Polarization
c. Both a. and b.
d. None of these

8. The equivalent capacitance of gate stack in case of negative capacitance TFET is:

a. Parallel combination of C_i and C_d
b. Series combination of C_i and C_d
c. Equal to C_i
d. Equal to C_d

9. What improvisations have been made in device performance by intervention of ferroelectric layer?

a. Stepper Subthreshold Swing
b. Improved I_{ON}/I_{OFF} ratio
c. Smaller I_D
d. All of the above

10. The negative capacitance of NCTFET is due to intervention of:

a. Ferroelectric Layer
b. Oxide Layer
c. Substrate
d. None of these

11. The 1T DRAM presented in this chapter has how many gate(s):

a. 1
b. 2
c. 3
d. 4

12. The optimized device presented in this chapter has:

a. Raised Source
b. Negative Capacitance
c. Shifted Gate Stack
d. All of the Above

13. Steepeness of subthreshold swing depicts:

a. Faster Switching of the device
b. Lesser Power Consumption
c. Lesser Leakage Current
d. None of these

14. Which parameter controls the tunneling rate of TFET device:

a. Source Material
b. Doping Profile
c. Terminal Resistance
d. None of these

15. Thicknes of Fe material used in gate stack has an impact on:

a. V_{FE}
b. Band to Band Generation Rate
c. Both a. and b.

d. None of these

ANSWER KEY

1. (c)

2. (a)

3. (c)

4. (a)

5. (d)

6. (b)

7. (c)

8. (b)

9. (d)

10. (a)

11. (b)

12. (d)

13. (a)

14. (b)

15. (c)

ACKNOWLEDGEMENT

This work is supported by the Science and Engineering Research Board (SERB), Department of Science & Technology, Government of India, CRG/2020/006229, dated: 05/04/2021.

REFERENCES

[1]   S. Salahuddin and S. Datta, "Use of Negative Capacitance to provide Voltage Amplification for Low Power Nanoscale Devices," *Nano Letters*, Vol. 8, Issue No. 2, pp. 405–410, 2008. [http://dx.doi.org/10.1021/nl071804g]

[2]   A. Biswas, and A. M. Ionescu, "1T capacitor-less DRAM cell based on asymmetric tunnel FET design", *IEEE Journal of the Electron Devices Society,* vol. 3, no. 3, pp. 217-222, 2015. [http://dx.doi.org/10.1109/JEDS.2014.2382759]

[3]   S. Chander, S.K. Sinha, and R. Chaudhary, "Comprehensive review on electrical noise analysis of TFET structures", *Superlattices Microstruct..* p. 107101, 2021.

[4]   C. Hu, T.-J. King, and C. Hu, "A capacitorless double-gate DRAM cell", *IEEE Electron Device Letters,* vol. 23, pp. 345-347, 2002.
[http://dx.doi.org/10.1109/LED.2002.1004230]

[5]   S. Chander, and S.K. Sinha, "Effect of raised buried oxide on characteristics of tunnel field effect transistor", *Silicon,* vol. 14, no. 14, pp. 8805-8813, 2022.
[http://dx.doi.org/10.1007/s12633-022-01681-z]

[6]   D. O. Kim, D.-I. Moon, and Y.-K. Choi, "Optimization of bias schemes for long-term endurable 1T-DRAM through the use of the biristor mode operation", *IEEE Electron Device Letters,* vol. 35, pp. 220-222, 2014.
[http://dx.doi.org/10.1109/LED.2013.2295240]

[7]   N. Navlakha, J. Lin, and A. Kranti, "Improved Retention Time in Twin Gate 1T DRAM with Tunneling Based Read Mechanism", *IEEE Electron Device Letters,* vol. 37, pp. 1127-1130, 2016.
[http://dx.doi.org/10.1109/LED.2016.2593700]

[8]   S.K. Sinha, K. Kumar, and S. Chaudhury, "Si/Ge/GaAs as channel material in nanowire :FET structures for future semiconductor devices", *11th IEEE International Conference on Electron Devices and Solid State Circuits..* Nanyang Executive Centre, pp. 527-530, 2015.
[http://dx.doi.org/10.1109/EDSSC.2015.7285167]

[9]   A.I. Khan, C.W. Yeung, C. Hu, and S. Salahuddin, "Ferroelectric negative capacitance MOSFET: Capacitance tuning & antiferroelectric operation", In: *Proceedings IEEE International Electron Devices Meeting.* pp. 11.3.1–11.3.4, 2011.
[http://dx.doi.org/10.1109/IEDM.2011.6131532]

[10]   A. Giovanni, "Demonstration of subthrehold swing smaller than 60mV/decade in Fe-FET with P(VDF-TrFE)/SiO2 gate stack", *IEEE International Electron Devices Meeting..* pp. 1-4, 2008.
[http://dx.doi.org/10.1109/IEDM.2008.4796642]

[11]   I. A. Pindoo, "Improvement of electrical characteristics of sige source based tunnel fet device", *Springer, Silicon.* Vol. 13, Issue No. 9, pp. 3209-3215, 2020.

[12]   A. Singh, and S.K. Sinha, "Performance analysis of device characteristics in negative capacitance field effect transistor", *2nd IEEE International Conference on Electronics and Sustainable Communication Systems at Hindusthan Institute of Technology.* Coimbatore, India, 4-6, 2021.
[http://dx.doi.org/10.1109/ICESC51422.2021.9532660]

[13]   A. Singh, and S.K. Sinha, "Comparative analysis of various ferroelectric materials used in negative capacitance field effect transistor (NCFET)", *International Conference on Recent Development on Materials, Reliability, Safety and Environmental Issues (IMRSE),* 2021. 2021.

[14]   S. K. Sinha, and S. Chander, "Investigation of dc performance of ge-source pocket silicon-on-insulator tunnel field effect transistor in nano regime inderscience", *Inderscience, International Journal of Nanoparticles.* Vol. 13, Issue No. 1, pp. 13-20, 2021.

[15]   T. Tanaka, E. Yoshida, and T. Miyashita, "Scalability study on a capacitorless 1T-DRAM: From single-gate PD-SOI to double-gate FinDRAM", *IEDM Technical Digest. IEEE International Electron Devices Meeting.* pp. 919–922, 2004.
[http://dx.doi.org/10.1109/IEDM.2004.1419332]

[16]   E. Yu, S. Cho, H. Shin, and B.-G. Park, "A band-engineered one-transistor DRAM with improved data retention and power efficiency", *IEEE Electron Device Letters.* Vol. 40, Issue No. 4, pp. 562–565, 2019.
[http://dx.doi.org/10.1109/LED.2019.2902334]

[17]   K. Chandrasekar, S. Goossens, C. Weus, M. Koedam, B. Akesson, N. When, and K. Soossens, "Exploiting expendable process-margins in DRAMs for run-time performance optimization", *Design,*

Automation & Test in Europe Conference & Exhibition.. pp. 1-6, 2014.
[http://dx.doi.org/10.7873/DATE.2014.186]

[18] S. Chander, S. K. Sinha, R. Chaudhary, and A. Singh, "Ge-source based l-shaped tunnel field effect transistor for low power switching application", *Silicon.* 2021.
[http://dx.doi.org/10.21203/rs.3.rs-874009/v1]

[19] A. Z. Badwan, Q. Li, and D. E. Ioannou, "On the nature of the memory mechanism of gated-thyristor dynamic-RAM cells", *IEEE J. Elec. Dev. Soc..* Vol. 3, Issue No. 6, pp. 468–471, 2015.
[http://dx.doi.org/10.1109/JEDS.2015.2480377]

[20] S. Chander, S. K. Sinha, R. Chaudhary, and R. Goswami, "Effect of noise components on L-shaped and T-shaped heterojunction tunnel field effect transistor", *IOP Science Semicond. Sci. Technol.,* vol. 37, no. 7, pp. 1-11, 2022.
[http://dx.doi.org/10.1088/1361-6641/ac696e]

[21] J. Wan, C. Le Royer, A. Zaslavsky, and S. Cristoloveanu, "Progress in Z2-FET 1T-DRAM: Retention time, writing modes, selective array operation, and dual bit storage", *Solid-State Electron.,* vol. 84, pp. 147-154, 2013.
[http://dx.doi.org/10.1016/j.sse.2013.02.010]

[22] A. Z. Badwan, Z. Chbili, Y. Yang, A. A. Salman, Q. Li, and D. E. Ioannou, "SOI field-effect diode DRAM cell: Design and operation", *IEEE Electron Device Letters.* Vol. 34, Issue No. 8, pp. 1002–1004, 2013.
[http://dx.doi.org/10.1109/LED.2013.2265552]

[23] S. K. Sinha, S. Chander, and R. Chaudhary, "Investigation of noise characteristics in gate-source overlap tunnel field-effect transistor", *Silicon.* 1-8, 2022.
[http://dx.doi.org/10.1007/s12633-022-01806-4]

[24] M. G. Ertosun, P. Kapur, and K. C. Saraswat, "A highly scalable capacitorless double gate quantum well single transistor DRAM: 1T-QW DRAM", *IEEE Electron Device Letters.* Vol. 29, Issue No. 12, pp. 1405–1407, 2008.
[http://dx.doi.org/10.1109/LED.2008.2007508]

[25] N. Rodriguez, S. Cristoloveanu, and F. Gamiz, "Novel capacitorless 1T-DRAM cell for 22-nm node compatible with bulk and SOI substrates", *IEEE Transactions on Electron Devices.* Vol. 58, Issue No. 8, pp. 2371–2377, 2011.
[http://dx.doi.org/10.1109/TED.2011.2147788]

[26] R. Narang, M. Saxena, R.S. Gupta, and M. Gupta, "Assessment of ambipolar behaviour of a tunnel FET and influence of structural modifications", *J. semicond. tech. sci..* Issue No. 12, pp. 482-491, 2012.

[27] S. Jin, J-H. Yi, J.H. Choi, D.G. Kang, Y.J. Park, and H.S. Min, "Modeling of retention time distribution of DRAM cell using a Monte-Carlo method", *IEDM Technical Digest. IEEE International Electron Devices Meeting* pp. 399–402, 2004.

[28] V.P-H. Hu, H.H. Lin, Y.K. Lin, and C. Hu, "Optimization of negative-capacitance vertical-tunnel FET (NCVT-FET)", *IEEE Trans. Electron Dev.,* vol. 67, no. 6, pp. 2593-2599, 2020.
[http://dx.doi.org/10.1109/TED.2020.2986793]

[29] K. Yamaguchi, "Temperature dependence of anomalous currents in worst-bit cells in dynamic random-access memories", *J. Appl. Phys..* Vol. 87, Issue No. 11, pp. 8064–8069, 2000.
[http://dx.doi.org/10.1063/1.373498]

[30] C. Santos, P. Vivet, D. Dutoit, P. Garrault, N. Peltier, and R. Reis, "System-level thermal modeling for 3D circuits: Characterization with a 65nm memory-on-logic circuit", *IEEE International 3D Systems Integration Conference (3DIC).* pp. 1-6, 2013.
[http://dx.doi.org/10.1109/3DIC.2013.6702379]

[31] C. G. Shirley, and W. R. Daasch, "Copula models of correlation: A dram case study", *IEEE*

Transactions on Computers. Vol. 63, Issue No. 10, pp. 2389-2401, 2014.
[http://dx.doi.org/10.1109/TC.2013.129]

[32]   S.T. Bu, D.M. Huang, G.F. Jiao, H.Y. Yu, and M.F. Li, "Low frequency noise in tunneling field effect transistors", *Solid-State Electron.,* vol. 137, pp. 95-101, 2017.
[http://dx.doi.org/10.1016/j.sse.2017.08.008]

[33]   N. Kamal, A. K. Kamal, and J. Singh, "L-Shaped tunnel field-effect transistor-based 1t dram with improved read current ratio, retention time, and sense margin", *IEEE Transactions on Electron Devices..* Vol. 68. Issue No. 6, 2021.
[http://dx.doi.org/10.1109/TED.2021.3074348]

[34]   M. Halid, T.B. Evelyn, M. Thomas, and S. Stefan, "Ferroelectric fets with 20-nm thick hfo2 layer for large memory window and high performance", *IEEE Trans. Electron Dev.,* vol. 66, no. 7, pp. 3828-3833, 2019.

[35]   D. Shafizade, M. Shalchian, and F. Jazaeri, "Ultrathin junctionless nanowire fet model, including 2-D quantum confinements", *IEEE Trans. Electron Dev.,* vol. 66, no. 9, pp. 4101-4106, 2019.
[http://dx.doi.org/10.1109/TED.2019.2930533]

CHAPTER 6

Impact of Electrode Length on I-V Characteristics to Linearity of TFET With Source Pocket

Prajwal Roat[1], **Prabhat Singh**[1,*] and **Dharmendra Singh Yadav**[1]

[1] *National Institute of Technology, Hamirpur, Himachal Pradesh, India*

Abstract: In this chapter, the author demonstrates a triple metal double gate TFET with a uniformly doped source pocket (TMG-SP-DG-TFET) to investigate the impact of triple metal length variation (Length of an electrode implanted above the oxide region) on the device performance. When the electrode length near the drain and source region varies, the electrostatic potential and electric field near the source-channel (SCi) and drain-channel interface (DCi) may vary accordingly. Due to these deviations, the tunneling improves or reduces for a moderately doped drain and a highly doped source region. Therefore, the I_{ON} (ON-state current) has shown significant functionalities with electrode length variation. This extensive study was carried out for the investigation of analog parameters, including EBD (ON/OFF state), E_{field}, Potential, g_m (Trans-conductance), C_{gs} and C_{gd} (Gate-to-source and Gate-to-Drain capacitance), Maximum cut-off frequency (f_t), Gain bandwidth product (GBP), Transit Time (τ), with Linearity figure of merit that includes, g_{m2}, g_{m3}, VIP_2, VIP_3, IIP_3, IMD_3, and 1dB compression point. This comprehensive study shows that varying the length of the metal electrode with a fixed doping level of the source pocket will improve the overall performance of TMG-SP-DG-TFET.

Keywords: BTBT (Band-to-band-tunneling), EBD (Energy-Band-Diagram), Work-function (WF), Capacitance, Electric field (E_{field}), Linearity FOMs.

INTRODUCTION

Metal oxide semiconductor FETs play a vital role in laying the foundation of electronic industries with a backbone in transistors. The advancement in technologies allows users to have more functionality, a high-speed device, and low supply voltage with low power dissipation. However, MOSFETs have limitations like low subthreshold swing, high leakage current, and high power dissipation. This problem should be considered to improve device efficiency, and that is where Tunnel FETs come into the picture [1 - 3]. Therefore, TFETs are regarded as an alternative to MOSFETs with high sub-threshold swing and low

---

* **Corresponding author Prabhat Singh:** National Institute of Technology, Hamirpur, Himachal Pradesh, India;
E-mail: prabhat@nith.ac.in

Gopal Rawat & Aniruddh Bahadur Yadav (Eds.)

leakage when it comes to aggressive downscaling of the device. But as for TFETs, it has limitations like low I_{ON} and high ambipolar conduction (I_{amb}) [4 - 7]. For TFETs, when it comes to I_{ON} current, it is decently equated to MOSFETs due to lessen BTBT at SCi (source-channel-interface). Henceforth, it is necessary to suggest a TFET design with a high I_{ON} current and overcome the restraint of TFET in various aspects comparable to high ambipolar current/behavior [8 - 10]. Regarding this, various engineering techniques are explored for similar double-gate TFETs (DG-TFET) [11 - 17], lower bandgap materials in source area [18 - 22], DG-TFETs with hetero-oxide, high-k dielectrics in gate oxide, Multiple gates over oxide region, *etc* [23, 24]. With the high-k dielectric material of higher permittivity, the leakage current of the device is reduced, and metal gates simultaneously screen the electrons to the high-k material [25].

In TFETs, the introduction of more than two gate designs leads to improved electrostatic control of the potential over the channel section leading to an effective increase in the tunneling region. Thus, it directly leads to an overall improvement in the I_{ON}. The combined effect of multiple gates and double gate TFETs with source pockets helps enhance the general state of the device [26 - 32].

SIMULATION DATA WITH DEVICE SPECIFICATIONS

The graphic of the TMG-SP-DG-TFET is exhibited in Fig. (1). It comprises the triple metal gate with hetero-Oxide of high-k oxide (HfO_2 and SiO_2) as a gate dielectric. The device is founded on silicon with the uniformly doped region of drain, channel, source, and source pocket. The n-type drain doping (N_D) is $5 \times 10^{18} cm^{-3}$, channel (N_C) is $1 \times 10^{17} cm^{-3}$, p-type source doping (N_S) set to 1×10^{20} cm^{-3}, p-type source pocket 1×10^{18} cm^{-3} correspondingly. The device parameters are enumerated in Table 1. The length of dual metal gates (L_{SE}, for M_1 and M_3) varies for 12nm, 15nm, and 18nm while keeping the middle metal gate (M_2) constant at 20nm, with work function varying from 4.0eV to 4.8eV for M_1 (Highest WF_{M1}), M_2 and M_3 (Lowest WF). The sum of M_1 and M_3 is set to 30nm. The performance analysis of the device has been studied and equated with this carried out work concerning linearity and analog constraints of the device.

DETAILS AND EQUATIONS OF ANALOG/RF FOMS

The explanation of analog/RF variables is the main focus of this section, like gate Capacitance ($C_{gg} = C_{gd} + C_{gs}$), g_m, GBP, f_t, transit time (ô), Transconductance Generation factor, and Frequency Product (TGF and TFP) with the mathematical equations.

$$C_{gd} = \frac{\partial Q_s}{\partial V_{gs}} \qquad (1)$$

$$C_{gs} = \frac{\partial Q_s}{\partial V_{gd}}$$

(2)

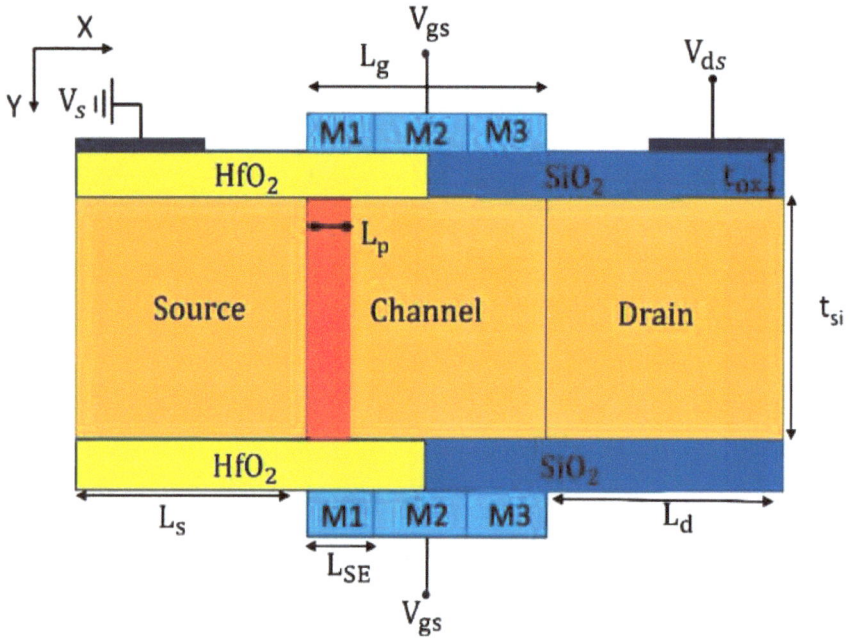

Fig. (1). Graphical illustration of TMG-SP-DG-TFET.

Table 1. Summary of simulation design variables.

S.No.	Parameters	Abbreviations	Values
1.	Gate-Length	Lg	50 nm
2.	Source-Length	L_S	100 nm
3.	Drain-Length	L_D	100 nm
4.	Source-Pocket-Length	L_p	5 nm
5.	Source-Doping-Level	N_S	10^{20} cm^{-3}
6.	Drain-Doping-Level	N_D	5×10^{18} cm^{-3}
7.	Channel-Doping-Level	N_C	10^{17} cm^{-3}
8.	Source-Pocket-Doping	N_P	10^{18} cm^{-3}
9.	Gate WF (M1)	WF_{M1}	4.2 eV
10.	Gate WF (M2)	WF_{M2}	4.8 eV
11.	Gate WF (M3)	WF_{M3}	4.0 eV

(Table 1) cont.....

S.No.	Parameters	Abbreviations	Values
12.	Thickness of body	t_{si}	10 nm
13.	Thickness of HfO$_2$	t_{ox}	3 nm

For the superior performance of the TFET device, both the parasitic capacitance, as mentioned in equations (**1 & 2**) (C$_{gd}$ and C$_{gs}$), should be lowest because they play an important role as it determines device performance at various operating frequencies for parasitic oscillations. Along with this, g$_m$ is responsible for the switching speed of the device and is mathematically represented by equation (**3**). A greater range of g$_m$, increases the enhancement in switching efficacy of the device [19, 20, 33].

$$g_m = \frac{\partial I_d}{\partial V_{gs}} \tag{3}$$

The next important parameter is GBP for the optimum effectiveness of the device. We require a significantly larger GBP value. It is contrariwise correlated to C$_{gd}$ and linearly proportionate to the g$_m$ [21, 22, 34]. The GBP's arithmetic equivalent is given by equation (**4**).

$$GBP = \frac{g_m}{2\pi C_{gd}} \tag{4}$$

The f$_t$ helps to determine the highest frequency for which a device operates without output distortions, such that at various operating frequencies, short circuit gain turns equal to unity [35, 36]. The relation intended for f$_t$ is:

$$f_t = \frac{g_m}{2\pi (C_{gd}+C_{gs})} \tag{5}$$

The other critical parameters are TGF and TFP, which determines the efficiency of the device and give a comparison of power dissipation and bandwidth. For optimum device performance, the value of TGF and TFP must be high [37 - 40]. The mathematical expression for both factors is mentioned in equations (**6 & 7**), respectively.

$$TGF = \frac{g_m}{I_d} \tag{6}$$

$$TFP = \frac{g_m f_t}{I_d} \tag{7}$$

Last, in analog/RF analysis, another critical parameter is τ (given by equation (8)), which is described as the time taken by carriers to be moved from a channel or defines how much delay the device produces [41].

$$\tau = \frac{1}{2\pi f_t} \tag{8}$$

RESULTS AND DISCUSSIONS

A very high potential barrier at the SCi interface presented, no BTBT occurs and hence no flow of electrons as depicted in Fig. (2) for OFF-state condition. Now for the EBD Fig. (3) for ON-state, with band alignment, relatively little potential separation exists between the source's valence band and the channel's conduction band, hence enhanced BTBT phenomena with the high flow of carriers and current starts to flowing. Fig. (4a) illustrates E_{field} intensity is very high near SCi due to the increase in charge accumulation, and less E_{field} near DCi because of low charge accumulation.

Fig. (4b) for potential, with the increase in gate voltage (V_{gs}), potential for L_{SE} = 12nm, increases with high E_{field} and hence contribute to the functionality optimization of the device in term of higher current. Fig. (5) reveals the transfer characteristics of TMG-SP-DG-TFET. It has been depicted from I_{ds} -V_{gs} curve for L_{SE} = 12nm; it shows the improved result in terms of steeper slope and enhanced I_{ON} when compared to L_{SE} =15nm and L_{SE} = 18nm with the triple metal gate. It happens because of the best possible electrostatic controllability over the channel region and increases in the effective tunneling area when L_{SE} = 12nm. Hence, it aids in enhancing the device's overall efficiency and effectiveness.

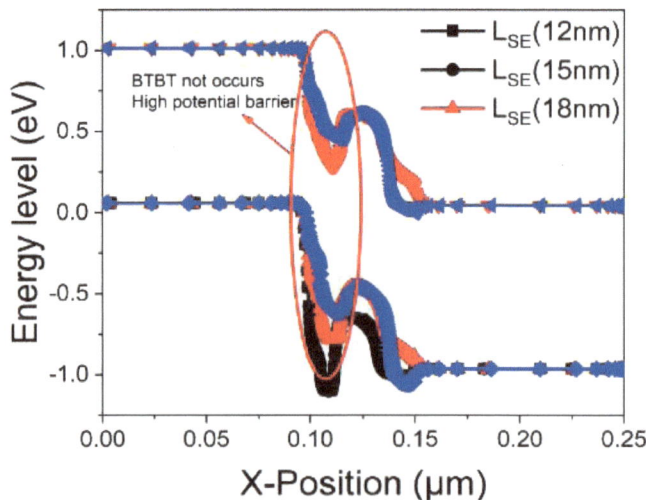

Fig. (2). EBD for OFF-state for structure in concurrence with Fig. (1).

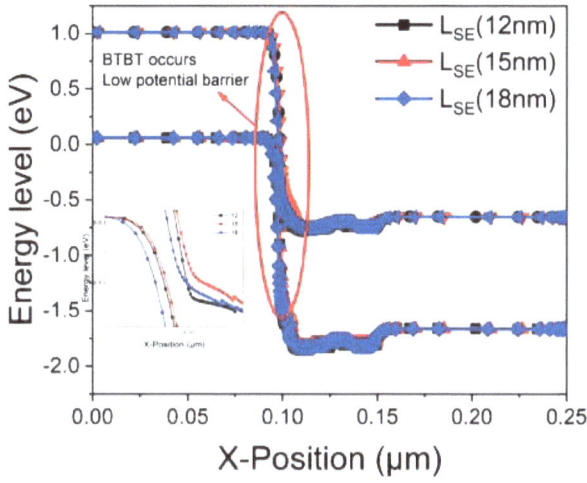

Fig. (3). EBD for ON-state for structure in consensus with Fig. (**1**).

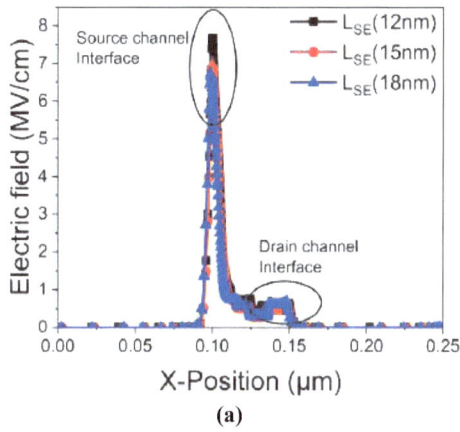

(a)

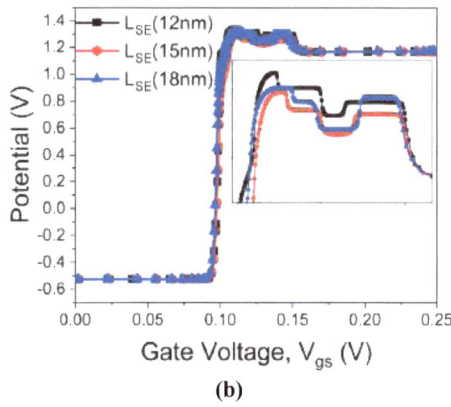

(b)

Fig. (4). (**a**) E_{field} and (**b**) investigation of Potential for structure in accordance with Fig. (**1**).

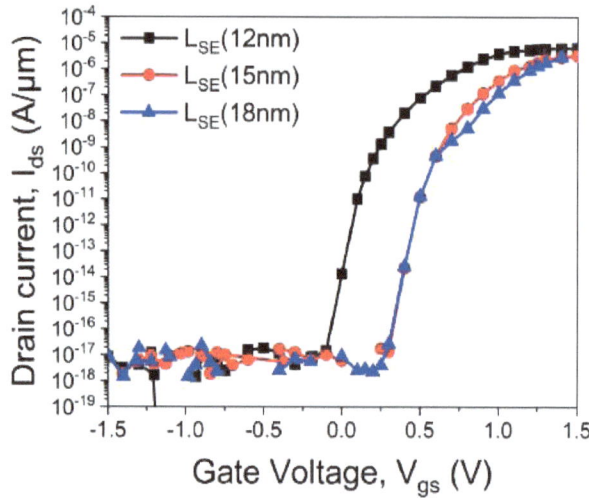

Fig. (5). Structure concurrence with Fig. (**1**), inspection of I_{ds} -V_{gs} curve.

The parasitic capacitance (C_{gd} and C_{gs}) plays a vital role to analyze the device's performance. From Fig. (**6**), the C_{gs} (for L_{SE} = 12nm) are reduced due to reduced charge accumulation and high tunneling rate near the SCi. Similarly, C_{gd} (for L_{SE} = 12nm) increases with the respective gate voltage, as per Fig. (**7**). The increasing C_{gd} is not good for device performance. The lowest C_{gd} values achieve for the L_{SE} = 15nm because, for this value, the charge inversion process near the DCi is very less.

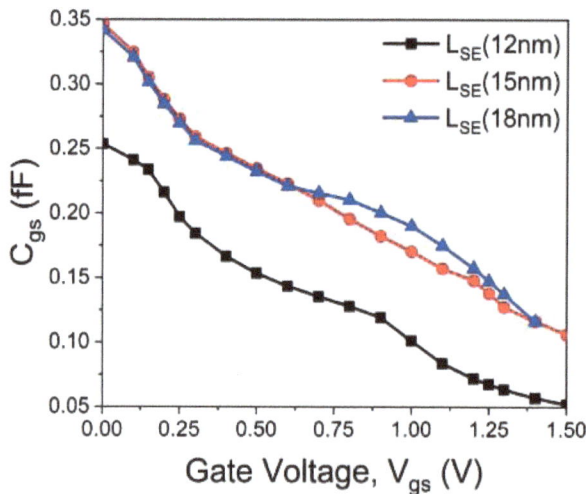

Fig. (6). Data plot of C_{gs}.

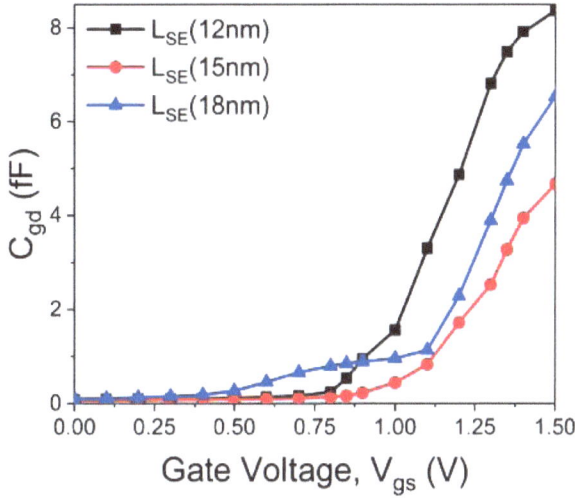

Fig. (7). Data plot of C_{gd}.

The g_m, which is the first derivative of I_{ds} to V_{gs}, is directly related to I_{ds} and hence for L_{SE} = 12nm, shows higher values as described in Fig. (**8**). From Fig. (**9**), f_t is decreasing with increment in L_{SE} because it is negatively proportional to parasitic capacitances and positively related to g_m. From equation (**4**), GBP is proportional to the g_m and contrarywise related to C_{gd}. The value of GBP improves with a high in g_m, as depicted in Fig. (**10**). For L_{SE} = 12 nm, GBP is in the range of 1.5 GHz because g_m is very high and dominates in the ratio of g_m to C_{gd}.

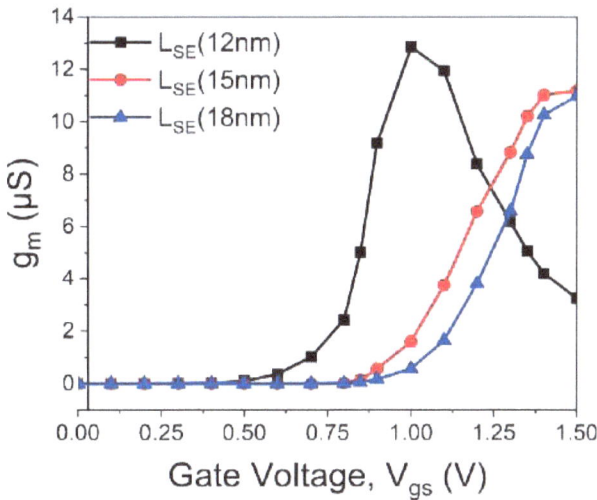

Fig. (8). Variation in g_m for different values of L_{SE}.

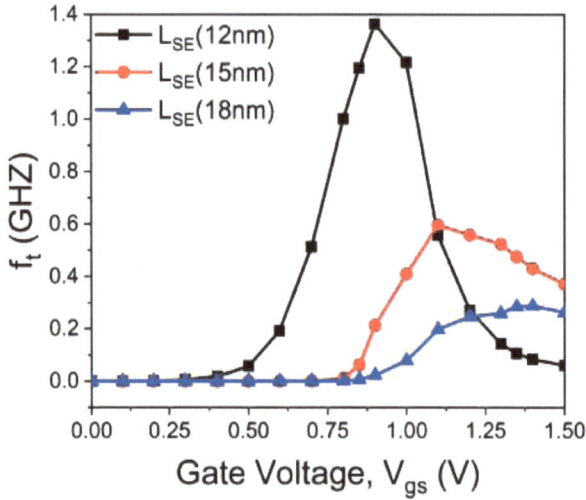

Fig. (9). Variation in f_t for different values of L_{SE}.

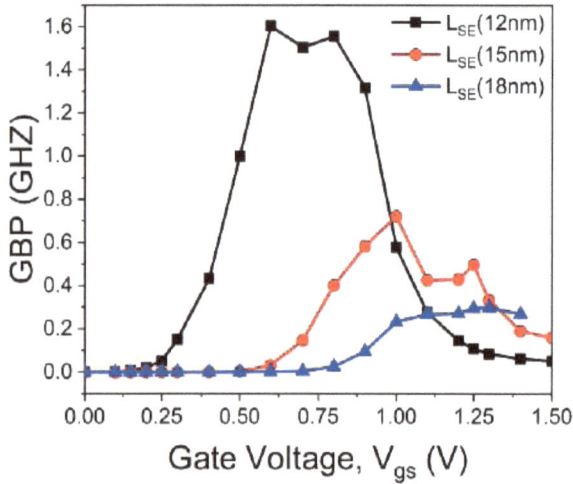

Fig. (10). Variation in GBP for different values of L_{SE}.

Figs. (**11** & **12**) illustrate the TFP and TGF (Transconductance Generation product and Transconductance Generation Factor) data deviations. Both of these variables are proportional to g_m. As a result, both parameters have a high value when L_{SE} is set to 12 nm. From Fig. (**13**), it is depicted that the value of ô is low as it is contrariwise related to the f_t. The lowest value helps to enhance the device's efficiency as measured by decreased device latency.

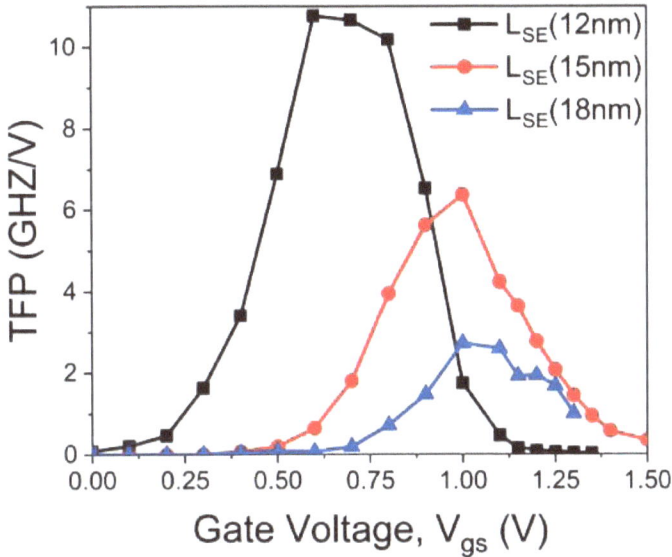

Fig. (11). Graph of TFP for different values of L_{SE}.

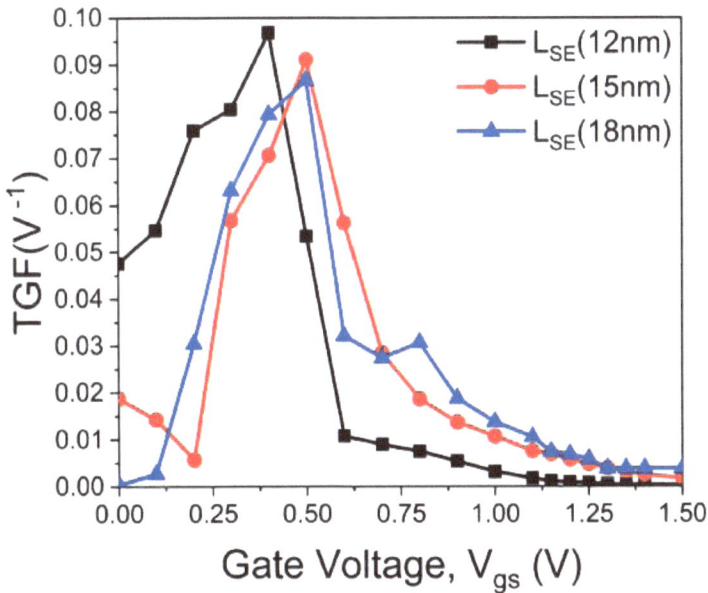

Fig. (12). Graph of TGF for dissimilar values of L_{SE}.

Linearity Analysis

The linearity features of the device, which include g_{m2} and g_{m3} (2nd and 3rd order derivative of g_m), VIP_2 and VIP_3 (2nd and 3rd order Voltage-Intercept-Point), IIP_3 and IMD_3 (3rd order Input-Intercept-Point and Intermodulation-point), 1db compression point ($1dB_{cp}$) are examined under various L_{SE} near the source region. For improved linearity, values produced by g_{m2} and g_{m3} must be low as possible for superior device performance with suppressed distortion in output [24, 25, 42 - 46].

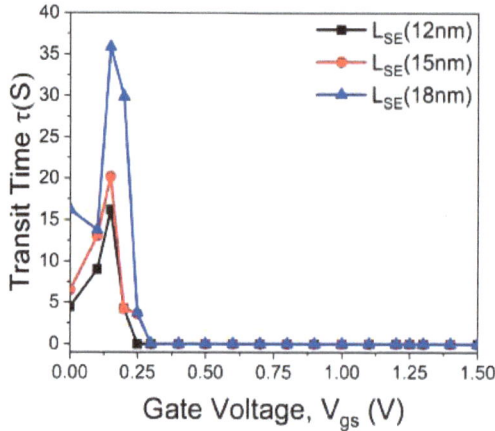

Fig. (13). Graph of ô for different values of L_{SE}.

The mathematical expression for g_{m2} and g_{m3} are specified by equation **(9 & 10)**, correspondingly [27, 47, 48]. From Figs. **(14 & 15)**, it can be seen for both the parameter's value initially increases and attain the maximum peak when $L_{SE} = 15nm$, then decreases with gate voltage. Hence, distortion of the device is reduced at lower bias voltage with lower L_{SE}, *i.e.*, better linear behavior can be obtained at low voltages and its degraded linearity for high V_{gs}.

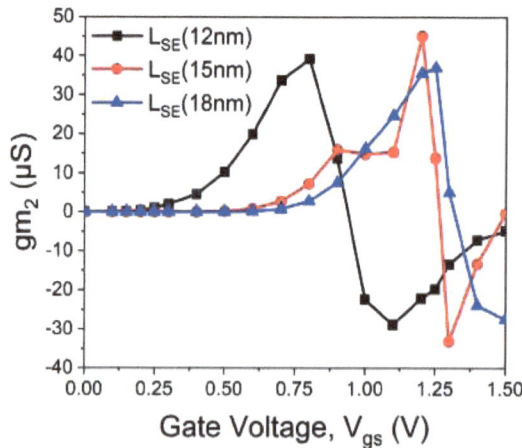

Fig. (14). Graph of g_{m2}.

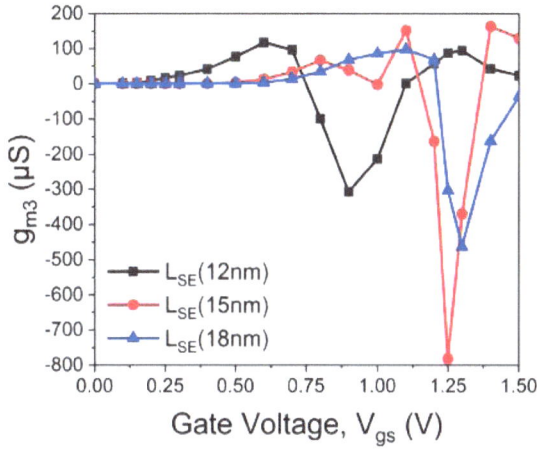

Fig. (15). Graph of g_{m3}.

$$g_{m2} = \frac{\partial^2 I_d}{\partial^2 V_{gs}} \tag{9}$$

$$g_{m3} = \frac{\partial^3 I_d}{\partial^3 V_{gs}} \tag{10}$$

The VIP_2, VIP_3, and IIP_3 are the linearity FOMs shown in Figs. (**16-18**), respectively. The expressions of these FOMs constraints are specified by equations (**11-13**). These parameters should be high for improved linearity characteristics [27, 49, 50].

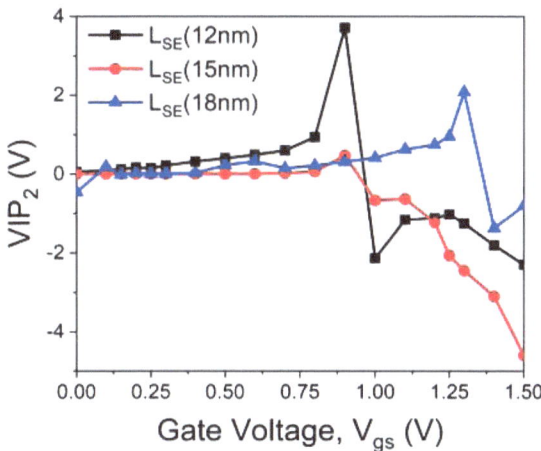

Fig. (16). Graph of VIP_2.

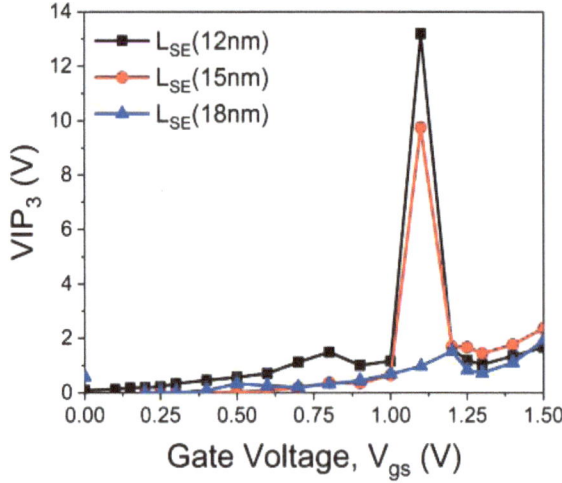

Fig. (17). Graph of VIP$_3$.

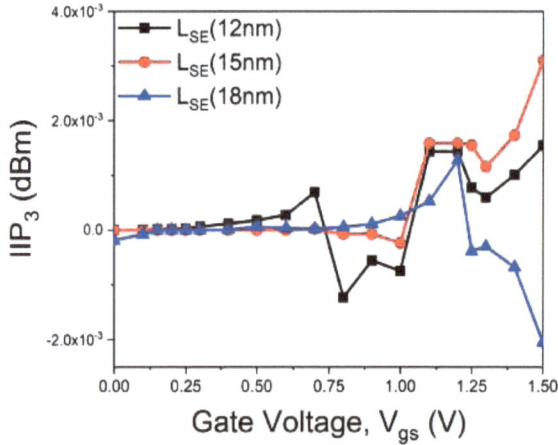

Fig. (18). Graph of IIP$_3$.

$$VIP_2 = 4\left(\frac{g_m}{g_{m2}}\right) \tag{11}$$

$$VIP_3 = \sqrt{24\left(\frac{g_m}{g_{m3}}\right)} \tag{12}$$

$$IIP_3 = \frac{2}{3}\left(\frac{g_m}{g_{m3}R_s}\right) \tag{13}$$

For distortion-less and significant linearity behavior, VIP$_2$ and VIP$_3$ would be high [29, 48]. It can be seen in Fig. (**16**); With varying V$_{gs}$, the curve for VIP$_2$ increases initially (For V_{gs} less than 0.75V) and then after starts decreasing (For V_{gs} higher than 0.8V) because in this region, the value of g_{m2} low as well as negative. The VIP$_2$ attains the highest peak when L_{SE} = 12nm and when L_{SE} = 15nm, attains the lowest range of VIP$_2$ because at the time, the significant controllability over the channel starts shifting to DCi and it will further increase when L_{SE} = 18nm (electrostatic potential of channel controlled by the DCi region electrode M$_3$). For VIP$_3$, it shows a similar trend as VIP$_2$ as portrayed in Fig. (**17**), it increases initially, but after (when V_{gs} = 1.15 V), it decreases as g_{m3} from V_{gs} = 1.0 V to V_{gs} = 1.50 V with negative values in this region for L_{SE} = 18nm. For IIP$_3$ (Fig. **18**), it shows a conversant trend with VIP$_3$, and it increases up to V_{gs}=1.15 V and then decreases afterward. All these three parameters attain the maximum values when L_{SE} is set to 12nm, and its value degrades as well as L_{SE} increases.

The two other critical linearity parameters are IMD$_3$ and 1dB$_{cp}$, and their mathematical relation with g_m and g_{m3} is given by equations (**14 & 15**), respectively [25, 27, 49]. IMD$_3$, with the lowest value, infers amended immunity to third-order intermodulation-harmonics and helps avoid power consumption. IMD$_3$ improves with increasing L_{SE}, as shown in Fig. (**19**), resulting in device reduction in performance.

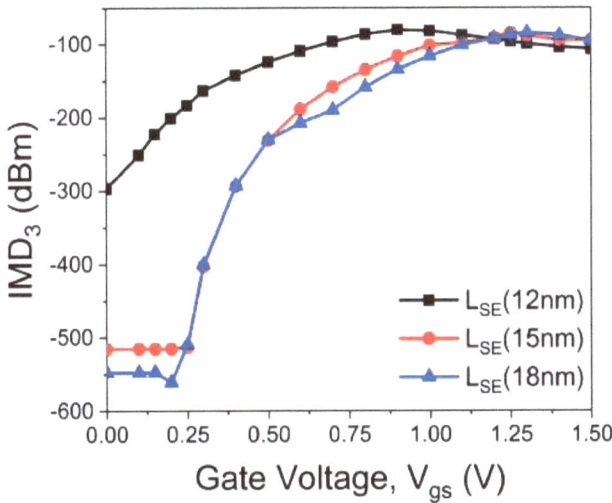

Fig. (19). Graph of IMD3.

$$IMD_3 = \left[\frac{9}{2} (VIP_3)^2 g_{m3} \right]^2 R_s \qquad (14)$$

$$1dB \ Compression \ Point = 0.22 \sqrt{\frac{g_m}{g_{m3}}} \qquad (15)$$

At last, investigated linearity parameter is $1dB_{cp}$. As it directs the input power rating where the output power reduces to 1dB first from the linear unity gain domain, a higher range of $1dB_{cp}$ is desired to hold the linear behavior of the device concerning input and output [26, 27, 41]. $1dB_{cp}$ increases as L_{SE} increases from 12nm to 18nm, as depicted in Fig. (**20**), which is a favorable indication of the enhanced linear performance of the device at lower L_{SE}.

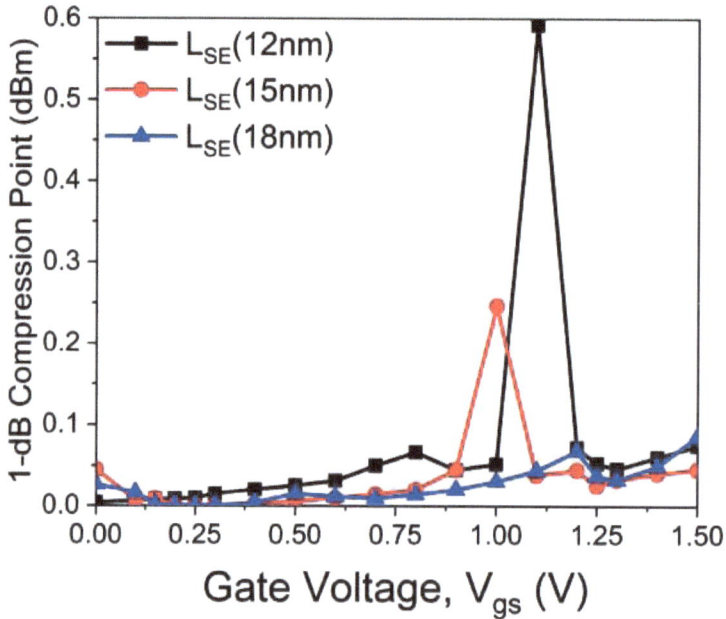

Fig. (20). Graph of 1-dB compression point.

CONCLUSION

From the perspective of analog/Rf and linearity characteristics, the improvement of TMG-SP-DG-TFET is investigated. The comprehensive study is accomplished with variable lengths of the metal gate (L_{SE}) with their fixed work function. It has been concluded that I_{ON} increases, which leads to an improved I_{on}/I_{off} ratio; hence it is imitated in better device outcomes with the suppression of ambipolar behavior. High ON-state current is achieved by employing lower band gap material in the source region so that the device is employed hetero-junction for better steepness in the transfer characteristics and improved SS value. Further, analog/RF and linearity parameters show significantly enhanced results for L_{SE} = 12nm. As a result, the TMG-SP-DG-TFET is appropriate for low-power applications for memory devices and next-generation VLSI circuitry.

LEARNING OBJECTIVES

• The impact of drain/source side gate electrode length variation was investigated with fixed electrode length at the middle.

• The length of the source/drain side gate electrode affects the tunneling rate at both interfaces, which may lead to a significant change in Ion and ambipolar current.

• This extensive study was carried out for investigation of energy band profile (ON/OFF state conditions) with deviation in electric field and parasitic capacitances of simulated structure.

• Whenever the length of the metal gate adjacent to the source region is reduced, the analog/RF functionality of the circuit exhibiting excellent steepness in the IDS-VGS curve is attained.

• The device reliability also improved when the length of the metal-gate adjacent to the source region is lower and higher near the channel-drain interface.

• All the device performance constraints are investigated carefully in this chapter with proper explanations with readable/colorful graphs.

MULTIPLE CHOICE QUESTIONS

1. Tunneling phenomena occur at which interface?

  a. Source-Channel interface
  b. Drain-Channel interface
  c. Neither source nor drain channel interface
  d. At both interface

2. Tunneling at the drain-channel interface contributed to:

  a. Improve ON-state current
  b. Decrease OFF-sate current
  c. Increase Ambipolar behavior
  d. Enhance I_{ON}/I_{OFF} ratio

3. Tunneling at source-channel interface contributed to:

  a. Improve ON-state current
  b. Enhance I_{ON}/I_{OFF} ratio

c. a and b both
d. Decrease Ambipolar behavior

4. For less parasitic oscillations:

a. Low C_{gs}, Low C_{gd}
b. High C_{gs}, Low C_{gd}
c. Low C_{gs}, High C_{gd}
d. High C_{gs}, High C_{gd}

5. The change in length of gate electrode may affect the:

a. Energy-Band-Profile
b. Electric Field
c. Potential
d. All of the above

6. For the higher Gain Bandwidth Product (GBP) value:

a. Low g_m, Low C_{gd}
b. High g_m, Low C_{gd}
c. Low g_m, High C_{gd}
d. High g_m, High C_{gd}

7. If the $(C_{gd} + C_{gs})$ is high, what should happen?

a. High GBP
b. High f_t
c. Both GBP and f_t low
d. Both GBP and f_t high

8. The correct relation between TFP and TGF:

a. TFP $\propto$ TGF
b. TFP $\propto$ 1/ TGF
c. TFP $\cong$ TGF
d. None

9. The device delay increases with:

a. As L_{SE} increases, f_t increase
b. As L_{SE} increases, f_t decrease
c. As L_{SE} decreases, f_t decrease
d. As L_{SE} decreases, f_t increase

10. The linearity characteristics of the device explain about:

a. Reliability
b. Harmonic Distortions
c. Intermodulation distortion
d. All of the above

11. The VIP_2, VIP_3, and IIP_3 are the linearity paraments, it should be:

a. Low, Low, Low
b. Low, Low, High
c. High, Low, Low
d. High, High, High

12. The VIP_3, and IIP_3 can be maximum when:

a. Low g_m, Low g_{m3}
b. Low g_m, High g_{m3}
c. High g_m, Low g_{m3}
d. High g_m, High g_{m3}

13. The relation between 1dB compression point and g_{m3} given by:

a. $0.22\sqrt{\dfrac{g_m}{g_{m3}}}$

b. $0.22\sqrt{\dfrac{g_{m3}}{g_m}}$

c. $0.32\sqrt{\dfrac{g_m}{g_{m3}}}$

d. $0.22\sqrt{g_{m3} * g_m}$

14. The relation between IMD_3 and g_m is:

a. $IMD_3 \propto (g_{m3})^{-2}$
b. $IMD_3 \propto (g_{m3})^{-1/2}$
c. $IMD_3 \propto (g_{m3})^{-1/4}$
d. $IMD_3 \propto (g_{m3})^{-1}$

15. What is the significant impact of more than one number of gates/electrodes on overall device performance?

a. SS and V_{th} both improve
b. Degradation in SS and V_{th}
c. SS improve
d. V_{th} improve

ANSWER KEY

1. (d)

2. (c)

3. (c)

4. (a)

5. (d)

6. (b)

7. (c)

8. (a)

9. (b)

10. (d)

11. (d)

12. (c)

13. (a)

14. (a)

15. (a)

ACKNOWLEDGEMENTS

The authors would like to thank the Department of Electronics and Communication Engineering, National Institute of Technology, Hamirpur, Himachal Pradesh, India, for providing valuable support to carry out this study.

REFERENCES

[1]   S. Kumar, and D.S. Yadav, "Assessment of interface trap charges on proposed tfet for low power high-frequency application", *Silicon,* vol. 14, pp. 9291-9304, 2022.

[2]   Byung-Gook Park, B-G. Park, J.D. Lee, and T-J.K. Liu, "Tunneling field-effect transistors (tfets) with subthreshold swing (ss) less than 60 mv/dec", *IEEE Electron Device Lett.,* vol. 28, no. 8, pp. 743-745, 2007.
        [http://dx.doi.org/10.1109/LED.2007.901273]

[3]   D.S. Yadav, D. Sharma, S. Tirkey, D. Soni, D.G. Sharma, S. Bajpai, and N. Sharma, "A comparative study of gap/sige hetero junction double gate tunnel field effect transistor", *IEEE International Symposium on Nanoelectronic and Information Systems (iNIS).,* pp. 195-199, 2017.
        [http://dx.doi.org/10.1109/iNIS.2017.48]

[4]   P. Singh, D.P. Samajdar, and D.S. Yadav, "Doping and dopingless tunnel field effect transistor", *6th International Conference for Convergence in Technology (I2CT).,* pp. 1-7, 2021.
        [http://dx.doi.org/10.1109/I2CT51068.2021.9418076]

[5]   W.M. Reddick, and G.A.J. Amaratunga, "Silicon surface tunnel transistor", *Appl. Phys. Lett.,* vol. 67, no. 4, pp. 494-496, 1995.
        [http://dx.doi.org/10.1063/1.114547]

[6]   K.K. Bhuwalka, S. Sedlmaier, A.K. Ludsteck, C. Tolksdorf, J. Schulze, and I. Eisele, "Vertical tunnel field-effect transistor", *IEEE Trans. Electron Dev.,* vol. 51, no. 2, pp. 279-282, 2004.
        [http://dx.doi.org/10.1109/TED.2003.821575]

[7]   S. Blaeser, S. Glass, C. Schulte-Braucks, K. Narimani, N. Driesch, S. Wirths, A. Tiedemann, S. Trellenkamp, D. Buca, and Q. Zhao, "Novel sige/si line tunneling tfet with high ion at low vdd and constant ss", *IEEE international electron devices meeting (IEDM).,* pp. 22-23, 2015.

[8]   P. Singh, D.P. Samajdar, and D.S. Yadav, "A low power single gate l-shaped tfet for high frequency application", *6th International Conference for Convergence in Technology (I2CT),* pp. 1-6, 2021.
        [http://dx.doi.org/10.1109/I2CT51068.2021.9418075]

[9]   Q. Huang, R. Huang, C. Wu, H. Zhu, C. Chen, J. Wang, L. Guo, R. Wang, L. Ye, and Y. Wang, "Comprehensive performance re-assessment of tfets with a novel design by gate and source engineering from device/circuit perspective", *IEEE International Electron Devices Meeting,* pp. 13-3, 2014.
        [http://dx.doi.org/10.1109/IEDM.2014.7047044]

[10]   G. Zhou, R. Li, T. Vasen, M. Qi, S. Chae, Y. Lu, Q. Zhang, H. Zhu, J-M. Kuo, and T. Kosel, "Novel gate-recessed vertical inas/gasb tfets with record high ion of 180 ua/um at v ds= 0.5 v", *International Electron Devices Meeting,* pp. 32-6, 2012.

[11]   K. Boucart, and A.M. Ionescu, "Double gate tunnel fet with ultrathin silicon body and high-k gate dielectric", *European Solid-State Device Research Conference,* pp. 383-386, 2006.
        [http://dx.doi.org/10.1109/ESSDER.2006.307718]

[12]   C. Li, X. Zhao, Y. Zhuang, Z. Yan, J. Guo, and R. Han, "Optimization of L-shaped tunneling field-effect transistor for ambipolar current suppression and analog/RF performance enhancement",

Superlattices Microstruct., vol. 115, pp. 154-167, 2018.
[http://dx.doi.org/10.1016/j.spmi.2018.01.025]

[13]   C. Anghel, P. Chilagani, A. Amara, and A. Vladimirescu, "Tunnel field effect transistor with increased on current, low-k spacer and high-k dielectric", *Appl. Phys. Lett.,* vol. 96, p. 122104, 2010.
[http://dx.doi.org/10.1063/1.3367880]

[14]   S. Tirkey, D.S. Yadav, and D. Sharma, "Controlling ambipolar behavior and improving radio frequency performance of hetero junction double gate tfet by dual work-function, hetero gate dielectric, gate underlap: Assessment and optimization", *International Conference on Information, Communication, Instrumentation and Control (ICICIC).,* pp. 1-7, 2017.
[http://dx.doi.org/10.1109/ICOMICON.2017.8279132]

[15]   K.K. Jha, and M. Pattanaik, "Analysis of pocket double gate tunnel fet for low stand by power logic circuits", *I. J. VLSI Design & Commun. Sys.,* vol. 4, no. 6, p. 27, 2013.
[http://dx.doi.org/10.5121/vlsic.2013.4603]

[16]   P. Singh, D.P. Samajdar, and D.S. Yadav, "A low power single gate l-shaped tfet for high frequency application", *6th International Conference for Convergence in Technology (I2CT).,* pp. 1-6, 2021.
[http://dx.doi.org/10.1109/I2CT51068.2021.9418075]

[17]   S. Anand, and R.K. Sarin, "Performance investigation of InAs based dual electrode tunnel FET on the analog/RF platform", *Superlattices Microstruct.,* vol. 97, pp. 60-69, 2016.
[http://dx.doi.org/10.1016/j.spmi.2016.06.001]

[18]   Seema, and S.S. Chauhan, "A new design approach to improve DC, analog/RF and linearity metrics of vertical TFET for RFIC design", *Superlattices Microstruct.,* vol. 122, pp. 286-295, 2018.
[http://dx.doi.org/10.1016/j.spmi.2018.07.036]

[19]   J. Madan, and R. Chaujar, "Gate drain underlapped pnin-gaa-tfet for comprehensively upgraded analog/rf performance", *Superlattices Microstruct.,* vol. 102, pp. 17-26, 2017.
[http://dx.doi.org/10.1016/j.spmi.2016.12.034]

[20]   D.S. Yadav, D. Sharma, R. Agrawal, G. Prajapati, S. Tirkey, B.R. Raad, and V. Bajaj, "Temperature based performance analysis of doping-less tunnel field effect transistor", *International Conference on Information, Communication, Instrumentation and Control (ICICIC).,* pp. 1-6, 2017.
[http://dx.doi.org/10.1109/ICOMICON.2017.8279131]

[21]   N. Parmar, P. Singh, D.P. Samajdar, and D.S. Yadav, "Temperature impact on linearity and analog/RF performance metrics of a novel charge plasma tunnel FET", *Appl. Phys., A Mater. Sci. Process.,* vol. 127, no. 4, p. 266, 2021.
[http://dx.doi.org/10.1007/s00339-021-04413-z]

[22]   C. Li, X. Zhao, Y. Zhuang, Z. Yan, J. Guo, and R. Han, "Optimization of L-shaped tunneling field-effect transistor for ambipolar current suppression and Analog/RF performance enhancement", *Superlattices Microstruct.,* vol. 115, pp. 154-167, 2018.
[http://dx.doi.org/10.1016/j.spmi.2018.01.025]

[23]   D.S. Yadav, D. Sharma, B.R. Raad, and V. Bajaj, "Impactful study of dual work function, underlap and hetero gate dielectric on TFET with different drain doping profile for high frequency performance estimation and optimization", *Superlattices Microstruct.,* vol. 96, pp. 36-46, 2016.
[http://dx.doi.org/10.1016/j.spmi.2016.04.027]

[24]   S. Anand, and R.K. Sarin, "Performance investigation of InAs based dual electrode tunnel FET on the analog/RF platform", *Superlattices Microstruct.,* vol. 97, pp. 60-69, 2016.
[http://dx.doi.org/10.1016/j.spmi.2016.06.001]

[25]   P. Singh, and D.S. Yadav, *Assessing the impact of drain underlap perspective approach to investigate dc/rf to linearity behavior of l-shaped tfet.,* vol. 14, pp. 11471-11481, 2022.*Silicon,* vol. 14, pp. 11471-11481, 2022.

[26]   A.K. Gupta, A. Raman, and N. Kumar, "Design and investigation of a novel charge plasma-based

core-shell ring-tfet: Analog and linearity analysis", *IEEE Trans. Electron Dev.,* vol. 66, no. 8, pp. 3506-3512, 2019.
[http://dx.doi.org/10.1109/TED.2019.2924809]

[27] P. Singh, and D.S. Yadav, "Impact of temperature on analog/RF, linearity and reliability performance metrics of tunnel FET with ultra-thin source region", *Appl. Phys., A Mater. Sci. Process.,* vol. 127, no. 9, p. 671, 2021.
[http://dx.doi.org/10.1007/s00339-021-04813-1]

[28] G. Han, Y. Wang, Y. Liu, C. Zhang, Q. Feng, M. Liu, S. Zhao, B. Cheng, J. Zhang, and Y. Hao, "GeSn quantum well p-channel tunneling FETs fabricated on Si (001) and (111) with improved subthreshold swing", *IEEE Electron Device Lett.,* vol. 37, no. 6, p. 1, 2016.
[http://dx.doi.org/10.1109/LED.2016.2558823]

[29] K. Tomioka, and T. Fukui, "Tunnel field-effect transistor using InAs nanowire/Si heterojunction", *Appl. Phys. Lett.,* vol. 98, no. 8, p. 083114, 2011.
[http://dx.doi.org/10.1063/1.3558729]

[30] C. Li, Z.R. Yan, Y.Q. Zhuang, X.L. Zhao, and J.M. Guo, "Ge/Si heterojunction L-shape tunnel field-effect transistors with hetero-gate-dielectric", *Chin. Phys. B,* vol. 27, no. 7, p. 078502, 2018.
[http://dx.doi.org/10.1088/1674-1056/27/7/078502]

[31] S. Yadav, R. Madhukar, D. Sharma, M. Aslam, D. Soni, and N. Sharma, "A new structure of electrically doped TFET for improving electronic characteristics", *Appl. Phys., A Mater. Sci. Process.,* vol. 124, no. 7, p. 517, 2018.
[http://dx.doi.org/10.1007/s00339-018-1930-9]

[32] Upasana, R. Narang, M. Saxena, and M. Gupta, "Modeling and TCAD assessment for gate material and gate dielectric engineered TFET architectures: Circuit-level investigation for digital applications", *IEEE Trans. Electron Dev.,* vol. 62, no. 10, pp. 3348-3356, 2015.
[http://dx.doi.org/10.1109/TED.2015.2462743]

[33] G. Betti Beneventi, E. Gnani, A. Gnudi, S. Reggiani, and G. Baccarani, "Optimization of a pocketed dual-metal-gate TFET by means of TCAD simulations accounting for quantization-induced bandgap widening", *IEEE Trans. Electron Dev.,* vol. 62, no. 1, pp. 44-51, 2015.
[http://dx.doi.org/10.1109/TED.2014.2371071]

[34] C. Rajan, D.P. Samajdar, J. Patel, A. Lodhi, S.K. Agnihotri, D. Sharma, and A. Kumar, "Linearity and reliability analysis of an electrically doped hetero material nanowire TFET", *J. Electron. Mater.,* vol. 49, no. 7, pp. 4307-4317, 2020.
[http://dx.doi.org/10.1007/s11664-020-08143-5]

[35] N. Paras, and S.S. Chauhan, "Insights into the DC, RF/Analog and linearity performance of vertical tunneling based TFET for low-power applications", *Microelectron. Eng.,* vol. 216, p. 111043, 2019.
[http://dx.doi.org/10.1016/j.mee.2019.111043]

[36] Upasana, R. Narang, M. Saxena, and M. Gupta, "Exploring the applicability of well optimized dielectric pocket tunnel transistor for future low power applications", *Superlattices Microstruct.,* vol. 126, pp. 8-16, 2019.
[http://dx.doi.org/10.1016/j.spmi.2018.12.005]

[37] B.V. Chandan, K. Nigam, D. Sharma, and S. Pandey, "Impact of interface trap charges on dopingless tunnel FET for enhancement of linearity characteristics", *Appl. Phys., A Mater. Sci. Process.,* vol. 124, no. 7, p. 503, 2018.
[http://dx.doi.org/10.1007/s00339-018-1923-8]

[38] D. Soni, D. Sharma, S. Yadav, M. Aslam, and N. Sharma, "Performance improvement of doped TFET by using plasma formation concept", *Superlattices Microstruct.,* vol. 113, pp. 97-109, 2018.
[http://dx.doi.org/10.1016/j.spmi.2017.10.012]

[39] M. Verma, S. Tirkey, S. Yadav, D. Sharma, and D.S. Yadav, "Performance assessment of a novel vertical dielectrically modulated TFET-based biosensor", *IEEE Trans. Electron Dev.,* vol. 64, no. 9,

pp. 3841-3848, 2017.
[http://dx.doi.org/10.1109/TED.2017.2732820]

[40]   N. Upasana, R. Narang, M. Saxena, and M. Gupta, "Linearity and analog performance realization of energy-efficient TFET-based architectures: an optimization for RFIC design", *IETE Tech. Rev.,* vol. 33, no. 1, pp. 23-28, 2016.
[http://dx.doi.org/10.1080/02564602.2015.1043153]

[41]   B.V. Chandan, S. Dasari, S. Yadav, and D. Sharma, "Approach to suppress ambipolarity and improve RF and linearity performances on ED□Tunnel FET", *Micro & Nano Lett.,* vol. 13, no. 5, pp. 684-689, 2018.
[http://dx.doi.org/10.1049/mnl.2017.0814]

[42]   Narang Rakhi, Saxena Manoj, R.S. Gupta, and Gupta. Mridula, "Linearity and analog performance analysis of double gate tunnel FET: effect of temperature and gate stack", *I. J. VLSI Design & Commun. Sys.,* vol. 2, no. 3, pp. 185-200, 2011.
[http://dx.doi.org/10.5121/vlsic.2011.2316]

[43]   J. Lee, R. Lee, S. Kim, K. Lee, H.M. Kim, S. Kim, M. Kim, S. Kim, J.H. Lee, and B.G. Park, "Surface Ge-rich p-type SiGe channel tunnel field-effect transistor fabricated by local condensation technique", *Solid-State Electron.,* vol. 164, p. 107701, 2020.
[http://dx.doi.org/10.1016/j.sse.2019.107701]

[44]   A. Lemtur, D. Sharma, P. Suman, J. Patel, D.S. Yadav, and N. Sharma, "Performance analysis of gate all around GaAsP/AlGaSb CP-TFET", *Superlattices Microstruct.,* vol. 117, pp. 364-372, 2018.
[http://dx.doi.org/10.1016/j.spmi.2018.03.049]

[45]   J.C. Lee, T.J. Ahn, and Y.S. Yu, "Si/Ge hetero tunnel field-effect transistor with junctionless channel based on nanowire", *J. Nanosci. Nanotechnol.,* vol. 19, no. 10, pp. 6750-6754, 2019.
[http://dx.doi.org/10.1166/jnn.2019.17109] [PMID: 31027023]

[46]   S. Chen, S. Wang, H. Liu, T. Han, H. Xie, and C. Chong, "A novel dopingless fin-shaped sige channel tfet with improved performance", *Nanoscale Res. Lett.,* vol. 15, no. 1, p. 202, 2020.
[http://dx.doi.org/10.1186/s11671-020-03429-3] [PMID: 33068207]

[47]   S.M. Turkane, and A.K. Kureshi, "Review of tunnel field effect transistor (TFET)", *Int. J. Appl. Eng. Res.,* vol. 11, no. 7, pp. 4922-4929, 2016.

[48]   S. Gupta, K. Nigam, S. Pandey, D. Sharma, and P.N. Kondekar, "Effect of interface trap charges on performance variation of heterogeneous gate dielectric junctionless-TFET", *IEEE Trans. Electron Dev.,* vol. 64, no. 11, pp. 4731-4737, 2017.
[http://dx.doi.org/10.1109/TED.2017.2754297]

[49]   K. Vanlalawmpuia, and B. Bhowmick, "Linearity performance analysis due to lateral straggle variation in hetero-stacked TFET", *Silicon,* vol. 12, pp. 955-961, 2019.

[50]   S. Kumar, K.S. Singh, K. Nigam, V.A. Tikkiwal, and B.V. Chandan, "Dual-material dual-oxide double-gate TFET for improvement in DC characteristics, analog/RF and linearity performance", *Appl. Phys., A Mater. Sci. Process.,* vol. 125, no. 5, p. 353, 2019.
[http://dx.doi.org/10.1007/s00339-019-2650-5]

CHAPTER 7

II-VI Semiconductor-based Thin-Film Transistor Sensor for Room Temperature Hydrogen Detection From Idea to Product Development

Sukanya Ghosh[1,*] and **Lintu Rajan**[1]

[1] *National Institute of Technology, Calicut, Kerala, 673601, India*

Abstract: Implementing gas sensors incorporating nanoelectronic devices to detect pollution and improve the safety control of industrial, medical, and domestic sectors has opened up a novel world with immense interest. As a promising renewable energy carrier and a potential replacement for fossil fuels, there is the paramount importance of hydrogen gas storage at extensive facilities worldwide. The sustainable production of hydrogen is increasing owing to its enormous energy per mass of any fuel. Nevertheless, due to its extreme flammability, simple and highly accurate sensors with promising sensing materials are required to detect the slightest traces of timely leak detection for developing a hydrogen economy. Various hydrogen detectors already exist, but expensive cost, large size, sluggish response, and high temperature limit their potential for widespread applications. The integral objective of the present chapter is to focus on a systematic investigation of Pd-Ti/ZnO Schottky TFT-based room temperature hydrogen sensors excluding any heating element. With high chemical and thermal stability, ZnO is a promising candidate for sensors in a hazardous atmosphere. The developed sensor exhibited room temperature detection with a maximum response of 33.8% to 4500 ppm H_2 in dry air. The selectivity analysis toward H_2 in the presence of other reducing and oxidizing gas species has also been investigated to ensure the real-time applicability of the sensor. Reliable operation of the sensor in a wide range of 500 ppm to 4500 ppm H_2 has been confirmed from the linear behavior of the sensor. The hydrogen sensing mechanism of the proposed sensor in terms of Schottky barrier height reduction at the interface of Pd-Ti/ZnO has also been detailed in this chapter. Room temperature detection of the hydrogen sensor presented here competes favorably with the existing studies. This study can be extended in exploring new routes to realize hydrogen sensing applications at room temperature for commercialization with precise control over film thickness and target gas concentrations.

Keywords: Hysteresis, Room temperature, Pd-Ti/ZnO Schottky TFT, hydrogen sensor, Repeatability, Selectivity, Reproducibility.

---

* **Corresponding author Sukanya Ghosh:** National Institute of Technology, Calicut, Kerala, 673601, India; E-mail: gsukanya66@gmail.com

Gopal Rawat & Aniruddh Bahadur Yadav (Eds.)

INTRODUCTION

Nanoelectronics, the core foundation of next-generation electronic science and technology, has emerged rapidly and vigorously in recent years owing to various possible applications. In addition, nanosensors, particularly gas sensors composed of nanomaterials, have made increasing developments due to their dramatic advantages and specific surface states. Variations of physical or chemical properties of several gases are transformed into standard electrical signals through a gas sensor in the field of gas analysis and safety applications. In this regard, the extending demands for hydrogen sensors are not only restricted to industrial process control and environmental welfare but also extend to glassmaking, metal smelting, space flights, propulsion fuels, coal mines, nuclear reactors, petroleum extraction, semiconductor manufacturing, food, medical and chemical industries, *etc* [1 - 3]. The importance of hydrogen as a clean energy source was strongly recommended at the Conference of Parties in Paris in 2015. National Renewable Energy Laboratory in the U.S published a document concentrating on hydrogen-specific application domains [4]. There are growing demands for carbon dioxide-free hydrogen vehicles worldwide. Broadly combustible concentration range (4%-75%), high burning velocity, high ignition heat (142 kJ/g H_2), low combustion energy (0.017 mJ), large diffusion coefficient (0.61 cm_2/s), colorless, odorless, and tasteless nature are some of the several inherent characteristics of hydrogen. The mixtures of hydrogen and air are highly flammable. Hence, accurate and timely leak detection of hydrogen is essential and the need of the hour during hydroge production and storage in industrial and domestic sectors [5, 6].

However, existing solutions based on the concepts of the electrochemical, optical, acoustic wave, and calorimetric undergo the drawbacks of higher operational temperature, poor lifetime, slow response, large size, costly fabrication, design complications, high dissipation of power, and so on. FET-based gas sensors incorporating their respective electrical characteristics have been widely investigated in recent years to overcome such barriers [7, 8]. The sensing layer is the heart of a gas sensor. In this context, metal oxide semiconductors for example, TiO_2 SnO_2, V_2O_5, WO_3, Nb_2O_5 *etc*., have been broadly employed as sensing layers in FET-based gas sensors due to their noticeably reasonable cost, reduced fabrication complexities, high sensing response, lower dissipated power and quick response time. ZnO, an n-type semiconductor, has been extensively used as a hydrogen sensor. It possesses several advantages, *e.g.*, the high bandgap energy (3.37 eV at 300 K), high mobility, huge exciton binding energy (60 meV at 300 K), higher thermal conductivity (116 Wm-1K-1), superior chemical and thermal stability, near UV emission, strong surface adsorption potential, d10 electronic configuration, and so on [9, 10]. Such properties ensure ZnO is a spectacular material for gas sensing. Several gas adsorbing sites determine a gas sensor's performance; hence, a high surface-to-volume ratio sensing element is favorable. Nanocrystalline materials ensure a high surface-to-volume ratio compared to bulk. Exposure of the surface increases with decreasing crystal size. There is also an increment of a fraction of atoms at the grain boundary with reducing crystal size. Surface conductivity increases with the increasing number of nodes at the surface. ZnO is bio-safe and bio-compatible. The hexagonal wurtzite structure and a high degree of flexibility of ZnO in growth geometries are popularly investigated in sensing characteristics at minimum processing temperature. ZnO thin film associated (Pd/ZnO/p-Si and Pd/ZnO/Zn) efficiently selective hydrogen sensor is reported in the literature with 2000-20,000 ppm H_2 concentration [11]. A maximum 7.8% sensing response is achieved by synthesizing a porous ZnO thin film using the sol-gel technique to detect 250 ppm CO at 350°C [12]. Pd/ZnO Schottky diode-based hydrogen sensor is investigated by incorporating sol-gel techniques and thermal evaporation at RT [13]. Mainly produced from the zincite mineral, the ionicity of II-VI compound semiconductor ZnO falls at the borderline between ionic and covalent semiconductors [14]. It should be noted that ZnO thin films have been deposited by several methods, including CVD [6], plasma-enhanced CVD [15], spray pyrolysis [16], a sol-gel technique [17], plasma laser deposition [18], *etc.*, for gas sensing applications. Although each method has its advantages, several issues still need to be overcome that may affect their gas sensing function. These relate to their poor repeatability, high synthesis time, and high operating temperature. RF sputtering is one of the best thin film deposition techniques. The advantage of this specific technique relates to uniformity in thickness, high purity, and high deposition rate; therefore suitable for large-area deposition, better controllability of deposition parameters, lower substrate temperature, and excellent film adhesion [19]. In sputtering, the material to be deposited forms the target electrode. Upon introduction inside the vacuum chamber, an inert gas, *e.g.*, argon, is ionized through an electron beam. These high-energy ions bombard the target electrode and remove the target material. A vapor composed of positively charged ions is created. These vapor atoms are then deposited, covering the substrate required to be coated. RF magnetron sputtering uses magnetic fields to trap electrons in front of the target so that ionization of the inert gas increases, allowing for increasing deposition rates. At each cycle of RF sputtering, the target material surface is cleaned by the charge formation process as the electrical potential gets altered. When the positive cycle arrives, the target material comes under negative bias as electrons attract it. Striking of the target electrode continues during the negative cycle of the RF sputtering process. RF sputtering can sustain lower pressure requirements throughout the chamber. As a result, the target material can be efficiently deposited due to the lower collisions of ionized gases. Also, plasmas are usually scattered throughout the chamber instead of localizing around the target electrode. RF sputtered ZnO thin film-based hydrogen sensor exhibited outstanding sensing response throughout a wide concentration range at an optimum temperature [20]. Noble catalytic metals (*e.g.*, Pd, Pt, Au, *etc.*) exhibit excellent hydrogen sensing responses owing to their catalytic activity and high hydrogen solubility [21, 22]. Among them, Pd has been most reported in effectively enhancing sensing response and selectivity of the sensing material toward hydrogen, also known as hydrogen collector. Pd has excellent hydrogen molecules adsorption capability and the dissociation of these hydrogen molecules into hydrogen atoms [23, 24]. Acceleration of the diffused hydrogen atoms occurs all over the surface, resulting in increased interaction with the surface-active sites. At room temperature, Pd can consume 900 times its equivalent volume of hydrogen [25]. Pd consumes hydrogen following an exothermic reaction in a hydrogen-enriched environment, displacing hydrogen molecules approaching the Pd contact. The interaction between hydrogen molecules and Pd atoms takes place by incorporating Van der Waals forces. Adsorbed H_2 gas molecules dissociate into H atoms and diffuse within the metal [26, 27]. ZnO thin film exhibits a superior Schottky contact with Pd [28]. Ti promotes adhesion as well as improves top electrical contacts. Ultrasmall Pd nanowire network-based hydrogen sensor with outstanding sensing performance and response times has been explored in the literature [29].

Existing FET-based gas sensors must be highly sensitive and reproducible for commercial applications [30 - 32]. Therefore, to obtain enhanced sensing response and expand the functional domain of semiconducting oxide-based TFTs, there is a need to develop a gas sensor by taking into account the target gas-consuming potentiality of the sensing material and the current regulating capability of TFT. The channel layer of the device serves as the target gas-reactive surface while operating at room temperature. The bottom gate top contact-based TFT hydrogen sensors are worth investigating because of their reduced cost, lower fabrication complexity, reduced power consumption, enhanced sensing response, fast response, and accurate detection. Such device structures are preferred to achieve effective performance due to reduced top metal-semiconductor contact resistance compared to a bottom contact-based design [33, 34]. Unlike the conventional FETs, the developed TFT structure consists of top metal/ZnO Schottky contacts instead of Ohmic contact. Such a design also preserves the regular gate-stacked MOS framework while including an Ohmic gate below the top metal contacts and the sensing material contrary to the MESFET.

This chapter briefly describes the developed sensor TFT's hydrogen sensing mechanism. Apart from discussing the fabrication procedure in detail, this chapter aims to highlight the structural analysis of the RF sputter deposited ZnO thin film with the electrical characterization of the fabricated TFT. The present chapter also investigates the dynamic response of the sensor TFT in the presence of different H_2 concentrations, along with repeatability, selectivity, and reproducibility analysis under room temperature.

SCHOTTKY BARRIER TFT (SB-TFT) AS HYDROGEN SENSOR

Upon depositing a metal over a lightly doped semiconductor, a Schottky contact is formed where $\varphi_m > \varphi_s$, φ_m and φ_s are the metal and the semiconductor work functions, respectively. As the metal and the n-type semiconductor come into contact, the flow of electrons from the semiconductor toward the metal lower energy states until the Fermi levels are aligned. Positively charged fixed donor ions form a space charge region in the n-type semiconductor while the electrons proceed toward the metal side. Hence the direction of electrons from the semiconductor to the metal gets opposed. So, to maintain the flat Fermi level, a band bending across the junction is initiated. It is considerably challenging to control ZnO Schottky barriers properly when the barrier height is calculated from the same metal on a given ZnO surface due to its wide variation. Schottky barrier heights depend on various extrinsic parameters such as preparation of the surface, crystal quality, and the conditions to form the contact. Moreover, for higher work function metals, *e.g.*, Pd, Pt, and Au, the n-type barrier heights (φ_{SB}^{n}) are usually lower than expected with several air-exposed treatments.

The formation of a TFT, a particular type of field-effect transistor, occurs when thin films of the dielectric, an active layer of the semiconductor, and metal contacts are placed over the substrate. There are specific differences between TFT and MOSFET. First, TFT operates in an accumulation region, whereas MOSFET operates in an inversion region. Next, the characteristic of TFT is amorphous, while it is crystalline MOSFET. TFT is undoped in comparison to MOSFET, which is mainly Si-doped. TFTs' structural configuration varies depending on the positions of the source, drain, and gate electrodes. Fig. (**1a**) depicts a typical bottom gate top contact TFT where the sensing layer is exposed to a particular analyte on one side. On the other end, the gate contact is isolated from the sensing layer (oxide semiconductor) through the gate dielectric. Such staggered geometry imparts a large area for injecting charge carriers into the semiconducting layer compared to the coplanar structure [35]. As a result, contact resistance gets lowered, and device performance improves since the charge carriers injected towards the interface probably involve the entire contact surface area instead of the contact edges similar to the coplanar structure. The channel's conductivity in TFT configuration depends on the number of free charge carriers controlled by the gate potential. When a gate voltage is induced, the polarization of the dielectric occurs. It leads to a charge distribution on the insulator surface of the same polarity as the gate potential. Also, a thin accumulation layer of mobile charge carriers of opposite polarity gets distributed within the oxide semiconductor region at the dielectric-semiconductor interface. Depending upon the gate bias polarity, dielectric capacitance, majority carriers (majority carriers for n-type semiconductors: electrons, and for p-type semiconductors, they are holes), and relative carrier concentration, this accumulation layer of mobile charge carriers can be either electrons or holes. Apart from these mobile charge carriers, deep trapped charges are also considered to be the origin of the threshold voltage. Once the effective gate bias is applied ($V_{EFF}= V_G$-V_{TH}), mobile charges contribute to the drain to source current. Under the application of zero gate voltage, TFT is treated as a chemiresistor helpful in measuring the conductivity change of the film when exposed to some chemical species. It goes to an equilibrium state without any gate bias applied. Under the application of an electric field across the gate dielectric, the gate electrode controls the drain-to-source current and the interaction between the analyte and the sensing layer. The transfer of electrons between analyte molecules and semiconductor varies the semiconductor work function, concentration, and mobility of free charge carriers. When a positive gate bias is applied, mobile electrons accumulate near the bottom side of the oxide semiconductor and dielectric, resulting in an accumulation region and a conducting channel. TFT can function in enhancement mode or depletion mode controlled by the requirement of gate bias to form a channel. When a negative bias is applied, electron depletion occurs from the channel, and the channel conductance gets reduced. Uniform and homogeneous charge carrier concentrations are present in the channel region without source-drain bias. In the channel region, a linear gradient of charge density forms in the presence of a small bias ($V_{DS} \ll V_{GS}$) across the source-drain terminal. In addition, the drain current becomes linearly proportional with drain bias and exhibits Ohmic behavior. A point is reached with increasing source-drain bias where $V_{DS}= V_{GS}$-V_T; at this pinch-off point (V_{SAT}), the depletion region becomes closer to the drain electrode, and the drain current no longer increases with increasing V_{DS}. If the source-drain potential increases, the depletion region expands, effective channel length shortens, the drain current no longer increases, and the transistor reaches its saturation zone. TFTs provide more measurable parameters and depth regarding the analyte for precise detection than conventional chemiresistors, hence more selective than the latter. TFT sensors offer better-sensing limits and response over those conventional chemiresistors due to the signal amplification of the TFT-based devices.

The related TFT hydrogen sensing mechanism based on Schottky barrier height reduction across the Pd-Ti/ZnO interface on the adsorption of target hydrogen molecules is presented in Fig. (**1b**). Carriers inserted across the Schottky barrier at the source are mainly regulated by the integral carrier transport mechanism T.E. Under hydrogen exposure, drain current increases with the work function variation between Pd-Ti/ZnO. At low H_2 concentrations, there is a significant role of Schottky contacts. The Schottky barrier height changes as the hydrogen molecules get adsorbed at the Schottky contact. The interaction of hydrogen molecules with Pd-Ti initiates this change in work function. Hydrogen dipoles are formed at the Pd-Ti/ZnO interface when hydrogen molecules are dissociated into hydrogen atoms. An electric field is created, allowing the electrons to flow toward the channel; hence the Schottky barrier height gets reduced between the source and oxide semiconductor [36, 37], represented in Fig. (**1b**). Schottky barrier height (φ_B) can be evaluated as per the TE model [38] in Equation. (**1**):

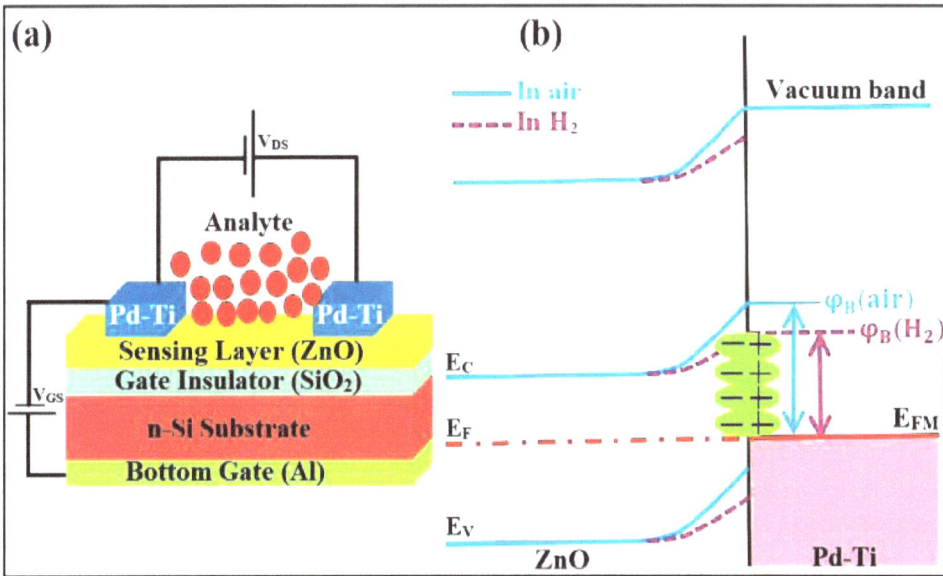

Fig. (1). (**a**) Cross-sectional schematic of a TFT gas sensor. Red circles depict the analyte of interest (hydrogen). (**b**) Energy band diagram of the sensor at source contact when zero bias is applied.

$$\varphi_B = -\frac{kT}{q}\ln\left(\frac{I_S}{AA^{**}T^2}\right) \qquad (1)$$

Where T= absolute temperature in Kelvin, q= electronic charge in coulomb, k= Boltzmann constant, A= area of Schottky contact in square meter, A**= effective Richardson constant of ZnO= 32 A em $2K^{-2}$ (for $me^* = 0.27_{m0}$) [39], IS= determined saturation current in ampere. The variation in Schottky barrier height (SBH) can be calculated as [38]:

$$\Delta\varphi_B = \varphi_{B,air} - \varphi_{B,H_2} \qquad (2)$$

where $\varphi_{B,\,air}$ and $\varphi_{B,H2}$ are the measured Schottky barrier heights in electron volts in air and H_2, respectively. Reduction in SBH occurs as hydrogen concentration increases which leads to a significant Schottky barrier height variation $(\Delta\varphi_B)$. The drain current is enhanced and considered a sensing metric; therefore, sensing response under room temperature improves. The work function and hydrogen solubility of the noble metals vary under hydrogen exposure, resulting in a significant change in sensing response even for a small barrier height variation across metal/oxide semiconductor Schottky junction. Pd-Ti catalytic contacts prevent any Pd-to-PdH$_X$ phase transformation. It is well established in the literature that, in the air atmosphere, stable oxygen molecules capture free electrons from the oxide semiconductor conduction band and get adsorbed on the semiconducting material surface to form chemisorbed O_2^- ions. It results in a depletion region at the ZnO surface, leading to an increase in Schottky barrier height, which limits the electron concentration; consequently, the drain current of the sensor decreases. The reaction kinematics is shown below:

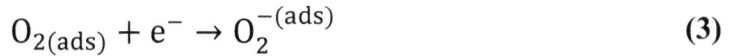

$$O_{2(ads)} + e^- \rightarrow O_2^{-(ads)} \qquad (3)$$

Hydrogen, being a reducing gas, interacts with adsorbed O_2^- at the ZnO surface and releases electrons to the oxide semiconductor conduction band upon oxidization. Consequently, the barrier height reduces owing to the decrease in depletion layer width; hence drain current of the sensor increases. It leads to enhanced sensing response and selectivity performance of the device under lower operating temperatures. The reaction kinematics follows the equation below:

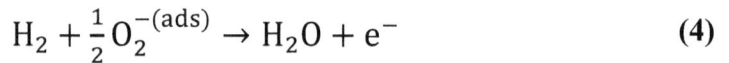

$$H_2 + \frac{1}{2}O_2^{-(ads)} \rightarrow H_2O + e^- \qquad (4)$$

Such chemisorbed reactions are highly dependent on hydrogen concentration. When hydrogen is removed, the depletion region reappears, and the drain current reduces. The increased hydrogen concentration enhances the reaction with chemisorbed oxygen species, ultimately improving the sensing response.

SENSOR TFT FABRICATION

The fabrication process flow of sensor TFT starts with the cleaning of 3" Si substrate with <100> orientation (thickness: 380 µm, resistivity: 1-10 Ω-cm.) using standard RCA-1 (NH$_4$OH: H$_2$O$_2$: DI water in 1:1:5) and RCA-2 (HCl: H$_2$O$_2$: DI water in 1:1:6) techniques at 75°C followed by 30 s of H$_2$O: H.F. (50:1) in ambient for removing native oxides.

Next, a 100 nm thick SiO_2, a dielectric gate layer, is grown on the substrate through dry oxidation (1100°C). The deposition obtains a 50 nm thick layer of ZnO for a period of 286 s using the RF magnetron sputtering technique at room temperature. Thickness inspection of the sputtered ZnO thin film is implemented with an ellipsometer to ensure measurements are within tolerance limits at the center, top, and edge. RF sputtering technique includes ZnO target (iTASCO, 99.99% pure), base pressure ~1.36×10^{-6} Torr, target to substrate distance ~7.5 cm, deposition pressure ~7.15×10^{-3} Torr, RF power ~80 W and sputtering gas Ar concentration ~50 sccm. The sample is then cleaned with acetone and IPA (C_3H_8O) sequentially, followed by dehydration at 110°C for 10 min, spin coating with HMDS solution at 6000 rpm for 40 s, and soft baking for 1 min at 110°C to make it ready for lithography. A positive PR AZ5214 is spin-coated on the top side of the wafer at 6000 rpm for 40 s, followed by soft baking for 1 min at 110°C. In optical lithography, a mask writer is prepared to pattern the required layer (Heidelberg Instruments, Model: µPG-501). Next, the mask is transferred to the ZnO layer using a UV dose (45 mV/cm^2, wavelength ~395 nm) for 55 ms in vacuum along with a 0.5 µm based mask aligner (SUSS MicroTech, Model: MJB4). The patterned layer was developed using MF26A surfactant developer in 18 s. The sample is dipped in a developer solution and then dried. Then, 20 nm thick Ti and 130 nm thick Pd as source/drain metal (dimension: 200×200 µm) is deposited using electron beam evaporation at RT with a base pressure of ~2.8×10^{-6} Torr, deposition rate ~2A°/s, deposition time ~33 min followed by lift-off process at chemical wet bench (FELCON). The lift-off process is more precisely achieved by incorporating 30 s sonication. Thickness inspection Pd-Ti was carried out using Dektak profilometry. The wafer front side is spin-coated with positive PR S1813, followed by hard baking for 3 min at 110°C. Thereon, SiO_2 removal is performed from the wafer backside using a wet-etching process that includes buffered HF with an etch rate of 68.67 nm/min followed by PR strip in acetone, IPA, and DI water before Al (thickness: 150 nm) gate metal deposition using DC sputtering with a deposition time of 567 s).

The electrical performance of the fabricated device is enhanced by an RTA process executed at 350°C for 1 min in Ar. A magnified representation of the developed TFT under study fabricated on 50 nm ZnO thin film post-annealing process is shown in Fig. (**2**).

THIN FILM CHARACTERIZATION

ZnO thin film crystallinity is investigated by XRD analysis through Rigaku, Smartlab High-Resolution X-Ray diffractometer using Cu-Kα (1.54 A°) radiation with 2θ range: 20°-80°. The XRD pattern of the ZnO thin film obtained using R.F. sputter deposition has been shown in Fig. (**3**). According to Joint Committee on

Power Diffraction Standards (JCPDS, number: 36-1451), the wurtzite structure of c-axis oriented ZnO thin film and alignment along (002) crystallographic plane

Fig. (2). Magnified version of the studied device (channel width: 200 µm, channel length:6 µm) is presented using a yellow circle from the micrograph of the fabricated devices.

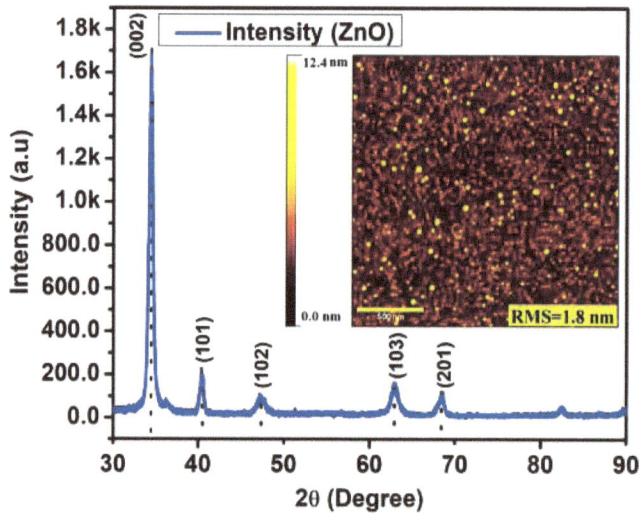

Fig. (3). XRD pattern of 50 nm ZnO thin film. Inset: AFM topographic image of the thin film.

is evident. Descending intensities are observed in other reflection peaks corresponding to (101), (103), (201), and (102) planes, respectively. The diffraction peaks of the ZnO thin film are observed at 34.49°, 40.41°, 47.26°, 62.89°, and 68.51° corresponding to (002), (101), (102), (103), and (201) crystal planes respectively. The pattern's peak broadening is due to the nanometer-sized thickness of the deposited film. The average crystallite size is 11.07 nm, evaluated using the Scherrer formula by measuring the broadening of (002) peak represented as:

$$D= \frac{0.9\lambda}{\beta \cos\theta} \qquad (5)$$

where D is the crystallite size in nm, β is the full width at half maximum of the intense diffraction peak in radians, λ is the X-Ray wavelength (0.154 nm), and θ is Bragg's diffraction angle in radians. The average crystallite size of the deposited thin film indicates its high surface area to volume ratio. There is a significant influence of surface area in gas sensing as it is a surface phenomenon. More active sites are provided for adsorption-desorption to occur by a high surface area, which enhances the sensing performance at low target gas concentrations under room temperature [40]. The surface morphology of the fabricated film has been analyzed using AFM (Bruker, Dimension Icon), shown as an inset of Fig. (**3**). The RMS roughness of the thin film is estimated at 1.8 nm. Surface roughness is another critical approach in increasing sensing response. A rougher surface improves the film's surface area-to-volume ratio by providing more active sites for adsorbing the target hydrogen atoms. It leads to the accumulation of hydrogen dipoles with a higher density at the interface of Pd-Ti and ZnO; hence an improved sensing performance can be expected.

ELECTRICAL CHARACTERIZATION OF THE TFT

The electrical properties of the proposed device have been characterized by Semiconductor Parameter Analyzer (Agilent B1500A) incorporating DC Probe Station. The operating characteristics of the fabricated TFT are presented in Fig. (**4**) Under zero gate bias, TFT is non-operational as charge carriers are absent from the oxide semiconductor. Charge carrier density from the interface between the Pd-Ti source and ZnO further diminishes, increasing drain bias. It indicates that the drain current becomes independent irrespective of the positive drain to source voltage. The channel starts conducting under a positive gate bias, and TFT starts operating. Developed TFT is observed to reach a saturation region at lower voltages owing to the presence of the Schottky barrier; thus, power consumption gets drastically reduced. Transfer and output characteristic curves estimate the electrical performance of the TFT. For positive drain to source voltage, almost ideal transistor output characteristics are evident, as an inset of Fig. (**4**). For lower values of V_{DS}, the output curve is practically linear. Still, the drain current starts deviating from linearity with increasing V_{DS} due to the reduced charge carriers by semiconductor potential near the drain electrode. The channel current enters the saturation region when V_{DS} exceeds the pinch-off point (V_{SAT}). The smaller I_D-V_{DS} slope in the saturation region confirms the strict saturation behavior of the developed TFT. The crucial electrical parameters of the device are calculated from I_D *vs.* V_{GS} plot when its operating in saturation regime ($V_{DS} > V_{GS}$-V_T) and the output current follows the Equation given below:

$$I_{DS} = \frac{W}{2L}\mu_{FE}C_{OX}(V_{GS} - V_T)^2 \tag{6}$$

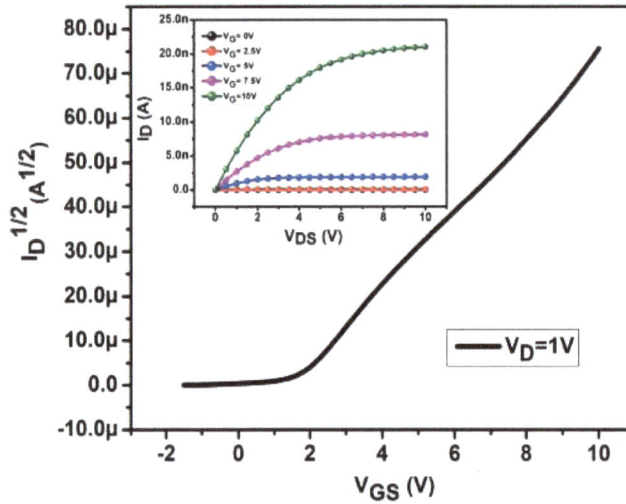

Fig. (4). TFT operating characteristics: $\sqrt{I_D}$ vs. V_{GS} plot. The inset shows output characteristics.

Where L and W are the channel length and channel width between the source and drain electrodes in μm, C_{OX} is the dielectric capacitance/unit area and has a value of 3.45×10^{-8} F/cm^2, μ_{FE} is the field effect mobility in cm^2/Vs, V_T is the threshold voltage in volts. The square root of the drain current in equation **(6)** exhibits linear dependency on gate bias. The transconductance parameter can be evaluated from the slope of $I_D^{1/2}$ *vs.* V_{GS} plot Threshold voltage can be extracted by extrapolating the transfer plot down to the V_{GS} axis. The on-off ratio is the ratio of drain currents when the device is in on and off states, respectively. Sub-threshold slope and on-off ratio are determined from the logarithmic plot of the transfer curve. Sub-threshold slope provides information on the extent of gate voltage required to increase I_{DS} by a factor of 10. The extracted values of the field effect mobility, threshold voltage, transconductance, and sub-threshold slope are 13.68 cm^2/Vs, 1.6 V, 7.87 μs, and 0.73 V/dec, respectively. The on-off ratio of the device is found to be in the order of 4, ensuring satisfactory performance in switching applications. The hopping transport of charge carriers is mainly responsible for the significantly high value of field effect mobility, which exceeds the value of finely-grained ZnO films grown using the PVD process at room temperature [33]. As a result, the effective separation between electrons and holes increases and resistance decreases; hence the sensing surface is expected to provide an amplified response.

RT HYDROGEN SENSING CHARACTERISTICS

For gas sensing measurements, two mass flow controllers (MFCs, Alicat Scientific, Inc. USA) are utilized to pass controlled gas flow and hydrogen concentration to the sensor chamber *via* the gas inlet. A uniform mixture of the synthetic air and target gas H_2 is required to obtain the analyte under study with appropriate concentrations through the static mixer in MFC. A computer-interfaced Keithley SMU 2450 incorporating mechanical simulation is used to monitor the drain current, which serves as the sensing metric on exposure to 500 ppm to 4500 ppm target hydrogen concentrations at room temperature. The Matlab program performs the data acquisition every second. A drain voltage of 1V and a gate voltage of 0.2 V are used to bias the sensor mounted inside the volumetric gas chamber. Initially, synthetic air is introduced inside the chamber for baseline current stabilization. Next, target H_2 is introduced to record the sensing metric. In the end, the sensor returns to the initial baseline current.

TRANSIENT SENSING CHARACTERISTICS AT DIFFERENT H_2 CONCENTRATIONS AND REPEATABILITY ANALYSIS

The drain current-time analysis is performed to investigate the transient response of the fabricated hydrogen sensor. The transient sensing characteristics are recorded at gate voltage (V_{GS}) of 0.2V and drain voltage (V_{DS}) of 1V for H_2 concentrations ranging from 500 ppm to 4500 ppm at room temperature. An elevation trend in drain current characteristics is noticed as the H_2 concentration increases due to the increasing hydrogen dipoles at the Pd-Ti/ZnO interface shown in Fig. (**5a**). The sensor returns to its initial baseline value as synthetic air is purged, indicating its reversible and reusable characteristics. The drain current values obtained are 48.5 µA, 52 µA, 54.6 µA, 57.5 µA, and 60.5 µA, respectively, for 500 ppm, 1500 ppm, 2500 ppm, 3500 ppm, and 4500 ppm, at room temperature. A small drain current drift indicates incomplete hydrogen desorption during the evacuation from the Pd-Ti electrodes [41]. The optimum sensing performance is achieved with a maximum 7.58 meV Schottky barrier height variation at 4500 ppm H_2 at room temperature.

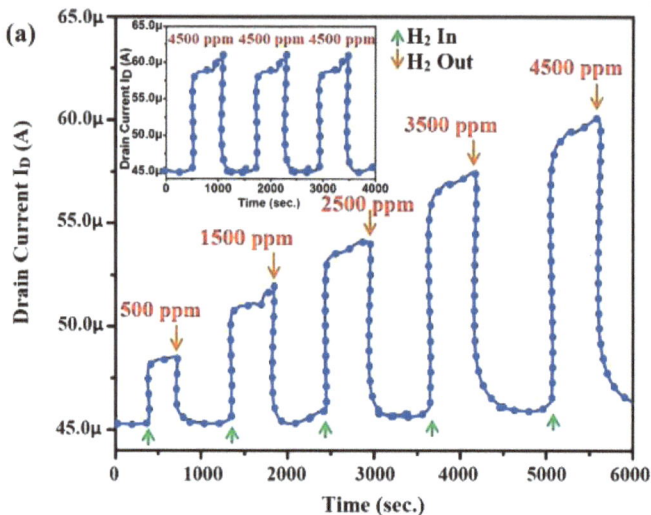

Fig. (5a). Transient response plot for different H_2 concentrations (500 ppm to 4500 ppm) at room temperature. Inset: Dynamic sensing characteristics for three cycles of 4500 ppm H_2 at room temperature.

Repeatability is a critical aspect of ensuring the practicability of a sensor. The consistent response to three successive cycles of H_2 at 4500 ppm illustrated as an inset of Fig. (**5a**) reveals the highly repeatable performance of the developed sensor. The applied drain and gate voltages are kept at 1V and 0.2V, respectively. The baseline current in the air is found at 45.2 μA, whereas the stable sensing current in 4500 ppm H_2/air is 60.5 μA.

VARIATION OF SENSING RESPONSE WITH HYDROGEN CONCENTRATION

The sensing response of the developed sensor can be expressed as [42]:

$$\text{Sensing Response (\%)} = \frac{I_{D(H_2)} - I_{D(air)}}{I_{D(air)}} \times 100 \qquad (7)$$

Where $I_{D(H2)}$ is the drain current value on exposure to H_2 and $I_{D\,(air)}$ is the drain current value when exposed to synthetic air in an ampere. The expected increasing trend in sensing response with increasing H_2 concentration is noticeable in Fig. (**5b**). The response of the device was noticed to increase from 7.3% to 33.8% at increasing hydrogen concentrations from 500 ppm to 4500 ppm at RT. It can be seen that the sensing response variation exhibits linear nature (R^2=0.987) with the variation in H_2 concentration, making it suitable for reliable operation under a wide range of concentrations at room temperature. The surface morphology of the fabricated thin film can probably be one of the reasons for the elevated sensing response in the room temperature. The ultrathin nature of the film and its inherent high surface-to-volume ratio satisfies this reasoning.

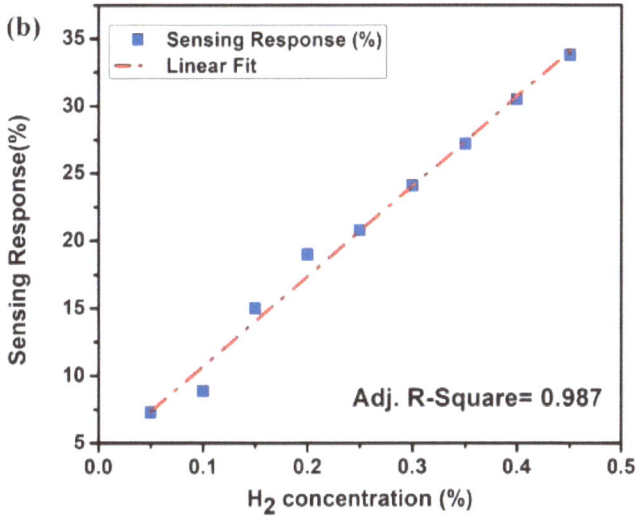

Fig. (5b). Linear fitting of sensing response *vs.* H_2 concentration plot.

RESPONSE-RECOVERY PROPERTIES

The response and recovery times of the sensor have been calculated as the time needed to acquire 90% of the steady-state response on exposure to target gas and the required time to reach 10% of the baseline value on the removal of the target gas at room temperature, respectively. While evaluating the corresponding sensing speed, reasonable response and recovery time values are obtained and recorded at room temperature, as depicted in Table **1**. The increase of drain current under the exposure of H_2 also represents the gas response characteristics of the sensor; when it reaches the saturation point, H_2 flow is stopped to detect its recovery characteristics. Response time values are observed to be reduced from 115 s to 50.3 s, whereas recovery time values are observed to be increasing from 72 s to 130 s with an increase of H_2 concentration from 500 ppm to 4500 ppm. It implies faster diffusion of H_2 and high surface interaction rates with chemisorbed oxygen on oxide semiconductors when exposed to increasing H_2 concentration as Schottky barrier height reduces, including narrow barrier width. Even though drain current values are observed to decrease to their initial values, the data of a little longer recovery time at increased H_2 concentration indicate that hydrogen desorbs slowly from the sensing material surface and metal/semiconductor interface at room temperature.

Table 1. Response-Recovery Parameters.

Concentration (ppm)	500	1000	1500	2000	2500	3000	3500	4000	4500
Response Time (sec.)	115	104	95	89	81	75	68	57	50.3
Recovery Time (sec)	72	80	88	94	101.6	110	116.2	124	130

SELECTIVITY AND REPRODUCIBILITY ANALYSIS

Selectivity is another important figure of merit that defines the responding capability of a sensor toward a specific target gas when exposed to several other gases. The selectivity of the fabricated sensor is tested upon exposure to H_2 as well as three industrially relevant gas species, such as methane, acetone, and nitric oxide, for 500 ppm to 4500 ppm concentrations at room temperature, presented in Fig. (**6**).

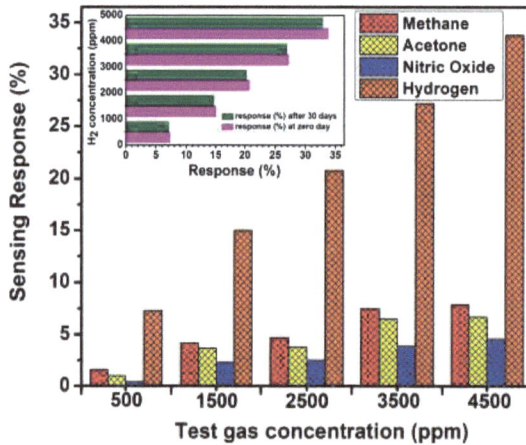

Fig. (6). Selectivity histogram of the sensor toward several gas concentrations. Inset: Stability test of the sensor after a month.

The response of the proposed sensor is observed to be much more dominant for H_2 compared to other tested gas species, rendering superior selectivity of the sensor toward H_2. The responses observed for methane, acetone, nitric oxide, and hydrogen are 1.6%, 1%, 0.4%, and 7.3%, respectively, for 500 ppm minimum detection limit and 7.9%, 6.7%, 4.6%, and 33.8% for 4500 ppm maximum detection limit. The sensing response for 1500 ppm, 2500 ppm, and 3500 ppm H_2 is also the highest in the presence of other interfering gases. It illustrates that the sensing limit for other tested gases is higher than hydrogen at a minimum gas concentration. It is known that the lowest activation energy and smallest molecular size of hydrogen compared to other test gases strengthen sensor surface reactivity, leading to more significant variation in the space charge region of the developed oxide semiconductor-based sensor. The sensing response of the target gases decreases with their increasing activation energy owing to the minimal change in space charge region. Smaller molecular size effectively influences the hydrogen molecules to dissociate swiftly over the Pd-Ti contact, and eventually, fast diffusion of hydrogen occurs to the Pd-Ti/ZnO interface.

The reproducibility or stability of a sensor is its ability to indicate constant response over a long duration. Reproducibility analysis of the developed sensor is performed after a month for different H_2 concentrations at room temperature. Notably, the sensing response is almost consistent over time, as shown in Fig. (**6**) For 500 ppm H_2 concentration, the response is found to decrease a little from 7.3% to 7.1%; for 1500 ppm, it dropped from 15% to 14.5%; for 2500 ppm, it fell from 20.8% to 20.3%; for 3500 ppm it decreased from 27.2% to 26.9% and for 4500 ppm H_2 concentration, response decreased from 33.8% to 32.9% when tested after a month. It ensures a somewhat favorable and stable response of the developed sensor.

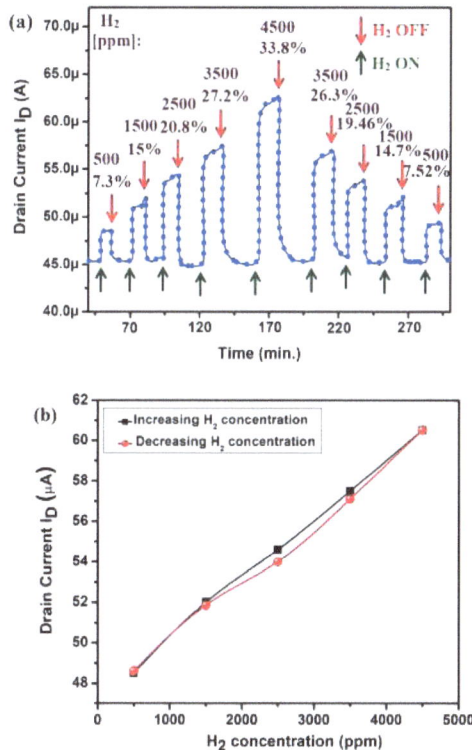

Fig. (7). (**a**) Variations in drain current from low to high concentration of H_2 and *vice versa.* (**b**) Associated hysteresis plot.

HYSTERESIS EFFECT

One of the most challenging factors of highly precise gas sensors is their hysteresis effect. Fig. (**7a**) exhibits the variation in drain current response while approaching low to high concentration values of H_2 and *vice versa, i.e.*, returning to the initial concentration from a high H_2 concentration. The associated hysteresis is plotted in Fig. (**7b**). The following formula is used to determine the hysteresis error and can be expressed as:

$$\delta H(\%) = \pm \frac{\Delta I_{max}}{2F_{FS}} \times 100 \qquad (8)$$

Where ΔI_{max} is the value obtained from the difference in drain currents in ampere (A) when H_2 concentration increases from 500 ppm to 4500 ppm (rise in drain current occurs from 48.5 μA to 60.5 μA) and when it reduces back to 500 ppm from 4500 ppm (drain current value reaches 48.6 μA). Full-scale output F_{FS} represents the difference in maximum drain current values obtained when the maximum (4500 ppm) and minimum input (500 ppm) concentrations are applied. The computed hysteresis error is found as ±2.58%, illustrating the fabricated sensor's superior sensitive and selective nature.

CONCLUSION

Room temperature hydrogen sensing characteristics of a Pd-Ti/ZnO Schottky thin-film transistor has been demonstrated in this chapter. Apart from the hydrogen sensing mechanism of the developed sensor, this chapter also explores detailed fabrication methods, thin film characterization techniques, and electrical characterization of the fabricated device. Experimentally, a minimum and maximum sensing response of 7.3% and 33.8% have been obtained under 500 ppm H_2/air and 4500 ppm H_2/air, respectively. Hence the developed device exhibits a wide concentration range for hydrogen sensing. The developed sensor is highly repeatable when tested under a constant hydrogen concentration. In addition, a decent response time (50.3 s at 4500 ppm H_2) and recovery time (130 s at 4500 ppm H_2) have been achieved. Notably, it demonstrates reasonably good selectivity and reproducibility at room temperature. The hysteresis error of the sensor is observed to be negligible (±2.58%). Such characteristics of the developed sensor make it a potential candidate for its application in room temperature sensing technology for H_2 detection and can open up the opportunity for a wide range of gas sensing applications.

LEARNING OBJECTIVES

By the end of this chapter, readers will be able to:

• Develop a clear understanding of potential nanoelectronic device design and conduct experiments considering relevant standards with practical constraints.

• Develop a step-by-step concrete idea of nanoelectronic device fabrication, material, and device characterization with various tools, techniques, and processes.

• Develop an ability to analyze, interpret, and synthesize data.

• Understand the need to develop gas sensors and their performance-determining factors.

• Develop adaptability to sensing device applications.

MULTIPLE CHOICE QUESTIONS

1. To enhance the performance of a gas sensor, the surface-to-volume ratio:

   a. Must decrease
   b. Must increase
   c. Should remain constant
   d. Should reduce at first, then it must improve.

2. Which of the following is not a property of hydrogen:

   a. Colorless, odorless, and tasteless gas.
   b. Highly flammable and explosive.
   c. Toxic but not combustible.
   d. Eco-friendly next-generation energy carrier.

3. Gas sensor:

   a. Converts the physical or chemical properties of the test gas into an electrical signal.
   b. Converts the electrical signal into chemical or physical quantities.
   c. Both a. and b.
   d. Can't generate any electrical signal, only allows electric current to flow through it.

4. In the presence of hydrogen for the Pd-Ti/ZnO TFT, the depletion region width at the ZnO surface and Schottky barrier height at the metal/semiconductor interface:

a. Increases and decreases, respectively.
b. Decreases and increases, respectively.
c. Both increases.
d. Both decreases.

5. Rougher surface:

a. Reduces the active sites for target gas adsorption.
b. Increases the active sites for target gas adsorption.
c. First increases then decreases the target gas adsorbing sites.
d. None of the above.

6. The Crystallite size of the ZnO thin film is evaluated from the:

a. XRD
b. XPS
c. UV-Vis Spectroscopy
d. All of the above

7. As deposition time increases, ZnO film thickness:

a. Decreases
b. Increases
c. Remains constant
d. First increases, then decreases

8. Strict saturation behavior of the TFT can be confirmed from:

a. Higher I_D-V_{DS} slope in the saturation region.
b. Smaller I_D-V_{DS} slope in the saturation region.
c. Higher I_D-V_{DS} slope in the linear region.
d. Smaller I_D-V_{DS} slope in the linear region.

9. Baseline current stabilization inside the volumetric gas chamber can be achieved through the following:

a. Introducing target gas.
b. Introducing synthetic air.
c. Introducing a uniform mixture of synthetic air and target gas.
d. Maintaining synthetic air flow off before introducing the target gas.

10. According to Joint Committee on Power Diffraction Standards (JCPDS, number: 36-1451), the angular peak of ZnO situates at:

a. 38.28°
b. 34.36°
c. 47.26°
d. 68.51°

11. Faster hydrogen adsorption and slower hydrogen desorption through ZnO and metal/ZnO interface make the response and recovery time:

a. To increase and decrease, respectively.
b. To decrease and increase, respectively.
c. To increase.
d. To decrease.

12. As the activation energy of the target gas increases, the sensing response:

a. Increases
b. Decreases
c. Remain constant
d. First increases, then decreases

13. The capability of a sensor toward a specific target gas, when exposed to several other gases, is termed as:

a. Stability
b. Transient Response
c. Repeatability
d. Selectivity

14. As hydrogen is removed, the sensor returns to its initial baseline value, indicating its:

a. Superior sensing response.
b. Negligible hysteresis error.
c. Reversible and reusable characteristics.
d. Quick response and recovery time.

15. The slight drift of the drain current in the transient response indicates:

a. Incomplete desorption of hydrogen during the evacuation from the Pd-Ti electrodes.
b. Superior selectivity of the sensor.
c. Negligible hysteresis error of the sensor.
d. Reversible and reusable characteristics of the sensor.

ANSWER KEY

1. (b)

2. (c)

3. (a)

4. (d)

5. (b)

6. (a)

7. (b)

8. (b)

9. (b)

10. (b)

11. (b)

12. (b)

13. (d)

14. (c)

15. (a)

ACKNOWLEDGMENTS

The present research is supported at the Centre for Nano Science and Engineering (CeNSE), Indian Institute of Science, Bangalore, funded by the Ministry of Human Resource Development (MHRD), Ministry of Electronics and Information Technology (MeitY), and Nanomission, Department of Science and Technology (DST), Govt. of India.

REFERENCES

[1]   R. Moradi, and K.M. Groth, "Hydrogen storage and delivery: Review of the state of the art technologies and risk and reliability analysis", *Int. J. Hydrogen Energy,* vol. 44, no. 23, pp. 12254-12269, 2019.
[http://dx.doi.org/10.1016/j.ijhydene.2019.03.041]

[2]   W.J. Buttner, M.B. Post, R. Burgess, and C. Rivkin, "An overview of hydrogen safety sensors and requirements", *Int. J. Hydrogen Energy,* vol. 36, no. 3, pp. 2462-2470, 2011.
[http://dx.doi.org/10.1016/j.ijhydene.2010.04.176]

[3]   V. Goltsov, and T.N. Veziroglu, "A step on the road to hydrogen civilization", *Int. J. Hydrogen Energy,* vol. 27, no. 7-8, pp. 719-723, 2002.
[http://dx.doi.org/10.1016/S0360-3199(01)00122-7]

[4]   W. Buttner, R. Burgess, M. Post, and C. Rivkin, "Summary and findings from the NREL/DOE hydrogen sensor workshop (June 8, 2011)," *National Renewable Energy Lab. (NREL), Golden, CO (United States), Tech. Rep.*, 2012.
[http://dx.doi.org/10.2172/1048994]

[5]   S. Ghosh, and L. Rajan, "Room temperature hydrogen sensing investigation of Zinc oxide Schottky thin-film transistors: Dependence on film thickness", *IEEE Trans. Electron Dev.,* vol. 67, no. 12, pp. 5701-5709, 2020.
[http://dx.doi.org/10.1109/TED.2020.3032084]

[6]   G. Korotcenkov, S.D. Han, and J.R. Stetter, "Review of electrochemical hydrogen sensors", *Chem. Rev.,* vol. 109, no. 3, pp. 1402-1433, 2009.
[http://dx.doi.org/10.1021/cr800339k] [PMID: 19222198]

[7]   H. Gu, Z. Wang, and Y. Hu, "Hydrogen gas sensors based on semiconductor oxide nanostructures", *Sensors,* vol. 12, no. 5, pp. 5517-5550, 2012.
[http://dx.doi.org/10.3390/s120505517] [PMID: 22778599]

[8]   C. Wang, L. Yin, L. Zhang, D. Xiang, and R. Gao, "Metal oxide gas sensors: Sensitivity and influencing factors", *Sensors,* vol. 10, no. 3, pp. 2088-2106, 2010.
[http://dx.doi.org/10.3390/s100302088] [PMID: 22294916]

[9]   N.H. Al-Hardan, M.J. Abdullah, and A.A. Aziz, "Sensing mechanism of hydrogen gas sensor based on RF-sputtered ZnO thin films", *Int. J. Hydrogen Energy,* vol. 35, no. 9, pp. 4428-4434, 2010.
[http://dx.doi.org/10.1016/j.ijhydene.2010.02.006]

[10] L. Rajan, C. Periasamy, and V. Sahula, "An in-depth study on electrical and hydrogen sensing characteristics of ZnO thin film with radio frequency sputtered gold schottky contacts", *IEEE Sens. J.,* vol. 19, no. 9, pp. 3232-3239, 2019.
[http://dx.doi.org/10.1109/JSEN.2019.2893025]

[11] S. Basu, and A. Dutta, "Room-temperature hydrogen sensors based on ZnO", *Mater. Chem. Phys.,* vol. 47, no. 1, pp. 93-96, 1997.

[http://dx.doi.org/10.1016/S0254-0584(97)80035-1]

[12]   H.W. Ryu, B.S. Park, S.A. Akbar, W.S. Lee, K.J. Hong, Y.J. Seo, D.C. Shin, J.S. Park, and G.P. Choi, "ZnO sol–gel derived porous film for CO gas sensing", *Sens. Actuators B Chem.,* vol. 96, no. 3, pp. 717-722, 2003.
[http://dx.doi.org/10.1016/j.snb.2003.07.010]

[13]   A.B. Yadav, C. Periasamy, P. Chakrabarti, and S. Jit, "Hydrogen gas sensing properties of Pd/ nanocrystalline ZnO thin films based Schottky contacts at room temperature", *Adv. Sci. Eng. Med.,* vol. 5, no. 2, pp. 112-118, 2013.
[http://dx.doi.org/10.1166/asem.2013.1199]

[14]   M.C. Carotta, A. Cervi, V. di Natale, S. Gherardi, A. Giberti, V. Guidi, D. Puzzovio, B. Vendemiati, G. Martinelli, M. Sacerdoti, D. Calestani, A. Zappettini, M. Zha, and L. Zanotti, "ZnO gas sensors: A comparison between nanoparticles and nanotetrapods-based thick films", *Sens. Actuators B Chem.,* vol. 137, no. 1, pp. 164-169, 2009.
[http://dx.doi.org/10.1016/j.snb.2008.11.007]

[15]   I. Eisele, T. Doll, and M. Burgmair, "Low power gas detection with FET sensors", *Sens. Actuators B Chem.,* vol. 78, no. 1-3, pp. 19-25, 2001.
[http://dx.doi.org/10.1016/S0925-4005(01)00786-9]

[16]   I. Rýger, G. Vanko, T. Lalinský, Š. Haščík, A. Benčúrová, P. Nemec, R. Andok, and M. Tomáška, "GaN/SiC based surface acoustic wave structures for hydrogen sensors with enhanced sensitivity", *Sens. Actuators A Phys.,* vol. 227, pp. 55-62, 2015.
[http://dx.doi.org/10.1016/j.sna.2015.02.041]

[17]   T. Usagawa, K. Ueda, A. Nambu, A. Yoneyama, Y. Kikuchi, and A. Watanabe, "Pt–Ti–O gate silicon–metal–insulator–semiconductor field-effect transistor hydrogen gas sensors in harsh environments", *Jpn. J. Appl. Phys.,* vol. 55, no. 6, p. 067102, 2016.
[http://dx.doi.org/10.7567/JJAP.55.067102]

[18]   Y. Halfaya, C. Bishop, A. Soltani, S. Sundaram, V. Aubry, P. Voss, J.P. Salvestrini, and A. Ougazzaden, "Investigation of the performance of HEMT-based NO, NO_2 and NH_3 exhaust gas sensors for automotive antipollution systems", *Sensors,* vol. 16, no. 3, p. 273, 2016.
[http://dx.doi.org/10.3390/s16030273] [PMID: 26907298]

[19]   H.S. Al-Salman, and M.J. Abdullah, "Fabrication and characterization of ZnO thin film for hydrogen gas sensing prepared by RF-magnetron sputtering", *Measurement,* vol. 46, no. 5, pp. 1698-1703, 2013.
[http://dx.doi.org/10.1016/j.measurement.2013.01.004]

[20]   L. Rajan, C. Periasamy, and V. Sahula, "Comprehensive study on electrical and hydrogen sensing characteristics of Pt/ZnO nanocrystalline thin film Schottky diodes on n-Si substrates using R.F. Sputtering", *IEEE Trans. Nanotechnol.,* vol. 15, no. 2, pp. 201-208, 2016.
[http://dx.doi.org/10.1109/TNANO.2015.2513102]

[21]   T. Hübert, L. Boon-Brett, G. Black, and U. Banach, "Hydrogen sensors : A review", *Sens. Actuators B Chem.,* vol. 157, no. 2, pp. 329-352, 2011.
[http://dx.doi.org/10.1016/j.snb.2011.04.070]

[22]   A. Pundt, "Hydrogen in nano-sized metals", *Adv. Eng. Mater.,* vol. 6, no. 12, pp. 11-21, 2004.
[http://dx.doi.org/10.1002/adem.200300557]

[23]   B.D. Adams, and A. Chen, "The role of palladium in a hydrogen economy", *Mater. Today,* vol. 14, no. 6, pp. 282-289, 2011.
[http://dx.doi.org/10.1016/S1369-7021(11)70143-2]

[24]   J. RaviPrakash, A.H. McDaniel, M. Horn, L. Pilione, P. Sunal, R. Messier, R.T. McGrath, and F.K. Schweighardt, "Hydrogen sensors: Role of palladium thin film morphology", *Sens. Actuators B Chem.,* vol. 120, no. 2, pp. 439-446, 2007.
[http://dx.doi.org/10.1016/j.snb.2006.02.050]

[25]   J. Li, R. Fan, H. Hu, and C. Yao, "Hydrogen sensing performance of silica microfiber elaborated with Pd nanoparticles", *Mater. Lett.,* vol. 212, pp. 211-213, 2018.
[http://dx.doi.org/10.1016/j.matlet.2017.10.095]

[26]   M. Dornheim, "Thermodynamics of metal hydrides: tailoring reaction enthalpies of hydrogen storage materials," In *Thermodynamics- Interaction Studies-Solids, Liquids, and Gases. Intech Open*, 2011.
[http://dx.doi.org/10.5772/21662]

[27]   M. Segard, "Ageing of palladium tritide: mechanical characterization, helium state, and modeling," *Ecole Nationale Superieure des Mines de Saint-Etienne (France), Tech. Rep.*, 2010.

[28]   H. von Wenckstern, G. Biehne, R.A. Rahman, H. Hochmuth, M. Lorenz, and M. Grundmann, "Mean barrier height of Pd Schottky contacts on ZnO thin films", *Appl. Phys. Lett.,* vol. 88, no. 9, p. 092102, 2006.
[http://dx.doi.org/10.1063/1.2180445]

[29]   X.Q. Zeng, M.L. Latimer, Z.L. Xiao, S. Panuganti, U. Welp, W.K. Kwok, and T. Xu, "Hydrogen gas sensing with networks of ultrasmall palladium nanowires formed on filtration membranes", *Nano Lett.,* vol. 11, no. 1, pp. 262-268, 2011.
[http://dx.doi.org/10.1021/nl103682s] [PMID: 21114299]

[30]   S. Han, X. Zhuang, W. Shi, X. Yang, L. Li, and J. Yu, "Poly(3-hexylthiophene)/polystyrene (P3HT/PS) blends based organic field-effect transistor ammonia gas sensor", *Sens. Actuators B Chem.,* vol. 225, pp. 10-15, 2016.
[http://dx.doi.org/10.1016/j.snb.2015.11.005]

[31]   E. Comini, "Metal oxide nano-crystals for gas sensing", *Anal. Chim. Acta,* vol. 568, no. 1-2, pp. 28-40, 2006.
[http://dx.doi.org/10.1016/j.aca.2005.10.069] [PMID: 17761243]

[32]   Y. Hong, C-H. Kim, J. Shin, K.Y. Kim, J.S. Kim, C.S. Hwang, and J-H. Lee, "Highly selective ZnO gas sensor based on MOSFET having a horizontal floating-gate", *Sens. Actuators B Chem.,* vol. 232, pp. 653-659, 2016.
[http://dx.doi.org/10.1016/j.snb.2016.04.010]

[33]   A.M. Ma, M. Gupta, F.R. Chowdhury, M. Shen, K. Bothe, K. Shankar, Y. Tsui, and D.W. Barlage, "Zinc oxide thin film transistors with Schottky source barriers", *Solid-State Electron.,* vol. 76, pp. 104-108, 2012.
[http://dx.doi.org/10.1016/j.sse.2012.05.005]

[34]   S. Ghosh, and L. Rajan, "Zinc oxide thin-film transistor with catalytic electrodes for hydrogen sensing at room temperature", *IEEE Trans. Nanotechnol.,* vol. 20, pp. 303-310, 2021.
[http://dx.doi.org/10.1109/TNANO.2021.3068994]

[35]   P.V. Necliudov, M.S. Shur, D.J. Gundlach, and T.N. Jackson, "Contact resistance extraction in pentacene thin film transistors", *Solid-State Electron.,* vol. 47, no. 2, pp. 259-262, 2003.
[http://dx.doi.org/10.1016/S0038-1101(02)00204-6]

[36]   I. Lundström, M. Armgarth, A. Spetz, and F. Winquist, "Gas sensors based on catalytic metal-gate field-effect devices", *Sens. Actuators,* vol. 10, no. 3-4, pp. 399-421, 1986.
[http://dx.doi.org/10.1016/0250-6874(86)80056-7]

[37]   W.P. Kang, Y. Gurbuz, J.L. Davidson, and D.V. Kerns, "A new hydrogen sensor using a polycrystalline diamond-based schottky diode", *J. Electrochem. Soc.,* vol. 141, no. 8, pp. 2231-2234, 1994.
[http://dx.doi.org/10.1149/1.2055094]

[38]   C.H. Chang, K.W. Lin, H.H. Lu, R.C. Liu, and W.C. Liu, "Hydrogen sensing performance of a Pd/HfO2/GaOx/GaN based metal-oxide-semiconductor type Schottky diode", *Int. J. Hydrogen Energy,* vol. 43, no. 42, pp. 19816-19824, 2018.
[http://dx.doi.org/10.1016/j.ijhydene.2018.08.213]

[39]   D. Somvanshi, and S. Jit, "Mean barrier height and Richardson constant for Pd/ZnO thin film-based Schottky diodes grown on n-Si substrates by thermal evaporation method", *IEEE Electron Device Lett.,* vol. 34, no. 10, pp. 1238-1240, 2013.
[http://dx.doi.org/10.1109/LED.2013.2278738]

[40]   J. Zhang, X. Liu, G. Neri, and N. Pinna, "Nanostructured materials for room-temperature gas sensors", *Adv. Mater.,* vol. 28, no. 5, pp. 795-831, 2016.
[http://dx.doi.org/10.1002/adma.201503825] [PMID: 26662346]

[41]   T.L. Poteat, and B. Lalevic, "Pd-MOS hydrogen and hydrocarbon sensor device", *IEEE Electron Device Lett.,* vol. 2, no. 4, pp. 82-84, 1981.
[http://dx.doi.org/10.1109/EDL.1981.25349]

[42]   S. Ghosh, and L. Rajan, "Room temperature hydrogen sensor using schottky contacted zinc oxide thin-film transistor: A comprehensive investigation", *IEEE Trans. Electron Dev.,* vol. 68, no. 9, pp. 4637-4643, 2021.
[http://dx.doi.org/10.1109/TED.2021.3097013]

CHAPTER 8

FinFET Advancements and Challenges: A State-of-the-Art Review

Rahul Ghosh[1], Tanmoy Majumder[1], Abhishek Bhattacharjee[1,*] and Rupanjal Debbarma[1]

[1] *Department of ECE, Tripura Institute of Technology, Narsingarh, Agartala-799015, India*

Abstract: A review of the electrical and physical characteristics of FinFETs is presented here. This work focuses on the latest structures of FinFET according to its classifications and three-dimensional schematics. Through studying the output I-V characteristics, the transfer characteristics, and the subthreshold current in the FinFET channel, the electrical characteristics of FinFETs have been analyzed. Considerations were made of coulomb, phonon, and surface roughness scattering to examine effective charge carrier mobility in the FinFET channel. Lastly, in this chapter, the impact of the Fin layer shape on device performance is studied.

Keywords: Body-tied FinFETs, Drain-induced barrier lowering (DIBL), Electromigration, Fin-height, FinFETs, Matthiesen's rule, Phonon scattering, Short channel effect (SCE), Sub-threshold slope, Spacer length.

INTRODUCTION

The FinFET structure is multigated and non-planar. With its back and front gates, the effect of the shorter channel length can be better controlled. Therefore, double-gated devices are suitable for low-power applications, since they enable a substantial reduction of reserve power while enhancing efficiency. Condensed channel impacts result from electrons floating in the direct and changing voltage caused by the contraction of channel lengths. SoI gadgets have a meager silicon film that prevents the off-state current (I_{OFF}) from flowing. It is necessary that the thickness of the silicon film is less than one-quarter of the length of the channel.

If we take HEMT (High Electron Mobility Transistor) as a counterpart of FinFETs, gate leakage is an important factor that influences their quality and presentation. By using a Gate oxide, MOS-HEMTs (Metal Oxide Semiconductors

* **Corresponding author Abhishek Bhattacharjee:** Department of ECE, Tripura Institute of Technology, Narsingarh, Agartala-799015, India; E-mail: abhishek89.187@gmail.com

Gopal Rawat & Aniruddh Bahadur Yadav (Eds.)

High Electron Mobility Transistors) can be made with improved Gate contact shape, a lower Gate current and a higher channel current. However, this results in a somewhat lower transconductance due to the larger Gate to channel detachment. A coordinated circuit's transistor count duplicates like clockwork, as Moore's law would have it. In the past, CMOS technology has been downsized using a variety of methods. This can be seen in multi-gate transistors. In a single gadget, MOSFETs are multi-gate transistors with more than one gate. As long as the cutoff and productivity are adequate, FinFET can be a fruitful apparatus. Current flow is reduced when high-κ dielectrics are used [1 - 15]. A MOSFET double gate device requires a different assembly method than a MOSFET single gate device, as shown in Fig. (**1**), which illustrates a schematic view of FinFET with geometric boundaries, such as Gate length (L_g), gate oxide thickness (t_{ox}) *etc.*

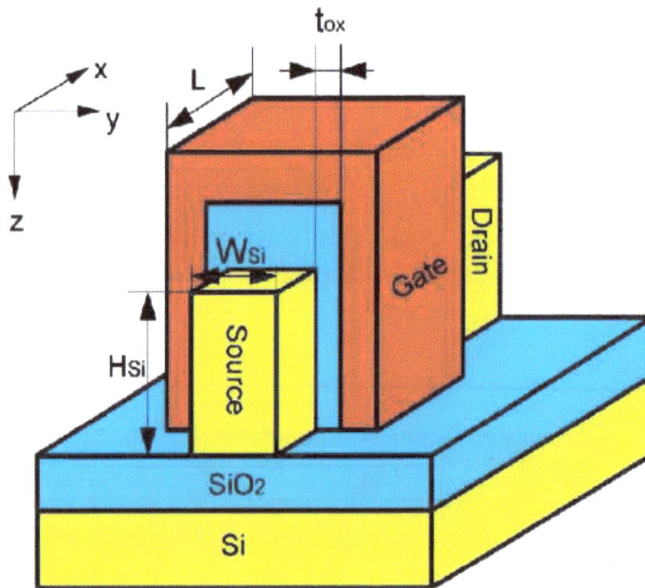

Fig. (1). Planar dimensions of FinFET [1].

On the nanometer scale, it may well be assumed that metal gates and high-κ dielectrics perform reasonably well. Due to their lower parasitic capacitances and robustness against random dopant behavior, double-gate FETs and tri-gate FETs are more appropriate for multi-gate devices [1 - 3]. Despite the reduced fringe capacitances, tri-gate FETs are complicated to fabricate. With less power consumption, immunity to SCE's, lesser area requirements, and a faster rate of operation, FinFETs are devices of choice in this era. There are several works available in the literature regarding the implementation of digital and analog circuits using symmetric and asymmetric FinFET structures. Although FinFETs cover most of the design hierarchy, many circuit-level implementations remain untouched. As a type of Multi-gate Field Effect Transistor (MGFET), FinFETs have been granted as one of the best devices for substituting bulk MOSFETs because of their improved slope, better stability, higher on-off current (I_{ON}/I_{OFF}) ratios, superior short-channel performance, and small intrinsic gate capacitances. In terms of SC characteristics, FinFET transistors with double gate (DG FinFET) and tri-gate gates have shown standard performance. These pros can be attributed to (i) thin Si-films, (ii) lighter doping of channels and (iii) double gates for better control of channels. Researchers can use this paper to evaluate FinFET devices and understand what changes are likely to happen in the next few years based on an in-depth understanding of FinFET technology and technical issues. Also relevant to the book chapter are the failure modes and reliability of FinFETs. For several generations into the coming future, High-Performance logic will continue to use FinFET architecture. The circuit design world is rapidly maturing as new materials are being introduced, and new devices like FinFETs are bringing significant changes to the circuit design industry.

ANALYSIS OF PHYSICAL ASPECTS OF FINFET

It was mostly in the later half of the 1990s and early 2000s that FinFETs were being made on SOI wafers, or basically SOI MOSFETs. It was mostly on SOI substrates that double-gate transistors were first demonstrated to overcome the short channel effects (SCE). Due to their shallow trench isolation (STI) process and absence of leakage paths adjacent to the junction depth, these devices are renowned for their easy fabrication and excellent scalability. The doping, Si film thickness, and conditions of bias of floating body SOI devices may affect floating body problems. Compared to bulk-Si wafers, SOI wafers are more costly and have a higher defect density [4 - 6]. It is important to note that the thick buried oxide in SOI FETs has a very low heat transfer rate, preventing the generated heat of the channel from dissipating into the substrate region. With an SOI FinFET, one can also switch from a conventional four-terminal (4-T) MOSFET to a three-terminal (body floating) device, as opposed to a conventional planar MOSFET. As a result, the three-terminal characteristics narrow the operating windows. Due to this, four-terminal FinFETs, which are connected directly to Si substrates, would be more appropriate. A bulk-Si FinFET with four terminals is called a body-tied FinFET or preferably a bulk FinFET. When they were initially reported, body-tied FinFETs were termed omega (Ω) MOSFETs because the cross-section of the body was similar to the Greek letter (Ω). When reported in 2002 *IEDM, F-L Yang et al.* coined their MOSFET as the *omega FET* since the gate structure was similar to the Greek letter Ω. Body-tied FinFETs were therefore called bulk FinFETs to separate them from SOI FinFETs and omega FETs. The fin bodies of bulk FinFETs and SOI FinFETs are floated separately and anchored to the substrate, as shown in Fig. (**2**). The three-dimensional schematics of a bulk Fin-FET are shown in Fig. (**3**). The gate electrode is represented by the letter "G". In comparison to three terminal FETs, bulk FinFETs are used more by IC designers, and their steps of fabrication are similar to those of conventional planar-channel (or 2-D) CMOS.

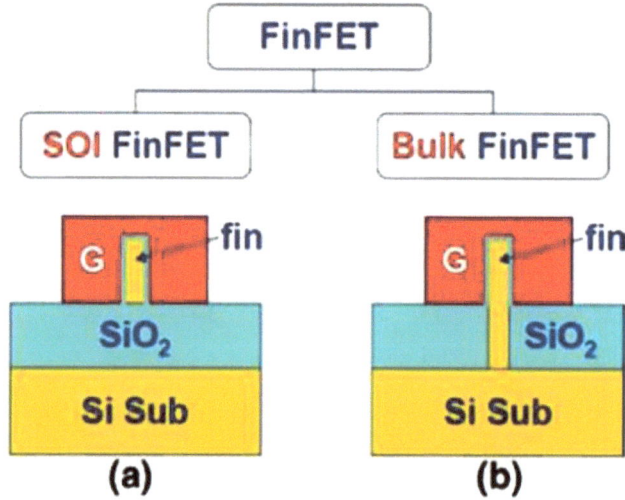

Fig. (2). Bulk & SOI FinFETS Structures [2].

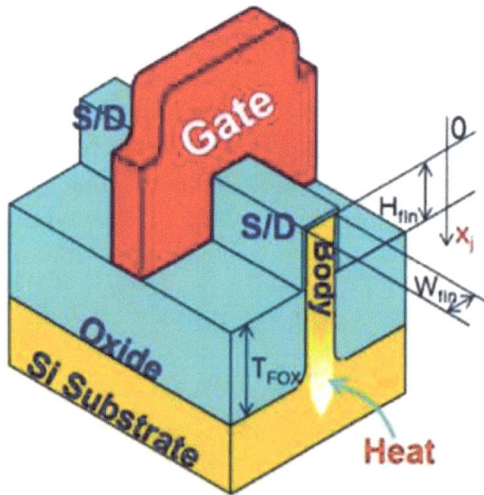

Fig. (3). Three-dimensionalschematic of bulkFinFET [2].

THE CHARACTERISTICS OF FINFETS FROM AN ELECTRICAL PERSPECTIVE

The application of the effective electric field is described as follows:

$$I_{ds} = 2\mu C_{ox}\frac{W}{L}\left(v_g - v_t - \frac{v_{ds}}{2}\right)v_{ds} \tag{1}$$

Unlike bulk planar MOSFETs, while discussing the device physics of FinFET's; we have to take into account multiple scattering mechanisms which exchange momentum between carriers and semiconductors to maintain mobility. Surface roughness and interface-trapped charges are major factors contributing to scattering mechanisms in semiconductor crystals as a result of lattice vibrations, ionized impurity atoms, and defects related to interfaces.

(a) Expression for coulomb scattering:

$$\mu_{col} = \mu_0^{ph}\left(1 + \frac{E_{eff}}{E_c^{col}}\right)^{v^{col}} \tag{2}$$

(b) Expression for phonon scattering:

$$\mu_{ph} = \frac{\mu_0^{ph}}{\left(1 + \frac{E_{eff}}{E_c^{ph}}\right)^{vph}} \tag{3}$$

(c) Expression for surface roughness scattering:

$$\mu_{sr} = \mu_0^{sr}\left(\frac{E_{eff}}{E_{effo}}\right)^{\gamma} \tag{4}$$

(d) Matthiesen's rule can be used to combine the three factors that contribute to total mobility:

$$\frac{1}{\mu_{eff}} = \frac{1}{\mu_{ph}} + \frac{1}{\mu_{col}} + \frac{1}{\mu_{sr}} \tag{5}$$

In the above equation, μ_{eff} represents the total effective mobility, while the terms on the right side describe the other factors of contribution. The above equation shows how the mobility of the inversion layer is influenced by the electric field (effective); note that the rule of Matthiessen is only an approximation and is valid in the event that there is no interdependence between scattering probabilities.

In the case of FinFETs, the threshold voltage (V_{th}) is given by:

$$V_{th} = \phi + n\frac{kT}{q}\ln\left(\frac{2C_{ox}kT}{q^2 n_i t_{si}}\right) + \frac{h^2\pi^2}{2m_{ds}\,W_{si}{}^2} \tag{6}$$

And electric field can be described by the following equation:

$$E_{eff} = \frac{1}{2}\frac{q}{\varepsilon}\left(\frac{N_{inv}}{2} + N_{sub} \times t_{si}\right) \tag{7}$$

Numerous scattering mechanisms are responsible for the exchange of momentum between carriers and semiconductors, and the scattering mechanisms can be explained by crystal defects of semiconductors, such as lattice vibrations and ionized impurities. Several renowned studies have been conducted on the mobility of the inversion channel log FinFETs.

ANALYSIS OF TRANSFER CHARACTERISTICS

Fig. (4) shows the Transfer characteristics with phonon scattering & surface roughness, while Figs. (5 & 6) show its transfer characteristics for a 10nm channel length, a Fin layer width of 150nm and a thickness of 30nm. Symbols depict experimental results, and solid lines show simulation results. Phonon scattering occurs when carriers interact with lattice vibrations. When the temperature rises, carrier-phonon interactions become more amplified, lowering phonon scattering mobility. Detached surfaces of SiO_2 display strong dependencies on effective fields as a result of surface roughness scattering. Surface roughness contributes to mobility degradation with strong fields due to strong fields pulling carriers toward the surface. Inversion of light is mainly dominated by phonon and Cloumb scattering at room temperature (300K). Phonon scattering and surface roughness dominate heavy inversion. Phonon scattering-induced device ON- current is reduced to 3360 µA/µm by using the relaxation time approximation. When phonon and surface roughness scattering are considered, the ON-current is further reduced to 3175 µA/µm, which is a reduction of approximately 10%. This study simulated the surface roughness of the interface of Si/SiO_2 along the channel region, including the regions of underlap between the gate and source [7]. In this study, a noncalibrated scattering model is applied to low-field mobility data. Additionally, the parasitic source/drain series resistance effect is not included in simulations. It would further reduce the ON-current if such a parasitic series resistance were included. ITRS-projected saturation threshold voltage is 192 mV, which is 140 mV based on simulations. To determine the threshold voltage, the drain current must reach 50 A/m (or 500 nA Wg/Lg, where Wg is the gate height [7]). For the considered optimized FinFET device, 82 mV/dec of subthreshold swing at 0.8 V and 46 mV/V of DIBL at 50 A/m were calculated.

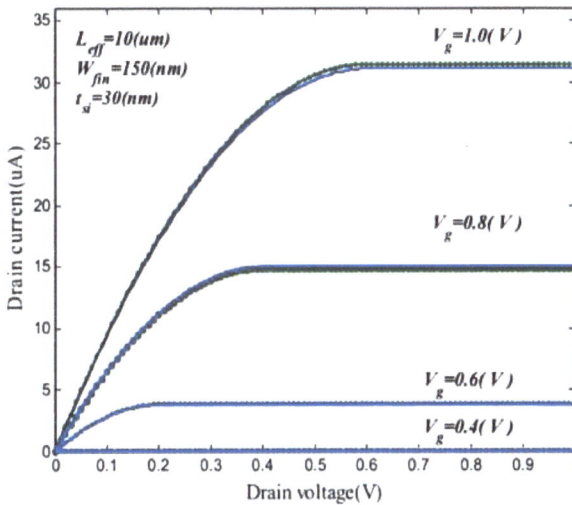

Fig. (4). Drain Characteristics [3].

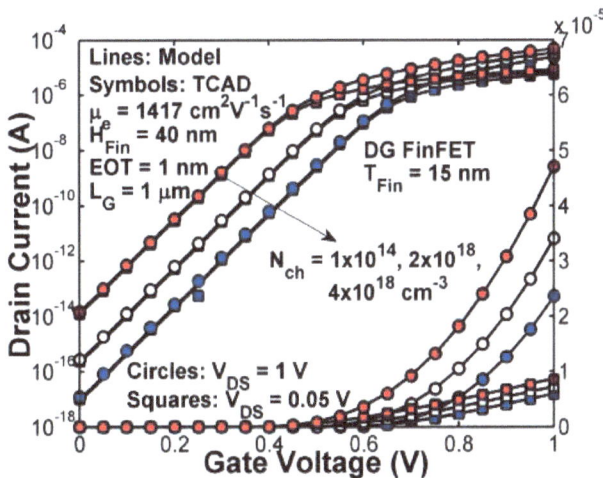

Fig. (5). Drain current vs gate voltage [4].

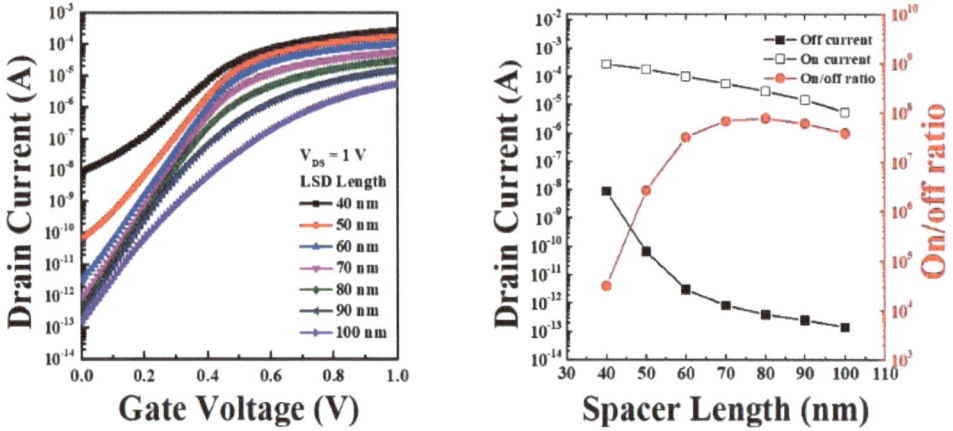

Fig. (6). I_d-V_g curve and on/off current variation with spacer length for N-type bulk FinFET [5].

In Fig. (**4**), as the drain voltage rises, there is an increase in drain current as well, until the pinch-off voltage is reached, at which point drain voltage and drain current are not related. In the presence of the drain-source voltage, the drain current increases. In the beginning, it increases linearly; then, it enters a saturation stage.

In contrast to Fig. (**5**), the channel area is A_{ch}, the amount of doping is N_{ch}, W is the channel width, and C ins is the insulator capacitance in units of length. As a result of solving the Poisson-carrier transport equation, the normalized drain current is represented by the following formula:

$$i_{DS} = \left[\frac{q_m^2}{2} - 2q_m - q_H \ln\left(1 - \frac{q_m}{q_H}\right)\right]\Bigg|_{q_{m,S}}^{q_{m,D}} \qquad (8)$$

It is important to note that FinFETs can be modeled with only four parameters. For example, DG FinFETs with simple cross-sections can be modeled accurately for various channel doping concentrations using the following parameters: A_{ch}, N_{ch}, W, and C_{ins}.

As the spacer length changes in an N-type FinFET, the on/off current changes accordingly, which can be seen in Fig. (**6**). The source and drain resistances increase with increasing spacer length. On-state current is degraded due to a rise in source resistance. In order to minimize power consumption, the spacer should be as small as possible. The leakage current increases when the spacer length is reduced below 60 nm, resulting in a rapid decrease in the effective channel length.

In this case, the on/off ratio is 7.73 x 10^{7}, while the spacer length is 80 nm. In addition to the increase in the spacer length, the increase in source resistance contributes to the lowering of the on-state current. DIBL and SS changes with spacer length are shown in Fig. (7). The effective channel length increases with increasing spacer length.Furthermore, at the channel-drain junction, a longer space-charge region is also formed, decreasing the DIBL. Conversely, as the electric field size as the channel volume rises, the problem of an increase in SS arises due to the decrease in the electric field from the drain to the channel [14, 15]. The lowest SS value of 64.29 mV/V is found when the spacer length is 90 nm, as shown in Fig. (8).

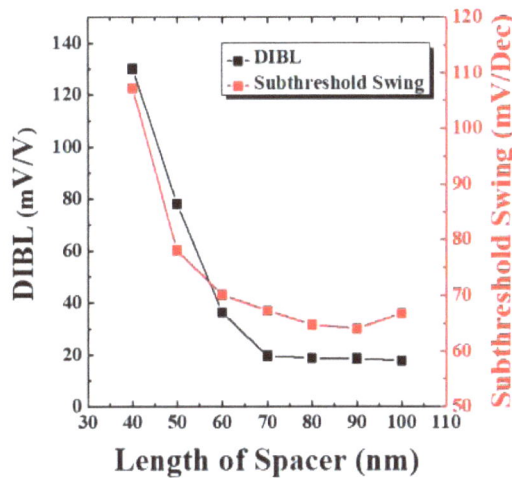

Fig. (7). Comparison of DIBL and S/S for n-type bulk FinFET according to spacer length [5].

LSD (nm)	On/Off Ration	DIBL (mV/V)	SS (mV/dec)
40	3.18×10^5	130.70	107.24
50	2.67×10^6	78.19	78.08
60	3.20×10^7	36.40	70.17
70	6.91×10^7	19.47	67.30
80	7.73×10^7	18.68	64.79
90	6.05×10^7	18.41	64.29
100	3.88×10^7	17.64	66.82

Fig. (8). Electrical properties of bulk FinFET w.r.t spacer length [5].

Device reliability is degraded by hot carriers produced by impact ionization [16 - 46]. When hot carriers are capable of passing through the gate insulator, they negatively impact the interface of Si-SiO$_2$, resulting in significant performance degradation of transistors. It is generally determined by examining the substrate current and how much device damage is caused by hot carriers. A change in spacer length will cause a change in substrate current. Due to impact ionization, the substrate current rises rapidly when the spacer length falls under 60 nm. As a result of impact ionization, hot carriers are generated that are created by the lattice temperature increasing, further speeding up the deterioration of the device. In relation to the change in spacer length, the substrate current correlates with the lattice temperature. It occurred when the p-doped and n-doped regions met at the junction, resulting in impact ionization. If a positive bias is given to the n-doped region (*i.e.*, drain region), the energy band in the n-type material will be lower as compared to the steady state. In the n-type region, the energy band will be lower when the bias is more positive. When the bias increases, electrons get attracted to the p-type region, where they are minority carriers.Upon colliding with the other ions, the incoming electrons are accelerated by the positive bias. A pair of electrons and holes (EHP) is created as a result [22 - 25]. A gate current is formed when the holes created at this time flow partly to the gate and pass through the substrate. As a result, the substrate current will increase when the impact ionization increases.

Fig. (**9**) shows the evolution of the transconductance of a FinFET device with different spacer lengths after channel hot electron injection. Drain and source voltages are ranged from 1.2 V to 2.0 V, while gate voltages are kept the same. A stress application time of 3800 seconds was considered.

A small drain voltage leads to significant degradation of the transconductance when the spacer length is short. According to Fig. (**10**), at large drain voltages, the deterioration increases with increasing spacer length. As we already have seen that even at low drain voltages, a large electric field can result in impact ionization when the spacer length is 40 nm. As a result, the deterioration rate is very rapid. High drain voltages, however, can lead to impact ionization deterioration in FinFETs even with 100-nm spacers.The electron interface trap area widens at high voltages, which explains why more deterioration occurs. A clear picture of this trend can be seen in the devices lifetime with HCE degradation from (Fig. **11**). A device's lifetime is calculated when its transconductance deteriorates by 10% over its lifetime [30 - 33]. From this experiment, it has been verified that, if the spacer is shorter, the degradation becomes easier, even at low drain biases. It is found that the voltage which fulfils the lifetime condition of 10 year for a 40 nm spacer is 1.11 V; for an 80 nm spacer it is 1.18 V; and for a 120 nm spacer, it is 1.32 V. As the spacer length increases, the driving voltage decreases, so the effective life should decrease when compared to an increase in applied voltage for same performance.

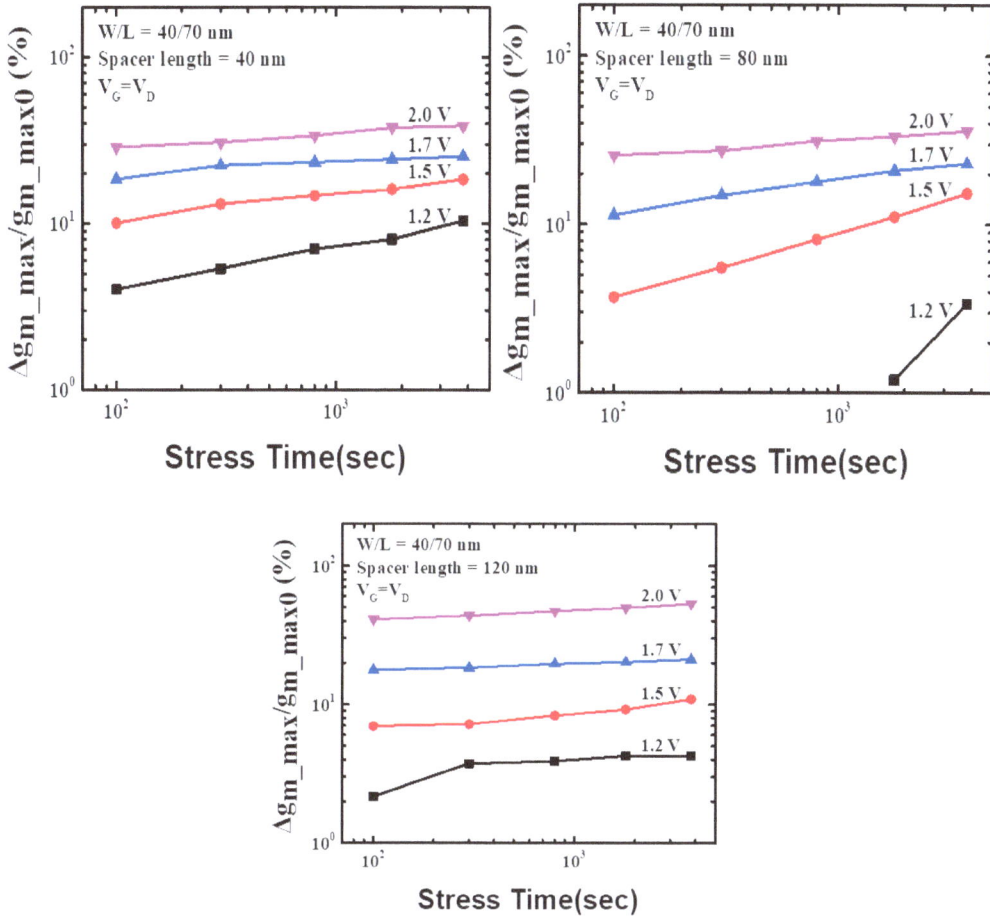

Fig. (9). Degradation of transconductance for different lengths of spacer under the stress conditions of LSD=40nm, LSD=80nm and LSD=120nm [5].

According to the change in spacer-length, Fig. (**11**) shows the change in substrate current. Shortening the spacer length below 60 nm causes impact ionization to increase the substrate current rapidly. In addition to accelerating the degradation of the device, the hot carriers created by impact ionization also increase the lattice temperature. Furthermore, the relationship between the substrate current and the lattice temperature changes with the length of the spacer is also shown. In order for that to occur, the N and P-doped regions must interact with one another by impact ionization.

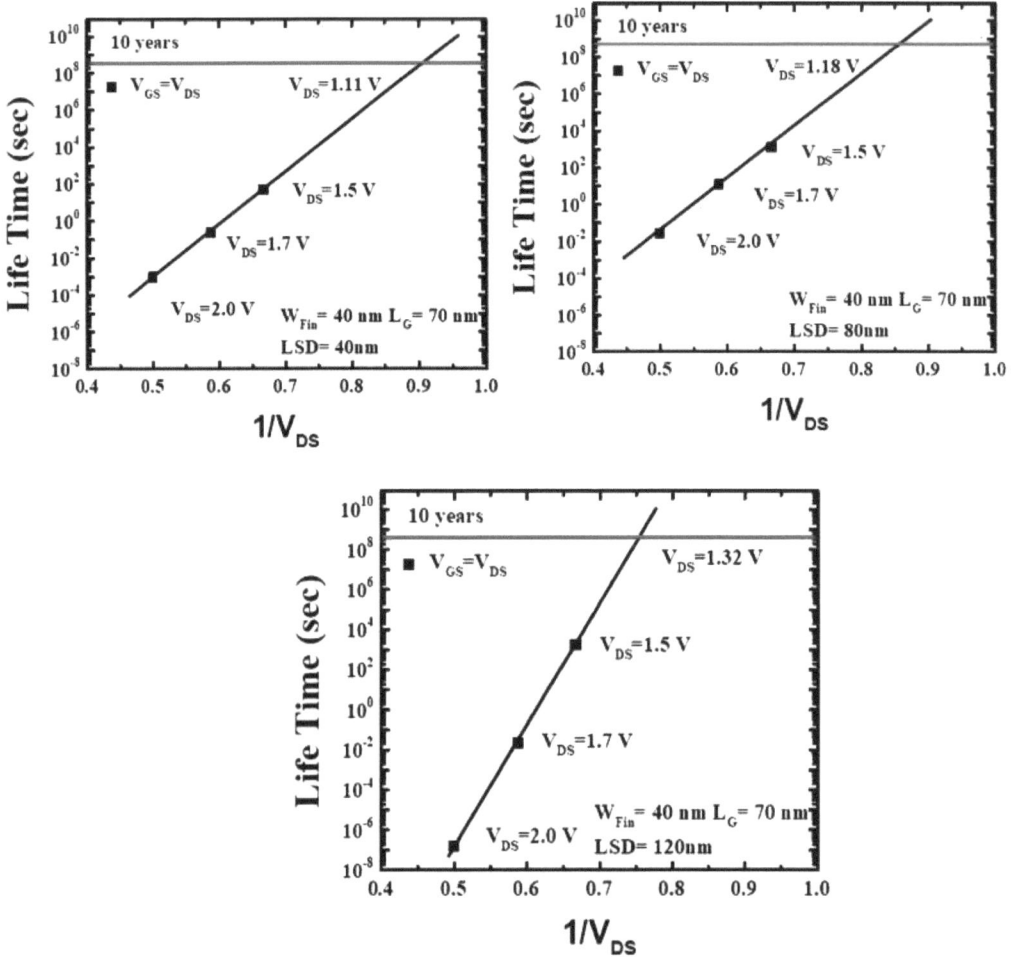

Fig. (10). Device lifetime for different stress bias conditions with different spacer lengths under the stress conditions obtained by LSD=40nm LSD=80nm LSD= 120nm [5].

NEW TRENDS IN FINFET STRUCTURES & OTHER LAYOUTS

Electronic devices have been scaled down from 320nm to 45nm over the past three decades due to the electronic revolution. Since CMOS technology evolved from micron devices to submicron devices, this scale-down process has major challenges, such as achieving low costs and high-performance devices. In addition, short channel effects will cause planar CMOS device OFF state leakage current to rise, so bulk CMOS devices will not adhere to the industry specifications to their lower transistor drive currents. The scaling factor also takes gate oxide thickness into account. In addition to those drawbacks, we also have the basic issue of short-channel effects, which increases the leakage current in bulk CMOS devices. Because of its lower transistor on current, bulk CMOS devices do not fulfill the specifications of industries. When considering the scaling factor, gate oxide thickness is also important.

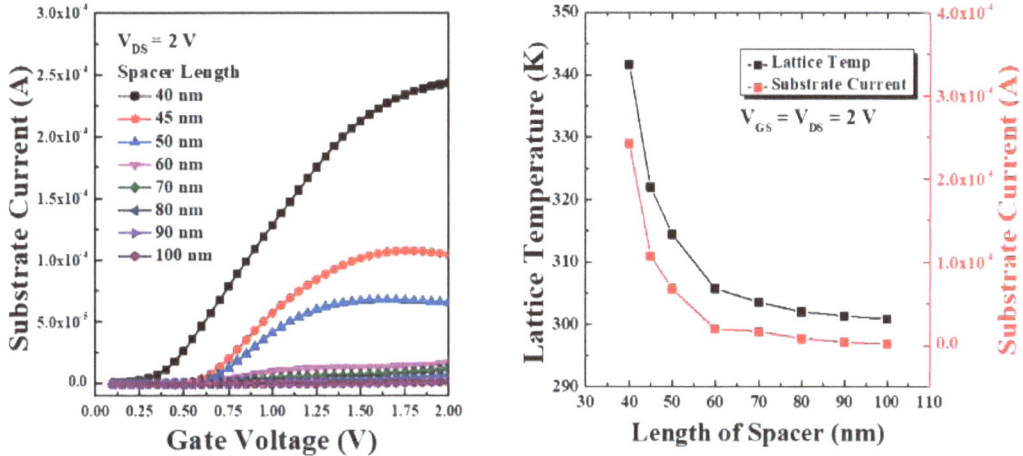

Fig. (11). Substrate current *versus* gate voltage characteristics and lattice temperature of N-type bulk FinFET for various spacer length [5].

As a solution to these problems, researchers tried high permittivity gate dielectrics (high-κ), keeping the sheet resistance low, which resulted in a high driving current. Consequently, CMOS devices gradually evolved to ultra-thin body transistors as the technology advanced. Thin silicon on the insulator reduces subsurface leakage paths in ultra-thin body transistors. There is, however, a very high parasitic resistance associated with thin-body MOS transistors. A comparison between bulk type Si MOS transistors and ultra-thin MOS transistors is shown in Figs. (**12 & 13**). Ultra-thin body MOS transistors have the least effective gate control.

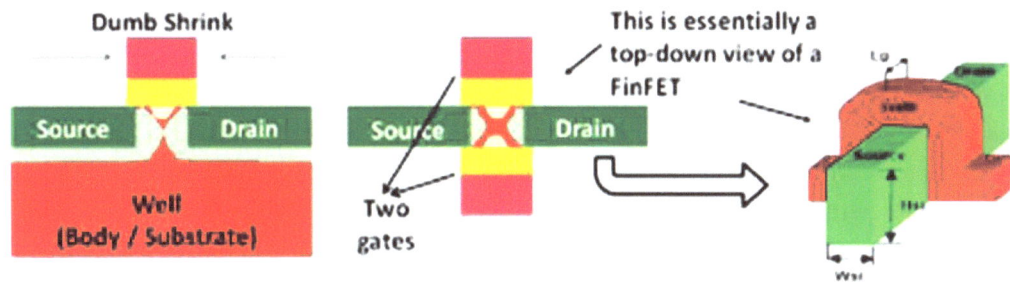

Fig. (12). Ultra-thin structure modification of finfet [42].

It is necessary to maintain the silicon substrate body as thin as possible in order to achieve high current through ultra-thin body devices. The parasitic resistance will be minimized by thicker source and drain terminals. Due to their high driving current, these systems are designed to drive large loads. The source and drain terminals are thickened using a low-pressure chemical vapor deposition (LPCVD) process. A decrease in the thickness of the ultra-thin body is achieved by increasing the threshold voltage of pmos transistors and nmos transistors. The thickness of the body affects the spacing between subbands. There is an inverse relationship between them both. It will result in a decrease in effective mass and an increase in mobility factor. Having a thin body with a high series resistance is one of the greatest challenges of ultra-thin MOS devices. With the FinFET devices, we will be able to overcome short-channel effects with threshold voltage variations found in CMOS. Non-planar devices are mostly multigate devices. Short gates, independent gates, and low power modes are some of the different modes available [36]. Multigate transistors with short gate modes are among those which are considered.To control the gate terminals, one gate of these transistors is shorted to another gate.There are two gates in these transistors, one of which is shorted to the other gate to provide additional control over the gate terminal. Fig. (**13**) shows the basic nonplanar FinFET devices.

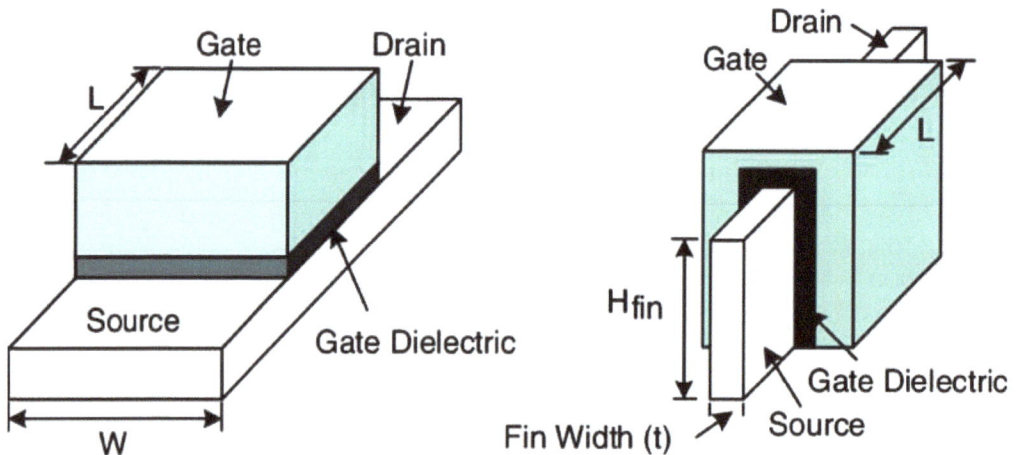

Fig. (13). Non-planar finfet devices [43].

In Fig. (**14**), the area occupancy of a FinFETmultigate transistor is depicted as a physical structure illustration. Using n-channel stacking of FinFET devices, the area overhead problem in CMOS bulky transistors can be overcome.

The three main modes are independent gated (IG), low power (LP), and hybrid IG-LP modes (low power), as can be seen in Fig. (**15**). In all FinFET models, one NAND gate logic design is considered. A simulation tool called Cadence Virtuoso is used with BSIM-CMG 45nm technology. All models based on NAND gates are compared. As compared to low-power gate and low-powered independent gate models, short gated model and independent gate model have high noise tolerance. Additionally, short gated mode and independent gate models have 40 percent less power leakage, 25 percent less power dissipation, and 20 percent less tolerance as compared to low-power gated models. In comparison to the short-gated model and independent model, the low-power model has a 10% delay with gates [21]. Last but not least, the low-power model has a higher speed than the other models. In the final model, the short-gated FinFET NAND gate design shows 35 to 45% less leakage power than the other FinFET NAND gate designs. It occupies less area compared to other models because it requires very few transistors and runs at very low power. However, it has some design limitations related to fin height and fin width.

Fig. (14). 3T FinFET physical structure [44].

Fig. (15). FinFet mode of operations [45].

CMOS and FinFET systems are analyzed using 32nm and 16nm models to differentiate static and dynamic power, and current through them.All simulation is carried out using 32nm CMOS and 32nm FinFET predictive technology model files.The results showed that 0.8V is the minimum voltage that can be applied to the CMOS and FinFET device at the 32nm model, according to the results. For 16nm models, the CMOS does not operate, and as for FinFET, it operates at 0.3V, as shown in Fig. (16). In 32nm model, dynamic power is 436.61 nW for the NAND gate and 439.99 nW for the NOR gate. Compared to 16nm, the NAND gate consumes 187.30nW and the NOR gate consumes 200nW.According to the author, when we reduce the channel length in a FinFET, we greatly reduce static and dynamic power. The static and dynamic power consumption of NAND gate-based FinFETs is lower than that of NOR gate-based FinFETs.In this study, a FinFETNAND gate based on independent gates is designed and a significant improvement in delay and power factors is observed when compared to previous cases.When compared to other FinFET gate models with low power, independent gate FinFETs achieved area reductions of up to 20%. Observations were made regarding timing, power and area constraints of the NAND gate in all three FinFET design models. FinFETs with independent gates operate slower than short-gated FinFETs due to their similarity to CMOS bulky transistors. Compared to independent gate FinFETs, shorted gate FinFETs consume very little power. There are, however, some limitations to it. In often cases, these limitations are observed in the form of:

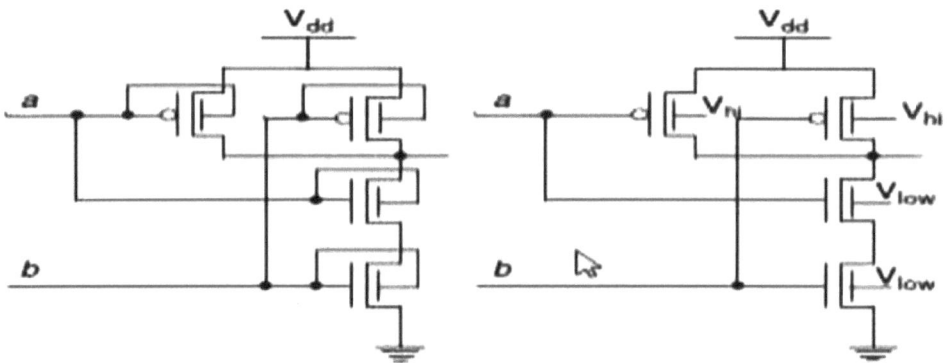

Fig. (16). FinFET based SG-mode NAND (shorted gate) & cascaded NAND [46].

• Due to the 3-D profile, parasitics are higher.

• Exceptionally high capacitances

• Whenever an electric field is present at a corner, it is amplified compared to the field at the sidewall. In corners, a nitrate layer can minimize this problem.

A 32nm technology library was verified by the author in Fig. (**17**), and the results were then compared with the power dissipation of the device; FinFET predicts technology delays better than the TCAD simulation tool using the predictive technology model PTM. However, it has significantly increased the number of rules-based layout designs. Routing problems may occur when regular layouts are used.XOR and XNOR circuits based on differential cascade voltage switching logic were designed by the author in independent gate mode and short-gated mode FinFET NANDs and NORs [16 - 19]. An independent gate FinFET device operating in short-gated mode is compared with one operating in dual threshold independent gate mode. FinFET-based static and DCVSL with dual threshold IG mode achieved lower power delay than FinFET-based short gated.

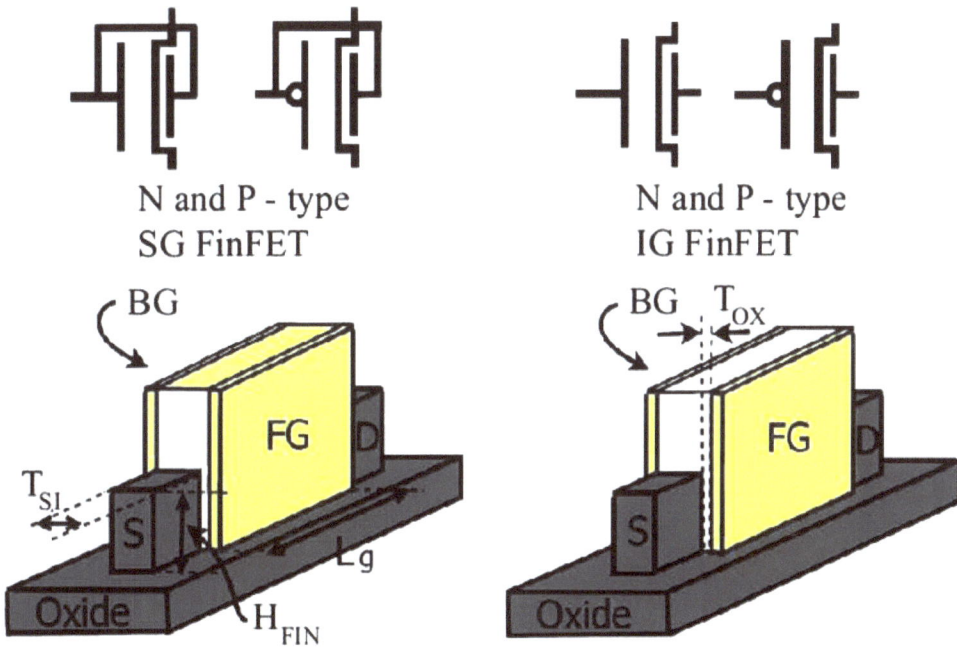

Fig. (17). FinFET-based SG, IG-mode [6].

Challenges in Designs Using FinFET

In a recent FinFET device, there is a short circuit gate mode with multiple gates and a 3D transistor with separate gates. The performance and power consumption of FinFETs is significantly better than those of CMOS bulky planar transistors. Furthermore, it has superior electrical properties and a significant reduction in static and dynamic power (leakage). For a technology as advanced as 10nm

FinFET, there are some major challenges in design. Aspects include analog, digital, parasitic, and signoff.

- In order to correctly print at below 20nm design technology, we need additional masking for the FinFET device fabrication.
- For layouts at 28nm or below, it becomes more challenging due to an increased number of nodes required.
- In order to measure the resistivity of metal layers, 50 times or more potentials must be applied.
- During the operation of low-process nodes, electromigration will increase dramatically.
- Design rules based on fabrication are more complicated.
- Our ability to design and fabricate millions of gates allows us to meet market demands more quickly.

Nevertheless, working at lower voltages can improve performance for IC designers.

Effect of Fin Design on Overall Structure & Performance

By introducing raised channels and multiple gates, FinFETs have become superior as compared to MOSFETs due to their improved channel width and gate control. Due to the fact that electrical current flows mainly along the fins, the fin height can have a direct impact on the device's drive current. In order to evaluate the effective channel width properly, it is necessary to use a taller fin, which gives a wider effective channel width:

$$Wch = 2Hfin + Wfin \qquad (9)$$

Fin height is represented by Hfin, and fin width is represented by Wfin. As a result, more carriers flow along the channel. The parasitic capacitance of 5 nm FinFETs is significantly affected by the narrow spacer and extremely thin oxide thickness due to the narrow critical dimension. According to Fig. (**18**), the drive current of the 15 nm, 19 nm, and 30 nm gate length devices increases linearly as fin height increases from 40 nm to 60 nm, boosted by 22.9%, 24.8%, and 26.1%, respectively. Due to the fact that the source/drain height does not change as the fin height approaches 60 nm, the current drive approaches saturation. While the fin height of device C_{gg} increases from 40 nm to 75 nm, the capacitance increases linearly. Due to the fact that the main contributor to the capacitance C_{gg} being the capacitance of gate to fin, source/drain height has little effect on capacitance. For 15 nm, 19 nm, and 30 nm gate lengths, the increase in fin height has led to 37.7%,

36%, and 39.6% increases in C_{gg}. C_{gg} increases more rapidly than drive current for change in fin height [23]. It requires challenging high aspect ratio etching and film deposition technologies to manufacture a tall fin while maintaining a fixed pitch. For the 5 nm drive current, which is more advanced than 16/14 nm and 7 nm, 70 μA/fin is the target for the 5 nm technology node. In order to maintain the drive current goal while not exceeding the C_{gg}, the fin height is optimally set between 50 nm and 60 nm.

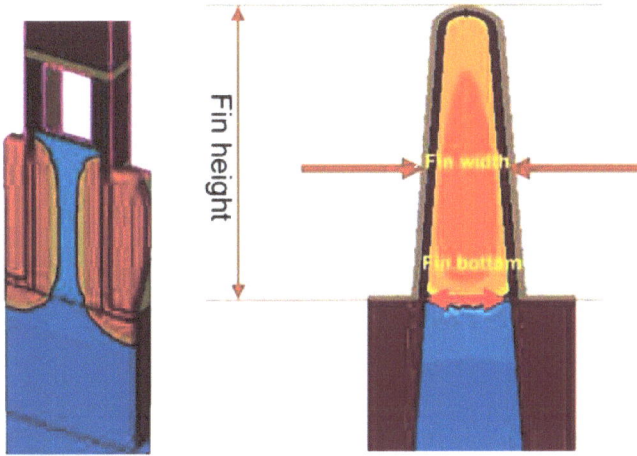

Fig. (18). 3D architecture FinFET half & fin profile slice across the channel. The top of the fin width is initially set to be 5 nm and the bottom at 5.5 nm. The fin height is set initially to 50nm [7].

Quantum Mechanical (QM) effects result in sub-band splitting as fin width shrinks to several nanometers. Due to the fact that the density of the inversion charge caused by the subband splitting is smaller than that of a classical (CL) channel, the QM channel potential is substantially lower than that of the CL channel, causing the V_{th} shift at a given gate voltage bias. V_{th} shift can be explained by the quantum confinement effect as follows:

$$\nabla V_{th} = \frac{SS}{(k_B T/q)\ln 10} \nabla \phi \qquad (10)$$

A subthreshold swing is represented by SS, a carrier charge by q, and a difference in channel potential by f between the CL and QM models is represented by f. According to Figs. (**19 & 20**), the device DIBL can be benefited from fin width shrink, but the V_{th} shift will worsen due to quantum confinement effects.

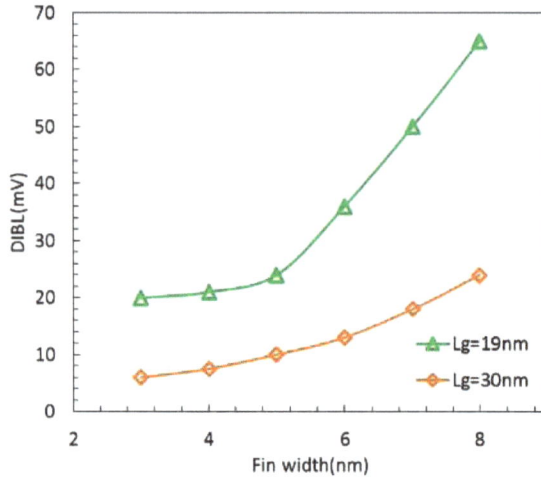

Fig. (19). FinFET device DIBL performance as fin width is lowering. Little DIBL due to control of gate [7].

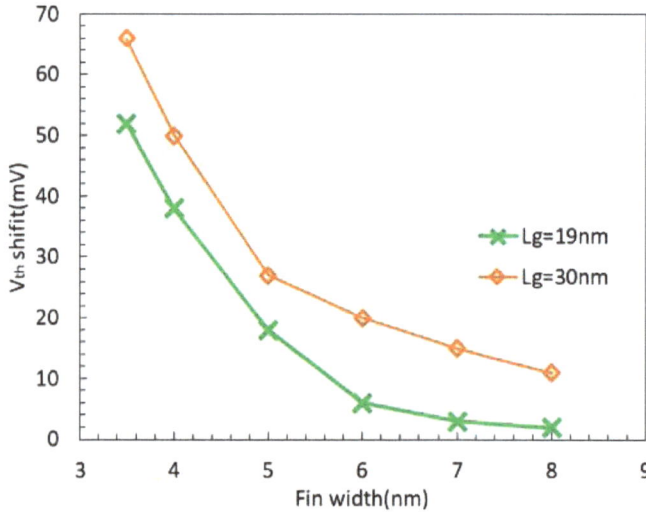

Fig. (20). Threshold voltage shift results due to quantum confinement effect [7].

V_{th} shifts are not significant when fin widths are large enough, such as 6 nm. The V_{th} shift can be significant when the fin narrows to 5 nm, which exhibits a steep rise as the fin shrinks, especially when the fin falls to 4 nm, where the V_{th} shift can become 50 mV (19 nm gate length). As a result, it is good to keep the fin width at or above 5 nm. Increasing fin height can boost device performance by an increasing drive current [31]. However, balancing the fast increase in capacitance may limit device speed if the fin height reaches a maximum. By making more vertical fins, we can reduce leakage current in addition to fin height, due to the leakage current occurring at the fin bottoms.

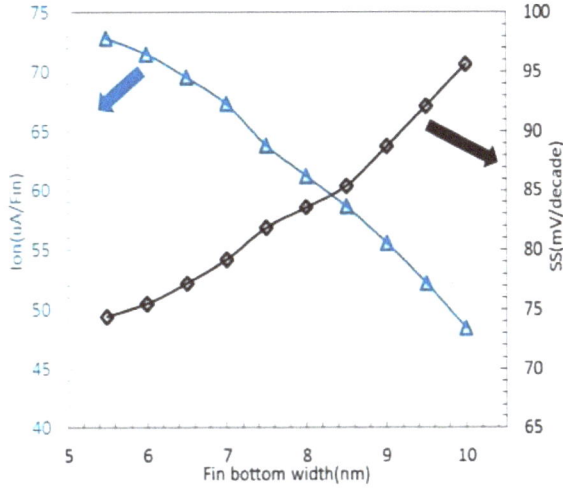

Fig. (21). Fin bottom width shrink results stronger on current (triangle with a through line) and lower *SS* (diamonds with a through line) with fin top fixed 5 nm [7].

Fig. (**21**) shows that the drive current increases as the fin bottom decreases from 10 nm to 5 nm, and the fin top is fixed at 5 nm, resulting in a 74 nm drive current with a fin bottom width of 5.5 nm; the current per fin is 5.5 mA. Further, when the slope of the fin becomes almost vertical, SS will drop down to 73 mV/Dec. In channels, the gate electric field is enhanced by shrinking the fin bottom. Currently, the current manufacturing process can accommodate a maximum sidewall angle of 89 degrees, which is not practical for manufacturability reasons. Due to narrow critical dimensions, fin profiles are important at the 5 nm technology node and probably beyond.The longer the fins, the stronger the drive current, but the more parasitic capacitance they cause. A 50 nm to 60 nm fin height was found to be optimal based on the above study. In addition to better DIBL performance, the narrow fin also increases V_{th} due to the quantum confinement effect. Fin widths below 5 nm will result in a significant V_{th} shift, which will increase steeply as fin widths decrease.Making more vertical fin profiles can result in better gate control, higher drive currents, and lower SS because of the steeper slope.

CONCLUSION

This chapter reviews FinFET made of different materials and various structures. The electrical and physical characteristics of FinFETs are being thoroughly discussed from considerable perspectives. The latest structures of FinFET and their relation to the performance parameters illustrated in I-V and transfer characteristics are also analyzed. Finally, the effect of fin shape on the overall performance is also brought into the limelight.

ACKNOWLEDGEMENT

The first author is highly grateful to his supervisor Mr. Rupanjal Debbarma as well as to Dr. Abhishek Bhattacharjee and Dr. Tanmoy Majumder from the Department of Electronics and Communication Engineering, TIT Narsingarh, without whose support and guidance this book chapter would have never been possible.

LEARNING OBJECTIVES

• This technological era is characterized by the development of emerging devices such as FinFETs, which have low power consumption, are immune from short-channel effects, require less space, and operate faster.

• Among multi-gate devices, FinFETs are much-worthy due to their lower parasitic capacitance and robustness against random doping behaviors.

• It is the material size that determines the area of the channel of FinFET devices, as it is wrapped around a silicon structure to shape the body of the integrated circuit.

• Due to their narrow critical dimension and extremely thin oxide thickness, FinFETs have significant parasitic capacitance. Manufacturing a tall fin while maintaining a fixed pitch requires challenging high aspect ratio etching and film deposition technologies.

• Compared to flat fins, more vertical fin profiles can deliver better gate control, larger currents, and lower SS because of the steeper slope caused by fin widths below 5 nm FinFETs.

MULTIPLE CHOICE QUESTIONS

1. Matthiesen's rule is only valid on the condition that:

  a. Total mobility is proportional to scattering components.
  b. Surface scattering and phonon scattering are a function of total mobility.
  c. There is no concept of dependency between the scattering probabilities.
  d. Collumb scattering is interdependent on overall mobility.

2. In HEMT (High Electron Mobility Transistor), which two factors influence the quality in FinFETs:

  a. The gate oxide thickness.

b. The channel dimensions.
c. Substrate thickness.
d. The gate leakage current and support current.

3. In the vicinity of the substrate below the gate, an N-region is formed as electrons accumulate in:

a. Source and drain regions.
b. Source and gate regions
c. Source and body regions.
d. Drain and gate regions.

4. As a potential alternative to bulk MOSFETs, FinFETs have been predicted to be one of the best devices with respect to which of the following aspects:

a. Improved slope.
b. Better stability.
c. Higher on-off current (I_{ON}/I_{OFF}).
d. All of the above.

5. FinFets on SOI wafers appeared in the timelines of which of the following years:

a. Early 1970's.
b. After 1900's.
c. Late 1990's & Early 2000's.
d. None of the above.

6. In comparison to bulk silicon wafers, SOI wafers are:

a. More costly and have a higher defect density.
b. Easily accessible.
c. More robust.
d. Cheaper.

7. Examples of Crystal defects in semiconductors that explains scattering mechanisms are:

a. Total mobility dispersions.
b. Surface jitters.
c. Lattice vibrations & ionized impurities.
d. None of the above.

8. At what voltage does the relationship of drain voltage and drain current ceases to exist:

a. Threshold voltage.
b. Pinch-off voltage.
c. Gate induced volatge.
d. Drain induced voltage.

9. The spacer length is directly proportional to the:

a. Substrate voltage.
b. Supply voltage.
c. The depletion layer width.
d. The drain and source resistances.

10. Device reliability is degraded by hot carriers which is produced by;

a. Impact ionization.
b. Short channel effect.
c. Narrow channel width.
d. The source and drain voltages.

11. In the case of short spacers, what voltage results in a degradation of transconductance?

a. Impact ionization.
b. Short channel effect.
c. Narrow channel width.
d. The source and drain voltages.

12. Which of the following insulating layer is used in Fabrication of FinFET?

a. Aluminium oxide.

b. Silicon Nitride.
c. Silicon dioxide.
d. None of the above.

13. Which of the following parameters decides the scaling factor in finfet:

a. Dielectric constant.
b. Gate oxide thickness.
c. Power supply voltage.
d. Gate to drain voltage.

14. High permittivity gate dielectrics (high-κ) keeps the sheet resistance low that resulted in:

a. High driving current.
b. Constant voltage.
c. Linear i-v characteristics.
d. No one of the above.

15. Why is it important to keep the silicon substrate body as thin as possible?

a. To maintain stability.
b. To achieve high drive through ultra thin body devices.
c. To widen the channel length.
d. To minimize short channel effects.

ANSWER KEY

1. (c)

2. (d)

3. (a)

4. (d)

5. (c)

6. (a)

7. (c)

8. (b)

9. (d)

10. (a)

11. (a)

12. (c)

13. (b)

14. (a)

15. (b)

REFERENCES

[1]   K. Bindu Madhavi, and S.L. Tripathi, "Strategic review on different materials for finfet structure performance optimization", *IOP Conf. Series Mater. Sci. Eng.,* vol. 988, no. 1, p. 012054, 2020. [http://dx.doi.org/10.1088/1757-899X/988/1/012054]

[2]   Yang F-L, Chen H-Y, Chen F-C, Huang C-C, Chang C-Y, Chiu H-K, Lee C-C, Chen C-C, HuangH-T, Chen C-J, Tao H-J, Yeo Y-C, Liang M-S, Hu C '25 nm CMOS omega FETs'. In: IEDM technical digest, pp 255–258, (2002).

[3]   Hossain M. Zakir, Hossain Md.Alamgir, Islam Md.Saiful, Rahman Md. Mijanur, and Chowdhury Mahfuzul Haque, "Electrical characteristics of trigate finfet", *Glob. J. research.eng.Electri.Electro. eng.,* vol. 11, no. 7, 2011.

[4]   Juan Duarte, and Navid Paydavosi, "Unified finfet compact model: Modelling trapezoidal triple-gate finfets", *International Conference on Simulation of Semiconductor Processes and Devices (SISPAD).,* 2013pp. 135-1358 Glasgow, UK [http://dx.doi.org/10.1109/SISPAD.2013.6650593]

[5]   Jinsu Park, Jaemin Kim, Sanchari Showdhury, and Changhwan Shin, "Electrical characteristics of bulk finfet according to spacer length", *Electronics,* vol. 9, no. 8, p. 1283, 2020.

[6]   V. Mahor, and M. Pattanaik, "Low leakage and highly noise immune finfet-based wide fan-in dynamic logic design", *J. Circuits Syst. Comput.,* vol. 24, no. 5, p. 1550073, 2015. [http://dx.doi.org/10.1142/S0218126615500735]

[7]   E. Shang, Y. Ding, W. Chen, S. Hu, and S. Chen, "The effect of fin structure in 5 nm finfet technology", *J. Microelectro.Manufac.,* vol. 2, no. 4, pp. 1-8, 2019. [http://dx.doi.org/10.33079/jomm.19020405]

[8]   C. Meinhardt, and R. Reis, "FinFET basic cells evaluation for regular layouts", *IEEE 4th Latin American Symposium on Circuits and Systems (LASCAS).,* pp. 1-4, 2013. [http://dx.doi.org/10.1109/LASCAS.2013.6519063]

[9]   X. Zhang, J. Hu, and X. Luo, "Optimization of dual-threshold independent-gate finfets for compact low power logic circuits", *IEEE 16th International Conference on Nanotechnology,,* pp. 529-532, 2016. [http://dx.doi.org/10.1109/NANO.2016.7751552]

[10]   N. Yadav, S. Khandelwal, and S. Akashe, "Design and analysis of finfet pass transistor based XOR and XNOR circuits at 45 nm technology", *International Conference on Control, Computing, Communication and Materials (ICCCCM).,* pp. 1-5, 2013.
[http://dx.doi.org/10.1109/ICCCCM.2013.6648909]

[11]   P.K. Mukku, S. Naidu, and D. Mokara, "recent trends and challenges on low-power finfet devices", In: *Smart Intelligent Computing and Applications . Smart Innovation, Systems and Technologies,* S. Satapathy, V. Bhateja, J. Mohanty, S. Udgata, Eds., vol. 160. springer: Singapore, 2020, pp. 499-510.
[http://dx.doi.org/10.1007/978-981-32-9690-9_55]

[12]   S.A. Tawfik, and V. Kursun, "FinFET technology development guidelines for higher performance, lower power, and strongerresilience to parameter variations", *52nd IEEE International Midwest Symposium on Circuits and Systems (MWSCAS),* pp. 431-434, 2009.
[http://dx.doi.org/10.1109/MWSCAS.2009.5236062]

[13]   D.D. Lu, C-H. Lin, A.M. Niknejad, and C. Hu, "Compact modeling of variation in finfet sram cells", *IEEE Des. Test Comput.,* vol. 27, no. 2, pp. 44-50, 2010.
[http://dx.doi.org/10.1109/MDT.2010.39]

[14]   J. Hu, Y. Zhang, C. Han, and W. Zhang, "An investigation of super-threshold finfet logic circuits operating on medium strong inversion regions", *Open Electr. Electron. Eng. J.,* vol. 9, no. 1, pp. 22-32, 2015.
[http://dx.doi.org/10.2174/1874129001408010263]

[15]   M. Rostami, and K. Mohanram, "Dual-vth independent-gate finfets for low power logic circuits", *IEEE Trans. Comput. Aided Des. Integrated Circ. Syst.,* vol. 30, no. 3, pp. 337-349, 2011.
[http://dx.doi.org/10.1109/TCAD.2010.2097310]

[16]   S.A. Tawfik, and V. Kursun, "Multi-threshold voltage finfetsequential circuits, ieee trans. onvery large scaleintegration (VLSI)", *Systems,* vol. 19, no. 1, pp. 151-156, 2011.

[17]   M.C. Wang, "Independent-gate finfet circuit design methodology", *IAENG Int. J. Comput. Sci.,* vol. 37, no. 1, 2010.

[18]   N. Paydavosi, S. Venugopalan, Y.S. Chauhan, J.P. Duarte, S. Jandhyala, A.M. Niknejad, and C.C. Hu, "Bsim—spice models enable finfet and UTB IC designs", *IEEE Access,* vol. 1, pp. 201-215, 2013.
[http://dx.doi.org/10.1109/ACCESS.2013.2260816]

[19]   C. Meinhardt, and R. Reis, "Finfet basic cellsevaluation for regularlayouts", *IEEE 4th Latin American Symposium on Circuits and Systems (LASCAS).,* 2013.
[http://dx.doi.org/10.1109/LASCAS.2013.6519063]

[20]   M. Alioto, "Analysis and evaluation of layoutdensity of finfetlogic gates", *IEEE International Conference on Microelectronics,* 2009.

[21]   Abhishek Bhattacharjee, Tanmoy Majumder, Rajib Laskar, Shradhya Kar, Tamanna Laskar, Nabadipa Dey, and Amanisha Chakraborty, "Design and Optimization of a 50 nm Dual Material Dual Gate (DMDG), High-κ Spacer, FiNFET Having Variable Gate Metal Workfunction", In: *MicroelectronicDevices, Circuits and Systems. ICMDCS 2021. Communications in Computer and Information Science.,* V. Arunachalam, K. Sivasankaran, Eds., vol. 1392. Springer, 2021.
[http://dx.doi.org/10.1007/978-981-16-5048-2_16]

[22]   A. Bhattacharjee, and S. Dasgupta, "Source/Drain (S/D) Spacer-Based Reconfigurable Devices-Advantages in High-Temperature Applications and Digital Logic", In: *Modelling, Simulation and Intelligent Computing. MoSICom 2020. Lecture Notes in Electrical Engineering,,* N. Goel, S. Hasan, V. Kalaichelvi, Eds., vol. 659. Springer: Singapore, 2020.
[http://dx.doi.org/10.1007/978-981-15-4775-1_48]

[23]   A. Bhattacharjee, and S. Dasgupta, "A Compact physics-based surface potential and drain current model for an S/D spacer-based DG-RFET", *IEEE Trans. Electron Dev.,* vol. 65, no. 2, pp. 448-455, 2018.

[http://dx.doi.org/10.1109/TED.2017.2786302]

[24]   Bagga. Navjeet, Kumar. Anil, A. Bhattacharjee, and S. Dasgupta, "Performance evaluation of a novel gaa schottky junction (gaasj) tfet with heavily doped pocket", *Superla. Microstruc.*, vol. 109, pp. 545-552, 2017.
[http://dx.doi.org/10.1016/j.spmi.2017.05.040]

[25]   A. Bhattacharjee, and S. Dasgupta, "Impact of Gate/spacer-channel underlap, gate oxide eot, and scaling on the device characteristics of a DG-RFET", *IEEE Trans. Electron Dev.*, vol. 64, no. 8, pp. 3063-3070, 2017.
[http://dx.doi.org/10.1109/TED.2017.2710236]

[26]   R.J.P. Lander, J.C. Hooker, J.P. van Zijl, F. Roozeboom, M.P.M. Maas, Y. Tamminga, and R.A.M. Wolters, "A tunable metal gate work function using solid state diffusion of nitrogen", *Proc. ESSDERC*, pp. 103-106, 2002.

[27]   Yeo. Yee-Chia, P. Ranade, King Tsu-Jae, and Hu. Chenming, "Effects of high-/spl kappa/ gate dielectric materials on metal and silicon gate workfunctions", *IEEE Electron Device Lett.*, vol. 23, no. 6, pp. 342-344, 2002.
[http://dx.doi.org/10.1109/LED.2002.1004229]

[28]   Y-K. Choi, K. Asano, N. Lindert, V. Subramanian, T-J. King, J. Bokor, and C. Hu, "Ultra-thin body SOI MOSFET for deep-sub- tenth micron era", *Int. Electron Devices Meeting Tech.*, pp. 919-921, 1999.

[29]   S-D. Kim, C-M. Park, and J.C.S. Woo, "Advanced model and analysis of series resistance for CMOS scaling into nanometer regime—part II quantitative analysis", *IEEE Trans.ElectronDevices*, vol. 49, pp. 467-472, 2002.

[30]   B. Doyle, R. Arghavani, D. Barlage, S. Datta, M. Doczy, J. Kava- lieros, A. Murthy, and R. Chau, "Transistor elements for 30 nm physical gatelengths and beyond", *Intel Tech. J.*, vol. 6, pp. 42-54, 2002.

[31]   B. Doris, M. Ieong, T. Kanarsky, Y. Zhang, R.A. Roy, O. Dokumaci, Z. Ren, F-F. Jamin, L. Shi, W. Natzle, H-J. Huang, J. Mezzapelle, A. Mocuta, S. Womack, M. Gribelyuk, E.C. Jones, R.J. Miller, H-S.P. Wong, and W. Haensch, "Extreme scaling with ultra-thin si channel MOSFETs", *Digest. International Electron Devices Meeting.*, pp. 267-270, 2002.

[32]   S. Zimin, L. Litian, and L. Zhijian, "Optimization of MOSFET's with polysilicon-elevated source/drain", *Proc. 5th Int. Conf. Solid- State and Integrated Circuit Technology*, pp. 188-189, 2013.

[33]   W.B. De Boer, D. Terpstra, and J.G.M. Van Berkum, Selective *versus* non-selective growth of Si and SiGe., *Mater. Sci. Eng. B*, vol. 67, no. 1-2, pp. 46-52, 1999.
[http://dx.doi.org/10.1016/S0921-5107(99)00208-1]

[34]   Yang-Kyu Choi, Daewon Ha, Tsu-Jae King, and Chenming Hu, "Nanoscale ultrathin body PMOSFETs with raised selective germanium source/drain", *IEEE Electron Device Lett.*, vol. 22, no. 9, pp. 447-448, 2001.
[http://dx.doi.org/10.1109/55.944335]

[35]   S.P. Ashburn, M.C. Öztürk, G. Harris, and D.M. Maher, "Phase transitions during solid-state formation of cobalt germanide by rapid thermal annealing", *J. Appl. Phys.*, vol. 74, no. 7, pp. 4455-4460, 1993.
[http://dx.doi.org/10.1063/1.354387]

[36]   S.P. Ashburn, M.C. Öztürk, J.J. Wortman, G. Harris, J. Honeycutt, and D.M. Maher, "Formation of titanium and cobalt germanides on Si (100) using rapid thermal processing", *J. Electron. Mater.*, vol. 21, no. 1, pp. 81-86, 1992.
[http://dx.doi.org/10.1007/BF02670924]

[37]   R. Chau, J. Kavalieros, B. Doyle, A. Murthy, and N. Paulsen, "A 50 nm depleted-substrate CMOS transistor (DST)", *in Int. ElectronDevices Meeting 2016 Tech. Dig.*, pp. 621-624, 2018.

[38] H. van Meer, and K. De Meyer, "The spacer/replacer concept: A viable route for sub-100 nm ultrathin-film fully-depleted SOI CMOS", *IEEE Electron Device Lett.,* vol. 23, no. 1, pp. 46-48, 2002. [http://dx.doi.org/10.1109/55.974808]

[39] H.S.P. Wong, D.J. Frank, P.M. Solomon, C.H.J. Wann, and J.J. Welser, "Nanoscale cmos", *Proc. IEEE,* vol. 87, no. 4, pp. 537-570, 1999. [http://dx.doi.org/10.1109/5.752515]

[40] L. Chang, "Extremely scaled silicon nano-CMOS devices", *Proceedings of the IEEE,* pp. 1860-1873, 2014. [http://dx.doi.org/10.1109/JPROC.2003.818336]

[41] Pavan Kumar Mukku, Sushmi Naidu, Divya Mokara, Puthi Pydi Reddy, and Kuppili Sunil Kumar, "Recent trends and challenges on low-power finfet devices", In: *Smart Intelligent Computing and Applications . Smart Innovation, Systems and Technologies,* S. Satapathy, V. Bhateja, J Mohanty, S Udgata, Eds., vol. 160. Springer: Singapore, 2019, pp. 499-510.

[42] P. Ajey, Ruilong Xie. Jacob, Sung Min Gyu, Lars Liebmann, Lee Rinus T. P., and Taylor Bill, "Scaling challenges for advanced cmos devices", *I. J. High Speed Electron. Sys.,* vol. 26, no. 01n02, p. 1740001, 2017. [http://dx.doi.org/10.1142/S0129156417400018]

[43] X. Guo, V. Verma, P. Gonzalez-Guerrero, S. Mosanu, and M.R. Stan, "Back to the future: Digital circuit design in the finfet era", *J. Low Power Electron.,* vol. 13, no. 3, pp. 338-355, 2017. [http://dx.doi.org/10.1166/jolpe.2017.1489]

[44] Y. Kim, J. Lee, G. Kim, T. Park, H. Kim, Y. Cho, Y. Park, and M. Lee, "Partial isolation type saddle-finfet(pi-finfet) for sub-30 nm dram cell transistors", *Electronics,* vol. 8, no. 1, p. 8, 2018. [http://dx.doi.org/10.3390/electronics8010008]

[45] Chen, Xianmin. "FinFET-based System Modeling and Low-Power System Design.", Doctoral thesis, Electrical Engineering Department, Princeton, NJ: Princeton-University, 2016. https://dataspace. princeton.edu/handle/88435/dsp016t053j389

[46] G. Saranya, and K. Kalarani, "Design and implementation of logic gates using finfet technology", *I. J. Advanced Tech. Eng,* vol. 3, no. 1, 2015.

CHAPTER 9

Optically Gated Vertical Tunnel FET for Near-Infrared Sensing Application

Vandana Devi Wangkheirakpam[1,*], Brinda Bhowmick[2], Puspa Devi Pukhrambam[2] and Ghanshyam Singh[3]

[1] *Department of Electronics and Communication Engineering, Indian Institute of Information Technology Senapati, Manipur, India*

[2] *Department of Electronics and Communication Engineering, National Institute of Technology Silchar, Assam, 788010, India*

[3] *Department of Electronics and Communication Engineering, Malaviya National Institute of Technology Jaipur, Rajasthan, 302017, India*

Abstract: This chapter presents a vertical tunnel FET (VTFET) designed for light sensing application to use in medical diagnosis and treatment, tracking of targets, analysis of the chemical composition, surveillance cameras, *etc*. Various aspects related to this optimized VTFET photosensor are analyzed to benchmark its performance among those available in the literature. A brief discussion on the conventional TFET geometry is presented to give a better understanding of the advantages of its working methodologies. The concept of sensing using optically gated VTFET is studied with a remarkable focus on design perspective and detection principle. The modified TFET geometry has a photosensitive gate called an optically gated VTFET to use in near-infrared sensing applications. The design approach based on Synopsys Technology Computer-Aided Design (TCAD), along with suitable physics-based models of simulation, is introduced in this chapter. A wavelength range of $0.7\mu m$ to $1\mu m$ is considered in the simulation process. Analyses of different sensing parameters, such as sensitivity, responsivity, *etc*., at low intensity of illumination, are brought to light with the main focus on the viability of the proposed sensor to be a superior one. Through such analysis, this chapter presents a low-power, highly sensitive, cost-effective, faster response time photodetector that may be applicable for next-generation photosensors.

Keywords: TFET, Tunneling, Photosensor, Wavelength, Near-infrared light, Illumination, Sensitivity.

* **Corresponding author Vandana Devi Wangkheirakpam:** Department of Electronics and Communication Engineering, Indian Institute of Information Technology Senapati, Manipur, India; E-mail: vannawang46@gmail.com

Gopal Rawat & Aniruddh Bahadur Yadav (Eds.)

INTRODUCTION

In modern times, sensors are omnipresent. Without automation, life would have been very difficult [1, 2]. A sensor may be defined as a device that converts one signal into another form that can be measured more conveniently, perhaps more accurately [3]. It receives a different form of signals, *i.e.*, physical, chemical and biological signals and transforms into electrical or optical signals [4]. The classification of sensors is done on the basis of the nature of the input, mechanism of conversion, applications, and various characteristics of the sensor, such as cost, range of detection or accuracy, properties of detection, *etc* [5]. There is an increasing demand for low-power, highly sensitive, faster response time, and low limit of detection optical sensors. Some commonly used photodetectors are photodiodes with PN junction, Avalanche photodiode, PIN photodiodes, FET-based phototransistors, *etc*. The presence of an intrinsic layer restricts the PIN diodes from using in applications involving high sensitivity [6]. Quantum efficiency is low in PN junction photodiodes, and Avalanche photodiodes are more prone to noise. Hence, FET photodetectors are favored over other photodetectors because of their CMOS compatibility and capability of downsizing [7]. But, the continued scaling down of the state-of-art MOSFETs experiences a power crisis problem; an issue arises due to the thermal limit of subthreshold swing (60 mV/dec). Moreover, after reaching its scaling limit, MOSFET induces short channel effects, limiting the device's performance. As a result, exploring advanced devices having steeper subthreshold slopes becomes a primary approach [8].

In recent years, many industries and researchers have proposed various devices whose working principle is different from the state-of-art MOSFETs [9, 10]. Band-to-band tunneling FET (TFET) promises a steeper SS since its current transport is governed by the tunneling of carriers from the valence band of the source to the conduction band of the channel instead of the thermionic emission as in the case of MOSFET [11]. An aggressive threshold voltage scaling can be performed in Tunnel FET, and thus it can provide a steeper sub-threshold swing (SS) [12]. However, the drain current ratio (I_{ON}/I_{OFF}) of TFETs is typically lesser when compared with MOSFETs as their ON state current depends on the tunneling probability [13]. Important research in developing the TFET from its primitive structure to its modern-day forms has been carried out to improve I_{ON} by adopting structural modifications as well as emerging materials [14]. TFET also shows another issue of current conduction in both negative and positive gate bias, known as ambipolar conduction. Till now, different modified TFET architectures having enhanced characteristics have been reported [15]. Some of the structures extensively used in the literature for various analyses include SOI TFET, hetero-junction TFET, nanowire TFET, gate-all-around TFET, dual metal gate TFET, circular gate TFET, *etc*. TFETs are very popular for their low-power digital circuits and memory applications [16]. In the recent past, the use of TFET in sensing has become a topic of interest. The application of TFET as a dielectric modulated label-free biosensor is widely studied for its higher sensitivity characteristics [17]. The implementation of TFET in a photo-sensing application is also emerging research and is yet to explore. This sensor can be used for the detection of visible to infrared wavelengths. A photosensitive n-type Si gate separated from the channel region by a dielectric behaves as an optically activated gate upon illumination of light depending on the charge carrier excited by a photon that is accumulated at the interface. This photogating technique is adopted to look after the conducting nature of the channel using the gate field or voltage induced by light [18]. The variation in the channel conductance due to the illumination of light in the gate causes drain current fluctuation, and the change is measured in terms of sensitivity [19].

Section 2 of this chapter discusses a brief survey of some of the works reported on FET-based photosensors. In Section 3, a review of the concept, geometry and working principle of conventional TFET is presented. Section 4 mentions the principle of optically gated tunnel FET technology. Section 5 studies the sensor design and near infra-red light detection process of vertical tunnel FET-based photosensor using the photogating effect. Different sensing parameters of VTFET photosensor are defined in Section 6. The conclusion of this chapter and the comments on the future scope are presented in Section 7.

BRIEF SURVEY

Optical sensors designed using Silicon (Si) have been widely produced for optical sensing applications as they are readily available in nature, occupy less volume, and possess a good signal-to-noise ratio. In 1990, Mishra *et al*. designed an optically controlled ion-implanted GaAs MESFET and found that the electron-hole pair generation due to the incident photon flux significantly produced a drain current higher than the current generated by ion-implantation only [20]. A. Jain *et al*. proposed a high sensitivity photosensor based on Cylindrical Surrounding Gate Metal Oxide Semiconductor Field Effect Transistor (CSG MOSFET) [21]. The sensitivity parameter was measured by studying the change in subthreshold current due to the substantial increase in conductance on being exposed to light. Enhanced photo-sensing characteristics can be obtained by utilizing advanced MOSFET geometries. However, P.S. Gupta and his group designed a novel optoelectronic transistor using band-to-band tunnel FET for the detection of multiple spectral lines at near-infrared spectrum ranging between the wavelengths 0.7µm and 1µm [22]. They made use of the steeper slope advantage of TFET and obtained a better spectral response, which could resolute spectral lines that are very close to each other (~100nm). The study was also extended by developing a drain current model of the photosensor based on a line tunneling approach to investigate the working of the device [23]. S. Joshi *et al.* designed a photosensor based on split gate TFET having a transition metal dichalcogenide (TMD) channel using a photogating effect for the detection of visible light. The results showed an improved sensing characteristic when compared to the sensitivity obtained by the dual material gate TFET photosensor [24].

TUNNEL FIELD EFFECT TRANSISTOR (TFET): CONCEPT, GEO-METRY AND WORKING PRINCIPLE

Conventional TFET

The conventional TFET is a three-terminal device having an insulator-gated reverse-biased p-i-n geometry. These terminals are the source, gate and drain. TFET can operate both in p channel and n channel mode based on the polarity of the applied voltage. For n-TFET, the source is degenerated p-type while the drain is degenerated n-type. Likewise, p-TFET has a highly doped n-type source and p-type drain. The channel is either intrinsic or lightly n/p doped and is kept between the source and drain, resulting in a p-i-n structure. Consequently, a narrow depletion layer is obtained at the junction, which provides a favorable condition for the carriers to tunnel from the valence band (E_v) of the source region to the conduction band (E_c) of the channel region. This particular device works under the principle of Zener tunneling [27]. Similar to MOSFET, the channel is separated from the gate electrode by a dielectric material. An n-TFET is presented in Fig. (**1**). In this TFET, the source is connected to the ground and the gate and drain electrodes are biased with a positive voltage.

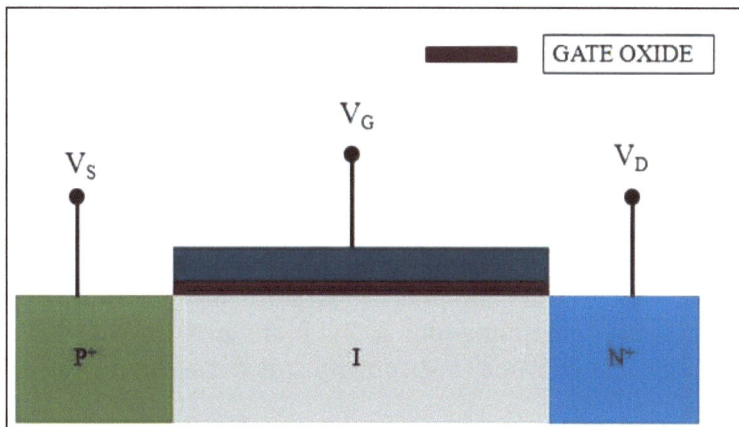

Fig. (1). 2D schematic of conventional TFET.

Working Principle of n-TFET

This section presents the biasing scheme of n-TFET. For such TFET, the bending of energy bands at the channel can be initiated by applying a positive voltage (V_{GS}) at the gate terminal. This bending of energy bands can increase by increasing VGS, thereby reducing the tunneling width, *i.e.*, the area between the EV of source and EC of channel where the bending is placed. The tunneling of carriers can be achieved only when V_{GS} reaches a limit known as the threshold voltage (V_{th}). This biasing condition is shown in Fig. (**2**), where the nature of E_C and E_V of the conventional TFET is plotted.

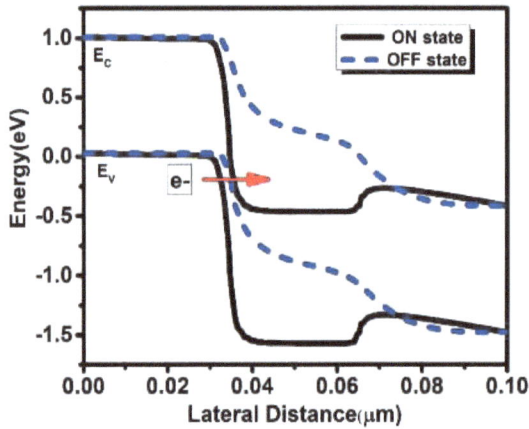

Fig. (2). Energy band diagram of nTFET illustrating the behavior at ON state and OFF state.

The tunneling probability of the energy barrier governs the current conduction of TFET. This probability is modeled through WKB approximation assuming a triangular barrier and is given by,

$$T(E) = exp\left(\frac{-4\lambda\sqrt{2m^*}E_g^{3/2}}{3q\hbar(E_g + \Delta\emptyset)}\right) \tag{1}$$

Where m* represents the effective mass, E_g denotes the forbidden energy gap, $\Delta\emptyset$ represents the energy of the area at the tunnel junction where bands are overlapped, λ is the tunneling width, q and $\hbar$ are the value of an electronic charge and reduced Planck's constant, respectively. λ is defined as,

$$\lambda = \sqrt{\varepsilon_s/\varepsilon_{ox}}t_{ox}t_s \tag{2}$$

Where ε_s and ε_{ox} correspond to the dielectric constants of substrate and oxide layer respectively, t_{ox} and t_s respectively represent the oxide and substrate thickness. It is evident from equations (**1 & 2**) that the drain current is dependent on the tunneling width, which is again a function of the material's dielectric constant (k). A higher k value will increase the gate and channel coupling and can increase the tunneling probability, thereby giving a higher value of I_D.

OPTICALLY GATED TFET TECHNOLOGY

Geometry

The objective of implementing a TFET-based photosensor requires the modification of the device geometry. The following are the basic requirements for such modification,

- The gate must be modified into a photosensitive region for being used in light-sensing applications. Such TFET is named optically gated TFET.
- An n-type Si can be used as the photosensitive gate which is separated from the channel region by a dielectric material.
- The photosensitive gate must be optically activated upon illumination of light depending on the photon excited charge carrier which are accumulated at the interface.
- Proper biasing must be applied to the gate electrode so as to induce sufficient gate-field or voltage due to the generated electron-hole pair. This photogating technique is adopted to monitor the channel conductance with the gate field or voltage induced by light.
- The TFETs must have significant doping concentrations. The designing of the sensor must be done in such a way that it should be able to give a good response even at low intensity of illumination.
- Also, the design of the TFET geometry should be made keeping in mind to suppress the ambipolar effect.

Simulation Strategy

The deviation in the channel conductance due to the illumination of light in the gate causes drain current fluctuation, and the change is measured in terms of sensitivity. The design approach based on Synopsys Technology Computer-Aided Design (TCAD), along with suitable physics-based models of simulation, is mentioned in this chapter [25]. The band-to-band tunneling mechanism of TFET is achieved in the simulation process by adopting predefined models, such as doping dependent model and a non-local BTBT model. The non-local BTBT model looks after the gradient of the energy band, and thus, it gives more realistic results. The effect of degenerated source and drain is incorporated by adding a band gap narrowing model. The ShockleyReadHall recombination model is activated to look after the quantization of the carriers' density gradient. In addition to these models, the Fermi Dirac Statistics and mobility models are also incorporated in simulating the TFET geometry. Furthermore, the use of the device in light sensing applications and optical generation from the incoming monochromatic light should take place. This is achieved in the device TCAD simulation by using the complex refractive index and optical generation model. The gate region is incident. The ray tracer method is adopted for the optical solver. The default optical parameters of Si provided in the simulation tool is utilized. The entire analysis is performed at low illumination intensity (I_0) of 0.5 Wcm^{-2}, and the performance of the sensor is studied in the wavelength range 0.7 μm–1 μm (near-infrared range).

DUAL MOSCAP VERTICAL TFET AS NEAR-INFRARED LIGHT SENSOR

This section presents a geometry of TFET named dual MOS capacitor (DMOS) Vertical TFET as an optical sensor and discusses some sensing parameters [26]. The sensor is designed for multiple real-time applications like tracking a target, medical diagnosis and treatment, chemical composition analysis, surveillance cameras, inter-chip data communication, *etc*. Various aspects related to this optimized VTFET photosensor are analyzed to benchmark its performance among those available in the literature [27]. The concept of sensing using optically gated VTFET is studied with a remarkable focus on design perspective and detection principle. The proposed VTFET consists of a dual MOS capacitor (MOSCAP) geometry with a vertically elevated channel/drain region sandwiched between these two MOSCAPs [28]. The structure gives an inverted-T geometry. A thin intrinsic epitaxial layer (epi-layer) between the gate-stack and the source region of MOSCAP acts as a part of the channel region and enables the band-to-band tunneling (BTBT) in a vertical direction and enhances the performance of the device [29, 30]. The electrical performance of the device is further improved by the incorporation of a δ-doped SiGe layer with a Germanium concentration of 40% at the source/channel junction, creating a heterojunction at this region. The modified TFET geometry has a photosensitive gate called optically gated VTFET for near-infrared sensing applications. Here, the metal gate is replaced by a photo-absorbing n-type Silicon gate with doping concentration, $N_g = 1 \times 10^{19}$ cm^{-3}. The sensor design is shown in Fig. (3). The elevated channel region (T_{ch}) has a thickness of 20nm. This thickness is maintained to make sure that the incident photon flux is completely absorbed inside the photosensitive gate.

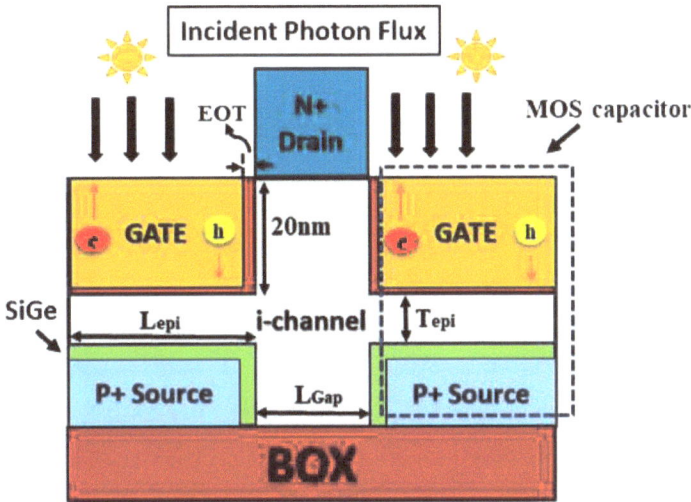

Fig. (3). 2D schematic image of DMOS Vertical TFET photosensor.

Illumination light of intensity 0.5 Wcm^{-2} and wavelength 0.7 μm is made to incident normally in this photosensitive gate from a monochromatic source through the illumination window. This optical incidence results in the creation of electron-hole pairs (EHP) in the photosensitive gate, and this is presented in Fig. (**4**). The plot is taken along the x-direction, and it is evident from this figure that the optical generation occurs only in this region and no such observation is seen in the channel region where there is no incident photon flux. This incident photon will excite those electrons having higher energy than that of the energy bandgap of Si. Such electrons at the gate region will get excited, causing band-to-band generation, which results in a phenomenon called photogeneration. Thus, excess electrons and holes are generated in this region. The motion of these photogenerated excess carriers is modulated by applying a gate bias (V_{GS}). After applying positive voltage V_{GS} at the gate region, electrons tend to move towards the gate electrode, whereas holes get attracted towards the interface of the gate and oxide layer. This effect of gate bias is termed the photogate effect. This effect perturbs the equilibrium of the gate region resulting in electron quasi-Fermi level (E_{Fn}) and hole quasi-Fermi level (E_{Fp}) separation, as shown in Fig. (**5**). This separation of Fermi level produces a photovoltage (V_{OP}) across the gate region and adds an extra voltage to the gate bias. Thus, V_{OP} reduces the amount of gate voltage needed for the device to start operating and increases the drain current. Figs. (**5a & b**) are respectively taken near the gate electrode and at the gate-oxide interface. It is evident from these figures that the area adjacent to the gate electrode has higher photogenerated electron density while the gate-oxide interface has higher hole density. However, the equilibrium is maintained, and there is a uniform distribution of electrons under zero-incident light conditions, as seen in Fig. (**5c**).

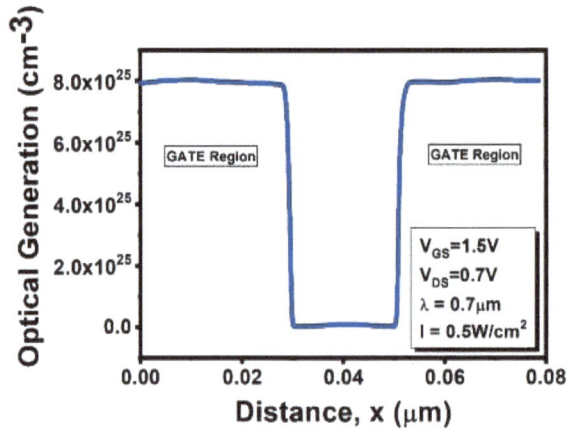

Fig. (4). Optical generation at the gate region due to incoming monochromatic light.

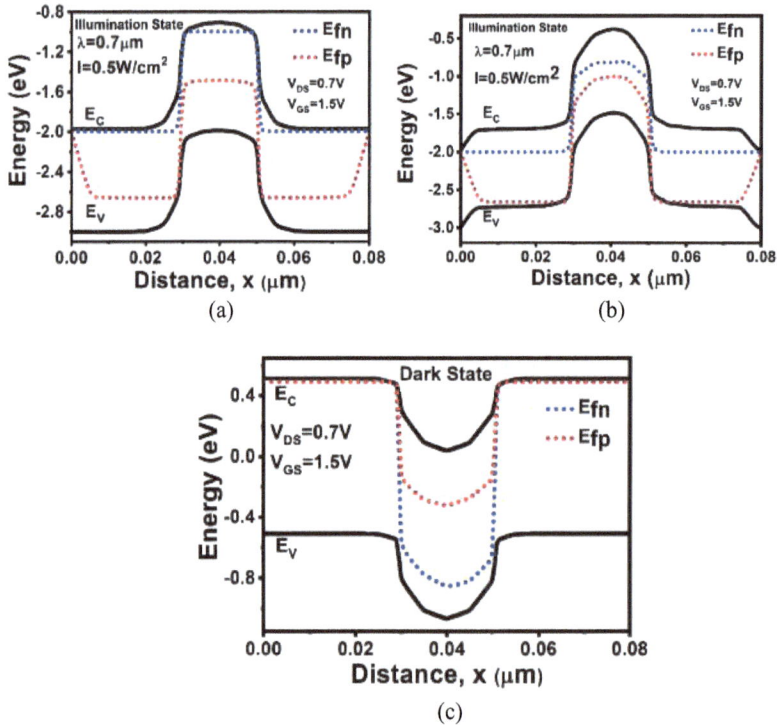

(a)

(b)

(c)

Fig. (5). Energy band diagram measured along x-direction under (**a**) illumination state near the gate electrode. (**b**) illumination state near the gate-oxide interface. (**c**) dark state.

SENSING PARAMETERS OF DMOS VTFET PHOTOSENSOR

Analyses of different sensing parameters, such as sensitivity, responsivity, *etc.*, at low intensity of illumination are brought to light with the main focus on the viability of the proposed sensor to be a superior one. The drain current characteristics of the D-MOS VTFET photosensor are studied under the illumination state as well as the dark state, *i.e.*, when there is no incident photon flux falling on the gate region. Fig. (**6**) shows the transfer characteristics curves measured in the wavelength range of 0.7μm to 1μm. The absorption coefficient of the incident light increases as the wavelength decreases, and due to this, photogeneration in the gate region increases. Thereafter, the optical voltage (V_{OP}) rises and lowers the gate voltage needed for the device to start operating. Hence, the drain current, along with the threshold voltage (V_{th}), is observed to be increasing as the wavelength of the incident light reduces, which can be seen in Fig. (**6**).

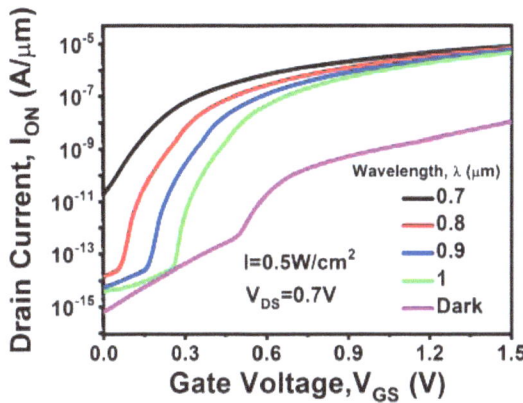

Fig. (6). TCAD simulated result of transfer characteristics of optically gated DMOS VTFET under illumination and dark state.

In this work, the quality of the photosensor is computed in terms of its spectral sensitivity (S_n), which decides how fast the device responds to the input. It is evaluated by calculating the shift in drain current when the value of λ of the incident light (λ) changes.

$$S_n = \frac{I_D(\lambda_1) - I_D(\lambda_2)}{I_D(\lambda_1)} \tag{3}$$

Where λ_1 and λ_2 indicate that the wavelength of the incident photon flux reduces from λ_2 to λ_1. Using equation (**3**), the spectral sensitivity is calculated and plotted in Fig. (**7**) for the considered wavelength pair. The trend of the plot shows an abrupt increase in S_n at lower gate voltage, reaches a peak value, and gradually declines. The reason behind the formation of such a trend is that the difference in the drain current value is higher in the subthreshold regime, and this difference slowly declines after reaching the super-threshold regime. For detecting spectral lines that are spaced very close to one another, a sensor having higher sensitivity is preferred. The peak sensitivity is obtained from each wavelength pair in Fig. (**7**). This peak spectral sensitivity (S_{nm}) is shown in Fig. (**8**). The S_{nm} values of 4.67 $\times$ 10^3, 3.3 $\times$ 10^3 and 9.77 $\times$ 10^2 is obtained for each considered wavelength pairs shown in the figure.

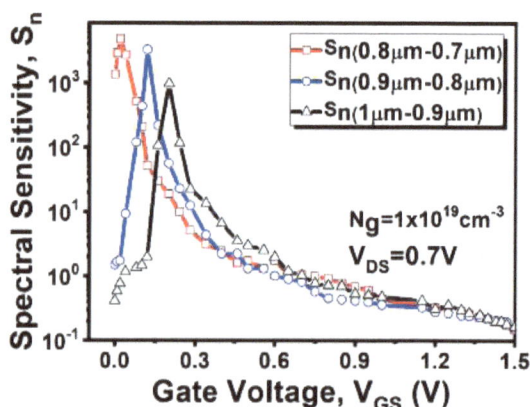

Fig. (7). Spectral sensitivity *vs.* gate voltage plot at various wavelength pairs.

The ratio of the output current to the input incident light power is defined as the responsivity (R) of the photosensor. It is also one of the sensing parameters and is calculated using the following relation,

$$R = \frac{I_D}{P_{opt}} \tag{4}$$

Where P_{opt} represents the optical power of the incoming monochromatic light. It is evident from equation (**4**) that responsivity is the direct function of the output drain current, and thus, a device having a higher I_D value will give better responsivity. Fig. (**9**) shows the responsivity plot for various near-infrared wavelengths, *i.e.*, from 0.7µm to 1µm. A higher sensitivity and better responsivity are observed for the proposed sensor, and hence it can be used effectively in sensing closely spaced spectral lines.

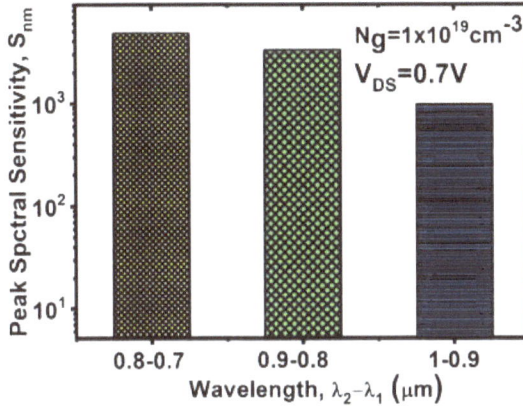

Fig. (8). Peak spectral sensitivity of D-MOS VTFET photosensor obtained from Fig. (7).

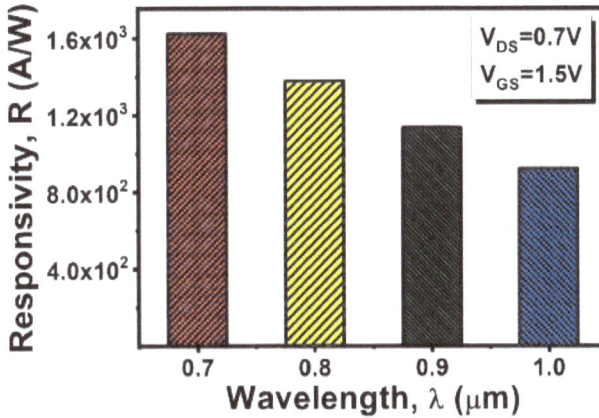

Fig. (9). Responsivity of D-MOS VTFET photosensor at steady state illumination.

CONCLUSION

This chapter has discussed an overview of Tunnel Field Effect Transistors (TFETs) as optically gated photosensors. TFETs have the ability to withstand aggressive downscaling without showing short channel effects, and thus, they are considered to be one of the most significant advanced FETs for low-power and high-frequency applications. Considering the steeper subthreshold swing characteristics of TFET, research on TFET-based sensors has become a new area of interest. This chapter has studied various features of TFET as an optically gated photosensor through TCAD simulation analyses focusing on its design and development. A novel modified geometry of TFET called dual MOS capacitor vertical TFET is presented in this chapter with the main aim to enhance its electrical characteristics andovercome the ambipolar effect of conventional TFET.

This device is then optically gated for being used in photo sensing applications, and it is observed to have an impressive sensitivity as well as responsivity. Through such analysis, this chapter presents a low-power, highly sensitive, cost-effective, with faster response time photodetector, which may be applicable for next-generation photosensors. The results of the TCAD tool can be validated by developing appropriate physics-based models.

LEARNING OBJECTIVES

By the end of this chapter, readers will be able to:

• Understand the issues related to the short channel state of art MOSFET and how it can be overcomed by Tunnel FET.

• Understand the importance of Vertical tunneling to improve the electrical characteristics of TFET.

• Understand how TFET can be used as light sensing device.

• Analyze and evaluate the sensitivity of TFET based photosensor.

• Understand the viability of TFET to use as low power, highly sensivitive, faster and cost-effective photodetector.

MULTIPLE CHOICE QUESTIONS

1. The power crisis issue is observed in short channel MOSFETs due to:

a. The mismatch between power supply scaling and threshold voltage scaling.
b. The higher leakage power.
c. The threshold voltage roll-off.
d. The hot carrier effect.

2. Which of the following parameter defines how fast the transistor can be turned OFF by reducing gate voltage?

a. Sensitivity
b. Response Time
c. Sub-threshold Swing
d. Responsivity

3. Which of the following photosensor has comparatively higher sensitivity?

  a. PIN photodiode
  b. Avalanche photodiode
  c. MOSFET based phototransistors
  d. TFET based phototransistor

4. In the presence of light, electron-hole pair generation takes place in the gate region and cause the Fermi level splitting. The reason behind the formation of Quasi Fermi Level is beacuse of the:

  a. Significant increase in majority carrier concentration.
  b. Significant increase in the minority carrier concentration.
  c. Significant increase in both majority and minority carrier concentration.
  d. None of the above.

5. At illumination condition, the photon induced shift in the gate voltage $(\Phi_p\text{-}\Phi_n)$ is .... throughout the gate region.

  a. Zero
  b. Unpredictable
  c. Constant
  d. Variable

6. TFET based photosensors are advantageous over the MOSFET because:

1. The current conduction mechanism is band-to-band tunneling.

2. The gate voltage required to turn ON the device is less.

3. TFET can have steeper sub-threshold swing.

  a. Statement 1 and 3 are correct.
  b. State 1 and 2 are correct.
  c. Only statement 1 is correct.
  d. All the statements are correct.

7. The senstivity of the DMOS Vertical TFET based photosensor depends on:

  a. The shift in drain current due to change in wavelength of incident light.
  b. Absorption co-efficient of the incoming light.

c. Both (a) and (b).
d. Only (a)

8. The design of the TFET must be done in such a way that it should be able to give good response at:

a. Low intensity of illumination light.
b. High intensity of illumination light.
c. Low supply voltage.
d. All of the above.

9. In DMOS VTFET based photosensor, the optical generation occurs:

a. Only in source region.
b. Only in gate region.
c. Both gate and source region.
d. None of the above.

10. As the wavelength of the incident light increases, the drain current:

a. Increases.
b. Decreases.
c. Doesn't depend on the wavelength of light.
d. Remains constant.

11. Under dark state,

Statement 1: The equlibrium is maintained.

Statement 2: There is uniform distribution of electrons.

a. Only statement 1 is correct.
b. Only statement 2 is correct.
c. Both statement 1 and 2 are correct.
d. Both statement 1 and 2 are incorrect.

12. The movement of the photogenerated carriers are controlled by,

a. Applied gate bias.
b. Applied drain bias.
c. Incident photon flux.

d. All of the above.

13. The induced photo voltage (V_{OP})

a. Reduces the drain current.
b. Increases the drain current.
c. Independent of drain current.
d. None of the above.

14. The photogeneration inside the gate region is a function of:

a. Wavelength of incident light.
b. Absorption co-efficient of the incident light.
c. Intensity of the incident light.
d. All of the above.

15. TFET based photosensor possesses higher sensitivity due to:

a. BTBT current conduction mechanism.
b. Steeper subthreshold swing.
c. Lower leakage current.
d. All of the above.

ANSWER KEY

1. (a)

2. (c)

3. (d)

4. (c)

5. (c)

6. (d)

7. (d)

8. (a)

9. (a)

10. (b)

11. (c)

12. (a)

13. (b)

14. (d)

15. (d)

REFERENCES

[1]   C.H. Lin, and C.W. Liu, "Metal-insulator-semiconductor photodetectors", *Sensors,* vol. 10, no. 10, pp. 8797-8826, 2010.
        [http://dx.doi.org/10.3390/s101008797] [PMID: 22163382]

[2]   H.W. Lin, S-Y. Ku, H-C. Su, C-W. Huang, Y-T. Lin, K-T. Wong, and C-C. Wu, "Highly efficient visible-blind organic ultraviolet photodetectors", *Adv. Mater.,* vol. 17, no. 20, pp. 2489-2493, 2005.
        [http://dx.doi.org/10.1002/adma.200401622]

[3]   G. Sarasqueta, K.R. Choudhury, J. Subbiah, and F. So, "Organic and inorganic blocking layers for solution-processed colloidal PbSe nanocrystal infrared photodetectors", *Adv. Funct. Mater.,* vol. 21, no. 1, pp. 167-171, 2011.
        [http://dx.doi.org/10.1002/adfm.201001328]

[4]   G. Konstantatos, I. Howard, A. Fischer, S. Hoogland, J. Clifford, E. Klem, L. Levina, and E.H. Sargent, "Ultrasensitive solution-cast quantum dot photodetectors", *Nature,* vol. 442, no. 7099, pp. 180-183, 2006.
        [http://dx.doi.org/10.1038/nature04855] [PMID: 16838017]

[5]   F. Gan, L. Hou, G. Wang, H. Liu, and J. Li, "Optical and recording properties of short wavelength optical storage materials", *Mater. Sci. Eng. B,* vol. 76, no. 1, pp. 63-68, 2000.
        [http://dx.doi.org/10.1016/S0921-5107(00)00400-1]

[6]   H. Zimmermann, and I.H. Speed, "High sensitivity photodiodes and optoelectronic integrated circuits", *Sens. Mater.,* vol. 13, no. 4, pp. 189-206, 2001.

[7]   A. Stoykov and R. Scheuermann, "Silicon avalanche photodiodes," *Lab. Muon Spin Spectrosc., Paul Scherrer Inst., Villigen, Switzerland, Tech. Rep.,* 2004.

[8]   R.A. Ismail, and W.K. Hamoudi, "Characteristics of novel silicon pin photodiode made by rapid thermal diffusion technique", *J. Electron Devices,* vol. 14, pp. 1104-1107, 2012.

[9]   Byung-Gook Park, B-G. Park, J.D. Lee, and T-J.K. Liu, "Tunneling field-effect transistors (TFETs) with subthreshold swing (SS) less than 60 mV/dec", *IEEE Electron Device Lett.,* vol. 28, no. 8, pp. 743-745, 2007.
        [http://dx.doi.org/10.1109/LED.2007.901273]

[10]   A.M. Ionescu, and H. Riel, "Tunnel field-effect transistors as energy-efficient electronic switches", *Nature,* vol. 479, no. 7373, pp. 329-337, 2011.
        [http://dx.doi.org/10.1038/nature10679] [PMID: 22094693]

[11]   K. Boucart, and A.M. Ionescu, "Double-gate tunnel FET with high-κ gate dielectric", *IEEE Trans. Electron Dev.,* vol. 54, no. 7, pp. 1725-1733, 2007.
        [http://dx.doi.org/10.1109/TED.2007.899389]

[12]   W.V. Devi, and B. Bhowmick, "Optimisation of pocket doped junctionless TFET and its application in digital inverter", *Micro & Nano Lett.,* no. April, pp. 1-5, 2018.
        [http://dx.doi.org/10.1049/mnl.2018.5086]

[13] V.D. Wangkheirakpam, B. Bhowmick, and P.D. Pukhrambam, "N+ pocket doped vertical tfet based dielectric-modulated biosensor considering non-ideal hybridization Issue: A Simulation Study", *IEEE Trans. Nanotechnol.*, vol. 19, pp. 156-162, 2020.
[http://dx.doi.org/10.1109/TNANO.2020.2969206]

[14] W.V. Devi, B. Bhowmick, and P.D. Pukhrambam, "N+ pocket-doped vertical tfet for enhanced sensitivity in biosensing applications: Modeling and simulation", *IEEE Trans. Electron Dev.*, vol. 67, no. 5, pp. 2133-2139, 2020.
[http://dx.doi.org/10.1109/TED.2020.2981303]

[15] V.D. Wangkheirakpam, B. Bhowmick, and P.D. Pukhrambam, "Modeling and simulation of optically gated tfet for near infra-red sensing applications and its low frequency noise analysis", *IEEE Sens. J.*, vol. 20, no. 17, pp. 9787-9795, 2020.
[http://dx.doi.org/10.1109/JSEN.2020.2991406]

[16] S. Basak, P.K. Asthana, and Y. Goswami, "Leakage current reduction in junctionless tunnel FET using a lightly doped source", *Appl. Phys. Lett.*, vol. 118, pp. 1527-1533, 2015.

[17] B.R. Raad, D. Sharma, P. Kondekar, K. Nigam, and D.S. Yadav, "Drain work function engineered doping-less charge plasma tfet for ambipolar suppression and rf performance improvement: A proposal, design, and investigation", *IEEE Trans. Electron Dev.*, vol. 63, no. 10, pp. 3950-3957, 2016.
[http://dx.doi.org/10.1109/TED.2016.2600621]

[18] S.B. Rahi, P. Asthana, and S. Gupta, "Heterogate junctionless tunnel field-effect transistor: Future of low-power devices", *J. Comput. Electron.*, vol. 16, no. 1, pp. 30-38, 2017.
[http://dx.doi.org/10.1007/s10825-016-0936-9]

[19] V.D. Wangkheirakpam, B. Bhowmick, and P.D. Pukhrambam, "Near infra-red photosensor using optically gated d-mos vertical tfet", *21st International Conference on Numerical Simulation of Optoelectronic Devices (NUSOD)*, pp. 47-48, 2021.
[http://dx.doi.org/10.1109/NUSOD52207.2021.9541496]

[20] S. Mishra, V.K. Singh, and B.B. Pal, "Effect of radiation and surface recombination on the characteristics of an ion-implanted GaAs MESFET", *IEEE Trans. Electron Dev.*, vol. 37, no. 1, pp. 2-10, 1990.
[http://dx.doi.org/10.1109/16.43794]

[21] A. Jain, S.K. Sharma, and B. Raj, "Design and analysis of high sensitivity photosensor using cylindrical gate MOSFET for low power applications", *Engineering Science and Technology, an International Journal*, vol. 19, no. 4, pp. 1864-1870, 2016.
[http://dx.doi.org/10.1016/j.jestch.2016.08.013]

[22] P.S. Gupta, S. Chattopadhyay, P. Dasgupta, and H. Rahaman, "A novel photosensitive tunneling transistor for near-infrared sensing applications: Design, modeling, and simulation", *IEEE Trans. Electron Dev.*, vol. 62, no. 5, pp. 1516-1523, 2015.
[http://dx.doi.org/10.1109/TED.2015.2414172]

[23] P.S. Gupta, H. Rahaman, K. Sinha, and S. Chattopadhyay, "An optoelectronic band-to-band tunnel transistor for near-infrared sensing applications: Device physics, modeling, and simulation", *J. Appl. Phys.*, vol. 120, no. 8, pp. 084510-084511, 11, 2016.
[http://dx.doi.org/10.1063/1.4961426]

[24] S. Joshi, P. K. Dubey, and B. K. Kaushik, "Photosensor based on split gate tmd tfet using photogating effect for visible light detection", *IEEE Sens. J.*, vol. 20, no. 12, pp. 6346-6353, 2020.
[http://dx.doi.org/10.1109/JSEN.2020.2966728]

[25] *Synopsys. Sentaurus Device User Guide. Synopsys* Inc.: Mountain View, 2011.

[26] V.D. Wangkheirakpam, B. Bhowmick, and P.D. Pukhrambam, "Near-infrared optical sensor based on band-to-band tunnel FET", *Appl. Phys., A Mater. Sci. Process.*, vol. 125, no. 5, p. 341, 2019.
[http://dx.doi.org/10.1007/s00339-019-2636-3]

[27]   R. Narang, M. Saxena, M. Gupta, and M. Gupta, "Comparative analysis of dielectric-modulated fet and tfet-based biosensor", *IEEE Trans. Nanotechnol.*, vol. 14, no. 3, pp. 427-435, 2015. [http://dx.doi.org/10.1109/TNANO.2015.2396899]

[28]   M. Ehteshamuddin, S.A. Loan, A.G. Alharbi, A.M. Alamoud, and M. Rafat, "Investigating a dual MOSCAP variant of line-TFET with improved vertical tunneling incorporating FIQC effect", *IEEE Trans. Electron Dev.*, vol. 66, no. 11, pp. 4638-4645, 2019. [http://dx.doi.org/10.1109/TED.2019.2942423]

[29]   V.D. Wangkheirakpam, B. Bhowmick, and P.D. Pukhrambam, "Investigation of a dual MOSCAP TFET with improved vertical tunneling and its near-infrared sensing application", *Semicond. Sci. Technol.*, vol. 35, no. 6, p. 065013, 2020. [http://dx.doi.org/10.1088/1361-6641/ab8172]

[30]   V.D. Wangkheirakpam, B. Bhowmick, and P.D. Pukhrambam, "Vertical Tunnel FET having Dual MOSCAP Geometry", In: *Sub-Micron Semiconductor Devices* CRC Press, 2022, p. 16. [http://dx.doi.org/10.1201/9781003126393-7]

CHAPTER 10

Self-Powered Photodetectors: Fundamentals and Recent Advancements

Varun Goel[1,*] and **Hemant Kumar**[1]

[1] *Department of Electronics and Communication Engineering, Jaypee Institute of Information Technology, Noida, Uttar Pradesh, India*

Abstract: This chapter focuses on the evolution of Self-powered Photodetectors, from single nanobelt to highly sophisticated Pyro-phototronic effect-assisted devices. The essentials of the self-powered photodetector, from material characterization to device engineering mechanism, are discussed in detail, such as Pyro-phototronic enriched devices. This study provides a state-of-the-art research trend of the Pyro-phototronic enriched self-powered photodetectors. Finally, a summary of various device structures with their figures of merit and conclusions, along with the research gap, is presented. This review focuses on providing valuable insights into improving self-powered photodetectors.

Keywords: Pyroelectric Effect, Pyro-phototronic Effect, Quantum Dots, Self-Powered Photodetector, Solution-Processed, ZnO.

INTRODUCTION

In today's era, imaging systems [1] and sensors make human life convenient. These systems do utilize photodetectors for their operation.

Applications like spectroscopy (study of light-matter interaction) [2], weather forecasting [3], remote farming [4], forest monitoring, astronomy (analysis of the universe) [5], self-driving vehicles [6], and night vision [7] utilizing infrared radiations are based on photodetectors only. Eye spy is a trending application that mimics the human eye for IoT-based applications [8].

Photodetectors discriminate among the photons of different wavelengths conditioned over the material's bandgap. High bandgap materials (> 3eV) like ZnO [9], TiO_2 [10], GaN [11], *etc.* sense UV radiations, while small bandgap materials (< 1eV) like PbS [12], PbSe [13], HgCdTe [14], InSb [15], *etc.* are uti-

---

[*] **Corresponding author Varun Goel:** Department of Electronics and Communication Engineering, Jaypee Institute of Information Technology, Noida, Uttar Pradesh, India; E-mail: varun.goel@jiit.ac.in

Gopal Rawat & Aniruddh Bahadur Yadav (Eds.)

lized to sense IR radiations. The materials that help visible light sensing are CuO [16], Cu_2O [17], Sb_2SeTe_2 [18], CdSe [19], *etc*.

In traditional working mechanisms, reversed-biased junctions are utilized for photodetection applications. Reverse-biased junction results in a wider depletion region and allows more light to fall over the depletion region. Suppose the incident photon's energy is higher than the material's bandgap; the photon imparts its energy to the covalent bond and generates electron and hole pairs after breaking the covalent bond. If the photon energy is less than the bandgap energy, then there is no generation of electron and hole pair resulting in the transmitted photon. The separation of charge carriers represents the external quantum efficiency of the detector achieved using an electric field set up by the external reverse potential.

The photodetection process can be picturized in four steps: optical absorption, charge excitation, charge separation, and charge collection, as shown in Fig. (**1**).

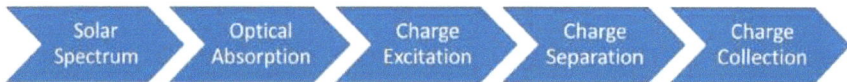

Fig. (1). Four steps of photodetection flow.

The working of the photodetector, as specified in Fig. (**1**), goes through the different losses associated with the operation of the photodetector are picturized as shown in Fig. (**2**) and discussed below:

Optical Loss: Some of the light incidents on the photodetector's surface are reflected, and some are transmitted through the photodetector. These reflected and transmitted components of light contribute to optical losses. To minimize the optical losses, anti-reflecting coatings and active layers are applied according to the wavelength of interest [20].

Thermalization Loss: This is the excess bandgap energy loss in the form of heat. This loss can be minimized using quantum dots and multilayer photodetectors [21].

Recombination Loss: The photogenerated charge carriers should be collected before energy loss or recombination. The generated charge carriers are limited by the lifetime and characteristic length of the device. These losses can be minimized by electron-transport and hole-transport layers. The transport layers are highly doped n-type and p-type semiconductors, which facilitate the collection and transport of charge carriers by improving the internal electric field across the photoactive area.

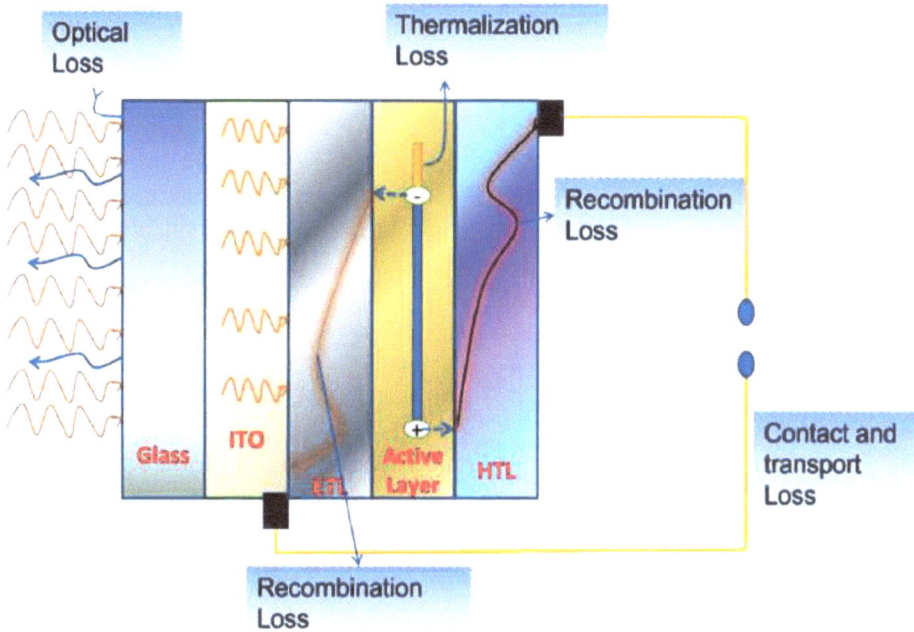

Fig. (2). Different losses associated with the Photodetector.

Contact and Transport Loss: These losses occur because of the resistance offered by contact paste and connecting wires. These losses can be minimized by using high-quality contact paste and low-resistance connecting wires.

The selection of materials for hole-transport-layer (HTL), electron-transport-layer (ETL), active area, and electrodes plays a vital role in the abovementioned losses. The performance of the photodetector accompanying these losses is generally characterized using the figures of merit, which are decided over the following parameters:

i. **Bias Voltage (volts)**: In a photodetector, a reverse bias voltage is used to implement the high electric field across the photoactive area required to separate the generated charge carriers [22]. It is desired that the applied bias voltage should be as low as possible to increase the device's viability for energy-saving applications.

ii. **Sensitivity**: Sensitivity defines the ability to detect the weak signals generated from the source. It is the ratio of current due to photogenerated electron-hole pairs (I_{ph}) to the current due to thermally generated electron-hole pairs under dark conditions (I_{dark}). The sensitivity of the photodetector should be as high as possible.

iii. **Rise Time (t_r) (seconds)**: The rise time indicates the delay required for the photodetector to moderate the output signal according to the source variation. The response time should be as minimum as possible for faster devices. The response time is the time taken to reach from an initial 10% to 90% of the saturated value of the output.

iv. **Reliability**: A Photodetector should produce desirable output when the input conditions are met. This defines the reliability of the photodetector, and therefore, the reliability of each component of the photodetector circuit should be as high as possible.

v. **Responsivity (A/W)**: The responsivity of a photodetector is a measure of the optical-to-electrical energy conversion efficiency of a photodetector. It is defined as the ratio of the photocurrent to the incident optical power intensity [23]. For a photodetector, responsivity should be as much high as possible.

$$Sensitivity = \frac{I_{ph}}{I_{dark}} \tag{1}$$

$$Responsivity = \frac{J_{ph}}{P_{opt}} \tag{2}$$

Where J_{ph} is the photocurrent density and P_{opt} is the optical power density.

vi. **External Quantum Efficiency (EQE)**: It is a measure of the number of electron-hole pairs collected at the terminals with respect to the number of photons incident. For a high-quality photodetector, EQE should be as high as possible.

$$EQE = 1240 \times \frac{Responsivity}{\lambda} \tag{3}$$

Where λ is the wavelength of the incident photon.

vii. **Detectivity (Jones or cmHz$^{1/2}$/W)**: The detectivity of a photodetector is the ability to detect weak signals (minimum number of photons). The noise current hinders the ability of the detector. Thus, the detectivity decreases with an increase in noise current and can be written as:

$$D^* = \frac{R\sqrt{A\Delta f}}{\sqrt{\langle I_n^2 \rangle}} \tag{4}$$

Where R is the responsivity, A is the area of the contact electrode, Δf is the bandwidth, and $\sqrt{(I_n^2)}$ is the noise equivalent power.

To improve the detectivity of any detector, the thermal current (noise) should be as minimum as possible.

viii. **Size**: The portability and integrability of a device depend on its size. Therefore, small-size photodetectors are preferred.

The photodetector's fabrication and device engineering process revolves around the abovementioned parameters and improving the same.

QUANTUM CONFINEMENT AND DIFFERENT DIMENSIONAL MATERIALS

Nanomaterials show changes in electrical and optical properties when the movement of electron-hole pairs is spatially confined. The thickness of the nanomaterials also helps with the confinement of electron-hole pairs, provided the particle size is less than the De-Broglie wavelength of the material, also called Bohr's radius. The confinement observed by the electron-hole pairs is also known as quantum confinement. The quantum confinement leads to the discretization of bandgaps, thus directly relating bandgap with the particle size of the nanomaterial [24].

$$E_g(QD) = E_g(bulk) + \left(\frac{h^2}{8R^2}\right)\left(\frac{1}{m_e} + \frac{1}{m_h}\right) - \frac{1.8e^2}{4\pi\epsilon\epsilon_0 R} \qquad (5)$$

The indicative variation of bandgap variation with QD particle size is shown in Fig. (**3**).

The classification of the nanomaterials based on confinement can be done as follows:

3-D Materials: In 3-D materials, the electron movement is not confined in any direction, *i.e.*, it is free to move in all three directions. An example of 3-D material is bulk materials. The depiction of bulk with continuum energy states is shown in Fig. (**4**).

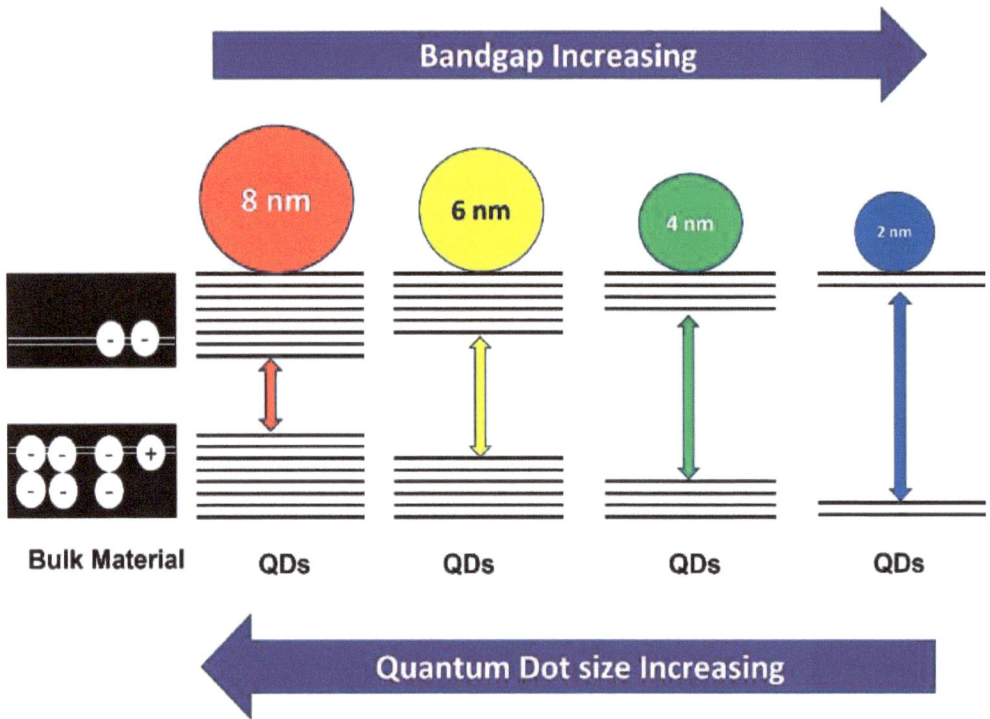

Fig. (3). Quantum Dot bandgap dependence on Size.

2-D Materials: In 2-D materials, the electron movement is restricted in one direction. However, electrons can move freely in two dimensions or along a plane. An example of 2-D material is quantum wells (nanosheets, nanoflakes), as indicated in Fig. (**4**).

1-D Materials: In 1-D materials, the electron movement is restricted in two directions, and the electron is allowed to move freely in one direction only or along the line. An example of 1-D material is quantum wires (nanowires, nanorods, nanotubes), as indicated in Fig. (**4**).

0-D Materials: In 0-D materials, the electron movement is restricted in all three dimensions. An example of 0-D material is quantum dots indicated in with discretized bandgap. QDs also show a tunable bandgap, which makes them widely valuable for displays, detectors, *etc*. QDs contain about 10 to 50 atoms and show a higher surface-to-volume ratio than other dimensional materials. The bandgap of the QDs increases with a decrease in its size, as shown in Fig. (**4**).

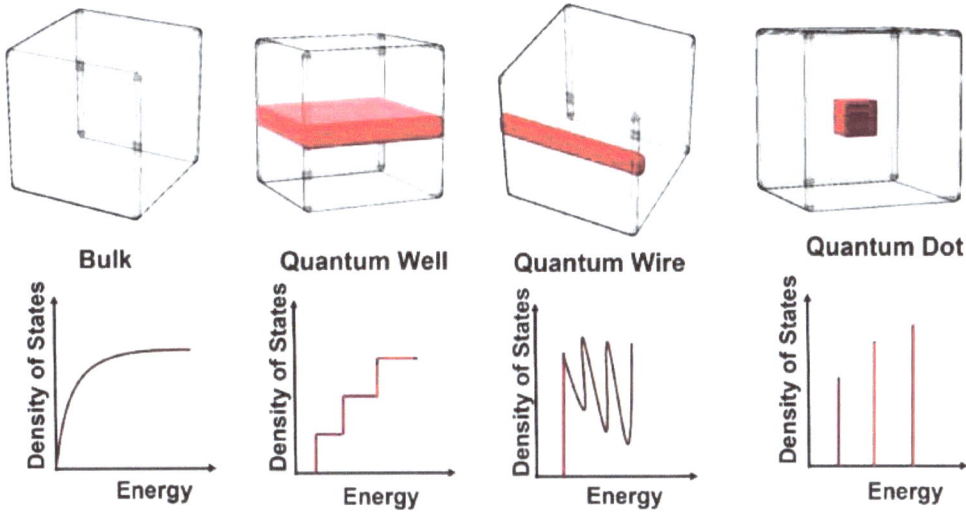

Fig. (4). Density of states *vs.* Energy for different dimensional materials.

Introducing quantum-confined materials in optoelectronics leads to opportunities in photodetection applications, such as tuning of spectrum by modifying particle size and extracting confined electron-hole pairs and energy generation, among others. Integrating the energy generation process with photon detection can lead to the development of self-powered photodetectors.

SELF-POWERED OPERATION

A self-powered photodetector utilizes the internal built-in electric field to separate charge carriers [25, 26]. Under dark conditions, the self-powered photodetector act as an open circuit, similar to a traditional photodetector, as shown in Fig. (5a). Ideally, there is no current through the device under dark conditions. However, a small amount of dark current flows through the device due to thermally generated carriers. During illumination, the charge extraction is performed under the influence of an electric field. Traditional photodiode's electric field is due to the influence of external bias. In contrast, the self-powered device utilizes an internal electric field for carrier separation and extraction. Thus, the working of a Self-Powered detector can be interpreted as a light-dependent source, as shown in Fig. (5b).

Self-powered photodetectors manufacturing gained a tremendous pace to shake hands with the mission of energy harvesting. This is possible because of the ease of synthesizing low-cost solution-processed materials like Nanowires (1-D) and QDs (0-D) [9, 27, 28]. The solution-processed techniques are low-cost, suitable for large-area fabrication, compatible with printing electronics, and suitable for flexible electronics [29]. Initially, researchers used inorganic materials for self-powered devices due to their high mobility and absorption [30]. However, the high processing temperature of inorganic materials restricts the applications, especially for flexible and printed electronics [31]. The inclusion of Low-dimensional solution-processed low-cost nanomaterials could extend the application spectrum of Self-Powered photodetectors toward flexible and printed electronics [28]. Further, the application of a self-powered photodetector can be extended for remotely located sites such as remote monitoring, weather forecasting, remote farming [4], and many others. The application-specific points and reducing energy dependence make Self-Powered photodetectors future-ready devices.

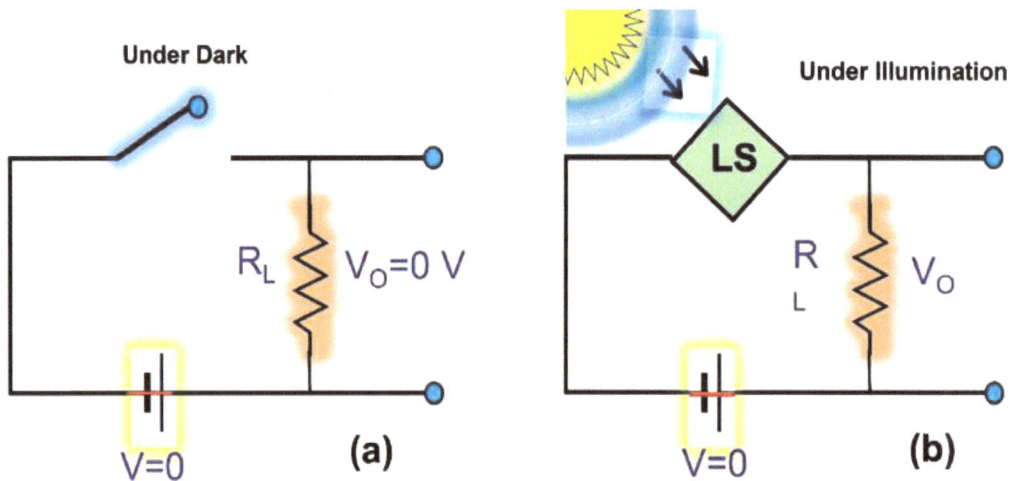

Fig. (5). The operation of a Self-Powered Photodetector under dark and illumination conditions (without external bias indicated as a short-circuited voltage source); the depiction is also valid for traditional detectors provided V≠0, and detector will behave as short-circuit under illumination.

The presented chapter can be broadly classified into different groups based on junction theory, like Schottky junction-based devices, heterojunction-based devices, and homojunction-based devices. Also, the novel nanostructure (QDs) based self-powered photodetectors are discussed.

ETL AND HTL EFFECTIVENESS IN SELF-POWERED PHOTODE-TECTORS

The incidence of photons over a photoactive material/device is responsible for generating charge carriers. The generated charge pairs should be separated and extracted (before recombination) under the influence of the internal field for the working of Self-Powered Photodetectors [26]. Generally, the improved electric field is achieved by sandwiching the photoactive layer between the electron transport layer (ETL) and hole transport layer (HTL) [30, 32]. The ETL and HTL are highly doped n-type and p-type materials and facilitate the transport of charge carriers before they recombine. The positive ions of ETL attract electrons, while the negative ions of HTL attract holes. Deepak *et al.* studied the comparative study of different organic HTLs for performance optimization of solar cells and found that Spiro-OMeTAD and PTAA are excellent organic HTLs [32]. Shen *et al.* [33] utilized Spiro-MeOTAD as HTL, $CsPbBr_3$ QDs as the active layer, and TiO_2 ETL and found an improved transient response with a rise time of 2.3 *msec*. Zhou *et al.* [34] fabricated a visible light self-powered photodetector. The device is based on GZO nanorods and $MAPBI_3$. This device shows an ultrafast response with a rise/fall time of 2 *msec*.

CLASSIFICATION OF SELF-POWERED PHOTODETECTORS BASED ON JUNCTION THEORY

Schottky Junction-Based Self-Powered Photodetectors

The most common, reliable, and easy way to fabricate a self-powered photodetector is by using suitable Schottky contacts [35, 36]. The choice of the metal electrode plays a crucial role in Schottky contact and hence influences the response of the self-powered photodetector [37, 38].

Graphene-Based Schottky Junction Devices

Graphene is a 2D material with high mobility that can be utilized for Schottky junction and is hence useful for self-powered photodetector operation. Jin *et al.* [39] fabricated a self-powered photodetector using a CdSe nanobelt. The device was prepared using CVD at around 1000°C and illuminated at 633 nm (visible-red light). This device shows high selectivity (peak at ~710 nm and FWHM <100 nm), excellent photosensitivity of 3.5×10^5 and photoconductive gain of 28 with the rise time (τ_r) 82 μsec and fall time (τ_f) 179 μsec. Gao *et al.* [40], extended the same research with the flexible substrate and found an improvement in transient behavior with 70 *μsec* and 137 *μsec* rise time and fall time, respectively. Graphene is also used with ZnO to provide ultrafast response and operation in the ultraviolet region of the self-powered photodetector. Boruah *et al.* [41] fabricated a device structure with a ZnO nanowire sandwiched in two graphene layers. This device shows high selectivity (with FWHM around 100 nm with a peak at ~350 *nm*) and thus works well under UV with t_r=3 s and t_d=0.47 s. Though the transient behaviour was not that good but it opened the path for other research to look after the ZnO based devices for self-powered operations. Duan *et al.* [35] fabricated graphene with Al-doped ZnO (AZO) nanorod array film as shown in Fig. (**6a**).

This device has shown good selectivity with a peak is around 350 *nm* and a very good transient response with a rise time of 37 *μsec* and a decay time of 330 *μsec* as shown in Fig. (**6b**). The device shows a significant improvement in UV-to-visible rejection ratio of 1×10^2 with a responsivity of 39 *mA/W*.

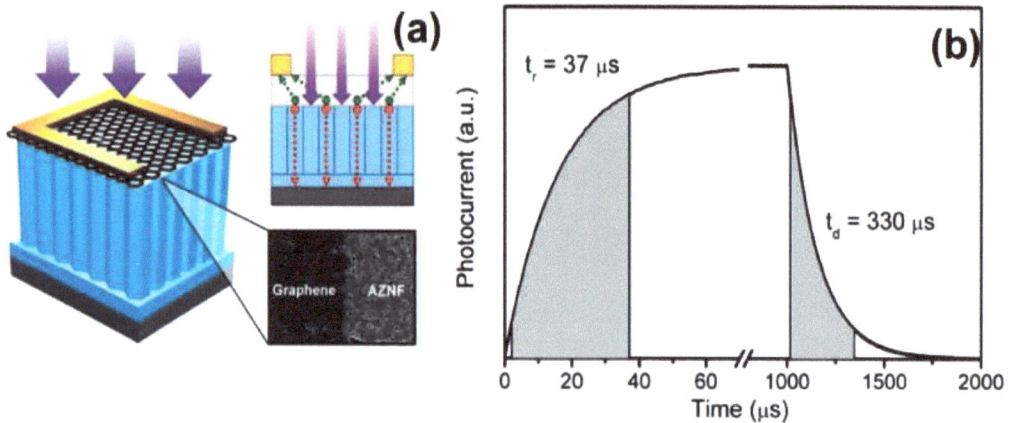

Fig. (6). (**a**) Graphene/Al: ZnO NR array film, and (**b**) Transient response of the DG/T-AZNF based device [35], Copyright 2017: American Chemical Society.

Gold-Based Schottky Junction Devices

Au electrode is commonly used for making Schottky contact-based photodetector. In 2014, Chen *et al.* [42] fabricated a UV selective asymmetric MSM (Metal-Semiconductor-Metal) structure-based device, and the Au electrode's width varies to make the asymmetric structure. This device has shown a responsivity of 20 *mA/W*. They have demonstrated the dependence of self-powered behavior on the asymmetric profile and achieved self-powered characteristics with an asymmetric ratio of 20:1 under UV illumination. Chen *et al.* [36] fabricated a MgZnO and asymmetric Au electrode Schottky junction-based self-powered UV photodetector. This device shows a spectral peak at around 320 *nm*, and narrow FWHM is achieved with a pure ZnO-based device. Mishra *et al.* [43] fabricated a ZnO/GaN Schottky junction with heterojunction with an Au electrode. In this structure, the ZnO active layer is fully depleted and improves the photocurrent to dark current ratio. They have studied structural variation. The performance of the devices is not that promising for self-powered applications but seems good under biased conditions. Ebrahimi and B Yarmand [44] fabricated nanostructured $Sn_xZn_{1-x}S$ composite material, with *x* can be changed from 0 to 1. This device is a Schottky contact-based self-powered device utilizing an Au electrode. They prepared the active layer of the device using a solvothermal process. Their main focus was on getting the bandgap tuning by varying the Sn concentration and they found the reduction in bandgap by increasing the Sn concentration. This device shows spectrum selective behavior with a peak, at ~400 nm and FWHM is less than 100 *nm*. Benyahia *et al*. [45] fabricated a ZnO-ZnS microstructure-based wideband self-powered photodetector using an Au electrode. The microstructure was prepared at a very high-temperature 5000OC to operate as an active layer. Such devices can be used in wideband (300 to 900 *nm*) applications with improved photo-response characteristics. The device shows a good response under UV, blue, green, and NIR and lacks spectrum selectivity (FWHM is ~300 *nm*).

Silver-Based Schottky Junction Devices

Silver metal is also widely used in Schottky-based devices. Purusothaman *et al*. [46] fabricated a flexible Schottky self-powered photodetector using floral-like ZnO NR and Ag-electrode on a PVDF substrate, as shown in Fig. (7). This device has shown a responsivity of 22.76 *mA/W*. Generally, the UV region is focused on devices utilizing ZnO. Researchers either look for a change in doping or the active material for other regions like visible and NIR.

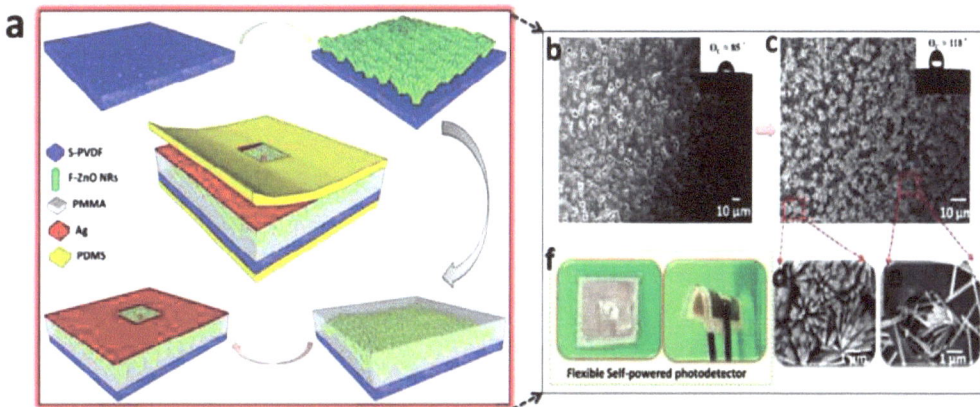

Fig. (7). (**a**) Schematic of self-powered photodetector utilizing floral ZnO NRs, (**b**) polyvinylidene difluoride (PVDF) untreated film, (**c**) SEM image of floral ZnO NRs, (**d**)-(**e**) magnified image floral ZnO NRs, and (**f**) fabricated self-powered device [46], Copyright 2018: American Chemical Society.

Other Metal/Alloy-Based Devices

Zhang *et al*. [47] introduced an MIS self-powered structure. They have used the Al_2O_3 layer between ZnO and Pt electrode, and found an efficiency improvement due to increased Schottky barrier height. The device's overall efficiency is improved by 2.77 times by introducing the Al_2O_3 layer. They have shown an improvement in the photoresponse property because of the Piezotronics Effect [48], with response and time being less than 0.1 *sec*. In 2019, Chang *et al*. [49]

used a Se microrod to achieve a broad spectrum covering from UV to visible, as shown in Figs. (**8a & b**). The Schottky junction is formed between Se and Ga-In alloy. This device has a very low dark current (200 *fA*), very high responsivity (408 *mA/W*), high detectivity $1.3\times10^{13}J$ and good transient response (t_r=124 μs and t_d=146 μs). But this device lacks the spectrum selectivity with FWHM is more than 300 *nm*.

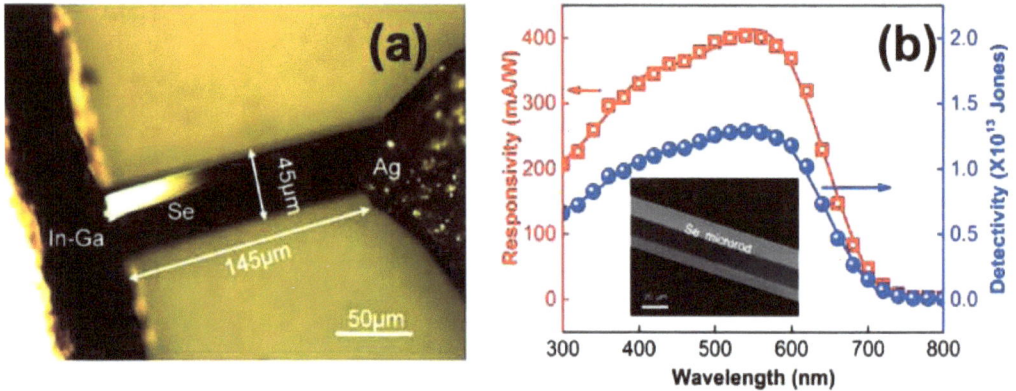

Fig. (8). (**a**) Photograph of the fabricated device with Se microrod, and (**b**) Responsivity and selectivity of Self-Powered device [49], Copyright 2019: American Chemical Society.

Quantum Dot-Based Devices

QDs are very useful for self-powered photodetectors because of their size-dependent tunable bandgap. Kumar *et al.* [28] reported the self-powered photodetector based on QDs. They have utilized CdSe QDs and Au for the Schottky junction and ZnO QDs layer as the charge transport layer, as shown in Figs. (**9a & b**). This device covers the visible region (350 to 750 *nm*) with a very fast response (τ_r=18 ms and τ_d=17.9 ms). This work reported the 17 times improvement in the photoresponse by incorporating the ZnO QDs layer. But this device lacks the spectrum selectivity. Kumar *et al.* [37] extended their research by replacing n-Si with ITO substrate and tried different metal electrodes to improve the photoresponse, as shown in Fig. (**10a**). They have implemented ITO/ZnO QDs/CdSe QDs based on a self-powered photodetector with Pd and Au electrodes. By using the Pd electrode, they found that the quantum efficiency increased 2.1 times, FWHM reduced from 190 to 61 nm, and transient response improved with τ_r=18.5 ms and τ_d=15.8 ms, as shown in Figs. (**10b & c**). In the [28] structure, light is incident from the top, while in the [37] structure, light is incident from the bottom side, thus utilizing the 100% illumination area.

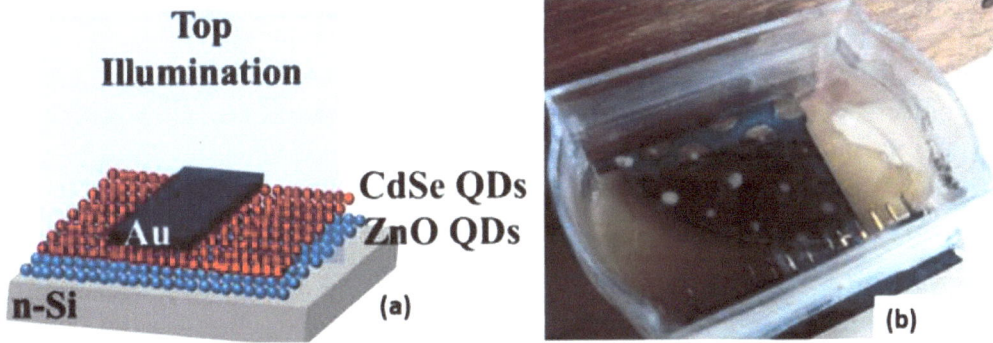

Fig. (9). (a) Schematic of QD-based self-powered PD, and (b) Image of the fabricated device [28], Copyright 2017: IEEE.

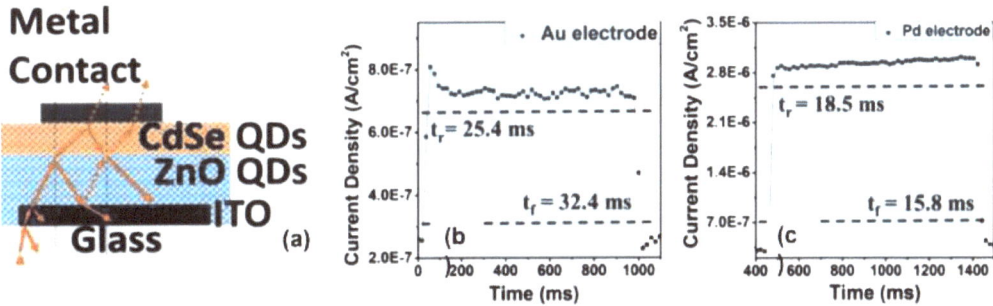

Fig. (10). (a) Schematic of the self-powered device, and (b)-(c) Transeint response of the device utilizing Au/Pd electrode [37], Copyright 2019: IEEE.

Heterojunction-Based Self-Powered Photodetectors

The heterojunction is formed between two dissimilar semiconductor materials with p-type and n-type doping. Bie *et al*. [50] reported an n-ZnO NW with a p-GaN-based self-powered UV selective device. Also, they have derived a CdSe NW self-powered photodetector. This device offers a good transient response with t_r=20 µs and t_d=219 µs and maximum out power of 1.1 μW. Hassan *et al*. [51] p-Si and n-ZnO NRs-based heterojunction self-powered photodetector with the transient response as t_r=25 ms and t_d=22 ms. These obtained results indicate poor transient response as compared to previously fabricated n-ZnO NW and p-GaN based device. Hatch el. al [52]. reported an n-ZnO and p-Copper Thiocyanate (CuSCN) based heterojunction self-powered UV selective photodetector, focusing on the improvement of transient behavior only. This device offers the ultrafast response t_r=500 ns and t_d=6.7 µs. They also extended their work by changing the annealing environment of ZnO [53] and reported transient responses with t_r=25 ns. Huang *et al*. [54] used ZnO NW, Graphene, CdS and electrolyte based heterojunction to fabricate UV self-powered photodetector. This device's spectral response extends to the visible blue region. The rise and fall time of the device are 5 ms each. Guo at. al [55]. fabricated ZIF-8@H:ZnO NR and Si heterojunction self-powered UV photodetector. The device characteristics are quite good under bias while not significant for self-powered operation. Ghamgosar *et al.* [56] reported p-Co_3O_4 and n-ZnO structure with an Al_2O_3 buffer layer, as shown in Fig. (**11a**). The comparative study is also performed utilizing nanowire and the thin film of ZnO, as shown in Figs. (**11b & c**). This structure shows improvement in photoresponsivity and the highest absorption at around 375 *nm*. The results obtained are quite promising under biased conditions but are not significant for self-powered operation. Yamada *et al.* [57] fabricated a transparent device utilizing p-CuI and n-In-Ga-Zn-O heterojunction. This device is transparent and works as a self-powered UV photodetector. Sinha *et al.* [58] reported a carbon dot and ZnO NRs on a paper substrate-based flexible self-powered photodetector, as shown in Fig. (**12**). The heterojunction is formed between ZnO NRs and graphite-coated paper substrate. Although the device shows good selectivity but has a feeble transient response of t_r=2 s and t_d=3.2 s.

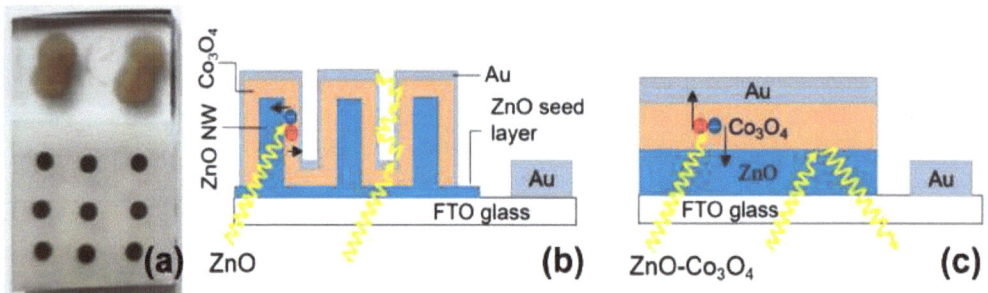

Fig. (11). (**a**) p-Co_3O_4 and n-ZnO structure with Al_2O_3 buffer layer self-powered device, Charge separation and light scattering comparison in **b**) NW and **c**) thin film geometry [56], Copyright 2019: American Chemical Society.

Agrawal *et al.* [59] reported a CZTS/MoS_2-based structure utilizing 2-D material with spectral response from 600 to 1200 *nm*. The photocurrent is more than 20 *nA* under self-bias conditions. The achieved results are not very promising for self-powered operations. Also, this device lacks spectrum selectivity. Wei *et al.* [60] fabricated p-NiO and AZO heterojunction-based devices using the chemical bath deposition technique (a low-cost technique). They have achieved a narrow spectral peak of around 400 with improved responsivity. Huang *et al.* [61] fabricated a heterojunction between Si NW and InGaZnO UV to a visible self-powered photodetector. This shows a spectral FWHM of less than 100 *nm* with a peak around 350 *nm* and a response time of 0.2 *msec*. Ahmed *et al.* [62] fabricated n-Si and p-NiO hetero- junction-based self-powered UV-NIR photodetector. The authors found that with an increase in annealing temperature from 300°C to 600°C, the size of crystallite increases, and the bandgap decreases. This device has shown an I_{ON}/I_{OFF} of 1.21×10^3 and a response time of less than 85 *msec*. This device lacks the spectrum selectivity. Zhou *et al.* [63] fabricated Cu-based device with $CsCu_2I_3$ and GaN heterojunction. The device exhibits self-powered operation with an FWHM of less than 100 nm. The device shows ultraviolet and red dual-wavelength photodetection and also shows an I_{ON}/I_{OFF} ratio of 97886 under UV and 31255 under red light at zero bias.

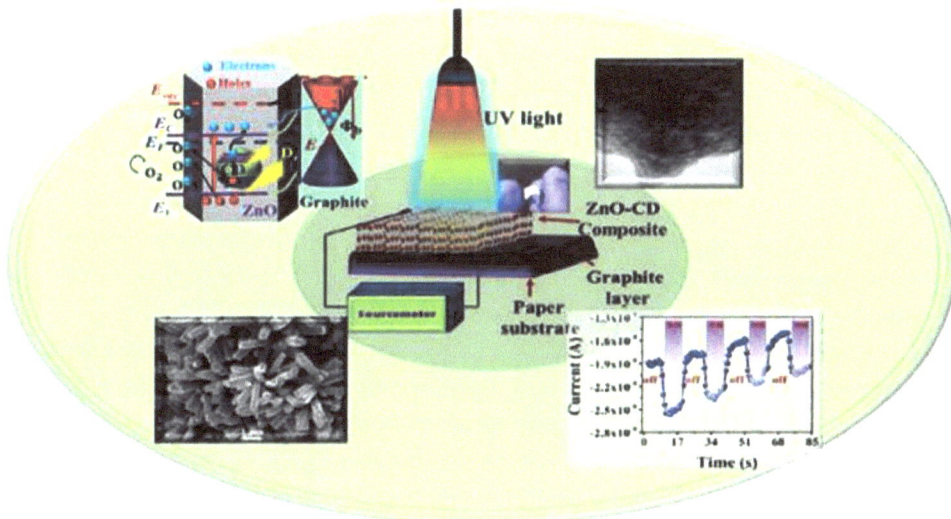

Fig. (12). Carbon dot-ZnO NR Graphite coated paper self-powered device [58], Copyright 2020: American Chemical Society.

Doping Based Devices

Many authors use doping as a tool to modify the optical and electrical properties of the materials. Chen *et al.* [64] reported alloying as doping. They have prepared n-$Mg_xZn_{1-x}O$ alloy with p-Si. They have found improvement in photoresponse because of the Piezotronics Effect. The piezo-potential changes and optical properties by varying the Mg content in ZnO. Zhou *et al.* [34] fabricated Ga-doped ZnO (GZO) NR with a perovskite-based self-powered device. The GZO NR was prepared using the water bath method, and ZnO NRs defects were reduced using Ga doping. This device has spectral coverage from 400 to 800 *nm* visible spectrum, which may be due to the incorporation of perovskite and Ga doping of ZnO, and lacks spectrum selectivity. The transient response is similar for the device with and without Ga doping. Saha *et al.* [65] prepared Ga-doped

ZnO NW with Ag coating using a p-Si substrate heterostructure device. This device has spectral coverage from 320 *nm* to 400 *nm* and excellent selectivity. Varying Ga doping leads to a redshift in the spectrum. Jiang *et al.* [66] changed the oxygen concentration in ZnO and fabricated SnS_2 and $ZnO_{1-x}S_x$ heterojunction-based self-powered devices. Though they have demonstrated spectral coverage from 365 to 850 *nm* with transient response of t_r=49.51 ms and t_d=25.93 ms but FWHM is less than 100 *nm*. They have also shown the transient behaviour with different illumination power under zero bias.

Hybrid Heterostructure Based Devices

Some researchers tried heterojunctions between organic and inorganic semiconductors to mitigate the mobility problem in organic semiconductors and get the advantage of organic semiconductors to fabricate flexible devices. Game *et al.* [67] fabricated an n-type nitrogen-doped ZnO and p-type spiro-MeOTAD hybrid heterojunction self-powered device. ZnO NRs are synthesized at 450°C to achieve the transient response of t_r=200 μs and t_d=950 μs. This device covers the UV-visible region and hence lacks spectrum selectivity. S. Sarkar and D. Basak [68] fabricated ZnO@CdS core-shell NR with PEDOT: PSS-based self-powered device. They used the aqueous chemical method for core-shell structure. The spectral coverage of the device is UV to visible and hence lacks spectrum selectivity with a transient response of 20 *msec*. Thermal treatment was utilized to reduce GO (Graphene Oxide) to rGO (reduced-Graphene Oxide) to dope ZnO by carbon atom and hence increase the spectral response in the visible region with higher spectral width. H. Kumar *et al.* [69] used quantum dots with organic material and fabricated PQT-12 and CdSe QD-based hybrid self-powered devices, as shown in Fig. (**13a**). They have demonstrated the transient response as t_r=15.32 ms and t_d=12.01 ms, as shown in Fig. (**13b**). They have shown a spectral peak around 400 *nm* but this device lacks spectrum selectivity. Li *et al.* [70] reported CdSe and p-$Sb_2(S_{1-x}Se_x)_3$-based self-powered NIR photodetector as shown in Figs. (**14a & b**).

This device can be utilized for heartbeat detection using 780 *nm* and 590 *nm* LEDs. Also, this device showed an FWHM of 35 *nm* without a filter with spectral coverage from 650 *nm* to 900 *nm*. Selectivity is achieved by changing the CdSe thicknesses and Se concentration, as shown in Figs. (**14c & d**).

The ZnO is doped with Al to enhance the optical characteristics of the self-powered device.

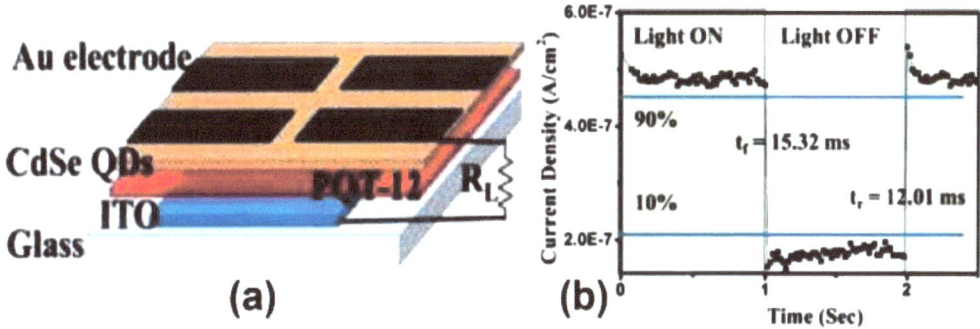

Fig. (13). (**a**) Schematic of QD/PQT-12 self-powered device, and (**b**) Transient response of the device [69], Copyright 2017: IEEE.

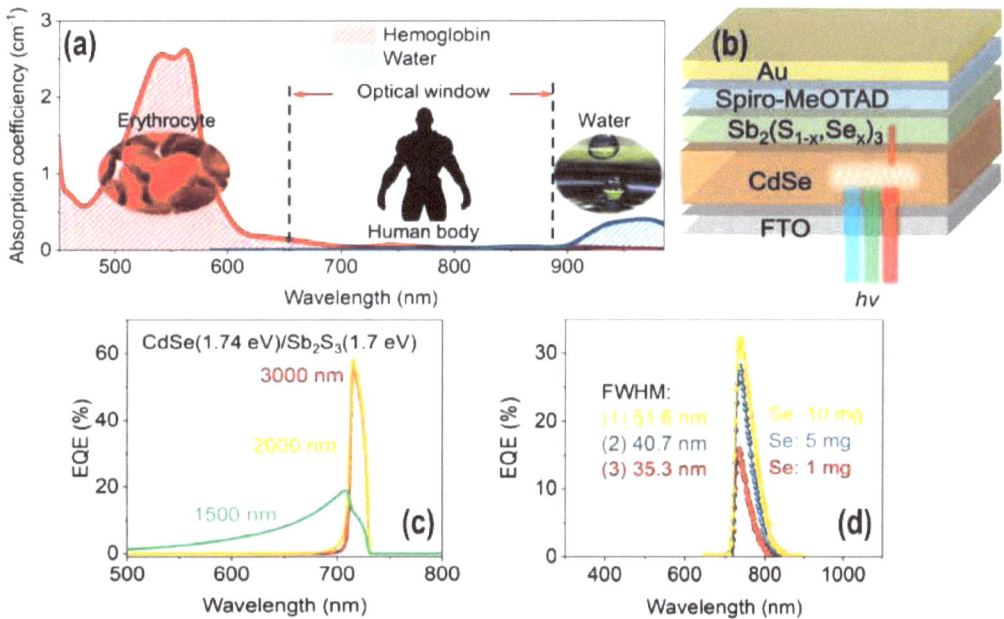

Fig. (14). (**a**) Absorption of haemoglobin and water, (**b**) Schematic of CdSe and p-$Sb_2(S_{1-x}Se_x)_3$-based device, (**c**) EQE for different CdSe thicknesses, and (**d**) EQE with different Se concentration [70], Published 2021 by Wiley as open access.

Homojunction Based Self-Powered Photodetectors

The homojunction is formed between two similar semiconductor materials with p-type and n-type doping. H. Shen *et al.* [26] reported a p-ZnO and n-ZnO-based

UV photodetector, as shown in Fig. (**15a**). They have used sapphire substrate and molecular beam epitaxy to deposit the ZnO layers (p-type and n-type). The n-ZnO layer acts as an active layer while the p-ZnO layer act as a filter layer, and hence detector shows a spectral width of 9 *nm,* as shown in Fig. (**15b**). Its optical response can be tuned from visible to NIR and can be used in multiple applications.

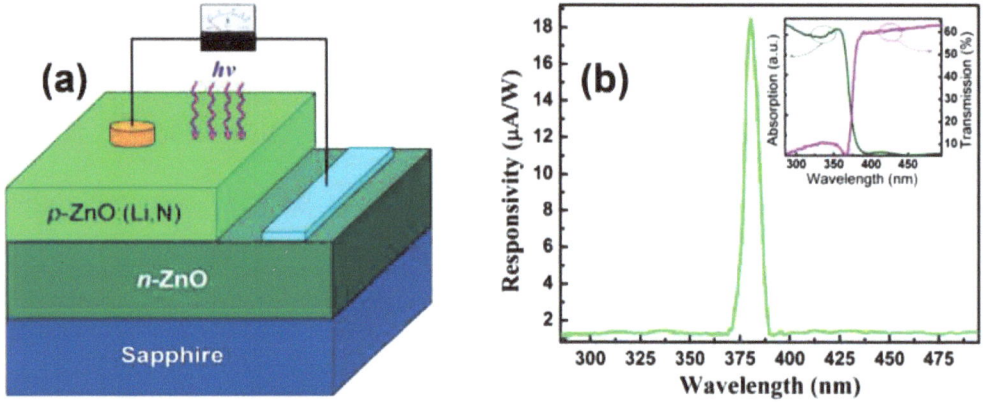

Fig. (15). (**a**) Schematic of n-ZnO and p-ZnO homojunction self-powered device, and (**b**) Responsivity of n-ZnO and p-ZnO based device [26], Copyright 2013: AIP Publishing.

PYROELECTRIC EFFECT

Some non-centrosymmetric crystals (10 out of 32 crystal structures) show spontaneous polarization along the polar axis, known as polar crystals. At constant temperature, spontaneous polarization is neutralized by internal or external conductivity or twinning. When the temperature changes, electrical changes can be observed across the crystal perpendicular to the polar axis [71]. The effect is called the pyroelectric effect. The pyroelectric effect is the temporary change in the polarization of the crystal with a temperature change. When the temperature increases $(dT / dt > 0)$, the thermal vibration increases, resulting in a reduction in the number of bound surface charges and hence the reduction in polarization and *vice-versa*. As a result, pyroelectric material shows electric potential under open-circuit conditions while electric current under short-circuit conditions [72].

Consider a Pyroelectric material with conducting electrode area (A), and the spontaneous polarization (P) is normal to the A. Then, bound surface charges on pyroelectric material are given by [71].

$$q = AP \tag{6}$$

Where

$$P = \chi E + P_s \tag{7}$$

P_s is the total polarization,

χ is the susceptibility and

E is the electric field inside the pyroelectric material

E may include any alternating electric field radiation and biasing field E_b. Therefore, the current flowing through the surface of pyroelectric material can be represented as:

$$i = A \frac{d}{dt}(P) \tag{8}$$

$$i = A \frac{d}{dt}(\chi E_b + P_s) \tag{9}$$

E_b is constant and independent variables χ and P_s varies with temperature. Now the current can be expressed as:

$$i = A \left(E_b \frac{d\chi}{dT} + \frac{dP_s}{dT} \right) \frac{dT}{dt} \tag{10}$$

Where, A $(dP_s/dT)(dT/dt)$ represents the true pyroelectric current while A $E_b(d\chi/dT)(dT/dt)$ represents the induced pyroelectric current. The induced current may be due to non-pyroelectric materials. So, in the absence of induced effects, the pyroelectric current can be defined as:

$$I_{py} = \rho A \frac{dT}{dt} \tag{11}$$

Where $\rho = dP_s/dT$ is the pyroelectric coefficient, which is a property of a material. So, pyroelectric current depends on the pyroelectric coefficient, electrode area, and temperature change rate with time.

The corresponding open-circuit voltage [73] is:

$$V_{py} = \rho A \frac{\Delta T}{C} \tag{12}$$

Where C is the equivalent capacitance of the device.

Table **1** shows the photodetectors commonly used as pyroelectric materials and their pyroelectric coefficient.

Table 1. Different Pyroelectric materials utilized in photodetectors.

S.No	Material	Pyroelectric coefficient ($\mu Cm^{-2} K^{-1}$)
1	CdS	3.5 [74]
2	CdSe	4 [74]
3	AlN	6-8 [72]
4	$KNbO_3$ nanowires	8 [72]
5	ZnO nanowires	12-15 [72]
6	$BaTiO_3$	225–259 [72]
7	$BiFeO_3$	90 [75]

Solution-processed techniques can prepare the above materials. The frequently used pyroelectric material is ZnO because of its comparably high pyroelectric coefficient and ease of synthesis.

The lower electric field and materials unavailability has threatened the application of self-powered photodetectors. But the utilization of the pyroelectric field in self-powered can be a game-changer.

Pyro-phototronic Effect

Initially, when there is no light, *i.e.*, the photodetector is under dark conditions, a built-in field exists due to the diffusion of electrons and hole across the junction, which restricts the flow of charge carriers across the junction. But thermally generated electron-hole pairs move across the junction and thus lead to I_{dark}.

Two effects are considered when photons are incident (Illumination is ON) on the pyroelectric material. First, there is the generation of electron-hole pairs. Second, there is a change in the temperature of the pyroelectric material w.r.t. time, which leads to the pyroelectric potential across the pyroelectric material (pyroelectric effect). This combined effect of photogeneration of electron-hole pair and the development of the pyroelectric field is known as the Pyro-phototronic effect. In the case of the self-powered photodetector, where there exists a problem of charge separation because of the lower field, this Pyro-phototronic effect plays a crucial role for photogenerated charge carriers due to an extra field.

Figs. (**16a & b**) show the generalized structure of the Pyro-phototronic effect-assisted self-powered photodetector. In this case, the light illumination is done from the bottom side, a transparent conducting oxide (TCO) coated glass substrate. TCO can be Indium-doped tin oxide (ITO) or Fluorine-doped tin oxide (FTO). Over TCOs, the deposition of a pyroelectric material is done. The most frequently used pyroelectric material is ZnO because of its higher pyroelectric coefficient, which leads to higher pyroelectric potential. The deposited ZnO material is inherently n-type. The active region for photogeneration of carriers is the depletion region of n-ZnO and p-type material (over ZnO) like PEDOT:PSS, p-GaN, *etc.*, as shown in Fig. (**16a**). An additional active layer can be used to provide a more photoactive region for the photogeneration of charge carriers as shown in Fig. (**16b**). In the latter case, the ZnO acts as an electron-transport layer, and there is a need for a hole transport laye over the active layer like SnOx, MoOx, *etc.*

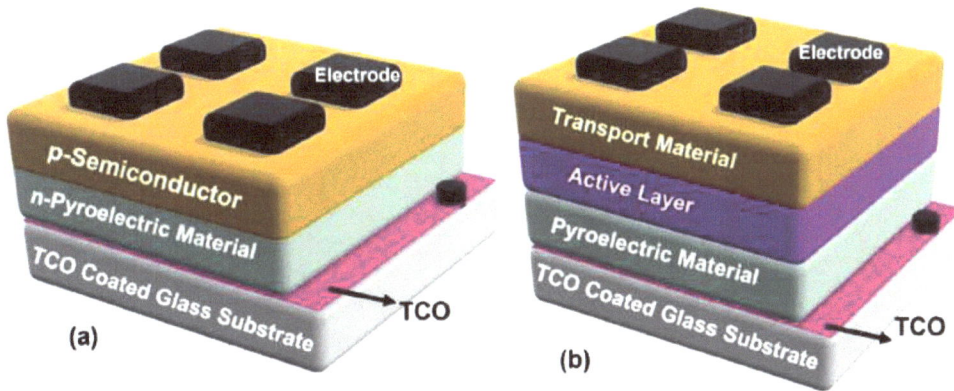

Fig. (16). Generalized structures of Pyro-phototronic effect assisted self-powered photodetector.

4-Stage Working Explanation

The generalized transient response of the pyro-phototronic effect-assisted self-powered photodetector is shown in Fig. (**17**). The illumination is ON and OFF for 1 sec each (frequency=0.5 Hz). Initially, there is no illumination of light, *i.e.*, the photodetector is under dark conditions, and the current through the photodetector is I_{dark}.

The operation of the Pyro-phototronic assisted self-powered photodetector can be explained in four stages as follows [76]:

1ˢᵗ Stage: At the time (t=0), the illumination is turned ON. If the incident photon's energy is higher than the bandgap of the active layer, this leads to the generation of electron-hole pairs. The built-in electric field separates these photogenerated charge carriers. At the same time, as the light is incident on pyroelectric material, there is a change in the temperature of the pyroelectric material w.r.t. time. This develops a temporary potential across the pyroelectric material. The electric field due to the pyroelectric effect (E_{py}) adds with the built-in field (E_{bi}) and hence there is an increase in an effective electric field responsible for faster charge separation and collection. At time (t_i), the maximum current ($I_{ph} + I_{py}$) is flowing where I_{ph} and I_{py} are the current because of the built-in electric field and the maximum pyroelectric current as indicated in Fig. (**17**) as 1st stage and also in Fig. (**18a**).

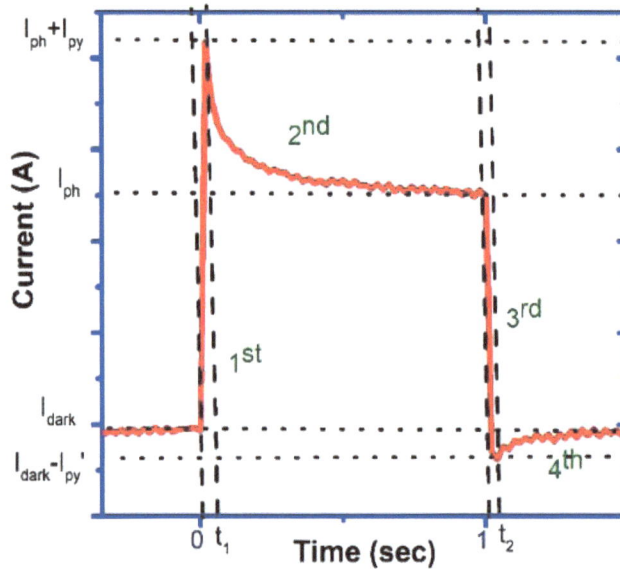

Fig. (17). Generalized Transient response of Pyro-phototronic effect assisted self-powered photodetector.

2nd Stage: The pyroelectric potential is temporary and dies out as the temperature gradient vanishes. Now the current through the photodetector saturates to a I_{ph} as shown in Fig. (**17**) (2nd stage) and Fig. (**18b**).

3rd Stage: At the time (t_{ph}), the illumination is turned OFF. The pyroelectric effect again plays a role but in the opposite direction. As the temperature decreases, the pyroelectric field acts opposite to the direction of the built-in electric field. This results in a sudden reduction in current. At the time ($t = t_1$), the maximum current starts flowing in the reverse direction (I_{dark}-I'_{py}). The resulting current variation is shown in Fig. (**17**) (3rd stage) and the device form in Fig. (**18c**).

4th Stage: Again, since the pyroelectric potential is temporary, therefore it dies out, and the current saturates to a dark current value (I_{dark}) as shown in Fig. (**17**) (4th stage) and Fig. (**18d**).

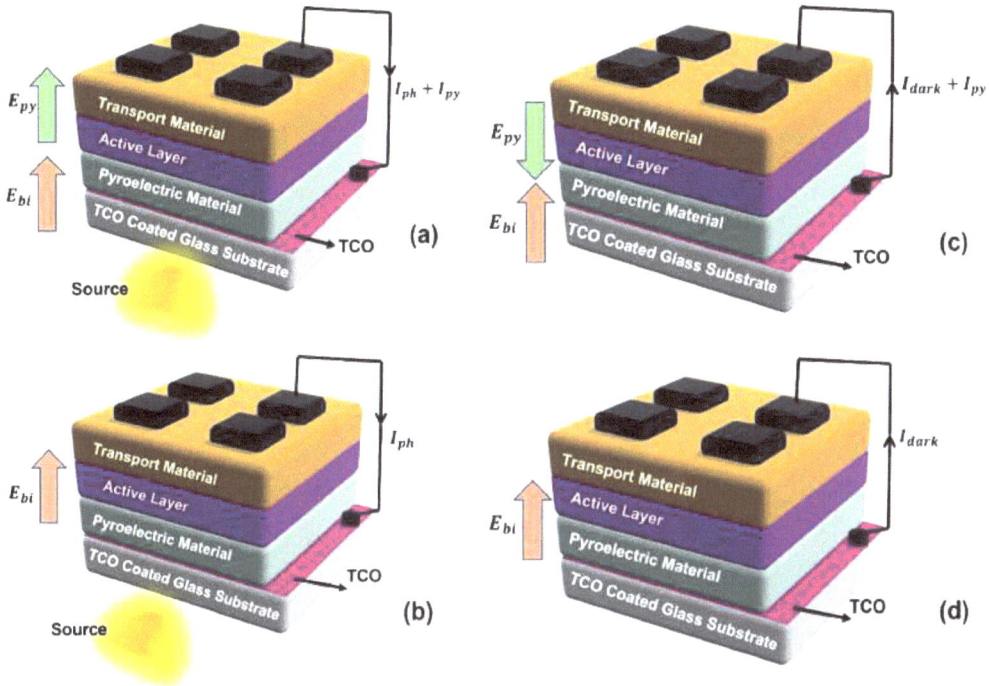

Fig. (18). Different stages working in Pyro-phototronic Effect assisted Self-Powered Photodetector.

RECENT ADVANCEMENTS

Boruah *et al.* [77] fabricated a self-powered photodetector utilizing the Pyro-phototronic effect. This device is based on ITO/Cl:ZNO NW/PEDOT PSS/Ag. The fabricated device has a rise time and fall time of 28 *msec* and 23 *msec,* respectively. In this structure, ZnO is doped by halogen to enhance the doping density in ZnO and hence result in a pyro-phototronic effect. The transient response indicates a pyro-phototronic effect. Chen *et al.* [78], fabricated graphene and ZnO Schottky contact-based self-powered photodetector. They have used the surface treatment of ZnO with H_2O_2 to enhance the transient behavior with a rise/decay time of 32 *msec* and responsivity of 50 *μA/W*. This device shows good selectivity with FWHM less than 100 nm. In this research, though the pyro-phototronic effect and built-in electric field interaction at the graphene/ZnO interface enhance the transient behavior, the increment is not significant compared to the previous result. To improve the responsivity, Chen *et al.* [79] fabricated a device based on ITO/ZnO NW/p-Si/Cu, as shown in Fig. (**19a**). This device has responsivity and detectivity of 164 *mA/W* and 8.78×10^{11} *Jones*. This device shows pyro-current approximately 6 times the saturated photocurrent, as shown in Fig. (**19b**). The same research group [73] studied the same device structure with changes in the frequency of light, power intensity, and different wavelengths of light and found enhanced detectivity and responsivity of 1300 mA/W and 1×10^{13} *Jones* for UV light 1 *kHz* frequency and light intensity of 3.7×10^{-3} mW/cm^2. Rana *et al.* [76] fabricated p-NiO and n-ZnO-based heterojunction-based devices. The fabricated device is a transparent and self-powered UV photodetector. It utilizes the pyro-phototronic effect, which improves responsivity and detectivity by 5.46 and 6.063 times in the UV region. Also, the device has a good transient response with t_r=3.92 μs and t_d=8.90 μs. Here narrowband semiconductor is used to enhance the spectral range. To improve the transient response, *Silva et. al* [80]. Utilized Si-substrate. They implemented ITO/ZnO/SnO$_x$/Si/Al-based devices. In this device, SnO$_x$ act as a hole-transport layer. The fabricated device shows rise and fall time of 2 *μsec* and 3 *μsec* respectively.

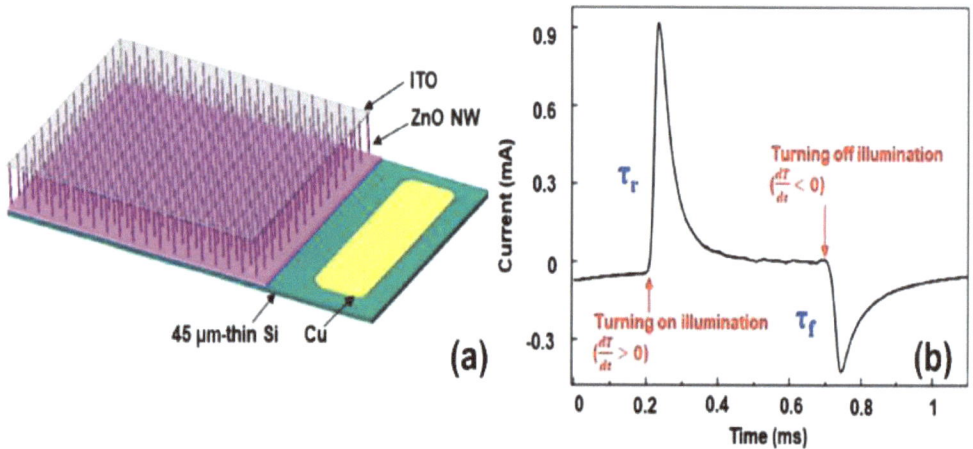

Fig. (19). (a) ITO/ZnO NW/p-Si/Cu based device, and (b) Transient response of the device [73], Published 2020 by Beilstein as open access.

SUMMARY

Table **2** shows the summary of different self-powered photodetectors indicating junction type and various figures of merit.

Table 2. Summary Table of Self-Powered Photodetectors.

Structure/ Materials	Structure Category	Response Time	Responsivity	Detectivity	Spectrum Region	EQE	References
Graphene/ZnO NW/graphene	Schottky Junction	3 s /0.47 s	~0.55 mA/W at zero bias	--	UV	-	[41]
Graphene/ZnO:Al nanorod-array	Schottky Junction	37 μs /330 μs	0.039 A/W at zero bias	--	UV	--	[35]
Graphene/H$_2$O$_2$-treated ZnO	Schottky Junction	32 ms	50 μA/W at zero bias	--	UV	--	[78]

(Table 2) cont.....

Au-ZnO-Au MSM structure	Schottky Junction	--	20 mA/W at zero bias	--	UV	--	[42]
Au/ZnO/GaN/Al$_2$O$_3$	Schottky+Heterojunction	20 ms	95.8 mA/W at zero bias	4.83×10^{13} Jones	UV	--	[43]
SnxZn1-xS nanostructures	Schottky Junction	--	0.0471 A/W at zero bias	--	UV	--	[44]
ZnO-ZnS Microstructured Composite	Schottky Junction	*22.5/45ms 22.3/40 ms 25.2/15 ms*	*3.34 mA/W 1.05 mA/W 0.47 mA/W at zero bias*	*8.9×10^{12} 2.8×10^{12} 1.4×10^{12} Jones*	*UV-Visible-NIR*	--	[45]
Floral-like F-ZnO nanorods on PVDF substrate	Schottky Junction	--	22.76 mA/W at zero bias	2.722×10^{10} cmH$^{1/2}$W^{-1}	UV	--	[46]
Se/In-Ga microrod	Schottky Junction	124 µs /146 µs	408 mA/W at zero bias	1.30×10^{13} Jones	UV-visible	--	[49]
Au/CdSe QDs/ZnO QDs/n-Si	Schottky Junction	17.9 ms /18 ms	10.23×10^{-3} A/W at zero bias	8.81×10^{9} cm Hz$^{0.5}$ W^{-1}	UV	--	[28]
Pd(Au)/CdSe QDs/ZnO QDs/ITO	Schottky Junction	17.15 ms (28.9 ms)	7.48 mA/W at zero bias	1.3×10^{10}	--	2.21% (0.87%)	[37]
ZnO/GaN	Heterojunction	20 µs /219 µs	--	--	UV	--	[50]
ZnO NR array/p-Si	Heterojunction	25 ms / 22 ms	--	--	UV	--	[51]
ZnO NR/ p-CuSCN	Heterojunction	4 ns/ 6.7 µs	0.0075 A/W at zero bias	--	UV	--	[52]
ZnO NR/ p-CuSCN	Heterojunction	~25 ns	0.05 A/W at near-zero bias	--	UV	--	[53]
ZnO NW Array/Graphene/CdS/Electrolyte	Heterojunction	5 ms	27.3 mA/W and 4.3 mA/W at zero bias	--	UV, Visible	--	[54]
ZIF-8@H:ZnO NRs/p-Si	Heterojunction	852µs /607 µs	~7.07×10^{4} mA/W	~2.14×10^{16} Jones	UV-NIR	--	[55]
ZnO-Co$_3$O$_4$ nanowire	Heterojunction	6s	21.80 mA/W at 0.1 V	4.12×10^{12} Jones	Visible	--	[56]
CuI/In-Ga-Zn-O	Heterojunction	2.5 ms /35 ms	0.6 mA/W at zero bias	-	UV	--	[57]
p-NiO/n-ZnO	Heterojunction	3.92µs / 8.9 µs	0.29 A/W	2.75×10^{11} Jones	UV	--	[76]
Carbon dot enhanced ZnO/graphite	Heterojunction	2s / 3.2 s	9.57 mA/W at zero bias	4.27×10^{8} Jones,	UV	--	[58]
Cu$_2$ZnSnS$_4$(CZTS) and MoS$_2$	Heterojunction	81 ms / 79 ms	141 mA/W at zero bias	--	--	--	[59]
InGaZnO/ p-silicon NW	Heterojunction	~0.1 ms	0.53 A/W at zero bias	--	UV	--	[61]
p-NiO/n-Si	Heterojunction	<85 ms	13.08, 46.02, 44.49, mA/W at zero bias	1.03×10^{11}, 3.65×10^{11}, 3.53×10^{11}, Jones	UV, Red, NIR	4.43%, 8.62%, 6.47%	[62]

(Table 2) cont.....

CuI/CsCu$_2$I$_3$/GaN	Heterojunction	--	71.7 mA/W 9.1 mA/W at zero bias	3.3×10^{12} Jones 4.2×10^{11} Jones	UV, Red	26.1%, 3.3%	[63]
SnS$_2$/ZnO$_{1-x}$S$_x$	Heterojunction	49.51 ms / 25.93 ms	8.28 mA/W at zero bias	5.09×10^{10} Jones	UV	--	[66]
N:ZnO–SpiroMeOTAD	Hybrid heterojunction	200 ms/ 950ms	17 mA/W at zero bias	--	--	--	[67]
ZnO@CdS Core-shell Nanorod Arrays	Hybrid heterojunction	20 ms	--	--	UV, Visible	--	[68]
CdSe QD/ PQt-12 polymer	Hybrid heterojunction	~12.01 ms/~15.32 ms	~3.3 mA/W at zero bias	5.4 × 109 cmHz$^{1/2}$W^{-1}	Visible	--	[69]
p-ZnO:(Li,N)/n-ZnO	Homojunction	--	~16 µA/W at zero bias	--	UV	--	[26]
Au–MgZnO–Au	Schottky	0.23 µs / 92 µs	2.22 mA/W at zero bias	4.4 × 10^{11} Jones	UV	--	[36]
n-NiO nanoflakes/AZO NRs	Heterojunction	~2 ms	85.12 mA/W at zero bias	1.737 × 10^{12} cm·Hz$^{1/2}$/W	UV	--	[60]
FTO/CdSe/ Sb2(S$_{1-x}$,Se$_x$)$_3$ /Spiro-OMeTAD/Au	Hybrid Heterojunction	--	0.19 A/W	--	NIR	32.2% at 735 nm	[70]
ITO / Cl:ZNO NW / PEDOT PSS / Ag	Hybrid Heterojunction	28 ms / 23 ms	2.2254 mA/W	1.54 × 10^{10}	UV	0.792	[77]
ITO/ZnO NW/p-Si/Cu	Heterojunction	15µs / 21µs	164 mA/W	8.78 × 10^{11}	NIR	-	[79]
ITO/ZnO/SnOx/Si/Al	Heterojunction	2 µs/3 µs	64.1 mA/W	2.40E+11	*UV-V* isible-*NIR*	-	[80]

CONCLUSION

The chapter discusses the necessity of self-powered photodetectors starting from the traditional photodiode to highly sophisticated Pyro-phototronic effect-assisted self-powered photodetectors. It has been observed that there is a need for a strong electric field for the photogenerated charge carrier separation and extraction. This electric field is generally provided by the inherent high bandgap materials of II-VI, III-V, and IV-IV group compound semiconductor materials. Pyroelectric materials significantly enhance the built-in electric field required for self-powered operation by developing temporary electric potential. It has been found that this pyroelectric potential varies with the frequency of light, the frequency set for transient analysis, and the intensity of light. Currently, the scope of these pyroelectric effect utilized (Pyro-phototronic effect) devices is limited to ultraviolet detectors because of the use of the ZnO nanomaterials. The involvement of pyro-electric materials focusing on other regions of the spectrum can also be emphasized in the future. Thus, a broad scope of research exists for self-powered operation in visible and Infrared regions.

LEARNING OBJECTIVES

• Photodetector applications, working and performance parameters.

• Self-Powered Photodetector working, benefits, and classification based on junction theory.

• Pyroelectric Effect and its utilization in Self-powered photodetectors.

• Recent Advancements in Self-powered photodetector.

MULTIPLE CHOICE QUESTIONS

1. The following material is *NOT* used for sensing UV radiations:

a. ZnO
b. TiO_2
c. PbS
d. GaN

2. A photodetector has a dark current of 10 *nA* while the current increases to 1 *μA* when light is illuminated. What is the sensitivity of the photodetector?

a. 10
b. 100
c. 1000
d. 10000

3. Nanowire is an example of which class of material:

a. 0-D
b. 1-D
c. 2-D
d. 3-D

4. The reflection from the surface of a device comes under:

a. Optical losses.
b. Thermalization losses.
c. Recombination losses.
d. Contact and transport losses.

5. The bandgap of a quantum dot (QD):

a. Increases with a decrement in size of QD.
b. Decreases with a decrement in size of QD.
c. Not change with the size of QD.
d. None of the above.

6. For the operation of a traditional photodetector:

a. A forward bias is required.
b. A reverse bias is required.
c. No potential is required.
d. None of the above.

7. For the operation of a self-powered photodetector,:

a. A forward bias is required.
b. A reverse bias is required.
c. No potential is required.
d. None of the above.

8. Electron-transport layer (ETL):

a. Attracts electrons towards itself.
b. Repels holes away from itself.
c. Both of the above.
d. None of the above.

9. The responsivity of a photodetector:

a. Increases with the applied reverse bias.
b. Decreases with the applied reverse bias.
c. Does not change with applied bias.
d. None of the above.

10. Hybrid heterojunction is a junction between:

a. Two n-type materials.
b. Two p-type materials.
c. N-type and p-type materials.

d. Organic and inorganic materials.

11. The pyroelectric current depends on:

a. Area of electrode.
b. Rate of change in temperature w.r.t. time.
c. Pyroelectric coefficient.
d. All of the above.

12. Which of the following has the largest Pyroelectric coefficient:

a. ZnO.
b. CdSe.
c. $BaTiO_3$.
d. CdS.

13. The Pyro-phototronic effect is explained in:

a. Two steps.
b. Three Steps.
c. Four Steps.
d. Five Steps.

14. On increasing the intensity of light, the photocurrent:

a. Increases.
b. Decreases.
c. Does not change.
d. None of the above.

15. The tunability of a self-powered photodetector can be achieved using:

a. Quantum Dots.
b. Nanowires.
c. Nano Wall.
d. Bulk materials.

ANSWER KEY

1. (c)

2. (b)

3. (b)

4. (a)

5. (a)

6. (b)

7. (c)

8. (c)

9. (a)

10. (d)

11. (d)

12. (c)

13. (c)

14. (a)

15. (a)

ACKNOWLEDGEMENT

Hemant Kumar thanks the Science and Engineering Department (SERB) through the Department of Science and Technology (DST), Government of India, under Grant SRG/2020/001282.

REFERENCES

[1]   M. El-Desouki, M. Jamal Deen, Q. Fang, L. Liu, F. Tse, and D. Armstrong, "CMOS image sensors for high speed applications", *Sensors,* vol. 9, no. 1, pp. 430-444, 2009.
[http://dx.doi.org/10.3390/s90100430] [PMID: 22389609]

[2]   M.B. Eyring, "Spectroscopy in forensic science", In: *Encyclopedia of Physical Science and Technology.* Elsevier, 2003, pp. 637-643.
[http://dx.doi.org/10.1016/B0-12-227410-5/00957-1]

[3]   Z.Q. Huang, Y.C. Chen, and C.Y. Wen, "Real-time weather monitoring and prediction using city buses and machine learning", *Sensors,* vol. 20, no. 18, p. 5173, 2020.
[http://dx.doi.org/10.3390/s20185173] [PMID: 32927855]

[4]   S. Pooja, D.V. Uday, U.B. Nagesh, and S.G. Talekar, "Application of MQTT protocol for real time weather monitoring and precision farming", *2017 International Conference on Electrical, Electronics, Communication, Computer, and Optimization Techniques (ICEECCOT)..* Mysuru, pp. 1-6, 2017.

[http://dx.doi.org/10.1109/ICEECCOT.2017.8284616]

[5]   M.E. Ressler, J.J. Bock, S.V. Bandara, S.D. Gunapala, and M.W. Werner, "Astronomical imaging with quantum well infrared photodetectors", *Infrared Phys. Technol.,* vol. 42, no. 3-5, pp. 377-383, 2001.
[http://dx.doi.org/10.1016/S1350-4495(01)00096-2]

[6]   C.I. Rablau, "Lidar: A new self-driving vehicle for introducing optics to broader engineering and non-engineering audiences", *Fifteenth Conference on Education and Training in Optics and Photonics: ETOP 2019..* Quebec City, Canada, p. 138, 2019.
[http://dx.doi.org/10.1117/12.2523863]

[7]   F. Rutz, "InGaAs SWIR photodetectors for night vision", *Proceedings Infrared Technology and Applications XLV,* vol. 11002, pp. 202-208, 2019.
[http://dx.doi.org/10.1117/12.2518634]

[8]   C. Choi, M.K. Choi, S. Liu, M. Kim, O.K. Park, C. Im, J. Kim, X. Qin, G.J. Lee, K.W. Cho, M. Kim, E. Joh, J. Lee, D. Son, S-H. Kwon, N.L. Jeon, Y.M. Song, N. Lu, and D-H. Kim, "Human eye-inspired soft optoelectronic device using high-density MoS2-graphene curved image sensor array", *Nat. Commun.,* vol. 8, no. 1, p. 1664, 2017.
[http://dx.doi.org/10.1038/s41467-017-01824-6] [PMID: 28232747]

[9]   Y. Kumar, H. Kumar, B. Mukherjee, G. Rawat, C. Kumar, B.N. Pal, and S. Jit, "Visible-Blind Au/ZnO quantum dots-based highly sensitive and spectrum selective schottky photodiode", *IEEE Trans. Electron Dev.,* vol. 64, no. 7, pp. 2874-2880, 2017.
[http://dx.doi.org/10.1109/TED.2017.2705067]

[10]   S. Li, T. Deng, Y. Zhang, Y. Li, W. Yin, Q. Chen, and Z. Liu, "Solar-blind ultraviolet detection based on TiO 2 nanoparticles decorated graphene field-effect transistors", *Nanophotonics,* vol. 8, no. 5, pp. 899-908, 2019.
[http://dx.doi.org/10.1515/nanoph-2019-0060]

[11]   Z. Fan, "An analysis of gan-based ultraviolet photodetector", *IOP Conf. Ser.: Mater. Sci. Eng.,* vol. 738, p. 012006, 2020.
[http://dx.doi.org/10.1088/1757-899X/738/1/012006]

[12]   X. Yin, C. Zhang, Y. Guo, Y. Yang, Y. Xing, and W. Que, "PbS QD-based photodetectors: Future-oriented near-infrared detection technology", *J. Mater. Chem. C Mater. Opt. Electron. Devices,* vol. 9, no. 2, pp. 417-438, 2021.
[http://dx.doi.org/10.1039/D0TC04612D]

[13]   M.C. Gupta, J.T. Harrison, and M.T. Islam, "Photoconductive PbSe thin films for infrared imaging", *Materials Advances,* vol. 2, no. 10, pp. 3133-3160, 2021.
[http://dx.doi.org/10.1039/D0MA00965B]

[14]   R.K. Bhan, and V. Dhar, "Recent infrared detector technologies, applications, trends and development of HgCdTe based cooled infrared focal plane arrays and their characterization", *Opto-Electron. Rev.,* vol. 27, no. 2, pp. 174-193, 2019.
[http://dx.doi.org/10.1016/j.opelre.2019.04.004]

[15]   S. Zhang, H. Jiao, X. Wang, Y. Chen, H. Wang, L. Zhu, W. Jiang, J. Liu, L. Sun, T. Lin, H. Shen, W. Hu, X. Meng, D. Pan, J. Wang, J. Zhao, and J. Chu, "Highly sensitive insb nanosheets infrared photodetector passivated by ferroelectric polymer", *Adv. Funct. Mater.,* vol. 30, no. 51, p. 2006156, 2020.
[http://dx.doi.org/10.1002/adfm.202006156]

[16]   H.J. Song, M.H. Seo, K.W. Choi, M.S. Jo, J.Y. Yoo, and J.B. Yoon, "High-performance copper oxide visible-light photodetector *via* grain-structure model", *Sci. Rep.,* vol. 9, no. 1, p. 7334, 2019.
[http://dx.doi.org/10.1038/s41598-019-43667-9] [PMID: 31089236]

[17]   L. Cong, H. Zhou, M. Chen, H. Wang, H. Chen, J. Ma, S. Yan, B. Li, H. Xu, and Y. Liu, "An ultra-fast, self-powered and flexible visible-light photodetector based on graphene/Cu 2 O/Cu gradient heterostructures", *J. Mater. Chem. C Mater. Opt. Electron. Devices,* vol. 9, no. 8, pp. 2806-2814,

2021.
[http://dx.doi.org/10.1039/D0TC05248E]

[18]   S.M. Huang, S.J. Huang, Y.J. Yan, S.H. Yu, M. Chou, H.W. Yang, Y.S. Chang, and R.S. Chen, "Extremely high-performance visible light photodetector in the Sb2SeTe2 nanoflake", *Sci. Rep.*, vol. 7, no. 1, p. 45413, 2017.
[http://dx.doi.org/10.1038/srep45413] [PMID: 28350014]

[19]   N.T. Shelke, S.C. Karle, and B.R. Karche, "Photoresponse properties of CdSe thin film photodetector", *J. Mater. Sci. Mater. Electron.*, vol. 31, no. 18, pp. 15061-15069, 2020.
[http://dx.doi.org/10.1007/s10854-020-04069-0]

[20]   T. Markvart, and L. Castañer, "Chapter IA-1 - principles of solar cell operation", In: *Practical Handbook of Photovoltaics.* Acadmic press, 2012, pp. 7-31.
[http://dx.doi.org/10.1016/B978-0-12-385934-1.00001-5]

[21]   H. Heidarzadeh, A. Rostami, and M. Dolatyari, "Management of losses (thermalization-transmission) in the Si-QDs inside 3C–SiC to design an ultra-high-efficiency solar cell", *Mater. Sci. Semicond. Process.*, vol. 109, p. 104936, 2020.
[http://dx.doi.org/10.1016/j.mssp.2020.104936]

[22]   G. Rawat, D. Somvanshi, Y. Kumar, H. Kumar, C. Kumar, and S. Jit, "Electrical and ultraviolet-a detection properties of e-beam evaporated n-TiO2 Capped p-Si nanowires heterojunction photodiodes", *IEEE Trans. Nanotechnol.*, vol. 16, no. 1, p. 1, 2016.
[http://dx.doi.org/10.1109/TNANO.2016.2626795]

[23]   E. Plis, J.B. Rodriguez, and S. Krishna, "6.06 - InAs", *GaSb Type II Strained Layer Superlattice Detectors," in Comprehensive Semiconductor Science and Technology.*, pp. 229-264, 2011.
[http://dx.doi.org/10.1016/B978-0-44-453153-7.00017-1]

[24]   CJ. Murphy, "Optical sensing with quantum dots", *Anal Chem.*, vol. 74, no. 19, pp. 520A-526A, 2002.
[http://dx.doi.org/10.1021/ac022124v]

[25]   A.P. Alivisatos, "Semiconductor clusters, nanocrystals, and quantum dots", *Science,* vol. 271, no. 5251, pp. 933-937, 1996.
[http://dx.doi.org/10.1126/science.271.5251.933]

[26]   H. Shen, C.X. Shan, B.H. Li, B. Xuan, and D.Z. Shen, "Reliable self-powered highly spectrum-selective ZnO ultraviolet photodetectors", *Appl. Phys. Lett.*, vol. 103, no. 23, p. 232112, 2013.
[http://dx.doi.org/10.1063/1.4839495]

[27]   H. Kumar, Y. Kumar, G. Rawat, C. Kumar, B. Mukherjee, B.N. Pal, and S. Jit, "Heating effects of colloidal zno quantum dots (QDs) on ZnO QD/CdSe QD/MoOx photodetectors", *IEEE Trans. Nanotechnol.*, vol. 16, no. 6, pp. 1073-1080, 2017.
[http://dx.doi.org/10.1109/TNANO.2017.2761785]

[28]   H. Kumar, Y. Kumar, B. Mukherjee, G. Rawat, C. Kumar, B.N. Pal, and S. Jit, "Electrical and optical characteristics of self-powered colloidal cdse quantum dot-based photodiode", *IEEE J. Quantum Electron.*, vol. 53, no. 3, pp. 1-8, 2017.
[http://dx.doi.org/10.1109/JQE.2017.2696487]

[29]   I. Lee, Y.H. Kim, J. Jang, K-H. Lee, J. Jang, Y-W. Lim, S-H.K. Park, C.B. Park, W. Lee, and B-S. Bae, "Flexible electronics: solution-processed, photo-patternable fluorinated sol–gel hybrid materials as a bio-fluidic barrier for flexible electronic systems (Adv. Electron. Mater. 3/2020)", *Adv. Electron. Mater.*, vol. 6, no. 3, p. 2070016, 2020.
[http://dx.doi.org/10.1002/aelm.202070016]

[30]   Q. Wang, Z. Lin, J. Su, Z. Hu, J. Chang, and Y. Hao, "Recent progress of inorganic hole transport materials for efficient and stable perovskite solar cells", *Nano Select,* vol. 2, no. 6, pp. 1055-1080, 2021.
[http://dx.doi.org/10.1002/nano.202000238]

[31] J.H. Lim, J.H. Shim, J.H. Choi, J. Joo, K. Park, H. Jeon, M.R. Moon, D. Jung, H. Kim, and H-J. Lee, "Solution-processed InGaZnO-based thin film transistors for printed electronics applications", *Appl. Phys. Lett.,* vol. 95, no. 1, p. 012108, 2009.
[http://dx.doi.org/10.1063/1.3157265]

[32] D.K. Jarwal, C. Dubey, K. Baral, A. Bera, and G. Rawat, "Comparative analysis and performance optimization of low-cost solution-processed hybrid perovskite-based solar cells with different organic HTLs", *IEEE Transactions on Electron Devices,* vol. 69, no. 9, pp. 5012-5020, 2022.
[http://dx.doi.org/10.1109/TED.2022.3194106]

[33] K. Shen, H. Xu, X. Li, J. Guo, S. Sathasivam, M. Wang, A. Ren, K.L. Choy, I.P. Parkin, Z. Guo, and J. Wu, "Flexible and self-powered photodetector arrays based on all-inorganic CsPbBr 3 quantum dots", *Adv. Mater.,* vol. 32, no. 22, p. 2000004, 2020.
[http://dx.doi.org/10.1002/adma.202000004] [PMID: 32319160]

[34] H. Zhou, L. Yang, P. Gui, C.R. Grice, Z. Song, H. Wang, and G. Fang, "Ga-doped ZnO nanorod scaffold for high-performance, hole-transport-layer-free, self-powered CH3NH3PbI3 perovskite photodetectors", *Sol. Energy Mater. Sol. Cells,* vol. 193, pp. 246-252, 2019.
[http://dx.doi.org/10.1016/j.solmat.2019.01.020]

[35] L. Duan, F. He, Y. Tian, B. Sun, J. Fan, X. Yu, L. Ni, Y. Zhang, Y. Chen, and W. Zhang, "Fabrication of self-powered fast-response ultraviolet photodetectors based on graphene/ZnO:Al nanorod-arra--film structure with stable schottky barrier", *ACS Appl. Mater. Interfaces,* vol. 9, no. 9, pp. 8161-8168, 2017.
[http://dx.doi.org/10.1021/acsami.6b14305] [PMID: 28240856]

[36] H. Chen, X. Sun, D. Yao, X. Xie, F.C.C. Ling, and S. Su, "Back-to-back asymmetric Schottky-type self-powered UV photodetector based on ternary alloy MgZnO", *J. Phys. D Appl. Phys.,* vol. 52, no. 50, p. 505112, 2019.
[http://dx.doi.org/10.1088/1361-6463/ab452e]

[37] H. Kumar, Y. Kumar, B. Mukherjee, G. Rawat, C. Kumar, B.N. Pal, and S. Jit, "Effects of optical resonance on the performance of metal (Pd, Au)/CdSe Quantum Dots (QDs)/ZnO QDs optical cavity based spectrum selective photodiodes", *IEEE Trans. Nanotechnol.,* vol. 18, pp. 365-373, 2019.
[http://dx.doi.org/10.1109/TNANO.2019.2907529]

[38] V. Goel, and H. Kumar, "Effect of electrodes on generation rate of cdte and pbse based optoelectronic devices", *6th International Conference on Signal Processing and Communication (ICSC).,* pp. 243-247, 2020.Noida, India
[http://dx.doi.org/10.1109/ICSC48311.2020.9182742]

[39] W. Jin, Y. Ye, L. Gan, B. Yu, P. Wu, Y. Dai, H. Meng, X. Guo, and L. Dai, "Self-powered high performance photodetectors based on CdSe nanobelt/graphene Schottky junctions", *J. Mater. Chem.,* vol. 22, no. 7, pp. 2863-2867, 2012.
[http://dx.doi.org/10.1039/c2jm15913a]

[40] Z. Gao, W. Jin, Y. Zhou, Y. Dai, B. Yu, C. Liu, W. Xu, Y. Li, H. Peng, Z. Liu, and L. Dai, "Self-powered flexible and transparent photovoltaic detectors based on CdSe nanobelt/graphene Schottky junctions", *Nanoscale,* vol. 5, no. 12, pp. 5576-5581, 2013.
[http://dx.doi.org/10.1039/c3nr34335a] [PMID: 23681339]

[41] B.D. Boruah, A. Mukherjee, and A. Misra, "Sandwiched assembly of ZnO nanowires between graphene layers for a self-powered and fast responsive ultraviolet photodetector", *Nanotechnology,* vol. 27, no. 9, p. 095205, 2016.
[http://dx.doi.org/10.1088/0957-4484/27/9/095205] [PMID: 26857833]

[42] H.Y. Chen, K-W. Liu, X. Chen, Z-Z. Zhang, M-M. Fan, M-M. Jiang, X-H. Xie, H-F. Zhao, and D-Z. Shen, "Realization of a self-powered ZnO MSM UV photodetector with high responsivity using an asymmetric pair of Au electrodes", *J. Mater. Chem. C Mater. Opt. Electron. Devices,* vol. 2, no. 45, pp. 9689-9694, 2014.

[http://dx.doi.org/10.1039/C4TC01839G]

[43] M. Mishra, A. Gundimeda, T. Garg, A. Dash, S. Das, Vandana, and G. Gupta, "ZnO/GaN heterojunction based self-powered photodetectors: Influence of interfacial states on UV sensing", *Appl. Surf. Sci.,* vol. 478, pp. 1081-1089, 2019.
[http://dx.doi.org/10.1016/j.apsusc.2019.01.192]

[44] S. Ebrahimi, and B. Yarmand, "Solvothermal growth of aligned SnxZn1-xS thin films for tunable and highly response self-powered UV detectors", *J. Alloys Compd.,* vol. 827, p. 154246, 2020.
[http://dx.doi.org/10.1016/j.jallcom.2020.154246]

[45] K. Benyahia, F. Djeffal, H. Ferhati, A. Bendjerad, A. Benhaya, and A. Saidi, "Self-powered photodetector with improved and broadband multispectral photoresponsivity based on ZnO-ZnS composite", *J. Alloys Compd.,* vol. 859, p. 158242, 2021.
[http://dx.doi.org/10.1016/j.jallcom.2020.158242]

[46] Y. Purusothaman, N.R. Alluri, A. Chandrasekhar, V. Vivekananthan, and S.J. Kim, "Direct In Situ hybridized interfacial quantification to stimulate highly flexile self-powered photodetector", *J. Phys. Chem. C,* vol. 122, no. 23, pp. 12177-12184, 2018.
[http://dx.doi.org/10.1021/acs.jpcc.8b02604]

[47] Z. Zhang, Q. Liao, Y. Yu, X. Wang, and Y. Zhang, "Enhanced photoresponse of ZnO nanorods-based self-powered photodetector by piezotronic interface engineering", *Nano Energy,* vol. 9, pp. 237-244, 2014.
[http://dx.doi.org/10.1016/j.nanoen.2014.07.019]

[48] K. Jenkins, V. Nguyen, R. Zhu, and R. Yang, "Piezotronic effect: An emerging mechanism for sensing applications", *Sensors,* vol. 15, no. 9, pp. 22914-22940, 2015.
[http://dx.doi.org/10.3390/s150922914] [PMID: 26378536]

[49] Y. Chang, L. Chen, J. Wang, W. Tian, W. Zhai, and B. Wei, "Self-powered broadband schottky junction photodetector based on a single selenium microrod", *J. Phys. Chem. C,* vol. 123, no. 34, pp. 21244-21251, 2019.
[http://dx.doi.org/10.1021/acs.jpcc.9b04260]

[50] Y.Q. Bie, Z.M. Liao, H.Z. Zhang, G.R. Li, Y. Ye, Y.B. Zhou, J. Xu, Z.X. Qin, L. Dai, and D.P. Yu, "Self-powered, ultrafast, visible-blind UV detection and optical logical operation based on ZnO/GaN nanoscale p-n junctions", *Adv. Mater.,* vol. 23, no. 5, pp. 649-653, 2011.
[http://dx.doi.org/10.1002/adma.201003156] [PMID: 21274914]

[51] J.J. Hassan, M.A. Mahdi, S.J. Kasim, N.M. Ahmed, H. Abu Hassan, and Z. Hassan, "High sensitivity and fast response and recovery times in a ZnO nanorod array/ p -Si self-powered ultraviolet detector", *Appl. Phys. Lett.,* vol. 101, no. 26, p. 261108, 2012.
[http://dx.doi.org/10.1063/1.4773245]

[52] S.M. Hatch, J. Briscoe, and S. Dunn, "A self-powered ZnO-nanorod/CuSCN UV photodetector exhibiting rapid response", *Adv. Mater.,* vol. 25, no. 6, pp. 867-871, 2013.
[http://dx.doi.org/10.1002/adma.201204488] [PMID: 23225232]

[53] S.M. Hatch, J. Briscoe, A. Sapelkin, W.P. Gillin, J.B. Gilchrist, M.P. Ryan, S. Heutz, and S. Dunn, "Influence of anneal atmosphere on ZnO-nanorod photoluminescent and morphological properties with self-powered photodetector performance", *J. Appl. Phys.,* vol. 113, no. 20, p. 204501, 2013.
[http://dx.doi.org/10.1063/1.4805349]

[54] G. Huang, P. Zhang, and Z. Bai, "Self-powered UV–visible photodetectors based on ZnO/graphene/CdS/electrolyte heterojunctions", *J. Alloys Compd.,* vol. 776, pp. 346-352, 2019.
[http://dx.doi.org/10.1016/j.jallcom.2018.10.225]

[55] T. Guo, C. Ling, X. Li, X. Qiao, X. Li, Y. Yin, Y. Xiong, L. Zhu, K. Yan, and Q. Xue, "A ZIF-8@H:ZnO core–shell nanorod arrays/Si heterojunction self-powered photodetector with ultrahigh performance", *J. Mater. Chem. C Mater. Opt. Electron. Devices,* vol. 7, no. 17, pp. 5172-5183, 2019.
[http://dx.doi.org/10.1039/C9TC00290A]

[56] P. Ghamgosar, F. Rigoni, M.G. Kohan, S. You, E.A. Morales, R. Mazzaro, V. Morandi, N. Almqvist, I. Concina, and A. Vomiero, "Self-powered photodetectors based on core-shell zno-co₃o₄ nanowire heterojunctions', *ACS Appl. Mater. Interfaces,* vol. 11, no. 26, pp. 23454-23462, 2019.
[http://dx.doi.org/10.1021/acsami.9b04838] [PMID: 31252456]

[57] N. Yamada, Y. Kondo, X. Cao, and Y. Nakano, "Visible-blind wide-dynamic-range fast-response self-powered ultraviolet photodetector based on CuI/In-Ga-Zn-O heterojunction", *Appl. Mater. Today,* vol. 15, pp. 153-162, 2019.
[http://dx.doi.org/10.1016/j.apmt.2019.01.007]

[58] R. Sinha, N. Roy, and T.K. Mandal, "Growth of carbon dot-decorated zno nanorods on a graphite-coated paper substrate to fabricate a flexible and self-powered schottky diode for uv detection", *ACS Appl. Mater. Interfaces,* vol. 12, no. 29, pp. 33428-33438, 2020.
[http://dx.doi.org/10.1021/acsami.0c10484] [PMID: 32573201]

[59] A.V. Agrawal, K. Kaur, and M. Kumar, "Interfacial study of vertically aligned n-type MoS2 flakes heterojunction with p-type Cu-Zn-Sn-S for self-powered, fast and high performance broadband photodetector", *Appl. Surf. Sci.,* vol. 514, p. 145901, 2020.
[http://dx.doi.org/10.1016/j.apsusc.2020.145901]

[60] C. Wei, J. Xu, S. Shi, Y. Bu, R. Cao, J. Chen, J. Xiang, X. Zhang, and L. Li, "The improved photoresponse properties of self-powered NiO/ZnO heterojunction arrays UV photodetectors with designed tunable Fermi level of ZnO", *J. Colloid Interface Sci.,* vol. 577, pp. 279-289, 2020.
[http://dx.doi.org/10.1016/j.jcis.2020.05.077] [PMID: 32485411]

[61] C.Y. Huang, C-P. Huang, H. Chen, S-W. Pai, P-J. Wang, X-R. He, and J-C. Chen, "A self-powered ultraviolet photodiode using an amorphous InGaZnO/p-silicon nanowire heterojunction", *Vacuum,* vol. 180, p. 109619, 2020.
[http://dx.doi.org/10.1016/j.vacuum.2020.109619]

[62] A.A. Ahmed, M.R. Hashim, T.F. Qahtan, and M. Rashid, "Preparation and characteristics study of self-powered and fast response p-NiO/n-Si heterojunction photodetector", *Ceram. Int.,* vol. 48, no. 14, pp. 20078-20089, 2022.
[http://dx.doi.org/10.1016/j.ceramint.2022.03.285]

[63] X. Zhou, C. Wang, J. Luo, L. Zhang, F. Zhao, and Q. Ke, "High-performance self-powered UV/red dual-wavelength photodetector based on CuI/CsCu2I3/GaN heterojunction", *Chem. Eng. J.,* no. Apr, p. 136364, 2022.
[http://dx.doi.org/10.1016/j.cej.2022.136364]

[64] Y.Y. Chen, C.H. Wang, G.S. Chen, Y.C. Li, and C.P. Liu, "Self-powered n-Mg Zn O/p-Si photodetector improved by alloying-enhanced piezopotential through piezo-phototronic effect", *Nano Energy,* vol. 11, pp. 533-539, 2015.
[http://dx.doi.org/10.1016/j.nanoen.2014.09.037]

[65] R. Saha, A. Karmakar, and S. Chattopadhyay, "Enhanced self-powered ultraviolet photoresponse of ZnO nanowires/p-Si heterojunction by selective in-situ Ga doping", *Opt. Mater.,* vol. 105, p. 109928, 2020.
[http://dx.doi.org/10.1016/j.optmat.2020.109928]

[66] J. Jiang, J. Huang, Z. Ye, S. Ruan, and Y.J. Zeng, "Self-Powered and Broadband Photodetector Based on SnS₂/ZnO₁₋ₓSₓ Heterojunction", *Adv. Mater. Interfaces,* vol. 7, no. 20, p. 2000882, 2020.
[http://dx.doi.org/10.1002/admi.202000882]

[67] O. Game, U. Singh, T. Kumari, A. Banpurkar, and S. Ogale, "ZnO(N)–Spiro-MeOTAD hybrid photodiode: an efficient self-powered fast-response UV (visible) photosensor", *Nanoscale,* vol. 6, no. 1, pp. 503-513, 2014.
[http://dx.doi.org/10.1039/C3NR04727J] [PMID: 24232600]

[68] S. Sarkar, and D. Basak, "Self powered highly enhanced dual wavelength zno@cds core–shell nanorod arrays photodetector: An intelligent pair", *ACS Appl. Mater. Interfaces,* vol. 7, no. 30, pp. 16322-

16329, 2015.
[http://dx.doi.org/10.1021/acsami.5b03184] [PMID: 26154060]

[69]   H. Kumar, Y. Kumar, G. Rawat, C. Kumar, B. Mukherjee, B.N. Pal, and S. Jit, "Colloidal CdSe quantum dots and PQT-12-based low-temperature self-powered hybrid photodetector", *IEEE Photonics Technol. Lett.,* vol. 29, no. 20, pp. 1715-1718, 2017.
[http://dx.doi.org/10.1109/LPT.2017.2746664]

[70]   K. Li, Y. Lu, X. Yang, L. Fu, J. He, X. Lin, J. Zheng, S. Lu, C. Chen, and J. Tang, "Filter-free self-power CDSE/Sb$_2$(S$_{1-x}$,Se$_x$)$_3$ nearinfrared narrowband detection and imaging", *InfoMat,* vol. 3, no. 10, pp. 1145-1153, 2021.
[http://dx.doi.org/10.1002/inf2.12237]

[71]   J. C. Joshi, and A. L. Dawar, "Pyroelectric materials, their properties and applications", *phys. stat. sol.,* vol. 70, p. 353, 1982.
[http://dx.doi.org/10.1515/9783112496282-001]

[72]   S.K. Ghosh, and D. Mandal, "Design strategy and innovation in piezo- and pyroelectric nanogenerators", In: *Nanobatteries and Nanogenerators.* Elsevier, 2021, pp. 555-585.
[http://dx.doi.org/10.1016/B978-0-12-821548-7.00022-1]

[73]   L. Chen, J. Dong, M. He, and X. Wang, "A self-powered, flexible ultra-thin Si/ZnO nanowire photodetector as full-spectrum optical sensor and pyroelectric nanogenerator", *Beilstein J. Nanotechnol.,* vol. 11, no. 1, pp. 1623-1630, 2020.
[http://dx.doi.org/10.3762/bjnano.11.145] [PMID: 33178547]

[74]   S.B. Lang, "Pyroelectricity: From ancient curiosity to modern imaging tool", *Phys. Today,* vol. 58, no. 8, pp. 31-36, 2005.
[http://dx.doi.org/10.1063/1.2062916]

[75]   Y. Yao, B. Ploss, C.L. Mak, and K.H. Wong, "Pyroelectric properties of BiFeO3 ceramics prepared by a modified solid-state-reaction method", *Appl. Phys., A Mater. Sci. Process.,* vol. 99, no. 1, pp. 211-216, 2010.
[http://dx.doi.org/10.1007/s00339-009-5499-1]

[76]   A.K. Rana, M. Kumar, D.K. Ban, C.P. Wong, J. Yi, and J. Kim, "Enhancement in performance of transparent p-nio/n-zno heterojunction ultrafast self-powered photodetector *via* pyro-phototronic effect", *Adv. Electron. Mater.,* vol. 5, no. 8, p. 1900438, 2019.
[http://dx.doi.org/10.1002/aelm.201900438]

[77]   B. Deka Boruah, S. Naidu Majji, S. Nandi, and A. Misra, "Doping controlled pyro-phototronic effect in self-powered zinc oxide photodetector for enhancement of photoresponse", *Nanoscale,* vol. 10, no. 7, pp. 3451-3459, 2018.
[http://dx.doi.org/10.1039/C7NR08125A] [PMID: 29393951]

[78]   D. Chen, Y. Xin, B. Lu, X. Pan, J. Huang, H. He, and Z. Ye, "Self-powered ultraviolet photovoltaic photodetector based on graphene/ZnO heterostructure", *Appl. Surf. Sci.,* vol. 529, p. 147087, 2020.
[http://dx.doi.org/10.1016/j.apsusc.2020.147087]

[79]   L. Chen, B. Wang, J. Dong, F. Gao, H. Zheng, M. He, and X. Wang, "Insights into the pyro-phototronic effect in p-Si/n-ZnO nanowires heterojunction toward high-performance near-infrared photosensing", *Nano Energy,* vol. 78, p. 105260, 2020.
[http://dx.doi.org/10.1016/j.nanoen.2020.105260]

[80]   J.P.B. Silva, E.M.F. Vieira, K. Gwozdz, A. Kaim, L.M. Goncalves, J.L. MacManus-Driscoll, R.L.Z. Hoye, and M. Pereira, "High-performance self-powered photodetectors achieved through the pyro-phototronic effect in Si/SnOx/ZnO heterojunctions", *Nano Energy,* vol. 89, p. 106347, 2021.
[http://dx.doi.org/10.1016/j.nanoen.2021.106347]

CHAPTER 11

Nanostructured Solar Cells as Sustainable Optoelectronic Device

Ankita Saini[1,*], **Sunil Kumar Saini**[1] and **Sumeen Dalal**[1]

[1] *Department of Chemistry, P.D.M. University, Bahadurgarh, Haryana, India*

Abstract: Owing to the strong interest in sustainable and renewable energy in the past recent years, the solar cell industry has grown vastly. Conventional solar cells are simply not efficient enough and are expensive to manufacture for large-scale electricity generation. There are potential and sustainable advancements in nanotechnology that have opened the door to the production of efficient nanostructured optoelectronics. Nanotechnology has depicted tremendous breakthroughs in the field of solar technology. Nanomaterials and quantum dots (QDs) have proved to be potential candidates in the field of solar cells. Nanotechnology is able to enhance the efficiency of solar cells, meanwhile helping in the reduction of manufacturing costs. Photovoltaics (PVs) based on inorganic, organic, and polymer materials are designed and synthesized with the aim of reducing cost per watt, even if it declines reliability and conversion efficiency. Such PVs absorb sunlight more efficiently with wider absorption spectra which also show better conversion of power to efficiency. Herein, we have highlighted nanoparticles based on inorganic, organic, or graphene-based functional materials, which exhibit enhanced physicochemical properties along with excellent surface-to-area ratio to be used as nanostructured thin layers coated with solar cell panels. Utilizing nanotechnology in developing low-cost and efficient solar cells would help to preserve the environment.

Keywords: III Generation Solar Cells, Nanomaterials, Nanotechnology, Optoelectronics, Photovoltaics, Renewable Energy, Sustainable Environment.

INTRODUCTION

The design, synthesis, and fabrication of functional materials as well as their use in various solar applications, are areas of study in the field of materials science. In recent years, nanosized polymers, nanoparticles, and carbon-based materials have been widely investigated and used as semiconductors in electronic applications such as nano-structured photovoltaics in place of conventional inorganic Silicon-

* **Corresponding author Ankita Saini:** Department of Chemistry, P.D.M. University, Bahadurgarh, Haryana, India; E-mail: ankitasaini53@gmail.com

Gopal Rawat & Aniruddh Bahadur Yadav (Eds.)

based photovoltaics (PVs). To address the growing global interest in energy usage, we need to develop an alternative approach and depart from conventional methods for capturing limitless and solar energy. In elaboration on the title of the book, in this chapter, we will discuss and highlight the importance of nanostructures in solar cell technology as nanostructured optoelectronic devices. A large pool of nanoparticles is being utilized in various nanostructured electronic innovations like light emitting diode (LED), dye-sensitized solar cells (DSSC) [1], organic solar cells (OSC), and sensors which renders their helpful electronic properties, for example, charge versatility, thermal, electrical stability, luminescence and film thinning capacity. From traditional inorganic microstructured solid-state devices to nanostructured devices, they offer the primary flexibility of semiconductors which considers the incorporation of different functionalities in molecules with the help of simple molecular and structural engineering. For more than the most recent 50 years, in nanoelectronics, compounds have been examined in the field of material science [2].

Solar cells are devices which light discharge electric charges so they can move unreservedly in a semiconductor and eventually course through the formation of electricity. They work on the principle of the photovoltaic effect where "photo means light and voltaic means electricity". The wavelength useful for conversion to electricity lies within the UV-visible range, 200-800nm (Fig. **1**). But all incoming light cannot be converted by solar cells into energy because some light gets out of the cell. It is known that energy is lost in the form of heat when it is greater than the band gap energy. However, if these excited electrons are not still captured, they will immediately recombine with the holes created at the interface, and the energy will be lost. A solar cell contains a layer of p-type Si which is kept close to the layer of n-type Si. The power acquired from PV is managed by area and type of material, sunlight's wavelength and its intensity; single crystal Si solar cells are limited to the conversion of 25% of solar energy into electricity.

The photovoltaic parameters of solar cells decide what will be the IPCE (Incident photon conversion efficiency) of the solar cell. Once the device is fabricated, analysis and characterization of the solar parameters are carried out. The parameters obtained from photocurrent voltage (J-V) estimation under reproduced daylight are IPCE. There are 3 photovoltaic parameters:

a. "Short circuit photocurrent density (J_{SC})"
b. "Open circuit voltage (V_{OC})"
c. "Fill factor (FF)".

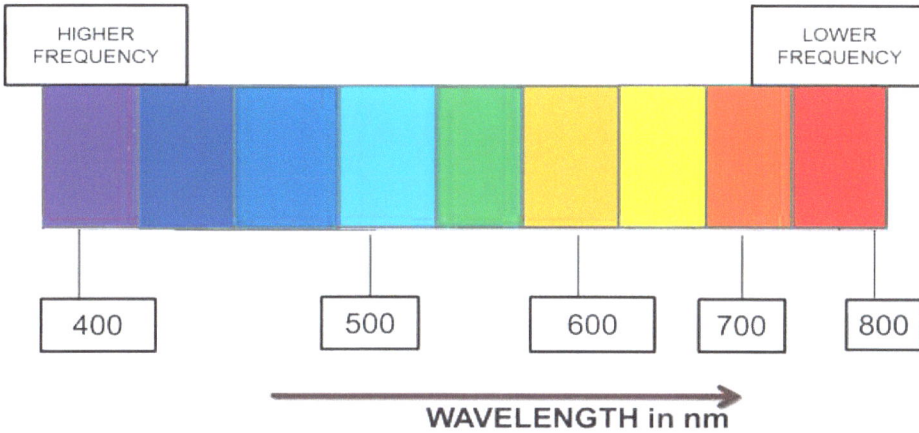

Fig. (1). Visible light spectrum.

With AM 1.5 illuminations and under standard circumstances of 100 mW/cm^2, *i.e.*, P_{in}, efficiency (η) of a cell can be calculated as:

"$\eta = (J_{SC} \times V_{OC} \times FF)/ P_{in}$" where FF is described as maximum power output, which is calculated from the equation "$FF = (J_{max} \times V_{max})/ (J_{SC} \times V_{OC})$".

There are 3 different types of solar panels available for usage in PV systems.

1. Monocrystalline

2. Polycrystalline

3. Amorphous thin film

Fig. (**2**) illustrates the 3 different types of solar panels available. The differences between these 3 types of panels lie in the expense and efficiency of the panels. Compared to mono-crystalline or polycrystalline solar panels, thin films, solar panels are less effective and have shorter lifespans. In contrast to crystalline solar panels, their costs are considerably cheaper because of the straightforward manufacturing processes. Modification is possible in thin-film solar panels to make them flexible, however, crystalline solar panels are rigid, brittle, and difficult to alter. Thin film panels are not advised for use in home solar systems due to their lower efficiency. To produce a certain amount of electricity, a user will require thin-film solar panels than a crystalline solar panel and hence require more area for its installation. Hence, these thin film solar panels are easily and frequently employed for commercial applications than by residential users.

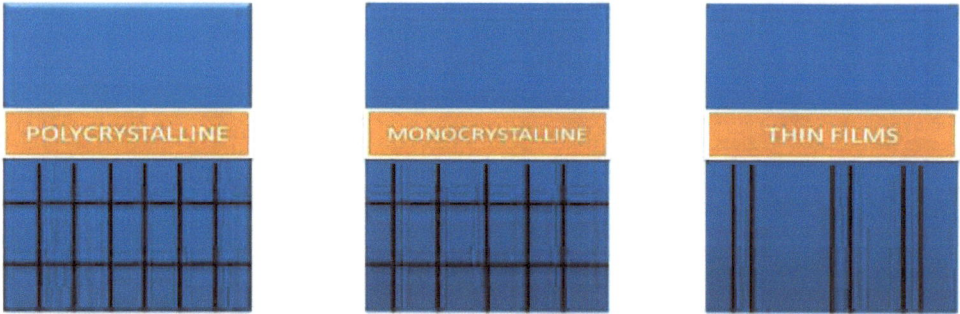

Fig. (2). Types of Solar panels.

There are three primary types of solar cells. (Fig. **3**) are as follows.

1. 1st Generation Solar Cells- Crystalline Silicon.

2. 2nd Generation Solar Cells- Thin Film.

3. 3rd Generation Solar Cells- Nanotechnology-based solar cells.

The first generation of solar cells makes use of superior crystalline silicon. In second-generation PV, such as Copper indium gallium selenide (CIGS) and cadmium telluride (CdTe), semiconducting semiconductors other than silicon are used in thin-film technology. These materials promise a higher theoretical efficiency than Si-based PV materials and are helpful in lowering process costs. Whereas, third-generation PV uses nanotechnologies for the generation of electricity, which mostly include tandem cells, quantum dots, nanostructured semiconductors, polymers, a nanosized active layer of amorphous silicon, organic sensitisers, nano-carbon materials, nanocrystals, and single-walled carbon nanotubes (SWCNTs).

Nanocrystals have gained interest for their ability to capture light in solar cells because they possess a larger surface area-to-volume ratio. These nanostructures support surface enhancement, which aids in the efficient light trapping achieved by conduction electron excitation at the interface. Moreover, the broader absorption from visible to near IR area, low cost, higher molar extinction coefficient, long-term stability, and ease of synthesis of the nanosized sensitizers that sensitise sunlight show their superiority over silicon and ruthenium-based dyes. Nanoparticles and quantum dots enhance the coupling of sunlight into the cell substrate.

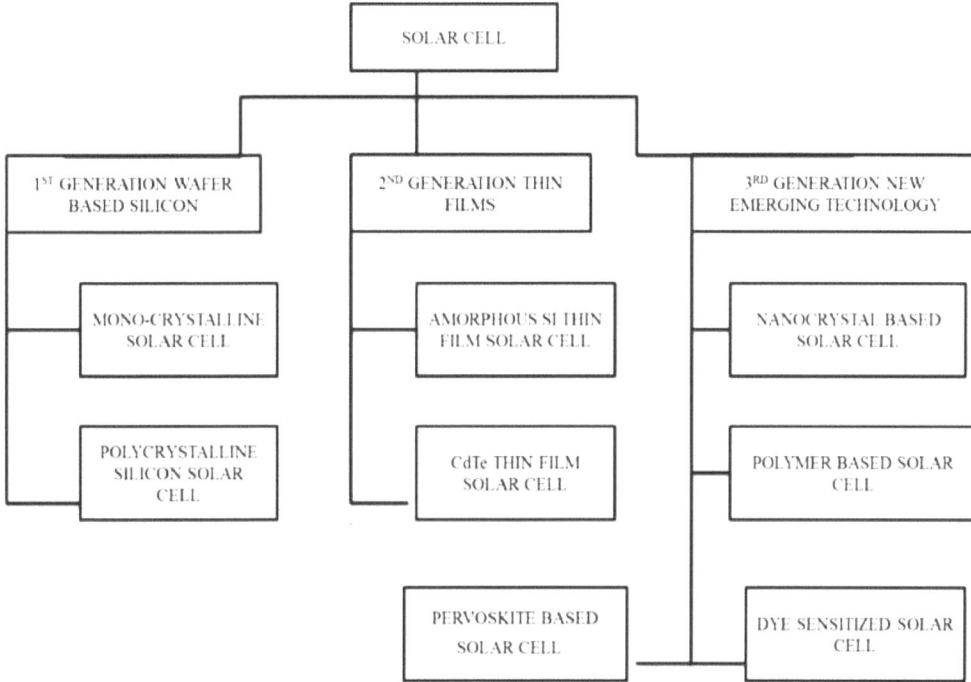

Fig. (3). Types of Solar Cells.

Table **1** depicts the comparison of some types of photovoltaic cells in terms of the thickness of cells. Using nanotechnology has made it possible to realize how the thickness of the solar cell has reduced from mm to nm while maintaining the optimum efficiency and various other solar power conversion parameters.

Table 1. Types of Photovoltaic Cells and Their Thickness.

Type of PV Cell	Category	Thickness	Efficiency
Mono-Crystalline Si	1st Generation	0.3 mm	15-20%
Thin Film CdTe	2nd Generation	10-15 μm	7-10%
Multi-Junction	3rd Generation	0.003 mm	33-35%
DSSC	3rd Generation	10 μm	>15%
Organic PV	3rd Generation	80-100 nm	20-25%

SOLAR NANOTECHNOLOGY AS A SUSTAINABLE ENERGY SOURCE

One of the fastest-growing industries in the world right now is renewable energy technology, which aims to fulfill future energy needs while simultaneously addressing environmental concerns and reducing climate change. Nanotechnology involves miniaturization and manipulation of the structure of materials in order to control their physical and chemical properties, which is difficult and different from the bulk properties. Although the large-scale application of nanostructured photovoltaics to capture solar energy has encountered several challenges, the most notable ones are the high cost of technology, environmental instability, and low power conversion efficiency. Apart from this, a major achievement in the field of materials science is the application of nanotechnology to the mass production of low-cost solar cells that are both effective and efficient. Thin-film solar cells with nanostructured layers have benefits over conventional solar cells. These are as follows (Fig. **4**).

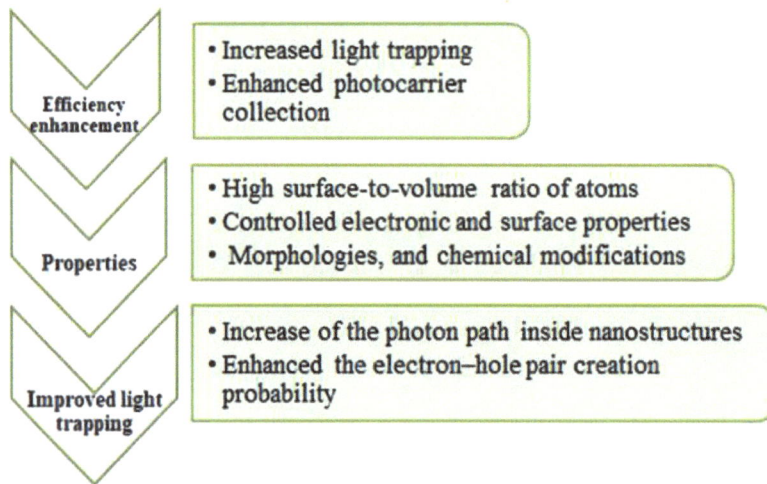

Fig. (4). Uses of Nano-Structured Layer in Solar Cells.

More Absorption: Optical path of light effective for absorption is larger than the actual thickness of the film corresponding to the active layer owing to the number of internal reflections. It facilitated increased light trapping and enhanced photo carrier collection.

Reduced Path: Recombination losses occur as a result of the necessity for photogenerated electrons (e⁻) and holes (h⁺) to traverse over a short distance. Improvement of the photon route within the nanostructure of the active layer increases the likelihood of the formation of electron-hole pairs.

Tailored Characteristics: With the tailored size of the nanoparticles with a high atom surface-to-volume ratio, controlled surface properties, morphology, and chemical modifications, the energy band gaps of different layers in the solar cell can be chemically adjusted to match the incident light. Large-sized nanocrystals absorb at higher wavelengths, while small-sized nanocrystals absorb at shorter wavelengths. The band gap energy level fluctuates along with the geometry of the nanomaterials. By improving the effective optical route and managing the energy band gap, nanotubes and quantum dots (QDs) are actively employed and researched to increase PV efficiency [3].

The goal of ongoing research and development is to increase efficiency so that solar energy is converted to electricity in a good percentage. As a result, the third generation of solar cells [4] utilises the newly developed semiconductor nanocrystals, QDs, CNTs, nanoparticles, and SWCT. Semiconductor dye-sensitized solar cells, semiconductor quantum dot solar cells, and semiconductor nanostructure polymer solar cells all outperform the traditional Shockley and Queisser limit of 32% for a single junction photovoltaic. Nanomaterials and nanostructures have the potential to enhance photovoltaic performance by improving optical absorption and photocarrier collection. Meanwhile, the new materials and structures can be manufactured at a low cost, allowing for cost-effective solar cell production. Solar cells can now be produced efficiently at a low cost owing to the new materials and structures that can be created. Various nanomaterials are used in solar cell technology owing to their size, physiochemical properties and usage (Fig. **5**).

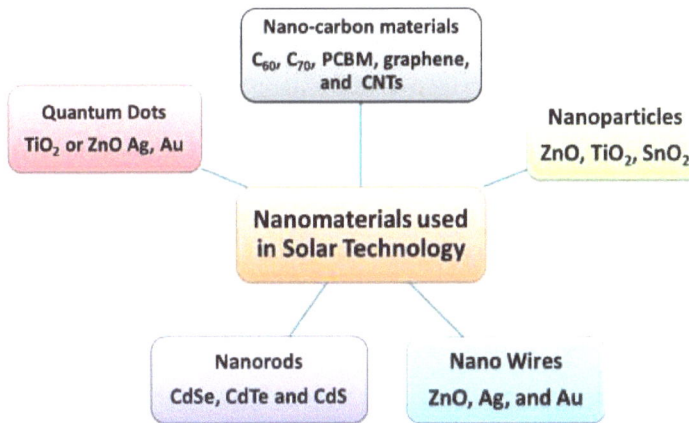

Fig. (5). Various nanomaterials used in solar cell technology.

By facilitating the flow of e⁻ to the surface of collecting electrodes in DSSCs, nanomaterials or CNT supports are used to anchor light-harvesting semiconducting nanoparticles. The Excitation of CdS nanoparticles makes it simpler to introduce charged CdS into SWCNTS. Charge transfer activities are induced when CNTs are connected to CdSe and CdTe and are exposed to visible light. The porous TiO_2 film's increased interconnectivity between the multiwalled CNTs and

TiO_2 particles improved the short circuit current density and efficiency overall. According to this, nanostructured semiconducting materials have the potential to obtain great efficiency at a low cost of production. The following are various types of nanomaterials that are used in PVs [5].

Nanoparticles: Over the past 20 years, Ti nanoparticles have been the subject of intensive research and development. For its application in solar cells, TiO_2 or ITO (Indium Tin Oxide) can absorb light in the visible light spectrum and transform solar energy into electricity. Nanomaterials offer a wide range of uses due to their physio-chemical characteristics, including paint, toothpaste, UV protection, photocatalysis, and photovoltaics.

Nano Wires: Nanowires are wires with lengths and widths of the order of 10 μm and 10 nm, respectively. Their width is measured in nanometers. These are particularly appealing for use in nanotechnology applications and studies. Nanowires enable us to significantly reduce the size of electrical gadgets while also enhancing their functionality. Due to the simplicity of charge transmission, ZnO, Ag, and Au are increasingly used in semiconductors and optoelectronics.

Nanorods: Nanorods are utilised with polymers as an active blend because they are able to produce electrons when they absorb light of a particular wavelength. These electrons travel through the nanorods until they come into contact with the electrode, where they are united to form a current and are converted into electrical energy. Such cells are easier to produce than conventional ones. The ability of these solar cells using nanorods, such as CdSe, CdTe, and CdS, to tune and absorb light throughout a broad spectrum, is one of its main features.

Quantum Dots: QDs are linked to a wide band gap nanomaterial like TiO_2 or ZnO for its application in the solar sector. Between the photoelectrode and the counter electrode lies a thin layer of liquid electrolyte, which contains a redox couple. The photoexcited electron-hole pairs that are present in the nanocrystals are produced as a result of QD absorbing the incident photons. To avoid charge recombination, the device then splits the charges that are photogenerated carriers into distinct sections of the solar cell.

Nano-carbon materials: In fullerene-based solar cells, fullerenes (C60, C70), PCBM, and PV are employed. In organic solar cells, C60 is widely employed as an electron acceptor [6]. It is anticipated that new fullerene acceptors, such as those other than C60 and other organic acceptor materials, will be employed to achieve greater efficiency. In DSSC, counter electrodes are made of nanocarbon materials: When substituting platinum as the electrode in III Generation solar cells, nanocarbon materials like graphene and CNTs (SW or MW) are employed to improve the counter electrodes in DSSC.

Single-walled carbon nanotubes: A desirable substance with distinctive structural, morphological, and electrical features is the semiconducting single-walled carbon nanotube. Semiconducting SWNTs exhibit intense photons from UV to IR and have a broad range of band gaps that match the whole solar spectrum. The carrier transport scattering is low, and the carrier mobility is great. Strong Coulomb interactions between electrons and holes are seen in semiconductors like SWNTs. They can produce additional electrons using incident photons that ordinary solar cells cannot absorb. SWNTs which are semiconductors, are hence the best material for PVs. Development in the research of SWNTS used in solar cells started in 2008 when the first fabricated SWNT-PV solar microcell was developed. A strong built-in electric field was generated in SWNTs, which is formed by directing an array of mono-layer SWNTs on two metal electrodes of different potentials. The separation of photogenerated electron-hole pairs was successful using this method. The efficiency rises to 12.6% when lit by solar energy. While a team of researchers at Cornell University created PV in 2009 using CNTs rather than conventional silicon. Multiple electron-hole pairs help to create and assemble in an NT p-n junction. The SWNT network was built on the surface of n-p junction Si solar cells by the Zhang group at Shanghai University, a year later. SWNTs exhibit great transparency while also lowering the surface's electrode resistance, which improves the collection efficiency and increases the efficiency by 3.92%. Solar Cells comprised of semiconductor-grade crystalline silicon and emerging thin-film PV technologies like CdTe, CIS, amorphous silicon, and CIGS materials offer significant cost savings for PV devices. Apart from this, from the thin film, nanotechnology is also used in several other components of solar cells, such as nanoelectrodes in addition to the film. Metal oxide nanoparticles are used for transparent electrodes based on ZnO, MoO_3, Graphene, and SnO_2. In the medium to long term, these nanoelectrodes have taken the position of ITO. CNT has been used as an ITO alternative for transparent electrodes in PV cells. These are discovered to be the upcoming replacements for transparent PV cell films as well as other organic or inorganic transparent conductors. Due to its superior mechanical quality, sheet resistance, and great optical transparency, graphene is also seen as a viable alternative to ITO films as electrodes. Thin film solar cells are covered with nanostructured layers made of nanoparticles because of their enhanced and promoted physicochemical quality and outstanding surface-to-area ratio. Although, fabricating solar cells with ultrathin layers lowers the material costs, it also degrades their efficiency. Such thin film PVs generally employ Si films that are between a size of a few nm and tens of micrometres thick instead of the 200 micrometre silicon wafers used in conventional PVs. A 100-fold reduction in Si might greatly increase the cost efficiency of such solar systems since it accounts for 10–20% of the total cost of any solar cell. Commercial versions of these devices presently convert sunlight into electricity at an efficiency of roughly 22%. Thin film PV systems are currently 7–13% less efficient than conventional systems, nevertheless. Table. **2** highlights the advantages and disadvantages of nanotechnology in photovoltaics.

Table 2. Advantages and disadvantages of Nanotechnology in PV.

ADVANTAGES	DISADVANTAGES
1. Environment-friendly energy	High upfront cost
2. Reduce your energy bills	Location sensitivity
3. Infinite energy	Less solar energy in winters
4. Less to maintain for years	High initial cost

Although nanotechnology has a large potential to significantly lower the cost of manufacturing solar cells, this is not the most promising use of the technology in the solar cell industry. In order to decrease the cost and improve the amount of energy produced, PV based on cadmium telluride, CIGS, CuInSe, and various other organic materials are continuously developed in research. Thin film solar cells are covered with nanostructured layers made of nanoparticles because they enhance and promote physicochemical quality and outstanding surface-to-area ratio. The use of nanostructures in photovoltaics has the potential for high efficiency by exploiting new physical mechanisms and by permitting solar cells with efficiencies closer to their theoretical maximum by modification in material properties. Simultaneously, nanostructures have the potential for reduced fabrication costs. Therefore, using nanotechnology to create and manufacture low-cost solar cells can undoubtedly contribute to environmental preservation [7].

REPORTED NANOSTRUCTURED SOLAR CELLS AS SUSTAINABLE OPTOELECTRONICS

It is no doubt to admire that nanotechnologies have the potential to improve the development of renewable energy sources [8, 9]. It has a high impact on the energy sector. Innovations in nanotechnology bring improvements in power and efficiency, higher performance in terms of lifetime, enhanced efficiency in solar power generation, and the reduction of costs. Regarding this, herein, we have reviewed some of the work done in the research field of nanostructured PVs. It has been discussed how nanostructure materials are now being used to enhance PV performance. Various strategies for lowering costs and boosting SC efficiency have been compiled.

Kang *et al.* showed that due to the efficient electron transportation that is orthogonal to the plane of the solar film, the use of CdTe nanorods increased the optical absorption and the dissociation of excitons [10]. CdTe nanorod/P3OT polymer was the photoactive layer employed in the construction of hybrid solar cells. Under AM 1.5 circumstances, the photovoltaic metrics measured were J_{SC} as 3.12 mA/ cm^2, V_{OC} of 0.714 V with FF of 47.7%, and overall efficiency of 1.06%. Another work by Kang *et al.* stated that the utilisation of well-aligned CdS nanorods and conjugated polymers, MEH-PPV, *i.e.*, poly-(2-methoxy-5-(2-ethylhexyloxy)-1,4-phenylene vinylene) solar cells have shown to operate as an effective channel for electron transport [11]. Under AM 1.5 circumstances, this polymer-based nano-structured PV displayed J_{SC} of 1.40 mA/cm^2, V_{OC} of 0.858 V, FF of 49.6%, and Power conversion efficiency of 0.6%. Hau *et al.* synthesized highly flexible air-stable inverted polymer PV using ZnO nanoparticles as an electron-selective layer. ZnO nanoparticle-based inverted devices with ITO on plastic substrates and low-temperature processing demonstrated high efficiency of 3.3% [12]. In contrast to the traditional one, which showed small photovoltaic activity after four days. The inverted device structure was found to be far more stable in ambient circumstances, preserving more than 80% of its initial conversion efficiency after 40 days. This is because the contact between poly-(3,4-ethylene-dioxythiophene) and poly(styrene-sulfonate) and Ag has a better stability. Temple *et al.* studied the performance of Si-solar cells based on localized surface plasmon excitation of Ag nanoparticles [13]. By evaporating 10 nm layers and then thermally annealing them, Ag nanoparticles have been loaded on silicon, glass, and silicon solar cells. Metal film islanding has been used to create Ag nanoparticles on the surface of solar cells, and the variations in reflectance and spectrum response that follow have been investigated. After installing a 10 nm Ag layer, the solar cell's efficiency rose from its initial 8.51% to 9.06%. On the addition of a coating of an antireflection on the surface of the layer of the solar cell, it resulted in broad reflectance minima with no increase in reflectance at any wavelength. Park *et al.* worked on organic solar cell doped graphene electrodes. Transparent electrodes for organic photovoltaic devices (OPV) were fabricated using graphene sheets with a predetermined layer of graphene produced by chemical vapour deposition [14]. It was discovered that PCE of devices using pure graphene electrodes is on par with that of those using ITO electrodes. However, the surface wetting issue between the hole-transporting layer (HTL) and the graphene electrodes made it more vulnerable to OPVs with pure graphene electrodes than those with ITO electrodes. After looking into several different options, it was discovered that doping AuCl$_3$ on graphene brought changes in the surface's wetting characteristics so that a homogeneous layer of a hole-transporting layer is obtained. They discovered that doping enhances the conductivity and alters the graphene electrode's work function, improving the PCE performance of the OPV devices overall. With this discovery, we can use transparent electrodes made of graphene instead of ITO in the future.

Zhang *et al.* developed inorganic solar cells using nanotechnology [15]. They discovered that inorganic solar cells, as a robust PV devices for capturing electric energy from sunlight, have seen exciting development because nanomaterials have distinctive optical and electrical features. Salvador *et al.* worked on metal nanoparticle electron accumulation in plasmon-enhanced OSC [16]. The efficiency of organic photovoltaics based on an active layer made up of polymer and fullerenes has been improved through the introduction of plasmonic metal nanoparticles. When in proximity to a photoexcited bulk hetero-junction layer composed of polymer P3HT, *i.e.*, poly(3-hexylthiophene) and C61, *i.e.*, phenyl-C61-butyric acid methyl ester, the accumulation of Ag nanoprisms displayed long-lasting negative charges. Chen *et al.* worked on the enhancement of broadband in thin-film amorphous Si solar cells enabled by the nucleated Ag nanoparticle [17]. Here, compared to reference cells without nanoparticles, broadband absorption enhancement and noticeably better performance were seen in solar cells coated with 200 nm Ag nanoparticles at 10% coverage density which resulted in 14.3% increase in photocurrent density and a 23% increase in efficiency. The maximum efficiency attained among the measured plasmonic solar cells was 8.1%. Lu *et al.* worked out the enhancement of the performance of polymer PV attributed due to the cooperative plasmonic effect of Ag and Au nanoparticles [18]. They displayed how polymer BHJSCs could perform better in the presence of a cooperative plasmonic effect. Dual nanoparticles exhibit superior behaviour on the enhancement of light absorption in comparison to single nanoparticles. Incorporating Ag and Au nanoparticles into the anode buffer layer resulted in efficiency of 8.66%, accounting for 20% improvement in the efficiency of a polymer solar cell. For solar cell applications, Au nanorods with Au nanoparticles were employed. The effect of plasmonic nanostructures on exciton-producing, dissociating, and recombining charges inside thin film devices has been thoroughly studied. By adding metal NPs to the PEDOT: PSS buffer layer, they increased the PCEs of BHJ solar cells. Devices containing Au Rods were tested for solar cell performance. A high PCE of 8.67% was achieved. Singh *et al.* synthesized 2, 6-bis-(1-methylbenzimidazol-2-yl)pyridine auxiliary ligand in order to create the thiocyanate-free heteroleptic ruthenium complex [19]. This compound produced an efficiency of 6.04% when used as a sensitizer in DSSC. The %age of the cell increased to 7.76% interestingly after the addition of 0.3 weight % of multiwall carbon nanotubes (MWCNTs) to the TiO_2 nanoparticles. Studies on the electrochemical impedance spectrum showed that the addition of a modest amount of MWCNTs boosted the cell's electron lifespan, producing a high level of V_{OC}. Park *et al.* [20] worked on an Ag-Nanowire-based polymer solar cell manufactured. The active layers are composed of low band-gap polymers such as poly [[(2,6-diyl-alt-(4-(2-ethylhexanoyl)-thieno [3,4-b]thiophene)-4,8-bis-(2-ethylhexyloxy)-benzo [1,2-b;4.5-b0]dithiophene]] -2,6-diyl] and acceptor as the methyl ester of phenyl-C61-butyric acid. Fabricated cells exhibited exceptional efficiencies of 5.27% (on glass) and 3.76% under AM 1.5 illumination (on PET). This study suggests that a substitute for the ITO electrode could be an Ag NW electrode made *via* electrohydrodynamic deposition (EHD) spraying. Li *et al.* investigated shape-dependent broadband plasmonic absorption in metallic nanoparticles to improve the efficiency of OSC [21]. Ag nanomaterials in a variety of forms, such as nanoparticles and nano-prisms, are combined. A straightforward wet chemical process is used to create the nanoparticles. Simultaneous activation of plasmonic low and high-order resonance modes, which vary in shape, size, and polarisation, causes the observed PCE enhancement. The high-order resonances have a greater impact on the absorption enhancement of OSCs than low-order resonances, especially for the Ag nanoprisms under study, due to a better overlap in the absorption spectrum of the active material. The wide-band absorption is improved, and Jsc increases by 17.91% with the addition of mixed nanomaterials to the active layer. In comparison to control OSCs that have already been tuned, PCE is improved by 19.44%. These findings point to a novel strategy for raising overall enhancement by enhancing broadband absorption. Wanninayake *et al.* worked on the enhancement of performance using Cu-oxide nanoparticles for polymer solar cells [22]. A sharp improvement in the photoabsorption of the active layer was observed in UV-visible absorption spectra with enhanced EQE, *i.e.*, the external quantum efficiency of P3HT/ PC70BM solar cells with various weight %ages of CuO nanoparticles. The use of nanoparticles increased the efficiency of the solar cells by 24% in comparison to the reference cell. The short circuit current increased from 5.23 mA/ cm^2 in the reference cell to 6.484 mA/ cm^2 in cells containing CuO NPs.and the FF went from 61.15 to 68.0%, demonstrating an improvement of 11.2%. Paramasivam *et al.* synthesized a series of four different nanostructured compounds with benzthiazole, fluorene-thiophene and benzo-carbazole as donors for solar use [23]. They discovered that the optical characteristics of these dyes would be suitable for use as donors in photovoltaic devices. These newly created dyes were used to create photovoltaic devices with PC60BM serving as the photoactive layer possessing the device architecture of ITO: PEDOT:PSS (38 nm): active layer: Ca (20 nm), and Al (100 nm). It was discovered that the best efficiency was 1.62% using BTPC61BM as an active layer. Saini *et al.* developed and studied two new chemical sensitizers as sensitizers for nanocrystalline TiO$_2$-based DSSCs [24]. These sensitizers have fluorenylidene decorated on a phenothiazine donor. The dye with fluorene-thiophene resulted in higher efficiency (3.31%) than the corresponding bithiophene analogs (2.83%) among DSSCs made using TiO$_2$ film, which was a 20 nm TiO$_2$ layer and a 300 nm TiO$_2$ scattering layer. Jin *et al.* worked to enhance the design and integration methods of CNTs and graphene [25]. Work has been done on the design and application of carbon nanomaterials for photoactive and charge transport layers in organic solar cells. Saini *et al.* [26] synthesized a series of four sets of D-A-π-A dyes, characterised by DSSCs and has a carbazole donor, a linker made up of naphthalimide as an auxiliary acceptor, spacers like phenyl, oligothiophene, dithienopyrrole, and cyanoacrylic acid as terminal acceptor. The nature of the spacer between cyanoacrylic acid and naphthalimide has a substantial impact on maximal absorption. Gradually, the absorption turns red as the conjugation lengthens and the electron richness increases. Dye energies are high-lying LUMO and low-lying HOMO, respectively, making it easier to inject electrons into TiO_2 conduction band and easily regenerate the oxidised dye by an electrolyte. A maximum current density of 3.61 mA cm^{-2} and voltage of 520 mV in DSSC was produced by a dye containing dithienopyrrole. All dyes had a small charge recombination resistance at the TiO_2/dye/electrolyte interface. This could be the cause of photoelectrochemical cell subpar performance. Additionally, the effect of rigidifying the linker increased the dye's total effectiveness by 10 times, from 0.13 to 0.92%. Kaçuş *et al.* highlighted the function of Au and Ag nanoparticles in P3HT: PCBM-based OPV [27]. They created a solar cell with an undoped P3HT:PCBM active layer, an active layer doped with 0.5 weight% AgNPs, and an active layer doped with 1.5 weight% AuNPs that were synthesised with chlorobenzene as solvent. In order to achieve the high PCE of organic polymer solar cells, the authors discovered that the rough areas resulting from nanoparticles act as charge carrier trap centres during carrier movement. Doped cells with the usage of 1.5% AuNPs and 0.5% AgNPs demonstrated improved efficiencies of 3.11 and 3.20% above undoped cells with 0.48% efficiency when compared to undoped PV. Ho *et al.* used an interconnecting layer with a tandem structure in the development of high-performance tandem organic solar cells, which offer a useful method for realising high-efficiency organic photovoltaic cells [28]. It is employed to increase the range of wavelengths for light harvesting. Interconnection layer (ICL) between subcells is crucial for the performance and reproducibility of tandem PV, while processing ICL has proven to be difficult. Using a commercially available polymer material, PEDOT: PSS HTL, as a hole-carrying material used for ICL, the fabrication of efficient tandem solar cells is reported. In comparison to the common PEDOT:PSS. Better charge extraction capabilities of the ICL are provided by the solar cell's improved wettability on the underlying non-fullerene photoactive layer. The tandem solar cell achieves an efficiency of 14.7% and the maximum PCE obtained was 16.1%. The device simulation results indicated that a PCE > 22% can soon be achieved in tandem cells with additional donor polymer development for device tuning. Alkhalayfeh *et al.* worked on an enhanced polymer PV with embedded plasmonic nanoparticles [29]. The efficiency of this thin-film polymer solar cell has recently been improved using a number of techniques. To increase light absorption, hole-charge carrier production, and carrier transport, metal nanoparticles such as Au and/ or Ag NPs were inserted in the active/buffer layer or at the interface of these two layers in the active layer of the solar cell. This increased the photocurrent in PSCs. Pezhooli *et al.* worked on an interdisciplinary work with the goals of preserving the environment and generating clean energy, which involved the synthesis and use of composite TiO_2-ZnO QDs on hybrid nanostructure perovskite solar cells [30]. TiO_2-ZnO nanocomposite was tested to show how well the nanocomposite with various ratios of TiO_2 and ZnO quantum dots performed in solar applications. They discovered an efficiency of around 5% produced by the solar cell built of TiO_2-ZnO nanocomposites. However, the reported efficiency can be boosted by optimising variables composited with other materials. Bertrandie *et al.* worked on an air-processable hole transport layer for non-fullerene OSCs. They developed a hybrid solution-processable HTL made of poly-(3,--ethylenedioxythiophene): PSS, *i.e.*, poly(styrenesulfonate) and tantalum doped tungsten oxide (TaWOx) nanoparticles, which showed excellent wettability across the hydrophobic active layer [31]. Using polymer as donor and acceptor without fullerene with a combination TaWOx:PEDOT::PSS layer as the HTL, n-i-p-type OSCs with full ambient processing yield an efficiency of 8.6%. In comparison to the devices built on the previously published MoOx-PEDOT:PSS HTL, which were solution-processed, OSCs using the TaWOx-PEDOT:PSS HTL show increased thermal stability when heated to 85°C. When annealed up to 120°C, TaWOx-PEDOT:PSS HTL optimized the thermal stability. TaWOx-PEDOT::PSS promises to be a good HTL material for making stable solar cells with roll-to-roll compatible printing and coating techniques as it improves temperature stability, air-processability, and wettability on hydrophobic surfaces.

The literature survey done here depicted that solar nanotechnology makes it possible to create smaller, more efficient electronics that are more successful at creating, absorbing, and storing energy. Modern nanoscience technologies enable effective photovoltaic manufacturing procedures [32, 33]. The material synthesized may be immediately put on flexible plastic foil or substrate, which could reduce costs and enhance efficiency [34]. Herein, we have highlighted nanoparticle-based functional materials which exhibit enhanced physicochemical properties along with excellent surface-to-area ratio to be used as nano-structured thin layers. Proper utilization of nanotechnology in development of low-cost and efficient solar cells would help preserve the environment in a sustainable way.

CONCLUSION

The future looks bright for solar energy. In a decade, solar energy has advanced significantly. With the advancements in nanotechnology, solar energy is supposed to become even more affordable in the upcoming decades. It is possible that by 2030, solar energy will be the sole and primary source of energy for producing electric power on most parts of the surface of Earth. The environment and climate change will benefit from this as well. The next generation of PV can be developed using QD and carbon nanostructured solar cells, which are still in the early stages of research. The cost of solar energy should be cut in half by 2030, according to the solar industry's explicit cost-reduction roadmap. Higher-efficiency modules are already being developed, which can provide 1.5 times as much power as currently available large-sized modules. Nanowires, nanorods, and quantum dot structures are the nanostructured materials being researched and developed for solar cell applications since they make it possible to create devices that are both highly effective and inexpensive. These will have a significant effect on moving forward on the path of solar nanotechnology. The best way to incorporate solar energy into our homes, places of work, and power systems is the other significant breakthrough. Thus, herein we highlight the enhancement in the performance of photovoltaics by incorporation of nanomaterials which also meet the sustainable development of energy in the future. Better power electronics and more extensive usage of inexpensive digital technology are the results of this. We are still unable to provide solutions that are both effective and affordable, given the design of the materials we use today. Novel materials are being created with the help of better-designed materials in the field of nanoscience and engineering knowledge so that we can improve the quality of life for both the present and future generations.

LEARNING OBJECTIVES

• Photovoltaics (PVs) based on inorganic, organic and polymer materials incorporating nanotechnology is designed and synthesized with the aim of reducing the cost per watt even if it declines in reliability and conversion efficiency.

• With AM 1.5 illumination and under standard circumstances of $100 \ mW/cm^2 i.e.$ Pin, efficiency (η) of a cell can be calculated as $\eta = (J_{SC} \times V_{OC} \times FF)/ P_{in}$ where FF is described as maximum power output, which is calculated from the equation "FF = $(J_{max} \times V_{max})/ (J_{SC} \times V_{OC})$".

• There are three primary types of solar cells, 1st Generation Solar Cells-Crystalline Silicon, 2nd Generation Solar Cells- Thin Film and 3rd Generation Solar Cells- Nanotechnology based solar cells.

• In recent years, nanosized polymers, nanoparticles, and carbon-based materials have been widely investigated and used as semiconductors in electronic applications, such as nanostructured photovoltaics, in place of conventional inorganic Silicon-based photovoltaics (PVs).

• Broader absorption from visible to near IR area, low cost, higher molar extinction coefficient, long-term, stability and ease of synthesis of the nanosized sensitizers that sensitise sunlight show their superiority over silicon and ruthenium-based dyes. Nanoparticles and quantum dots enhance the coupling of sunlight into the cell substrate.

• Solar nanotechnology makes it possible to create smaller, more efficient electronics that are more successful at creating, absorbing, and storing energy.

• Nanoparticles-based functional material which exhibits enhanced physico-chemical properties along with excellent surface-to-area ratio to be used as nano-structured thin layers. Proper utilization of nanotechnology in the development of low-cost and efficient solar cells would help to preserve the environment in a sustainable way.

MULTIPLE CHOICE QUESTIONS

1. The term photovoltaic is used since _____.

a. 1840
b. 1844
c. 1849
d. 1850

2. The term photovoltaic comes from which language?

a. Spanish
b. Greek
c. German
d. English

3. Which of the following materials is used in the fabrication of solar cells?

a. Barium
b. Silicon
c. Silver
d. Selenium

4. A solar cell convert _____ Energy into electrical energy.

  a. Mechanical
  b. Magnetic
  c. Chemical
  d. None of the above

5. A solar cell unit is _____.

  a. Semiconductor diode
  b. Semiconductor triode
  c. Transistor
  d. None of above

6. _____ is one of the most important materials and known as solar grade silicon.

  a. Crushed silicon
  b. Crystalline silicon
  c. Powdered silicon
  d. Silicon

7. Quantum dot solar cells are based on _____.

  a. Gratzel cell
  b. Solar cell
  c. Voltaic cell
  d. Galvanic cell

8. The quantum dots generally used are _____.

  a. CdS
  b. CdTe
  c. PbO
  d. GaAs

9. Solar cells are made from bulk materials that are cut into a wafer of _____ thickness.

a. 120-180µm
b. 120-220µm
c. 180-220µm
d. 180-240µm

10. Quantum dots can be used in _____.

a. Crystallography
b. Mechanics
c. Optoelectronics
d. Quantum physics

11. The efficiency of a solar cell based on nanostructure may be in the range.

a. 2 to 5%
b. 10 to 15%
c. 30 to 40%
d. 15 to 20%

12. The current of solar cell is _____proportional to light/illumination.

a. Directly
b. Indirectly
c. Inversely
d. Exponentially

13. Full form of FF in the solar cell parameter is.

a. Form factor
b. Face factor
c. Fire factor
d. Fill factor

14. Calculate Fill factor using the data: Pmax=15 W, V_{oc}=15 V, Jsc=5 A.

a. .64
b. .40

c. .20
d. .38

15. Organic polymer solar cells are made from:

a. Polyphenylene
b. Vinylene
c. Carbon fullerenes
d. All of the above

ANSWER KEY

1. (c)

2. (b)

3. (b)

4. (d)

5. (a)

6. (b)

7. (a)

8. (a)

9. (d)

10. (c)

11. (d)

12. (a)

13. (d)

14. (c)

15. (d)

REFERENCES

[1]   B. O'Regan, and M. Grätzel, "A low-cost, high-efficiency solar cell based on dye-sensitized colloidal TiO2 films", *Nature,* vol. 353, no. 6346, pp. 737-740, 1991.
[http://dx.doi.org/10.1038/353737a0]

[2]   K.L. Kelly, E. Coronado, L.L. Zhao, and G.C. Schatz, "The optical properties of metal nanoparticles: The influence of size, shape, and dielectric environment", *J. Phys. Chem. B,* vol. 107, no. 3, pp. 668-677, 2003.
      [http://dx.doi.org/10.1021/jp026731y]

[3]   P. Singhal, and H.N. Ghosh, "Hot charge carrier extraction from semiconductor quantum dots", *J. Phys. Chem. C,* vol. 122, no. 31, pp. 17586-17600, 2018.
      [http://dx.doi.org/10.1021/acs.jpcc.8b03980]

[4]   S. Lu, H. Lin, S. Zhang, J. Hou, and W.C.H. Choy, "A switchable interconnecting layer for high performance tandem organic solar cell", *Adv. Energy Mater.,* vol. 7, no. 21, p. 1701164, 2017.
      [http://dx.doi.org/10.1002/aenm.201701164]

[5]   H. Sun, P. Dai, X. Li, J. Ning, S. Wang, and Y. Qi, "Strategies and methods for fabricating high quality metal halide perovskite thin films for solar cells", *Journal of Energy Chemistry,* vol. 60, pp. 300-333, 2021.
      [http://dx.doi.org/10.1016/j.jechem.2021.01.001]

[6]   Y. Huang, E.J. Kramer, A.J. Heeger, and G.C. Bazan, "Bulk heterojunction solar cells: Morphology and performance relationships", *Chem. Rev.,* vol. 114, no. 14, pp. 7006-7043, 2014.
      [http://dx.doi.org/10.1021/cr400353v] [PMID: 24869423]

[7]   Y.J. Cheng, S.H. Yang, and C.S. Hsu, "Synthesis of conjugated polymers for organic solar cell applications", *Chem. Rev.,* vol. 109, no. 11, pp. 5868-5923, 2009.
      [http://dx.doi.org/10.1021/cr900182s] [PMID: 19785455]

[8]   C.J. Brabec, M. Heeney, I. McCulloch, and J. Nelson, "Influence of blend microstructure on bulk heterojunction organic photovoltaic performance", *Chem. Soc. Rev.,* vol. 40, no. 3, pp. 1185-1199, 2011.
      [http://dx.doi.org/10.1039/C0CS00045K] [PMID: 21082082]

[9]   C. Duan, K. Zhang, C. Zhong, F. Huang, and Y. Cao, "Recent advances in water/alcohol-soluble π-conjugated materials: new materials and growing applications in solar cells", *Chem. Soc. Rev.,* vol. 42, no. 23, pp. 9071-9104, 2013.
      [http://dx.doi.org/10.1039/c3cs60200a] [PMID: 23995779]

[10] Y. Kang, N.G. Park, and D. Kim, "Hybrid solar cells with vertically aligned CdTe nanorods and a conjugated polymer", *Appl. Phys. Lett.,* vol. 86, no. 11, p. 113101, 2005.
      [http://dx.doi.org/10.1063/1.1883319]

[11] Y. Kang, and D. Kim, "Well-aligned CdS nanorod/conjugated polymer solar cells", *Sol. Energy Mater. Sol. Cells,* vol. 90, no. 2, pp. 166-174, 2006.
      [http://dx.doi.org/10.1016/j.solmat.2005.03.001]

[12] S.K. Hau, H.L. Yip, N.S. Baek, J. Zou, K. O'Malley, and A.K.Y. Jen, "Air-stable inverted flexible polymer solar cells using zinc oxide nanoparticles as an electron selective layer", *Appl. Phys. Lett.,* vol. 92, no. 25, p. 253301, 2008.
      [http://dx.doi.org/10.1063/1.2945281]

[13] T.L. Temple, G.D.K. Mahanama, H.S. Reehal, and D.M. Bagnall, "Influence of localized surface plasmon excitation in silver nanoparticles on the performance of silicon solar cells", *Sol. Energy Mater. Sol. Cells,* vol. 93, no. 11, pp. 1978-1985, 2009.
      [http://dx.doi.org/10.1016/j.solmat.2009.07.014]

[14] H. Park, J.A. Rowehl, K.K. Kim, V. Bulovic, and J. Kong, "Doped graphene electrodes for organic solar cells", *Nanotechnology,* vol. 21, no. 50, p. 505204, 2010.
      [http://dx.doi.org/10.1088/0957-4484/21/50/505204] [PMID: 21098945]

[15] Y. Zhang, H. Geng, Z. Zhou, J. Wu, Z. Wang, Y. Zhang, Z. Li, L. Zhang, Z. Yang, and H.L. Hwang, "Development of inorganic solar cells by nano-technology", *Nano-Micro Lett.,* vol. 4, no. 2, pp. 124-134, 2012.

[http://dx.doi.org/10.1007/BF03353703]

[16]   M. Salvador, B.A. MacLeod, A. Hess, A.P. Kulkarni, K. Munechika, J.I.L. Chen, and D.S. Ginger, "Electron accumulation on metal nanoparticles in plasmon-enhanced organic solar cells", *ACS Nano*, vol. 6, no. 11, pp. 10024-10032, 2012.
[http://dx.doi.org/10.1021/nn303725v] [PMID: 23062171]

[17]   X. Chen, B. Jia, J.K. Saha, B. Cai, N. Stokes, Q. Qiao, Y. Wang, Z. Shi, and M. Gu, "Broadband enhancement in thin-film amorphous silicon solar cells enabled by nucleated silver nanoparticles", *Nano Lett.*, vol. 12, no. 5, pp. 2187-2192, 2012.
[http://dx.doi.org/10.1021/nl203463z] [PMID: 22300399]

[18]   L. Lu, Z. Luo, T. Xu, and L. Yu, "Cooperative plasmonic effect of Ag and Au nanoparticles on enhancing performance of polymer solar cells", *Nano Lett.*, vol. 13, no. 1, pp. 59-64, 2013.
[http://dx.doi.org/10.1021/nl3034398] [PMID: 23237567]

[19]   S.P. Singh, K.S.V. Gupta, M. Chandrasekharam, A. Islam, L. Han, S. Yoshikawa, M. Haga, M.S. Roy, and G.D. Sharma, "2,6-Bis(1-methylbenzimidazol-2-yl)pyridine: a new ancillary ligand for efficient thiocyanate-free ruthenium sensitizer in dye-sensitized solar cell applications", *ACS Appl. Mater. Interfaces*, vol. 5, no. 22, pp. 11623-11630, 2013.
[http://dx.doi.org/10.1021/am4030627] [PMID: 24187913]

[20]   S.E. Park, S. Kim, D.Y. Lee, E. Kim, and J. Hwang, "Fabrication of silver nanowire transparent electrodes using electrohydrodynamic spray deposition for flexible organic solar cells", *J. Mater. Chem. A Mater. Energy Sustain.*, vol. 1, no. 45, p. 14286, 2013.
[http://dx.doi.org/10.1039/c3ta13204h]

[21]   X. Li, W.C.H. Choy, H. Lu, W.E.I. Sha, and A.H.P. Ho, "Efficiency enhancement of organic solar cells by using shape-dependent broadband plasmonic absorption in metallic nanoparticles", *Adv. Funct. Mater.*, vol. 23, no. 21, pp. 2728-2735, 2013.
[http://dx.doi.org/10.1002/adfm.201202476]

[22]   A.P. Wanninayake, S. Gunashekar, S. Li, B.C. Church, and N. Abu-Zahra, "Performance enhancement of polymer solar cells using copper oxide nanoparticles", *Semicond. Sci. Technol.*, vol. 30, no. 6, p. 064004, 2015.
[http://dx.doi.org/10.1088/0268-1242/30/6/064004]

[23]   M. Paramasivam, R.K. Chitumalla, S.P. Singh, A. Islam, L. Han, V. Jayathirtha Rao, and K. Bhanuprakash, "Tuning the photovoltaic performance of benzocarbazole-based sensitizers for dye-sensitized solar cells: A joint experimental and theoretical study of the influence of π-Spacers", *J. Phys. Chem. C*, vol. 119, no. 30, pp. 17053-17064, 2015.
[http://dx.doi.org/10.1021/acs.jpcc.5b04629]

[24]   A. Saini, K.R.J. Thomas, C.T. Li, and K.C. Ho, "Organic dyes containing fluorenylidene functionalized phenothiazine donors as sensitizers for dye sensitized solar cells", *J. Mater. Sci. Mater. Electron.*, vol. 27, no. 12, pp. 12392-12404, 2016.
[http://dx.doi.org/10.1007/s10854-016-5146-5]

[25]   S. Jin, G.H. Jun, S. Jeon, and S.H. Hong, "Design and application of carbon nanomaterials for photoactive and charge transport layers in organic solar cells", *Nano Converg.*, vol. 3, no. 1, p. 8, 2016.
[http://dx.doi.org/10.1186/s40580-016-0068-8] [PMID: 28191418]

[26]   A. Saini, K.R.J. Thomas, Y.J. Huang, and K.C. Ho, "Synthesis and characterization of naphthalimide-based dyes for dye sensitized solar cells", *J. Mater. Sci. Mater. Electron.*, vol. 29, no. 19, pp. 16565-16580, 2018.
[http://dx.doi.org/10.1007/s10854-018-9750-4]

[27]   H. Kaçuş, M. Biber, and Ş. Aydoğan, "Role of the Au and Ag nanoparticles on organic solar cells based on P3HT:PCBM active layer", *Appl. Phys., A Mater. Sci. Process.*, vol. 126, no. 10, p. 817, 2020.

[http://dx.doi.org/10.1007/s00339-020-03992-7]

[28] C.H.Y. Ho, T. Kim, Y. Xiong, Y. Firdaus, X. Yi, Q. Dong, J.J. Rech, A. Gadisa, R. Booth, B.T. O'Connor, A. Amassian, H. Ade, W. You, T.D. Anthopoulos, and F. So, "High-performance tandem organic solar cells using hsolar as the interconnecting layer", *Adv. Energy Mater.,* vol. 10, no. 25, p. 2000823, 2020.
[http://dx.doi.org/10.1002/aenm.202000823]

[29] M.A. Alkhalayfeh, A. Abdul Aziz, M.Z. Pakhuruddin, and K.M. M Katubi, "Spiky durian-shaped Au@Ag nanoparticles in PEDOT:PSS for improved efficiency of organic solar cells", *Materials,* vol. 14, no. 19, p. 5591, 2021.
[http://dx.doi.org/10.3390/ma14195591] [PMID: 34639989]

[30] N. Pezhooli, J. Rahimi, F. Hasti, and A. Maleki, "Synthesis and evaluation of composite TiO2@ZnO quantum dots on hybrid nanostructure perovskite solar cell", *Sci. Rep.,* vol. 12, no. 1, p. 9885, 2022.
[http://dx.doi.org/10.1038/s41598-022-13903-w] [PMID: 35701463]

[31] J. Bertrandie, A. Sharma, N. Gasparini, D.R. Villalva, S.H.K. Paleti, N. Wehbe, J. Troughton, and D. Baran, "Air-processable and thermally stable hole transport layer for non-fullerene organic solar cells", *ACS Appl. Energy Mater.,* vol. 5, no. 1, pp. 1023-1030, 2022.
[http://dx.doi.org/10.1021/acsaem.1c03378]

[32] L. Zhan, S. Li, T-K. Lau, Y. Cui, X. Lu, M. Shi, C-Z. Li, H. Li, J. Hou, and H. Chen, "Over 17% efficiency ternary organic solar cells enabled by two non-fullerene acceptors working in an alloy-like model", *Energy Environ. Sci.,* vol. 13, no. 2, pp. 635-645, 2020.
[http://dx.doi.org/10.1039/C9EE03710A]

[33] L. Zhan, S. Li, T-K. Lau, Y. Cui, X. Lu, M. Shi, C-Z. Li, H. Li, J. Hou, and H. Chen, "Over 17% efficiency ternary organic solar cells enabled by two non-fullerene acceptors working in an alloy-like model", *Energy Environ. Sci.,* vol. 13, no. 2, pp. 635-645, 2020.
[http://dx.doi.org/10.1039/C9EE03710A]

[34] F. Qi, L.O. Jones, K. Jiang, S.H. Jang, W. Kaminsky, J. Oh, H. Zhang, Z. Cai, C. Yang, K.L. Kohlstedt, G.C. Schatz, F.R. Lin, T.J. Marks, and A.K.Y. Jen, "Regiospecific N -alkyl substitution tunes the molecular packing of high-performance non-fullerene acceptors", *Mater. Horiz.,* vol. 9, no. 1, pp. 403-410, 2022.
[http://dx.doi.org/10.1039/D1MH01127H] [PMID: 34666341]

CHAPTER 12

Nanomaterials Applicability in Blended Perovskite Solar Cells: To commercialize Lead-free Content, Including Easy End of life Management in Solar Infrastructure

Bhavesh Vyas[1,*], Jayesh Vyas[2], Vineet Dahiya[1] and Puja Acharya[1]

[1] *Department of EEE & ECE, K.R. Mangalam University, Gurgaon, Haryana, India*

[2] *Department of Mechanical & Chemical Engineering, Indian Institute of Technology, Jammu, India*

Abstract: This chapter provides insights about the Perovskite type of Solar Cell (PVSC) that can be utilized as a probable substitute for existing solar panels. Pros of reduced lead participation in chemical structure and better end-of-life supports have boosted research explorations. Combinatorial-based explorations in perovskites have created a collection of various chemical designs with unique properties and improved functionalities. Investigations in terms of catalytic nature, environment adaptability, and spintronic properties provide vivid structural arrangements. But still, constraints based on application-specific composition, raw material utilization, problems of stability issues, and the reduction in lead content are the objectives yet to be achieved at the commercial level. Interlinking of chemical structural, the nature of multiple layers created for building the material can be improvised with nanomaterials integration as detailed in the chapter. Comparative lab results analysis and efforts over raised stability and reduced lead content are taken into the study to present PVSC as upcoming commercialized products. Moreover, future scope briefs about existing nanomaterials composites with improvised electron and hole transport layers. Also, findings related to back electrode-based placement to achieve better efficiency and stability are submitted to reveal suitable obtains of the experiments covered in the study.

Keywords: Interface Recombination, Responsivity, Atomic Layer Deposition, Deposition Sequence, Chromophore, Photostability and Photocatalytic activity, Work Functions, Plasmon Resonance, slip flow, Nucleation Promoter, thickness tailoring, Quantum Confinement effect.

---

* **Corresponding author Bhavesh Vyas:** Department of EEE & ECE, K.R. Mangalam University, Gurgaon, Haryana, India; E-mail: tonu567@gmail.com

Gopal Rawat & Aniruddh Bahadur Yadav (Eds.)

INTRODUCTION

Technological changes are the research upgrades that always occur over a while. After introducing new research and innovation, big targets and challenges of making zero carbon emissions require a significant impetus from each domain.

The present book deliberates on the role of nanotechnology in various fields. Similarly, this chapter relates the upcoming modifications in the sizeable solar infrastructure market. From the solar energy point of view, conventional silicon crystals have always faced challenges in providing fabricated products at lower rates. Building crystals and weighted panels require higher temperature facilities that all include the running, operating, and summation of all types of cost, which makes it expensive. Suppose the same conditions are compared with perovskite-type solar cells (PVSCs). In that case, it sounds more reliable as perovskites can be prepared at a laboratory scale with low/average temperatures. Chemical components handling based on accuracy is required with the desired knowledge. They can be fabricated on thin films, and among all, with better power conversion efficiency compared to conventional setups.

Creating suitable perovskite is selecting the right ingredients; the rest are the procedural step. Possibilities available are nearly half a million; based upon suitable atomic configurations, the testing arrangements of precursor, additive and thermal feasibility are acquired. A random search could have an infinite loop as multiple combinatorial constraint satisfaction applied to them.

ARTIFICIAL INTELLIGENCE-BASED SCALING UP OF THE SOLAR MANUFACTURING SECTOR

The PVSC's group identified a possible upcoming replacement of the existing conventional silicon construction. Replacing the earlier practices of high-degree wafers enrichment and conversion to crystalline structures through room temperature-based fabrication setup, having fewer contingencies converted to costing over the constraints of placing and installing are considered. But still, the way to market is half done.

In the present world, the blending of algorithms has improved the results, and similar practices will be preferred in the upcoming time. Creating PVSCs utilizes multiple changing parameters that may vary concerning creation practices. Hence, machine learning will aid the work by boosting the fabrication facility. As the commercial platform is considered, the E-level of 18.5% is achieved with specific prerequisites, the work proposed by Tonio Buonassisi *et al*. [1]. Structural layering, the composition is considered for PVSCs. There are multiple numbers of combinations based on compounds concerning structural design exists. Moreover, the creation facility need not depend only on the spin coating type of process. New practices to create better products from lab standards are still a part of experimentation. Substitute techniques are also used in rapid spray plasma processing, vacuum deposition, rolling surface ways, *etc*. It will be improvised for better outcomes in the upcoming time. Better obtaining of precursor inks on jet and spray-type techniques for roll-to-roll platforms. Time processing in the building of the silicon frame also acts as a constraint. In the case of PVSC's, it depends upon various parameters that need to be controlled with the help of machine learning. Conditions such as structure finalized, moisture, and temperature [2] . Identification of toxin-categorized elements can be submitted for data sets so that actual materials that need to be excluded can be categorized [3].

GAPS IDENTIFIED FROM THE INTERNATIONAL PROGRESS OF PVSC'S

Large-sized market cap of entrepreneurs is aimed at PVSC-based outlets globally, such as Microquanta has set up the world's first pilot manufacturing facility for perovskite products with a size of 5 GW capacity. GCL New Energy followed them up with a 100 MW-volume PVSC production setup in Kunshan. Market survey invokes that deploying an 18% efficient PVSC module can reduce 70% of the production cost of making conventional silicon PV's. The research study at Oxford PV achieved an efficiency of 29.52% in tandem structures added by Berlin's 125 MW- production setup [4, 5]. In the outdoor power generation segments: Wonder, Solar Limited fabricated printable film-type modules with a panel area of 110 m^2 are also running. [6 - 8].

The companies need to update and pass the testing/regulations in 2020. Microquanta company cleared the stability test in which a 20 cm^2 perovskite module underwent a 3000-hour damping heat test without degradation and provided an efficiency loss of less than 2% after a UV preconditioning test over a product lifetime of over 25 years. Utmo Light Ltd. followed a similar direction by passing IEC tests with a stabilized efficiency of over 20%. Recently, the Okinawa Institute of Science and Technology Graduate University (OIST, Japan) reported over an 1100-h operational lifetime for a 10×10 cm^2 solar module. Global data suggest the upcoming value of count from 84 to higher in solar cell-based startups such as:

Swift Solar in solar devices from the United States produces high power to weight share devices in the efficient PVSC field. They also utilize the ink of perovskite in energy-related services. Swedish Algae Industry – diatom frustules added dye-Sensitized based PVSCs made from algae in the direction of cell creation. Targets are set initially to raise the efficiency level to 4% higher than the present-day scenarios. Sunew – An Organic Photovoltaics (OPV) based manufacturer from Brazil. Freschfield: A Quantum Dot carry works in cost minimization of solar manufactured assemblies. It utilizes nano paints in support of additive materials placed over glass plates [4].

Deficiencies: Major challenge lies in scaling panel dimensions from a commercialization point of view. Moreover, the success rate of power output decreases to 19.6% with the increment in the aperture part. from 0.1 cm^2 to 10 cm^2, further drips to 17.9% for an area approaching 1000 cm^2. These values remain behind conventional crystalline silicon cells having 26.7% at 79.0 cm^2 and 24.4% at 13,177 cm^2 [11, 45, 46].

THE NEED FOR MACHINE LEARNING INSPIRED PVSC'S FABRICATION

The discovery of new combinations can be initialized by the initial search and looping properties of structures with stabilized PVSCs [18]. Then sorting based on attributed model creation will be proceed by further experimentation. The structure-based deep molecular relationship is established. Moreover, the prediction of new configurations can evolve in the field of PVSCs using machine learning tools. Specific properties or parameters can be set as a threshold, and better-predicted perovskite can be found [19 - 22]. Specificity could be channelized through machine learning [19 - 22] by satisfying the curves with reduced data and even arrangements per size. Advancement with neural network in machine learning [19 - 22] with merging Bayesian programs will be capable of satisfying limited data sets collected, thus contrasting with artificial intelligence. Fabrication is a function of Portraying PVSC's electrical traits. Constraint-specific (Band gaps, strength, *etc.*) series sequence can be initiated while processing the final product with reduced time sequence. Reduction in errors and increment in power conversion levels are obtained. It can predict the physical and role-based priority of materials utilized in the additive interfacing of chemicals [20].

The time elapsed in fabrication processes demands fast algorithms and training on the machine [21]. To balance the constraint parameters of the environment for judging the best possible combinations in terms of stability and efficiency with a fast rate of PVSC creation. Sample finding- data modeling- training -applied – optimization- set model created. These could be the sequence of algorithmic work functions [22].

The rolling time of the sheet and the space between the jet sheet subtract, and the curing ways used for the purpose need to be channelized based on feedback. Variations are required to modify the average room temperature as per the need for chemical equations. All could be better managed through a machine learning-guided experimental approach. A deficiency in machine learning is that raw case studies can be implemented, but human analytics or visual inspection-based requirements can't be easily implemented. The identified defect can be rectified through Bayesian- probability approach. The test scenarios created will be based on comprehensive data observations collected from earlier collected data sets. Making the code freely available to commercial companies can be utilized for improvement purposes, and changes in the model can be practiced. Consensus in using manufacturing methods, reducing human hours, processes involved with constraints, satisfaction, *etc.*, for PVSC is ambiguous at the commercial level.

PEROVSKITE JOURNEY AND CONTRIBUTION AS SOLAR CELL

Longer eras of crystal-silicone-type solar cells have dominated the present setup of the market. The database confirms the four percent solar contribution to world power (International Agency). They are considering the 25–30-year life span of these solar modules. Solar generation was 1.4 GW in 2000 and has grown to 760 GW by 2020. *Vice versa*, it can be estimated that if three million solar panels are required for one gigawatt of energy on average, then for 760 GW approx. A count of 7.6×10^{11} solar modules are already working, which will be out of service as per duration. These panels are already built in high-temperature conditions and not easy to discard or end of life success rate. However, 90% on record, the remaining 10% has a tough impact on the sustainability and health of the environment and earth because that 10% is either dumped or mixed with the environment in natural conditions.

Most solar cells are constructed with "single crystalline silicon" to cope with efficiency difficulties. Since there is no grain boundary in monocrystalline solar cells because of their continuous structure, excited electrons can travel freely across the silicon structure without being constrained by grain boundaries. This cell's efficiency is drastically reduced by up to 10-15% compared to polycrystalline cells because the latter has much more grain boundaries that prevent the continuous flow of exciting electrons in the semiconductor. So, according to this study, "thin film solar cells" were created to balance efficiency and cost. Due to the lack of structures with uniform crystals in place, even though this is made of amorphous silicon, it still needs to be more efficient.

WHY IS PEROVSKITE PREFERRED HIGHER THAN CONVENTIONAL

Processing costs: Compared to crystalline silicon solar cells, perovskite solar cells are relatively less expensive to produce because they don't need to go through the same extraction process as the latter, which is instead accomplished through a method known as "solution processing" (referring grossly), which is also employed in the printing of newspapers.

Scalability: Compared to other existing solar panel production technologies, production costs may be reduced because of the solution processing technique's high degree of scalability [9].

Energy conversion efficiency: Compared to silicon solar cells, perovskite solar cells have a comparatively longer carrier diffusion length, which leads to increased efficiency.

In addition to the benefits listed above, PSCs such as "methylammonium-lead halide (MALH) material with a perovskite structure" that operate with single junctions and have an efficiency of 25.2% and multijunction technologies that, when used in conjunction with highly efficient crystalline silicon photovoltaic cells, further increase efficiency through integrated photovoltaics, bringing it theoretically up to 40 percent [10, 11].

With time duration, three different generations have undergone advance. The present and upcoming era will be of Perovskites, the NREL's graph has elaborated the research carried out in a short time over PVSC, as provided in Fig. (1). Proceeded by the reference number in brackets.

1st Generation		2nd Generation	
1. Monocrystalline Si Cell 2. Polycrystalline		1. Thin film solar panels (TFSC) 2. Amorphous silicon solar panels (A-Si)	
3rd Generation			
1. Cadmium Telluride solar panels 2. Concentrated PV panels			
RESEARCH BASED MULTIPLE VARIANTS OBTAINED IN SOLAR CELLS (SC'S)			
A	Perovskite SC	E	Bifacial SC [23]	I	Intermediate Band SC [24]
B	Liquid Inks SC [25]	F	Photon Up & Down Conversion SC [26]	J	Light Absorbing Dyes SC [27]
C	Quantum Dots SC [28-30]	G	Organic/Polymer SC [31-33]	K	Adaptive Cells SC [34]
D	Surface Texturing SC [35-40]	H	Encapsulation SC [41-43]	L	Autonomous Maintenance SC [44-48]

Fig. (1). Multiple advancements in solar cell technology.

Fig. (1) shows multiple varieties with different structural combinations, yet one must be the best for commercialization among all variants provided. Updations in materials from photon up/down conversion, light absorbing dyes [16], quantum dots, organic polymers *etc.*, the list is still under updation due to combinatorial patterns generated from different compositions of materials [27 - 31]. For Mono-type, the efficiency lies at 17 to 19%, and the price is Rs 47 for 300 watts and higher. For poly-si type, it is a 1410-degree Celsius temperature-prepared silicon material with a longer work cycle. Able to work under extreme weather situations with 17% accuracy. Higher prices compared to the earlier one. Thin film types are multiple segregations of films, utilizing precursor and substrate, although unreliable but suitable for large space sections. Amorphous type of silicon cells is triple layered with 7% efficiency and low cost in low weight variety [10]. From here onwards, the utilization of organic/inorganic content varieties is raised. Like cadmium type, although better technical features end of life causes health hazards. The other concentrated type is the most reliable, with a 41% efficiency level. A curved outside and lens system placed in the cooling part raises energy conversion. The facility of multiple junctions and angular-based reception is higher in such facilities. The upcoming Table **1** will also compare some basic properties of multi-varieties of solar PVSCs.

Table 1. Comparison of Multiple Variants of PVSCs with Doping differences.

CRITERIA	C-SI	CHALCOGENIDE	ORGANIC	III-V	PEROVSKITE	PEROVSKITE-SI-TANDEM
Power Conversion Efficiency	An efficiency of 20.5%-23% is observed [50]. The current lab efficiency record of a cell is 26.7% [49].	The efficiency of the range of 16-17% . The current lab efficiency record of a cell is efficiency is Around 22-23% [49].	An efficiency of 6%- 8% is observed. The current lab efficiency record of a cell is 17.4%	An efficiency of 30% is observed. The current lab efficiency record of a cell is 38.8%. (For non-concentrated irradiation, 5-junction cell)	An efficiency of 18% has been recorded. The cell's current lab efficiency record is 25.2%.	A target module efficiency of around 30% is set.
Cost	Currently, a typical cell costs around 1650 INR/m², while a module costs about 3200 INR/m.	The price range of chalcogenide solar modules is comparable to that of c-Si modules².	Organic solar modules would be less expensive than c-Si technology.	III-V multijunction cells nowadays are 1-2 times greater expensive per Wp than crystalline Si.	After the first ramp-up phase, the module costs are lower than for cSi technology.	Prize greater than that of cSi, it is anticipated that the efficiency improvement will more than cover the extra price.
Long term Stability	Cells need to be protected from humidity. Or it may deteriorate due to different thermal expansion coefficients of the cell encapsulation. If there is a mechanical load, the c- Si cells could break.	Chalcogenide solar cells are a little more humidity sensitive. Hence, more work is needed than for cSi technology.	Organic solar cells have poor long-term stability, which requires significant efforts to stop deterioration. Along with humidity, cells also deteriorate when exposed to continuous UV light; to avoid it use UV filtering.	Incredibly reliable in space applications. The cells can withstand radiation and temperatures up to 150 °C without deteriorating. Humidity may cause corrosion.	Since moisture has a similar degrading impact on cell performance, long-term stability is equivalent to that of organic solar cells.	Similar effects apply to perovskite cells

(Table 1) cont.....

CRITERIA	C-SI	CHALCOGENIDE	ORGANIC	III-V	PEROVSKITE	PEROVSKITE-SI-TANDEM
Performance under lower irradiation and elevated temperature	Efficiency under low light is significantly decreased to about 90% At 100 W/m² irradiation; Higher temperatures drop the cell's output power by roughly 0.5 percent /K.	Different chalcogenide cell types may operate differently under lower irradiation and higher temperatures. While the efficiency drop for lower irradiation is smaller for CdTe but somewhat larger for CIGS, the temperature effect is slightly less than c-Si.	Lower irradiance and higher temperatures have a less pronounced impact than they do for c-Si solar cells.	III-V solar cells perform better than c-Si solar cells when exposed to low light and high temperatures. Lower dark saturation current and temperature coefficient are responsible for it.	For 100 W/m² the performance at higher temperatures and lower irradiance is around 90%. In comparison to other cell technologies, cell technology is, therefore less impacted.	The silicon bottom cell mainly drives the impact of low, light, and high temperatures. As a result, it could be comparable to c-Si cells.

Thus, the note from Table **1** suggests that research still needs to go on with patience to search for a better commercial product that can be implemented for a longer duration with better features and less harm to society and the earth. The upcoming section relates the utilization of the nanomaterials family in PVSC's building processes.

NANOPARTICLES UTILIZATION IN TRANSPORT LAYERS AND PROBABLE AS BETTER LEAD SUBSTITUTES

The family of materials, especially at the nano level, needs a specialized setup with a skill set of knowledge to create structural level changes, doping, *etc*., along with special measurement devices.

The current section briefly details the building of PVSCs by participating nanoparticles in some proportion to enrich the functionality or structural properties by mixing blend indirectly. It starts from aluminum oxide and its group members. It says that [12] utilizing silicon and germanium for creating vivid junctions-based lead content halide PVSCs is regular while fabricating photodetectors. But the kinematics of interfacing and trapping declines the performance and provides reverse effects. The following issue can be resolved by giving a new combination through an interfacial strategy. That is deploying an interlayer of aluminum oxide. To obtain better performance of photodetection in case of self-powered. It directly varies the interface trap states causing the increment of photocurrent. Thus, a higher photocurrent value per unit in terms of the detector's sensitivity is considered and obtained better per earlier records. SEM, Photo-luminance spectra (PLS), and XRD-type observations were carried out to judge the growth of microcrystals $[CH_3NH_3PbBr_3/Al_2O_3/p\text{-}Si]$ heterojunctions are modelized in the lab with supported crystallization processes at average room temperature. 0.39 A W−1, 8.45×10^{12} Jones at 0 bias voltage with rising time and falling time are 110 ms and 53 ms, also supported with all variations and values at different excitation power points 6.23 µW - 205.3 µW. Comparison can be carried with surface-controlled growth, and templated type growth [12, 14]. The illustration shown in Fig. (**2**) relates the utilization of variants available for nanoparticle compositions [13]; it also links the basics of vivid composite materials available in the field of PVSCs by integration at a nano level.

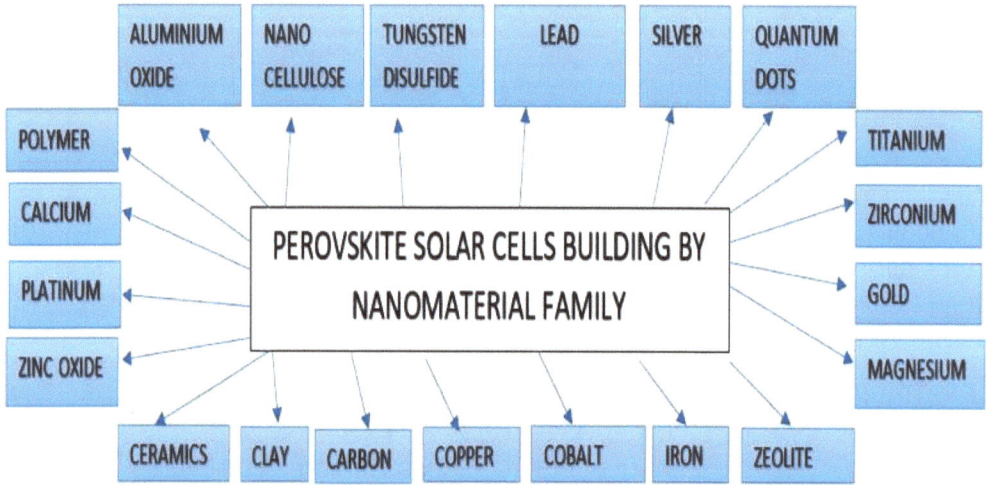

Fig. (2). Nanomaterial Family Description Utilized for PVSC's Creation.

Amidst the description of perovskite research with nanomaterials, various measuring instruments, knowledge, and multiple terminologies must be learned steps by step, such as identifying the nano-level structural changes by actions or series of lab experiments requires the utilization of advanced ways to obtain data for the better observance of the changes that occurred, to create a better model of the situation in a utility-specific direction. Thus, in the case of nanotechnology experimentation, the optical characterization ways like checking out absorbance or transmittance [47]. Temperature effects through photoluminescence, and primarily highly expertise is through Raman scattering measurement. The recognition of non-visible properties such as metamaterial behaviors in the case of nanorods of Ag [13]. Purity identification of multiwall carbon nanotubes using R-Scattering effects of temperature-based annealing through FT-IR, absorbance, and fluorescence spectroscopy. Various other examples include changes in optical traits, developments in the structural pattern, variations in optical and electrical characteristics, and optical transmission [14, 15]. The deposition of several electron carriage constituents continues the fabrication of thin films in the field of PVSCs. Utilization of two nanoparticles of aluminum, zinc oxide is mixed in combined as doped ones. Outcome and stability effects-based studies will be executed over the deposition sequence through doped nanomaterials on the fabricated films. Layer deposition sensitivity is checked using an X-ray spectroscopy based on photoelectron (PST). The blend of materials raises the temperature-facing capability of the MA (methylammonium) based layer of lead iodide. It also smooths the process of charge transferring at boundaries. Thus, further conversion-based testing provides 18.09% obtaining's and low hysteresis. They are moreover retaining efficiency to a level of 82% after a while of 100 h under normal conditions. Hence, approaching atomic layer-based engineering approaches could be carried out to obtain new variants of PVSC's target of better thermal strength is a primary demand.

[16, 17] PVSCs are also fabricated by taking the nanomaterial as the bottom electrode, and nanowires of silver sprayed to coat as the top electrode. The interfacial connection between the top and bottom electrode layers varies the outcomes. Cost-wise comparison of PVSC's electrode scales with the indium-ti--oxide type one with better results at the commercial level. But here, it has been substituted with thin aluminum layers over an added advantage of cost and conductivity in comparison to earlier ones. Utilizing the aluminum-based electrode concept targets the cost reduction in perovskites commercial cost. Experiments of using an aluminum electrode with poly(bis(4-phenyl)) (2,4,6-trimethylphenyl) amine (PTAA) results in a decline in performance in the transportation of holes. But the modification of the layers by placing a molybdenum (VI) oxide (MoO_3) layer between them resolves the issue and reduces the barrier, faster when hole injection is obtained. To obtain its best, thickness optimization needs to be maintained. It will obtain 7.09% converted efficient, low-cost commercial PVSC products.

With the merging of nanopillars [18] type of technology, nano participation is raised in each fold to support the solar product in any possible way of materials identification and combination. Carbon is an abundant material and is explained here by explored research work the glimpse of the structure shown in Fig. (**3b**). [19]. The utilization of cobalt oxide and readily available carbon are placed on building PVSCs through the printing process over the screen.

The hole transport layer will be practiced to replace carbon materials; here, the stabilization will rise due to the water-repellant nature of carbon. Spinel cobaltite oxide will try to raise the passing of holes by reducing combinations of charges. Reducing charge recombination will lead to redundancy in power conversion during interface. Nanoparticles are made through chemical precipitation and analysed *via* XRD/FESEM/XAS/TEM and UV-Vis spectroscopy. X-ray diffraction , X-ray absorption spectroscopy, field emission scanning electron microscopy, transmission electron microscopy, and UV-Vis spectroscopy. If the working part is considered more than such thin film, inorganic content-based PVSCs suggest good energy levels, and an improved efficiency level from 11.25% of earlier to 13.27%. If hours-wise stability is taken, then under ambient situations, 2500 hours working is obtained. Thus, the practicability of a Co_3O_4-based interlayer product with a 70 cm^2 area and a set conversion level of >11% without any hysteresis is found to be more stable. The continuous rise in research in the field of perovskites depicts the importance of materials and their implementation to obtain perfect combinatorial structures that survive for more extended periods and with better payback conditions. Over the last five [20, 21] years, noble metals have been used to consider electrode materials. As a result, carbon has also been coming into a role through various structures such as graphite, carbon black, graphene, and carbon nanotubes based on variations in its characteristics and structure combinations; these are taken into practice and scaled as laboratory products. Compared to gold, it has been prioritized over efficiency and stability bases. The water-repellent property makes it suitable to deploy in the electron transportation layer. Thus, it leads to obtaining a material having solid and flexible traits as substrate. It increases the probability of using carbon to build PVSCs with better state-of-art facilities. The stability of chemicals, temperature, and inherent better conductivity also promote its utilization level at the industrial scale.

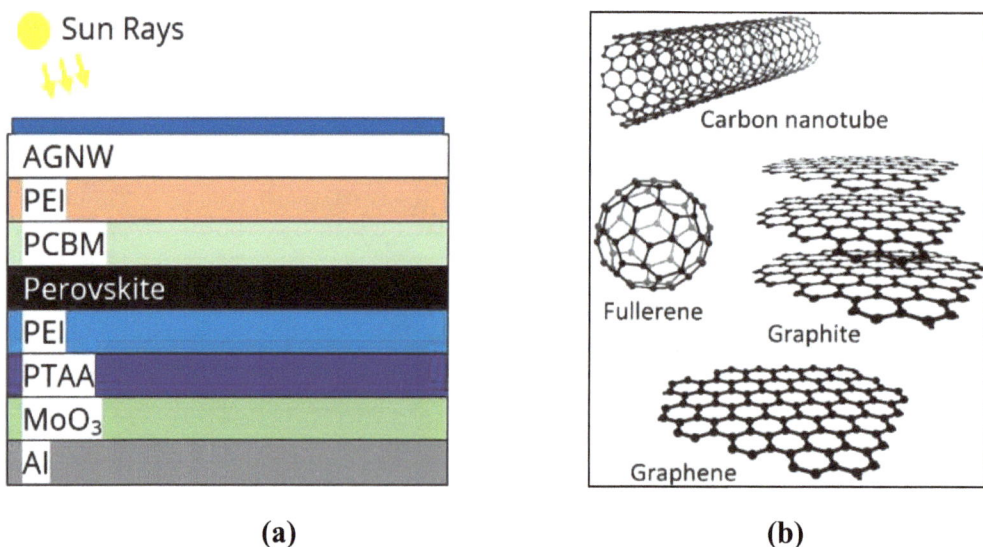

(a)   (b)

Fig. (3). (a)Al/MoO3 Based PVSC's Layering, (b) Carbon nanostructures utilized for PVSC's.

Inching [22], the efficiency level from single to double digits, involves deep research in building new materials. All along with it, stability always needs to be satisfied as the commercial purpose is the target. The following work briefs out the contribution of carbon and its various variants utilized in the development cycle of PVSCs. Many times the end-of-life step is deprived for the sake of more profits and larger life span utilizations, but it should be avoided if the product is in an action of commercialization. The complete universe has the most significant eternal source of power supply, that is, the sun, which must be utilized at full by encroaching on the flattest technological measures, and updating them promptly can contribute as a whole. To gain efficiency levels, modifications in the structure of perovskites by involving the organic, inorganic type of metal halides are in place to raise the efficiency level above 25%. The primary constraint of product stability in 15 to 20 years is the only part yet to be achieved usefully. Present research involves the utilization of nanomaterials *etc.*, to vary the different layers involved to strengthen them or to predict better layer composition with more correct additives, an interfaced engineering approach, *etc.* The work imparts knowledge about utilizing the most abundant and cheap materials, such as carbon, for PVSCs. Until today, the available materials need to be scrutinized based on pros and cons,. Then after perfect standardization and validation, some of the best combinations need to be practiced for commercialization product as a sample. Ceramics-based PVSC's are stepped in the direction of improving the efficiency of PVSC's by a stream of interface engineering, and the same is extended to various levels. In the order to optimize the optoelectronics character, separate the charge at identified electrodes, *etc.* The utilization of metal carbides (two-dimensional types, MXene, Ti_3C_2Tx) over a particular work function is experimented with along multiple terminating groups. To affect the absorber materials in perovskite and enhance the electron transportation layer to get a better variant. Checking by density function and spectroscopy provides information on dipole induction and the possibility of band arrangement amid layers. Utilizing 2-D members of the ceramics family and applying work function-based experiments along with interfacing variations over layers can provide an improved version of PVSCs. The achievement obtained by blending multiple engineering techniques is reduced hysteresis and conversion efficiency of 26%. It is far better than the perovskite cell not having ceramic content in MXene [23, 24]. Utilization and applications are emerging from every sector. Building and construction patterns and coloring ways have suggested better modifications that can be provided in perovskites for better power conversions. Deploying something new in existing models of power modules is the central theme. As a result, changing the color pattern of PVSC's module inspired by inkjet printable reflective stains is practiced. Moreover, zirconium-converted balls are used while preparing pigment at a definite temperature level. An excellent modification in the patterns of the sample created was obtained.

The concept that emerged among all practices is an invariant angle type of color perception that is outstanding amidst all available. Peculiar power conversion rates of 65% and 11% higher than the existing PVSCs (non-colored type) were obtained by introducing bright magenta and yellow shades. Various other patterns resembling the marble family, as well as corten steel surfaces, are also showcased. Inkjet printing is a low-budget, dynamic-adaptable technology that can be used over a short call with a bit of setup. A sample-sized material of 4 cm^2 is demonstrated with conversion to 14% under white marble type of optical characteristics [25], as shown in Fig. (**4**).

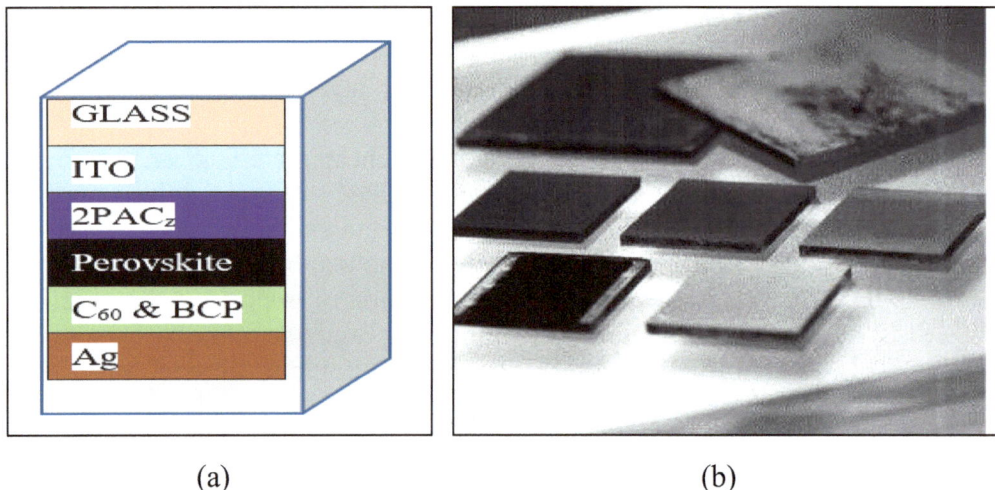

(a)   (b)

Fig. (4). (**a**) Ag-Based PVSC's Layering, (**b**) Sample materials for PVSC's.

By creating [26, 27], a thermal shock utilizing electrode and ionic powders and a calcination process followed by 40 hours of furnace perovskite blocks could be processed. The duration of high thermal conditions binds more strongly and fast among the constituents added. The structure utilized for electrolyte is of fluorite base type $(Y_2O_3)0.08(ZrO_2)0.92$ (YSZ) and $(Sm)0.2(Ce)0.8(O)1.95$(SDC). This process can create solid oxide cells in a short duration. By tailoring the carbon support size, multiple fast samples could be generated. Hence, if an economic point of view is considered, such processes can be utilized to obtain new ceramic powders for solid oxide cells, as shown in the Fig. (**5**).

[28] The scope of obtaining new composites may be based on reduced carbon emission contents to exit under the combination of perovskites with clay-type materials, as illustrated in the work hereby. The variety of organic and inorganic materials varies the outcome properties of substances, such as light-emitting materials that can be created as commercial products with better luminescence

and stability concerning external climate conditions. Similar experiments could be practiced at a laboratory scale by combining clay nanomaterials such as kaolinite, montmo- rillonite, and halloysite (aluminosilicates of clay) to prepare new composite clay substances that can enhance PVSC's external lighting products. One such base material that can be utilized in the nanocrystals category is methylammonium lead bromide, PVS. Clay additive experiments can raise external environment stability. When created with the above clay composites, it provides green emission values to 540 nm and an improvised stability level against temperatures up to 250°C. Moreover, luminescence is raised under moisture situations higher than a month. Thus, spontaneous products can be utilized to create commercial inside or outside lighting devices.

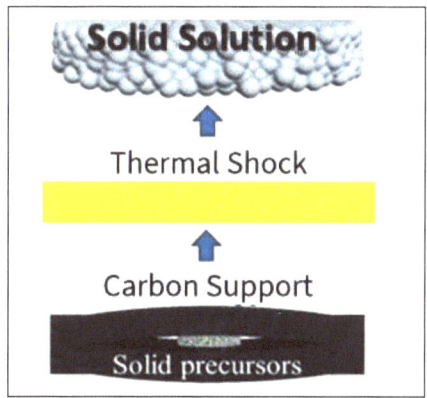

Fig. (5). Thermal Shock-Based PVSC's Creation.

[29, 30] Lead mixed halide types of PVScs are ready to place the product if commercial viability is a concern. Still, some manufacturing constraints need to be resolved with a better substitute (Cobalt as here one). Initially, these products are cost-wise, low, and have good efficiency. But the transportation layer formed at the hole level of the p-i-n type has ABXn formula. As a result, it requires pre-processing for a 15-minute duration at 300 degrees Celsius for better transportation effects. Such a process of heating with time duration needs a better replacement in terms of impactful quick boiler technology. Annealed material in 50-55 seconds is the objective that needs to be fulfilled. It is provided with an approach of using an infra-red lighting facility with a high-temperature range of about 2500 K for better-pressed facilities in a short duration. Experimentation studies were carried out, and the results suggest that infra-red-based setup suits are more suitable than conventional technologies for annealing during fabrication. Here, it utilizes the nanomaterial particles of cobalt from which it is doped. These new changes result in better power obtaining from 15.99% of earlier to an enhanced 17.77%. Due to the improved way of hole separation, reduced particle

gathering and reduction of voltage loss are reflected as a decline in resistance and more systematic tuning of the work functions involved. Thus, low cost and higher efficiency are added up with low-scaled technological equipment. [31] The utilization of cobalt is also characterized, and an experiment with PVSC's composition of more than 20-30 above combinations was tested at this moment in the work presented here. Multiple combinations of cobalt-based PVScs were tested to obtain the best combination in terms of long-span stable category, the overall best performance in terms of hole extraction, and photovoltaics action is considered. The inverted planer composition is well attached to the cobalt base and provides better hole separation and layer composition of ITO/ $CoOx/CH_3NH_3PbI_3/PCBM/Ag$. Even the conversion efficiency is attained at 14.5% compared to existing combinations. Photoluminescence (PL) decay concerning other variants is studied comparatively and measured with various measurement practices. The hole separation traits provide the time sequence as CoOx (2.8 ns) < PEDOT: PSS (17.5 ns) < NiOx (22.8 ns) < CuOx (208.5 ns), consistent with the trend of their photovoltaic performances. The duplicability and durable firmness are tried to be obtained for an extended time as the cobalt combined device retains a power conversion value of 12% till 1000 h different combinations (copper here) based upon organic and inorganic types of halide materials type PVSCs in the setting of layer selecting a hole during the photovoltaic experience. It has been collectively obtained those common issues of high cost, chemically active, temperature affected, and instability issues remain. But the research studies suggest that problems identified issues could be easily rectified by using carbon nanoparticles at room temperature during fabrication. Utilizing a vacuum-treated planer of PVSCs with copper phthalocyanine (CuPc), at the molecular level is practiced to obtain a solution for the above-identified problems. It has been obtained that the considered material has properties that make it a suitable material to face chemical and thermal-based variations during the fabrication process. The cheap rates of the identified materials make it more suitable along, with the advantage obtained from test results related to power energy conversion of 15.42% (OC volt.1.04, FF: 77.47%) and steady-state efficiency of 14.5%. The identified material composite suits to be well defined in terms of the selective hole process, which can be taken over larger-scale fabrication and higher life span with stability considerations too.

[32, 33] Lead removal will always be a target for every commercialized product before cost constraints because upcoming government agendas and policies are concerned with the country's sustainable growth. As shown in Fig. (**6**), the lead content needs to be removed as early as possible, and the utilization of copper nanoparticles can perform similar experiments. The highest conversion efficiency is of $CH_3NH_3PbI_3$ (MAPbI$_3$) composition, but removing lead and raising stability are two major bottlenecks. Hence, here copper plays its role by substituting Pb

quickly, and since it has the same set of optoelectronic characteristics, it is more eco-friendly, not expensive, and available. Thus, experiments are carried out by creating MACuxI3 ($1 \leq x \geq 2$), the complete parts need to be removed, but observations are carried out for partial as well as the mixed composition of both copper and lead. After which, they are used with carbon types of PVSC combinations. Higher absorbance rate of infrared causes better results in the case of the solar spectrum. The Cu penetration in MAPbI format provides a conversion efficiency of 12.85%, which is 0.42% higher than the mixed one. Other than this, if a complete replacement is obtained, it will be 4.0% only. While degradation of material till 1000h is negligible, that is, not so in the case of lead samples for the same duration. A minor modification or inclusion of the compound after lab experiments may solve the conversion efficiency issue. Cu material has a higher participation ratio for commercialization as it has better stability in terms of temperature and chemical interactions.

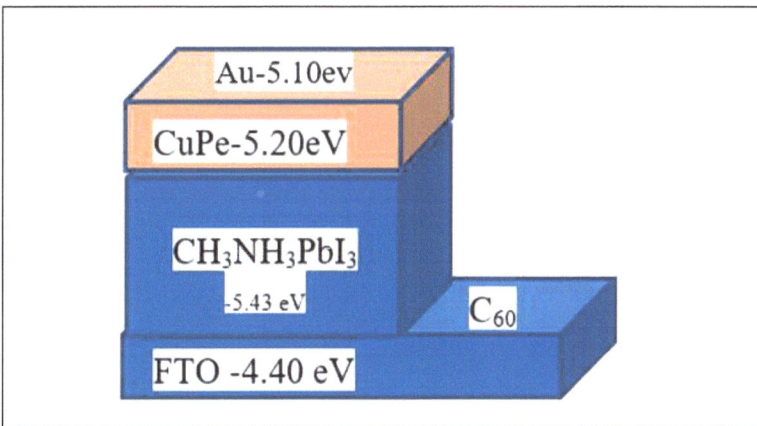

Fig. (6). CH3NH3PbI3 Based PVSC's Layering.

[34] Gold: Planer type of multijunction-based PVSC's composition of ITO/PEDOT: PSS/$CH_3NH_3PbI_3$/PCBM/Al are developed by chemical solutions carried out in cold conditions. Using gold as nanorods sprayed up with silica placed between polymer-based conductive layer of PEDOT: PSS, that is, transporting the holes to the other one $CH_3NH_3PbI_3$, works as the active layer. The experiment results and simulations also show a boost in the power conversion efficiency from 10.9% to 15.6% and finalized the product to 17.6%. These results are obtained due to the plasmon resonance of nanorods and having a localized surface. It advances the capturing system of light rays and increments sending-receiving of charge carriers. The pos replacement with metal can also be tested if cost considerations need to be considered. Creating an electron transport layer with higher efficiency is targeted by changing the structure by introducing gold

nanomaterials. As observed in the case of mesoporous TiO_2 photoelectrodes-based PVSC, its competence of power conversion is 15.19 %. The identified device is merged with a gold nanostructure of star shape, that is, having the structure of TiO_2-AuNSs now able to convert power to 17.72 %. This process is caused due to the improvement in light-collecting sensitivity and the curbing of charge recombination flow by adding Au. As the absorption of photons is raised, the output efficiency is increased automatically, as detailed in Table **2**.

Table 2. PV parameters based on different compositions.

PV Parameters of Devices Created				
ETL	Jsc (mA cm^{-2})	Voc [V]	FF	PCE [%]
TiO_2 Only	21.14	1.05	069	15.19
TiO_2 - AuNSs	22.97	1.08	.71	17.72

[35] Iron: PVSCs generally take the composition of TiO_2 to separate the charges that cause the decline of light-collecting traits under a light emission state [79]. Hence it maintains the photons collection property and keeps a better efficiency level. Simulated and lab-fabricated iron (III) oxide nanoisland electrodes can be utilized to build for extracting electrons in a better way during PVSC's workflow. A key finding is that the presence of iron oxide supports good conduction of light rays along with better charge withdrawal function. It will be reflected in a better power conversion efficiency obtained at 18.2%. Hence, the peculiarity obtained by utilizing iron composition is due to the low photocatalytic response, the rise in UV sensitivity, and the increment in long-duration stability measure [44]. Thus, this study raises the hidden space of blending nanoparticles with existing compositions. Then only highly efficient and long-term stable PVSCs will be obtained as a perfect commercialized product. Methyl ammonium leads tri-iodide PVSC's are having 20 % conversion rate. By utilizing polytriarylamine polymer (PTAA) and spiro-OMeTAD (Spiro). The issue exists at the absorber stage that is not available or not being fabricated. Moreover, the price of materials used in the hole transport layer is high. Hence, cost-based optimization studies need to be promoted. It is resolved by adjusting the requirements through a readily available material called to be mineral iron pyrite (FeS_2). It can be utilized as a layer for transporting holes in PVSCs. The structure consists of n and p layers on TiO_2 and FeS_2, respectively. Conversion efficiency was obtained to be 11.2%. Quenching types of research are carried out by observing the photoluminescence of the pyrite capable of collecting and sending holes at a higher rate than spiro. Thus, reflected in cost terms, it can be said as pyrite exists as 300 times less costly than a commercialized PVSC module of 1 m$_2$ [36]. The follow-up of research work after the discovery of perovskite is not so long as the revolution started after 2012. The

studies have confronted the importance of halide type PVSC's providing a higher impact on the upcoming chain of cells for the new era of green energy. Reports and tests have already been generated that, based on the efficiency level of lead constituents, PVSc's are on top. Still, as far as toxicity is considered, it also holds its position high. As a result, its future scope needs to be altered with a different material composition. The search for better substituent materials or modified structural compositions with better traits is in progress. The upcoming series of PVSCs need to be lead-free although the achievement level is not adjoined to $APbI_3$ substances. Along with it, improvements in stability level were also treated as significant constituents. The upcoming foundation of a lead deployed PVSCs as the cheapest material may have let everyone invest under the scheme with a reasonable scope [37]. The realization of toxicity that will be a part of the environment has forced the stoppage. Different compositions are put forward for the sake of stability concerns. But negative results were obtained in the direction of unhealthiness and leaching of lead. Certain types of research study in progress to reduce the damage caused by lead content through the better recovery of lead substances when the end of life is concerned. Such as by preparing a chemical composition that can sequestration the leakage of lead to 96-98% level in case of device demolition. Applying multiple chemicals sprayed on both sides, applying of film working as lead absorbent phosphonic family elongated. Moreover, lead-chelating substances on polymer films can be added between the electrode and packing film. All possible solutions are aimed toward the recovery of lead rather than its leakage due to environmental actions. The lead obtained will be in a solid structural form to be utilized for other industrial processes or as raw material later, rather than making it a part of the ecosystem in dissolved contents.

[38, 39] Utilization of magnesium is carried out with doping effects in other chemical compounds prepared from the same. Such as raising the motion of electron flow in the transport layer. As the issue obtained in zinc oxide, layered PVSC's magnesium nanoparticles are doped in 1–3 wt.% of ZnO by electrospinning applied to the identified transport layer. X-ray-based diffraction studies are carried out to assess the effects caused by such amalgamated composition—changes over structure, chemical, and morphological impacts. Identification of hexagonal wurtzite buildings with lattice arrangement of ZnO is obtained. It was also found that an inverse relation between band gap energy and magnesium doping exists. A raising difference of 0.56eV has been received with a rise in magnesium proportion. The mixture has improved the composition structure-wise. Reflected in terms of fill factor, its efficiency, and density of current. Penetrating from 1 to 3 wt. % efficiency rose from 8.5 to 13.53%. Hence, it can be coined as a better facile Mg-doped material. One industrial application outcome in terms of photocatalytic activity is the making of antibacterial textiles starts utilization.

[40, 41] Accepting PVSCs as the upcoming replacement for conventional silicon crystals is accepted worldwide. Similarly, the creation of new variants by varying the doping levels is in continued research now a day. The design of materials at low cost and availability of raw materials makes it more favorable to use, along with the higher amount of power obtained. The paradigm shift in bio-medical and electronics devices by merging PVSC's materials still needs to come on a large scale. But to promote green energy, we are unaware that it may raise white pollution, [42, 43] which must be checked in parallel by building up all such compounds (polymer substrates) and elements. Thus, to avoid such phases, nonexpensive type acrylic resin sprayed translucent nitrocellulose paper is created to build up flexible PVSCs. These are nature-friendly degradable materials having an efficiency of 4.25%. The ratio of power to device weight is at a larger scale of 0.56 W g–1. Improved material with stability feature and working still 80% of starting after bending 50 times. As an added advantage, such substrates can provide a large market in coverage and small bio-medical-based electronics devices.

Utilization of platinum materials in PVSC's preparation also exists. Considering Photon Downshifting layers, the platinum complex provides cell arrays of perovskite-type. A jetting process in an ultrasonic way that is inspired by automation and keeps track of control is used for the deposition procedure. Fabrication of Water-Repellent Platinum (II) Complex-Based Photon Down-shifting Layers for Perovskite Solar Cells is passed through Ultrasonic Spray, and deposition is executed. It is taking the technical inefficiencies of PVSC's under ultraviolet exposure to humidity and casual degradation. The problem is now dealt with through integrating water-repellant nanoparticles, *i.e.*, platinum groups. A luminescent photon downshifting layer (PDL) creates an identified sample in Pt-F composition. Afterward, the fabrication is layered over the glass type of substrate through jetting and deposition by ultrasonic. The obtained product possesses crystallinity and photoluminescence quantum yield (PLQY). It is different compared to materials processed by the spin coating mechanism. The addon material of PDL-based coating obtains the device-based output of 22.0%. Earlier issues of UV exposure and humidity are improved to a great extent. Now the new product created can be utilized as multipurpose applications over an extensive range of optoelectronics items.

Following the route of a lead halide of PVSC as the light-capturing property is a success. A proper understanding and deep learning are required to implement new materials with earlier racks available [21, 22]. Considering p-or- n-type materials, the selectivity and performance of the hole transport layer must be known. The work now deals with introducing polymers based on fluorene or carbazole leading chains with add-on character tics that can be easily modified per the requirements in PVSCs. The traits related to conductivity are monitored from orbitals energy segments. Temperature over glass transition and the wettability of the HTM effect. Overall, all these modify the overall performance of samples. Penetration of polymer into the hole transport layer and redesigning the architecture with new features and properties have been practiced. Hence, new products with new applications from earlier samples can be recreated to stabilize the existing issues of the economy. [44] Some experimental studies also found that the material utilized for hole transportation must be free from doping to obtain better stability and efficiency in PVSCs. Introducing conjugated polymer is practiced in this direction. But it's not easy to deploy its functioning traits; it cannot perform the desired reaction with absorber and charge extraction. At this moment, the polymer is mixed with the triphenylamine group to obtain more stability. It raises the interaction with PVSC, thus promoting the passivation feature. The transportation of charges is incremented by the higher number of face-on oriented crystallites. The material prepared has an efficiency level of 20.4% and better environment stability over a longer span than substitute materials working without addition (15.8%).

[45,49] Quantum dots are a type of nanomaterial utilized for various applications. They can be utilized for energy-saving purposes, not requiring initial purification; they can be directly used to create desirable substrates and mixed with materials like light, plastics, *etc*. In the case of PVSC's work as an absorber part and has various self-traits over energy stages and similar related quantum properties. These can preserve the properties of materials and which can be later utilized as advantages.

Different methods are discussed at this moment as per the chemical reactions involved, the passivation between ions and ligands, structure combinations related to existing, inverted type, dimension wise or planer else mesoscopic, *etc*. will lead to the creation of solid proportion materials termed as quantum dots as shown in Fig. (7).

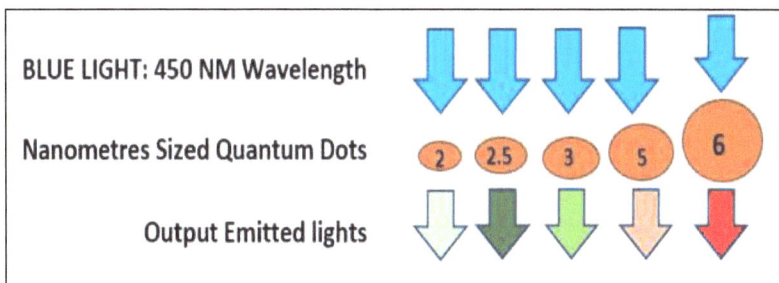

Fig. (7). Nanomaterials Sizing Depiction with Quantum Dots Concept for PVSC's.

CsPbIBr$_2$ types of PVSCs have gained focus as an industrial product output, in terms of its viability over applications related to electronics and optical-based hybrid devices. The market-launched product must be capable enough to handle exterior conditions and efficient working compared to old conventional devices, *etc.* In terms of structural composition, the energy orbitals transitions must be more stabilized with the high power of optical collection. Creation of material CsPbIBr$_2$ through deposition of processes multiple grains with its numerous boundaries initiates starting defects during formation. The identified issues when treated with silver iodide by additive engineering utilizing precursor, and then a new structure of PVSC is obtained. The unique composition will be CsI(PbBr$_2$)1-x(AgI)x. The addition of silver introduces a special effect in which the particles grow out of the merging process to create a uniform film with reduced grain boundaries and respective sizes. Simple PVSC's structure with an optimized quantity of AgI can provide a conversion efficiency of 7.2%. Thus, by utilizing such processes, the improved quality films of PVSCs have higher thermal and optoelectronic indexes to work upon.

[50] Basic studies are also practiced to judge the effects over layers formed through the titanium dioxide nanoparticle. Practices are taken by utilizing the paste of TiO$_2$ instead of fabricated units. Substitutes are made as such by replacing hole transportation layer with carbon material rather than using gold-based electrodes. The PVSC's structure layers consist of FTO glass/ compact TiO$_2$/ mesoporous TiO$_2$/ CH$_3$NH$_3$PbI$_3$/ carbon electrodes. A mesoporous layer was also created from anatase TiO$_2$ mixed with ethanol at different percent levels. A conversion efficiency of 0.44% was obtained through a mesoporous Tio$_2$ cell with a 1.27% of concentration. The strontium titanate type of nanoparticles is also used to prepare perovskites with better photocatalyst properties. Nano cubes of SrTi are designed using hydrothermal procedures using them (Sr(NO$_3$)$_2$) as precursors. A sizing of 18 nm with a cube-shaped perovskite structure is prepared under the tuned mole ratio of Sr and Ti. The substance was able to provide photocatalytic degradation of Methylene Blue (MB) and Tartrazine (TZ) dye under UV light irradiation. Thus, the fundamental property of photocatalytic nature is enhanced by the additive mixing of nanoparticles under fixed ratios.

Disulfide art of mixing and modifying layers of the hole and charge transportation at the nano level can suggest better power conversion efficiency as per the combinations prepared. Technological and structural change requires expertise and perfect tuning that perfect more combinations of chemical reactions need to be confirmed. Similar experiments performed here targeting the tungsten oxide process are created to develop a charge transport layer through the evaporation process in vacuum situations. Tungsten oxide is experimented to used with tin-oxide substrates. The fundamental relation lies in a sequence of thickness-tailoring crystallization that accelerates and channelizes the interface between the recombination process with the transferring of charges. Working over thickness optimization led to better homogeneity in the structure, and a 30 nm thick, dense transport layer of electrons was formed. It also reduces the resistance and improves the conduction process in perovskite layers. Stable operation till 1000h of the bending cycle is obtained. Experimentation at average room temperature and creating an electron transport layer prepares the process compatible with roll-to--the to-roll type of PVSC's manufacturing process.

The field of nanomaterials is a field of combination and corrections. Multiple materials with multiple properties are either suitable or nonsuitable or may be ideal for a third amidst layer to exist. Among all perfect combinations, satisfying multi-parameters must be finalized for a long-term goal of 15 to 20 years. At this moment, the work is to replace titanium dioxide material with zeolite family compounds. And further studies are performed on the new combination. Zeolites are introduced to the photoanodes of solar cells if those are dye-sensitized. Moreover, the utilization of a hydrothermal-based process is involved. Measurements using the characterization identified procedure are undertaken to view the polycrystalline structure bandgap of composites (4.14 eV) and surface morphology variations from rectangular to spherical. Results obtained by voltage values prove the creation of p-n type multi-variant materials having characteristics and patterns resembling solar cells.

CONCLUSION

This chapter's literature findings suggest comparing the multi-verse single junction of all available silicon cells. The prior one remains ahead by 16.3%, along with the high cost of building. The findings, as mentioned earlier, were found to be optimistic for PVSCs since technological upgradations by building multiple junctions' the type of PV in solution processing that uses acetonitrile/methylamine solvent rather than silicon type in the perovskites category. To get an efficiency of 26%, it increases market competition among small-sized multijunction-based PV cells. The hidden potential of PVSC's will dominate the market in the upcoming era. However, some processes are still underway to obtain a better solution in case of deposition or end-of-life management.

Studies over intermolecular forces blended nanomaterials amid multiple layers of cell composition, and bonding ways are still under experimentation and improvisation to provide a definite composite suite with both stability and functionality of better disposable after its utilization.

The high injection of moisture causes the break-up of perovskite; it becomes a chain reaction in similar conditions of high temperature, interaction with a nearby electric field, and UV encroachment. Various combinatorial conditions participate throughout external PVSC exposure. Looking inside, if the structural point is considered, the available vacancies start the exchanging process, and it inversely creates fluctuation or hidden dipole movement, thus hampering the focus of solar generation. Hence, internal chemistry needs better ways of maintaining the bond relationship.

Effects in terms of photocatalytic nature, the bounce-back process of current trap passivation, and practice of breaking product by UV exposure, afterward proposing a new different architecture of devices. Studies are in continuation over the selection of better-absorbing materials, options available through grain engineering, betterment of transportation layer, usages of additives, defect passivation, motivating studies working tacitly to provide a better commercial product that can be used for a long duration. Hence, aiming towards the high sustainability development goals for net zero emissions in the future. The research experiments are the way to create a sustainable upcoming future. The topic limits here.

LEARNING OBJECTIVES

• Basic Idea of Nanotechnology Materials and their role in solar Energy supplies.

• Interfacing and atomic layer combination availability in nanomaterial.

• Focus on existing gaps in Perovskite embedded material and related exploration to obtain lead substitutes in the present-day market.

• Processes involved in the commercialization of lead substitutes with improved composition.

• Role of artificial intelligence in better composition identification through balance characterization strategies.

• Lastly, to contribute to the existing perovskite-inspired solar economy followed by possible ways of commercialization, decomposition, and utilization.

MULTIPLE CHOICE QUESTION

1. The limit of visibility of the human eye exists to be:

a. 10 nm
b. 100 nm
c. 1000 nm
d. 10,000 nm

2. Based on specialized product, nanomaterials can be classified with the following assets:

a. Force
b. Pressure
c. Temperature
d. Friction

3. Nano-prefix originated from:

a. Billion - French
b. Dwarf - Greek
c. Particle - Spanish
d. Invisible - Latin

4. Substances having 1-D in an array of nanoscale and rest dimensions of two in a bulk category are termed as:

a. Quantum wire
b. Micro-material
c. Quantum well
d. Quantum dot

5. Coloring of the material having gold as constituents can be specified as:

a. Green
b. Blue
c. Black
d. Variable

6. Carbon Nanotube has a thickness of:

a. 1 nm
b. 1.3 nm
c. 1.55 nm
d. 2 nm

7. Top-down ways of creating nanomaterials include the following process:

a. An agglomeration using gas mode
b. Molecular self-assembly
c. Mechanical grinding
d. Molecular beam epitaxy

8. Identify the bottom-up ways to create nanomaterials from options:

a. Etching
b. Dip pen nano-lithography
c. Lithography
d. Erosion

9. Quantum confinement deals with:

a. Energy gap in the semiconductor is proportional to the inverse of the square root of the size.
b. Energy gap in a semiconductor is proportional to the inverse of the size.
c. Energy gap in a semiconductor is proportional to the square of size.
d. Energy gap in a semiconductor is proportional to the inverse of the square of the size.

10. Organic polymer solar cells are made from Polyphenylene:

a. True
b. False

11. In optimizing the cost of device fabrication for perovskite solar cells, the material can be used as a hole transporting material:

a. Mineral iron pyrite (FeS2)
b. TiO_2

c. TPPI
d. PCBM

12. The material which is not utilized in preparing the TCO layer is:

a. Al_2O_3
b. SnO_2
c. ITO
d. ZnO

13. Techniques used to determine the size of nanoparticles:

a. FPS
b. Dynamic Light Scattering
c. Inductively Coupled Plasma mass spectroscopy
d. UV-Vis

14. The size of the TCO layer can be found in:

a. $3\lambda/4 * 1/n(TCO)$
b. $\lambda/4 * 1/n(TCO)$
c. $\lambda/4 * 2/n(TCO)$
d. $6\lambda/4 * 1/n(TCO)$

15. What kind of ETM is utilized in perovskite solar cells:

a. Al_2O_3
b. TiO_2
c. ITO
d. PEDOT: PSS

ANSWER KEY

1. (a)

2. (d)

3. (b)

4. (c)

5. (d)

6. (b)

7. (c)

8. (b)

9. (d)

10. (a)

11. (a)

12. (a)

13. (b)

14. (b)

15. (b)

ACKNOWLEDGMENT

I am thankful to my family, friends, K R Mangalam University family members for providing support during the write-up of chapters.

REFERENCES

[1]   L. David, "Engineers enlist ai to help scale up advanced solar cell manufacturing mit news office", Available from: https://news.mit.edu/2022/ai-perovskite-solar-manufacturing-0413 Date: April 13. 2022.

[2]   Teaching AI to ask clinical questions. Available from: https://library.georgiancollege.ca/ c.php?g=720607&p=5151225 (Access on Jul.13, 2022).

[3]   X. Cao, G. Zhang, Y. Cai, L. Jiang, X. He, Q. Zeng, J. Wei, Y. Jia, G. Xing, and W. Huang, "All green solvents for fabrication of CsPbBr3 films for efficient solar cells guided by the Hansen solubility theory", *Sol. RRL,* vol. 4, no. 4, p. 2000008, 2020. [http://dx.doi.org/10.1002/solr.202000008]

[4]   M. Green, E. Dunlop, J. Hohl Ebinger, M. Yoshita, and N. Kopidakis, "Solar cell efficience tables (version 57)", *Prog. Photovoltaics,* vol. 29, no. 1, pp. 3-15, 2021.

[5]   R. Vidal, J.A. Alberola-Borràs, S.N. Habisreutinger, J.L. Gimeno-Molina, D.T. Moore, T.H. Schloemer, I. Mora-Seró, J.J. Berry, and J.M. Luther, "Assessing health and environmental impacts of solvents for producing perovskite solar cells", *Nat. Sustain.,* vol. 4, no. 3, pp. 277-285, 2020. [http://dx.doi.org/10.1038/s41893-020-00645-8]

[6]   D. Timothy, "The path to perovskite commercialization: A perspective from the united states solar energy technologies office", *ACS Energy Lett.,* vol. 7, no. 5, pp. 1728-1734, 2022. [http://dx.doi.org/10.1021/acsenergylett.2c00698]

[7]   G. Tong, D.Y. Son, L.K. Ono, Y. Liu, Y. Hu, H. Zhang, A. Jamshaid, L. Qiu, Z. Liu, and Y. Qi, "Scalable fabrication of >90 cm2 perovskite solar modules with >1000 h operational stability based on the intermediate phase strategy", *Adv. Energy Mater.,* vol. 11, no. 10, p. 2003712, 2021.

[http://dx.doi.org/10.1002/aenm.202003712]

[8]   H. Chen, F. Ye, W. Tang, J. He, M. Yin, Y. Wang, F. Xie, E. Bi, X. Yang, M. Grätzel, and L. Han, "A solvent- and vacuum-free route to large-area perovskite films for efficient solar modules", *Nature,* vol. 550, no. 7674, pp. 92-95, 2017.
       [http://dx.doi.org/10.1038/nature23877] [PMID: 28869967]

[9]   T. Wu, X. Liu, X. He, Y. Wang, X. Meng, T. Noda, X. Yang, and L. Han, "Efficient and stable tin-based perovskite solar cells by introducing π-conjugated lewis base", *Sci. China Chem.,* vol. 63, no. 1, pp. 107-115, 2020.
       [http://dx.doi.org/10.1007/s11426-019-9653-8]

[10]   X. Liu, Y. Wang, T. Wu, X. He, X. Meng, J. Barbaud, H. Chen, H. Segawa, X. Yang, and L. Han, "Efficient and stable tin perovskite solar cells enabled by amorphous-polycrystalline structure", *Nat. Commun.,* vol. 11, no. 1, p. 2678, 2020.
       [http://dx.doi.org/10.1038/s41467-020-16561-6] [PMID: 32472006]

[11]   X. Jiang, F. Wang, Q. Wei, H. Li, Y. Shang, W. Zhou, C. Wang, P. Cheng, Q. Chen, L. Chen, and Z. Ning, "Ultra-high open-circuit voltage of tin perovskite solar cells via an electron transporting layer design", *Nat. Commun.,* vol. 11, no. 1, p. 1245, 2020.
       [http://dx.doi.org/10.1038/s41467-020-15078-2] [PMID: 32144245]

[12]   X. Meng, Y. Wang, J. Lin, X. Liu, X. He, J. Barbaud, T. Wu, T. Noda, X. Yang, and L. Han, "Surface-controlled oriented growth of FASnI3 crystals for efficient lead-free perovskite solar cells", *Joule,* vol. 4, no. 4, pp. 902-912, 2020.
       [http://dx.doi.org/10.1016/j.joule.2020.03.007]

[13]   T. Nakamura, S. Yakumaru, M.A. Truong, K. Kim, J. Liu, S. Hu, K. Otsuka, R. Hashimoto, R. Murdey, T. Sasamori, H.D. Kim, H. Ohkita, T. Handa, Y. Kanemitsu, and A. Wakamiya, "Sn(IV)-free tin perovskite films realized by *in situ* Sn(0) nanoparticle treatment of the precursor solution", *Nat. Commun.,* vol. 11, no. 1, p. 3008, 2020.
       [http://dx.doi.org/10.1038/s41467-020-16726-3] [PMID: 32546736]

[14]   X. Liu, T. Wu, J.Y. Chen, X. Meng, X. He, T. Noda, H. Chen, X. Yang, H. Segawa, Y. Wang, and L. Han, "Templated growth of FASnI 3 crystals for efficient tin perovskite solar cells", *Energy Environ. Sci.,* vol. 13, no. 9, pp. 2896-2902, 2020.
       [http://dx.doi.org/10.1039/D0EE01845G]

[15]   K. Wang, J. Liu, J. Yin, E. Aydin, G.T. Harrison, W. Liu, S. Chen, O.F. Mohammed, and S. De Wolf, "Defect passivation in perovskite solar cells by cyano-based π-conjugated molecules for improved performance and stability", *Adv. Funct. Mater.,* vol. 30, no. 35, p. 2002861, 2020.
       [http://dx.doi.org/10.1002/adfm.202002861]

[16]   S. Xiong, J. Song, J. Yang, J. Xu, M. Zhang, R. Ma, D. Li, X. Liu, F. Liu, C. Duan, M. Fahlman, and Q. Bao, "Defect passivation using organic dyes for enhanced efficiency and stability of perovskite solar cells", *Sol. RRL,* vol. 4, no. 5, p. 1900529, 2020.
       [http://dx.doi.org/10.1002/solr.201900529]

[17]   Z. Yang, J. Dou, S. Kou, J. Dang, Y. Ji, G. Yang, W-Q. Wu, D-B. Kuang, and M. Wang, "Multifunctional phosphorus-containing lewis acid and base passivation enabling efficient and moisture-stable perovskite solar cells", *Adv. Funct. Mater.,* vol. 30, no. 15, p. 1910710, 2020.
       [http://dx.doi.org/10.1002/adfm.201910710]

[18]   Y. Jiang, L. Qiu, E.J. Juarez-Perez, L.K. Ono, Z. Hu, Z. Liu, Z. Wu, L. Meng, Q. Wang, and Y. Qi, "Reduction of lead leakage from damaged lead halide perovskite solar modules using self-healing polymer-based encapsulation", *Nat. Energy,* vol. 4, no. 7, pp. 585-593, 2019.
       [http://dx.doi.org/10.1038/s41560-019-0406-2]

[19]   Q. Tao, P. Xu, and M. Li, "Efficient and stable tin perovskite solar cells enabled by amorphous polycrystalline structure", *Nat. Commun,* vol. 11, no. 1, p. 2678, 2021.
       [http://dx.doi.org/10.1038/s41524-021-00495-8]

[20]   Y. Liu, W. Yan, S. Han, H. Zhu, Y. Tu, L. Guan, and X. Tan, "How Machine Learning Predicts and Explains the Performance of Perovskite Solar Cells", *Sol. RRL,* vol. 6, no. 6, p. 2101100, 2022.
[http://dx.doi.org/10.1002/solr.202101100]

[21]   P. Raccuglia, K.C. Elbert, P.D.F. Adler, C. Falk, M.B. Wenny, A. Mollo, M. Zeller, S.A. Friedler, J. Schrier, and A.J. Norquist, "Machine-learning-assisted materials discovery using failed experiments", *Nature,* vol. 533, no. 7601, pp. 73-76, 2016.
[http://dx.doi.org/10.1038/nature17439] [PMID: 27147027]

[22]   M. Srivastava, J.M. Howard, T. Gong, M. Rebello Sousa Dias, and M.S. Leite, "Machine Learning Roadmap for Perovskite Photovoltaics", *J. Phys. Chem. Lett.,* vol. 12, no. 32, pp. 7866-7877, 2021.
[http://dx.doi.org/10.1021/acs.jpclett.1c01961] [PMID: 34382813]

[23]   A. Luque, "Procedimiento para obtener celulas solares bifaciales" Spanish patent 458514, application, vol. 24, p. 24669, 1977.

[24]   A. Luque and A.Martí Increasing the Efficiency of Ideal Solar Cells by Photon Induced Transitions at Intermediate Levels", *Phys. Rev. Lett. 78,* vol. 30. 1997.
[http://dx.doi.org/10.1103/PhysRevLett.78.5014]

[25]   Spatial Element Distribution Control in a Fully Solution-Processed Nanocrystals-Based 8.6% Cu2ZnSn(S, Se)4" Device Wan-Ching Hsu, Huanping Zhou, Song Luo, Tze-Bin Song, Yao-Tsung Hsieh, Hsin-Sheng Duan, Shenglin Ye, Wenbing Yang, Chia-Jung Hsu, Chengyang Jiang, Brion Bob, and Yang Yang. ACS Nano Article ASAP.
[http://dx.doi.org/10.1021/nn503992e]

[26]   MA Hernández-Rodríguez, MH Imanieh, LL Martín, and IR Martín, "Experimental enhancement of the photocurrent in a solar cell using upconversion process in fluoroindate glasses exciting at 1480 nm", *Solar Energy Materials, and Solar Cells.* 116, 171-175.
[http://dx.doi.org/10.1016/j.solmat.2013.04.023]

[27]   P. Wang, S.M. Zakeeruddin, J.E. Moser, M.K. Nazeeruddin, T. Sekiguchi, and M. Grätzel, "A stable quasi-solid-state dye-sensitized solar cell with an amphiphilic ruthenium sensitizer and polymer gel electrolyte", *Nat. Mater.,* vol. 2, no. 6, pp. 402-407, 2003.
[http://dx.doi.org/10.1038/nmat904] [PMID: 12754500]

[28]   P. Yu, J. Wu, L. Gao, H. Liu, and Z. Wang, "InGaAs and GaAs quantum dot solar cells grown by droplet epitaxy", *Sol. Energy Mater. Sol. Cells,* vol. 161, pp. 377-381, 2017.
[http://dx.doi.org/10.1016/j.solmat.2016.12.024]

[29]   S.J. Moon, Y. Itzhaik, J.H. Yum, S.M. Zakeeruddin, G. Hodes, and M. Grätzel, "Sb 2 S 3 -Based Mesoscopic Solar Cell using an Organic Hole Conductor", *J. Phys. Chem. Lett.,* vol. 1, no. 10, pp. 1524-1527, 2010.
[http://dx.doi.org/10.1021/jz100308q]

[30]   P.K. Santra, and P.V. Kamat, "Mn-doped quantum dot sensitized solar cells: a strategy to boost efficiency over 5%", *J. Am. Chem. Soc.,* vol. 134, no. 5, pp. 2508-2511, 2012.
[http://dx.doi.org/10.1021/ja211224s] [PMID: 22280479]

[31]   https://www.kurzweilai.net/organic-polymers-create-new-class-of-solar-energy-devices

[32]   C. Guo, Y.H. Lin, M.D. Witman, K.A. Smith, C. Wang, A. Hexemer, J. Strzalka, E.D. Gomez, and R. Verduzco, "Conjugated block copolymer photovoltaics with near 3% efficiency through microphase separation", *Nano Lett.,* vol. 13, no. 6, pp. 2957-2963, 2013.
[http://dx.doi.org/10.1021/nl401420s] [PMID: 23687903]

[33]   "Miguel A., M. H. Imanieh, L. L. Martín and Inocencio Rafael Martín. "Experimental enhancement of the photocurrent in a solar cell using upconversion process in fluoroindate glasses exciting at 1480 nm."", *Sol. Energy Mater. Sol. Cells,* vol. 116, pp. 171-175, 2013.

[34]   P. Wang, S.M. Zakeeruddin, J.E. Moser, M.K. Nazeeruddin, T. Sekiguchi, and M. Grätzel, "A stable quasi-solid-state dye-sensitized solar cell with an amphiphilic ruthenium sensitizer and polymer gel

electrolyte", *Nat. Mater.,* vol. 2, no. 6, pp. 402-407, 2003.
[http://dx.doi.org/10.1038/nmat904] [PMID: 12754500]

[35] D. Sharma, R. Jha, and S. Kumar, "Quantum dot sensitized solar cell: Recent advances and future perspectives in photoanode", *Solar Energy Materials and Solar Cells*, vol. 155, pp. 294-322, 2016.
[http://dx.doi.org/10.1016/j.solmat.2016.05.062]

[36] O.E. Semonin, J.M. Luther, S. Choi, H-Y. Chen, J. Gao, A.J. Nozik, and M.C. Beard, "Peak External Photocurrent Quantum Efficiency Exceeding 100% via MEG in a Quantum Dot Solar Cell", *Science,* vol. 334, no. 6062, pp. 1530-1533, 2011.
[http://dx.doi.org/10.1126/science.1209845]

[37] P.K. Santra, and P.V. Kamat, "Mn-doped quantum dot sensitized solar cells: a strategy to boost efficiency over 5%", *J. Am. Chem. Soc.,* vol. 134, no. 5, pp. 2508-2511, 2012.
[http://dx.doi.org/10.1021/ja211224s] [PMID: 22280479]

[38] S.J. Moon, Y. Itzhaik, J.H. Yum, S.M. Zakeeruddin, G. Hodes, and M. Grätzel, "Sb 2 S 3 -Based Mesoscopic Solar Cell using an Organic Hole Conductor", *J. Phys. Chem. Lett.,* vol. 1, no. 10, pp. 1524-1527, 2010.
[http://dx.doi.org/10.1021/jz100308q]

[39] J. Du, Z. Du, J.S. Hu, Z. Pan, Q. Shen, J. Sun, D. Long, H. Dong, L. Sun, X. Zhong, and L.J. Wan, "Zn–Cu–In–Se Quantum Dot Solar Cells with a Certified Power Conversion Efficiency of 11.6%", *J. Am. Chem. Soc.,* vol. 138, no. 12, pp. 4201-4209, 2016.
[http://dx.doi.org/10.1021/jacs.6b00615] [PMID: 26962680]

[40] Peng Yu, Jiang Wu, Lei Gao, Huiyun Liu, Zhiming Wang, InGaAs and GaAs quantum dot solar cells grown by droplet epitaxy, Solar Energy Materials and Solar Cells, Vol. 161, p. 377-381, 2017.
[http://dx.doi.org/10.1016/j.solmat.2016.12.024]

[41] Alex C. Mayer, Shawn R. Scully, Brian E. Hardin, Michael W. Rowell, Michael D. McGehee, "Polymer-based solar cells, Materials Today, Vol. 10, (11), p. 28-33, 2007.
[http://dx.doi.org/10.1016/S1369-7021(07)70276-6]

[42] R.R. Lunt, and V. Bulovic, "Transparent, near-infrared organic photovoltaic solar cells for window and energy-scavenging applications", *Appl. Phys. Lett.,* vol. 98, no. 11, p. 113305, 2011.
[http://dx.doi.org/10.1063/1.3567516]

[43] R.R. Lunt, T.P. Osedach, P.R. Brown, J.A. Rowehl, and V. Bulović, "Practical roadmap and limits to nanostructured photovoltaics", *Adv. Mater.,* vol. 23, no. 48, pp. 5712-5727, 2011.
[http://dx.doi.org/10.1002/adma.201103404] [PMID: 22057647]

[44] P. Campbell, and M.A. Green, "Light trapping properties of pyramidally textured surfaces", *J. Appl. Phys.,* vol. 62, no. 1, pp. 243-249, 1987.
[http://dx.doi.org/10.1063/1.339189]

[45] J. Zhao, A. Wang, M.A. Green, and F. Ferrazza, "19.8% efficient "honeycomb" textured multicrystalline and 24.4% monocrystalline silicon solar cells", *Appl. Phys. Lett.,* vol. 73, no. 14, pp. 1991-1993, 1998.
[http://dx.doi.org/10.1063/1.122345]

[46] H. Hauser, B. Michl, V. Kübler, S. Schwarzkopf, C. Müller, M. Hermle, and B. Bläsi, Nanoimprint Lithography for Honeycomb Texturing of Multicrystalline Silicon, Energy Procedia, Vol. 8, p. 648-653, 2011.
[http://dx.doi.org/10.1016/j.egypro.2011.06.196]

[47] N. Tucher, J. Eisenlohr, H. Gebrewold, P. Kiefel, O. Höhn, H. Hauser, J.C. Goldschmidt, and B. Bläsi, "Optical simulation of photovoltaic modules with multiple textured interfaces using the matrix-based formalism OPTOS", *Opt. Express,* vol. 24, no. 14, pp. A1083-A1093, 2016.
[http://dx.doi.org/10.1364/OE.24.0A1083] [PMID: 27410896]

[48] Anastassios Mavrokefalos, Sang Eon Han, Selcuk Yerci, Matthew S. Branham, and Gang Chen,

"Efficient Light Trapping in Inverted Nanopyramid Thin Crystalline Silicon Membranes for Solar Cell Applications", *Nano Lett.* vol. 12, 6, p. 2792–2796, 2012.
[http://dx.doi.org/10.1021/nl2045777]

[49]   M.A. Green, E.D. Dunlop, J. Hohl-Ebinger, M. Yoshita, N. Kopidakis, and X. Hao, "Solar cell efficiency tables (version 56)", *Prog. Photovolt. Res. Appl.,* vol. 28, no. 7, pp. 629-638, 2020.
[http://dx.doi.org/10.1002/pip.3303]

[50]   Hameed, Mehvish & Bhat, Rouf & Singh, Dig & Mehmood, Mohammad. White Pollution: A Hazard to Environment and Sustainable Approach to Its Management, 2019.

CHAPTER 13

Nanomaterials and Their Applications in Energy Harvesting

Anup Shrivastava[1,*], **Shivani Saini**[1] and **Sanjai Singh**[1]

[1] *Computational Nano-Materials Research Laboratory (CNMRL), Indian Institute of Information Technology- Allahabad, India*

Abstract: The rapid advancement in technologies and a surge in the global population with the swift Industrialization led to severe challenges in fulfilling global energy demand. In the last few decades, researchers have been fiercely looking for the development of sustainable energy sources to achieve the goal of carbon neutrality and green energy generation. Among the various approaches to green energy generation, solar and thermoelectric conversions are the most lucrative. In both solar and thermoelectric means of energy generation, direct conversion of sunlight and temperature gradient into useful electricity is possible without involving heavy mechanical instruments or hazardous gases, which makes it more robust and prone to environmental degradation. Despite the several advantages, solar cell and thermoelectric power generation are still suffering from the challenges of low power conversion efficiency and long-term stability. In 2004, graphene was discovered, ushering in a new age of 2D materials research. The family of two-dimensional materials has been intensively explored in recent years as a reliable and effective alternative material for numerous applications, including thermoelectric power conversion and solar cell components, due to their distinctive material features. Geometric symmetry has a big impact on the electrical characteristics of 2D layered nanomaterials because they are highly sensitive to structural perfection. Numerous nanomaterials have recently been exfoliated, and computational predictions have been made for their potential applications in solar cells and thermoelectric generators. In this chapter, we will basically discuss the insight of emerging nano-materials, their key properties, and applications for designing renewable energy devices. The ultra-thin layer-based solar cells and nano-materials-based thermoelectric generators will also be introduced as a section of the chapter. The concept of emerging classes of nano-materials such as Janus monolayers, van-der Waals structures, and group-IV chalcogenides further piques the interest of the reader to learn about the different perspectives of nano-materials and nano-devices.

Keywords: 2D Materials, Optoelectronic Materials, Thermoelectric Generators (TEG), Solar Cells Nanomaterials, FoM, Energy harvesting, Janus Materials, VdW structures.

---

* **Corresponding author Anup Shrivastava:** Computational Nano-Materials Research Laboratory (CNMRL), Indian Institute of Information Technology- Allahabad, India; E-mail: rse2017002@iiita.ac.in

INTRODUCTION TO NANOMATERIALS

With the advent of technological growth and rapid industrialization, device dimensions are shrinking to ultra-small scales called nano-scales. Nanotechnology has revolutionized almost all fields of science, such as physics, chemistry, life sciences, engineering, photonics, *etc.* The idea and concepts of nano-sciences were started and popularized by a famous scientist of the late 20th century, Prof. Richard Feynman, in his talk entitled "There's plenty of rooms at the bottom," presented on December 29, 1959, at a conference of the American Physical Society at the California Institute of Technology (CalTech). During his presentation, Professor Feynman described how science would be able to regulate and govern specific atoms and molecules. Initially, people were not much connected with his ideas, until 1981, when a German and Swedish scientist invented the Scanning tunneling microscope (STM) at IBM Zurich. With the STM, scientists are able to see and manipulate individual atoms with high precision. In the meanwhile, Prof. Norio-Taniguchi of Japan, who was experimenting with high-precision machining methods, originally used the word "Nanotechnology" in 1974 to describe the processes of atom- or molecule-scale separation, deformation, and consolidation of materials [1]. Almost after twelve years (in 1986), an American scientist Kim Eric Drexler unintentionally used the phrase linked to nanotechnology to describe what would subsequently be recognized as molecular nanotechnology (MNT) [2] in his book "Engines of Creation: The Coming Era of Nanotechnology". Afterward, the last decades of the 20th century and the beginning of the new century came with some promising breakthroughs in the nano-regimes. Now, nano-technologies are being developed for extensive uses in biomedical, electronics, energy storage, and generations, water treatments, plasma physics, space technologies, *etc.* Owing to its extremely high precision and high accuracy, nano-sciences emerged as a key concept for all branches of science and engineering.

Nanomaterials are indispensable components of nanotechnology. Materials with at least one dimension smaller than 100 nanometers are generally referred to as nanomaterials. The manipulation of matter at the nanoscale (1–100 nanometers) length and the exploitation of noble characteristics and phenomena formed at this scale lead to the production of useful nanomaterials, devices, and systems. Owing to the change in the surface area (from micro to nano) due to nanostructuring results in an alteration of the physical, chemical, thermal, and morphological qualities of the materials. Henceforth, it has been observed that the materials are producing superlative performances at the nanoscale. It is also worth noting that, without altering the chemical makeup, it is feasible to change the basic characteristics of materials by patterning matter on the nanoscale.

Nanomaterials are found to be tremendously useful in all aspects of societal challenges, such as biomedical applications, implants, wearable electronics, IoT, energy harvesting, spintronics, data storage, water treatments, smart skins, and defense applications. Some of the key application areas of nanomaterials are supercapacitors, hydrogen toys for cars, and thermoelectric generators/coolers, as shown in Fig. (**1**). In this chapter, we will focus our discussion on the various aspects of energy harvesting techniques using nanomaterials.

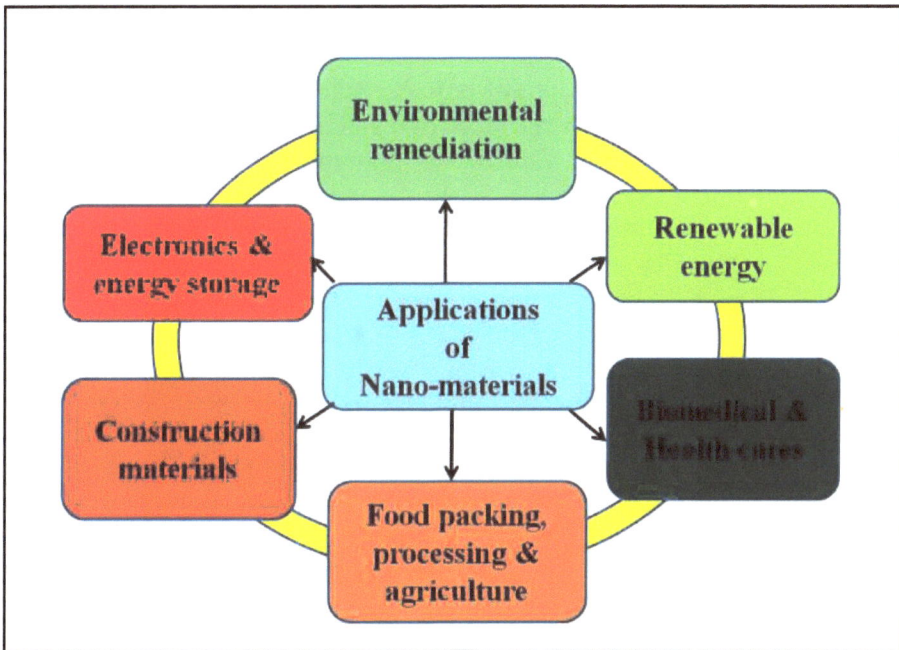

Fig. (1). Various applications of nanomaterials.

ENERGY HARVESTING

What exactly is energy? It is described as power obtained *via* the use of physical or chemical resources, particularly to give light and heat or to operate machinery. Energy is nothing more than a power that may be used to light things up, run machines, or accomplish anything else. Energy cannot be generated nor destroyed; it can only be moved from one form to another. Energy should be produced first and then stored for later use.

People have been using and storing energy in different forms since the beginning of civilization, but it attracted widespread consideration only after the industrial revolution in the 18th century. As machines and technologies have become prominent in day-to-day life, the demand for energy has also increased manyfold in the last few centuries. Over several hundred years, the world's energy consumption which relies on non-renewable energy sources such as fossil fuels, oil, coal, *etc.*, is now running out with the ever-increasing demand for energy consumption. Also, the combustion of fossil fuels is a threat to global climate change, caused by the production of large amounts of greenhouse gases. To fulfill the energy requirements of society, researchers are looking for non-conventional energy sources and energy harvesting through different approaches. The term "energy harvesting" or "scavenging of energy" refers to a technique or process of deriving energy from external sources (such as sunlight, thermal energy, wind energy, and kinetic energy, also called ambient energy), capturing and storing it for later use of a very small amount of power, such as wireless sensors, wearable electronics devices, and low-power electronics. They typically generate a small amount of energy, from microwatts to several milliwatts. Out of the various techniques, the four most popular energy harvesting approaches are (a) optoelectricity, (b) thermoelectricity, (c) piezoelectricity, and (d) RF energy harvesting. The detailed taxonomy for energy scavenging techniques is summarized in Fig. (**2**). Although the four types of energy scavenging techniques are shown in Fig. (**2**), we shall restrict our discussion to the first two techniques, such as thermoelectric and photovoltaic, only because of their paramount contribution compared to piezoelectricity and RF energy harvesting.

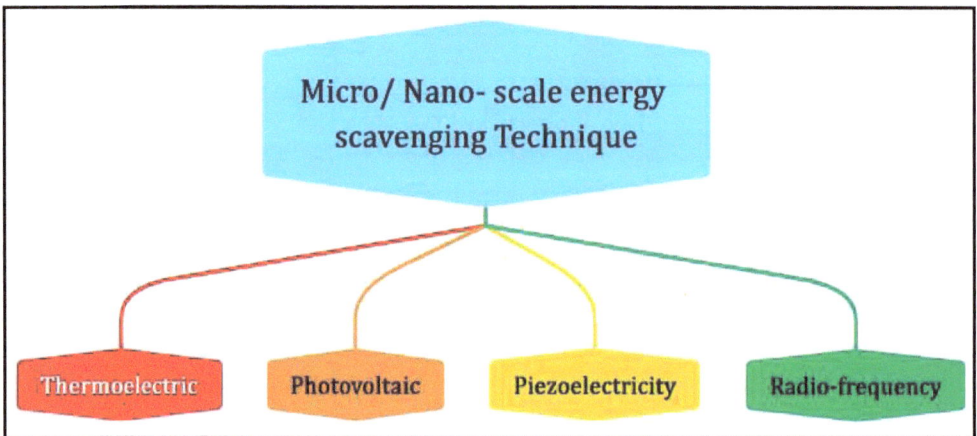

Fig. (2). Taxonomy for energy harvesting techniques.

Advantages and Applications of Energy Harvesting

- It is self-sustainable
- It has unlimited usage
- Environment friendliness
- Cost-effective

Applications

- Charging of low-power electronic devices
- Remote area sensing
- IoT applications
- Healthcare applications
- Automotive network,
- Industrial automation
- Eco management

Role of Nanomaterials in Energy Harvesting

In the last few decades, the advent of several nanomaterials and the implications of low-power electronics in diverse applications have instigated material scientists to explore the in-depth understanding of the role of nanomaterials in energy harvesting and their device design. Several attempts have been made for the all-around development of different potential materials for thermoelectric, photovoltaic, and piezoelectric energy conversion. Even after the discovery of graphene in 2004, a new era of 2D materials started, which exhibit outstanding performance in terms of their physical and chemical properties. Researchers across the globe are continuously making efforts to invent such materials through which we can improve the overall conversion efficiency in the case of all energy harvesting techniques. Numerous techniques, such as engineered 2D materials, nanowires, nanotubes, quantum dots, Janus monolayers, *etc.*, have been proposed and found to be very promising in energy harvesting applications.

Thermoelectric Energy Harvesting

In the last two decades, researchers across the world have shown an interest in thermoelectric energy harvesting. Thermoelectric energy harvesting might play a paramount role in clean and green energy solutions for the current global energy crisis. The main idea behind thermoelectric energy harvesting is to generate electrical energy from waste heat using the Seebeck theory. One of the surveys reports recently marked that over 2/3rd of the total energy across the world is wasted as heat. Although, a lot of research has been done and numerous materials have been invented that exhibit good transport and thermoelectric behavior, the efficiency of thermoelectric materials used today is still a challenge, which makes them uncompetitive with traditional power plants and refrigerators. The interest in the creation of a new generation of thermoelectric materials is being motivated by the rising demand for alternative energy sources. To compete with current technologies, new thermoelectric materials must increase efficiency by a factor of three over current values. Improving the efficiency of thermoelectric devices may enable the use of waste heat from previously untapped sources. Automotive waste heat sources, such as exhaust, brakes, and engines, can be used to improve the efficiency of future hybrid vehicles. A vital new source of energy could be provided by harnessing the heat produced by industrial plants, electrical plants, and nuclear power plants. The heat from radioactive decay could be used to power next-generation radioisotope thermoelectric generators, allowing us to reach further into space. In order to provide developing/under-developed nations with the necessary power in rural regions, thermoelectric power generation will play a very important role. For example, a wood-burning stove or open fire in a small town might function as a possible cell phone charging station or even supply LED illumination. Thermoelectric technology has the potential to act as a backup energy source and reduces our dependency on fossil fuels. To create the next generation of thermoelectric materials, scientific research in this area will need to be increased. Three fundamental physics processes, the Seebeck effect, the Peltier effect, and the Thomson effect, primarily describe thermoelectricity.

The Seebeck effects demonstrate that the electric current can be driven in a closed circuit by induced voltage as an application of temperature gradients. The temperature differential between the two junctions directly correlates with the induced voltage. Furthermore, Jean Charles Athanase Peltier, a French scientist, explained the phenomena of heat flux development across a substance when exposed to an electric current or current density in 1834. In 1851, Sir William Thomson exhibited that when an electric current passes through a conducting material with a temperature gradient, then a reversible heat is generated.

After the formulation of the three fundamental principles of thermoelectricity, systematic research has begun between 1885-1910. The German scientist Edmund Altenkirch evaluated the thermoelectric generators' potential efficiency and specified the material parameters needed to build usable devices in 1910. To express the qualitative performance of the thermoelectric materials, he developed a relationship known as the thermoelectric "Figure of Merit" [3, 4]. Unfortunately, metallic conductors were the only materials readily available at the time, making it impossible to create thermoelectric generators with an efficiency higher than roughly 0.5%.

Later, in 1950, a 5% efficient semiconductor-based generator was developed. In 1949, using the concept of the "Figure of Merit zT," A.F. Ioffe laid the foundation for the modern theory of thermoelectricity, according to his classic works on semiconductor thermoelements and thermoelectric cooling (1956) [5]. In order to measure results and improve performance, Ioffe promoted the use of semiconductors in thermoelectricity and semiconductor physics.

For the following several decades, increases in the efficiency of thermoelectric power generation remained modest despite significant research and development. The closing decades of the previous century saw a surge in new ideas, which rekindled interest in thermoelectricity. The hope that designed structures may increase zT, especially at the nanoscale has renewed interest in the study of thermoelectric materials [6, 7]. Some of the many theories that have been put forth over the past few decades have turned out to be unsuccessful, while others have led to the creation of totally new classes of complicated thermoelectric materials [8]. Bell cites how the need for alternate energy sources has reignited interest in commercial applications [9] and stimulated the development of low-cost, environment friendly thermoelectric materials.

A major breakthrough in thermoelectric research was made by Dresselhaus and Hicks in 1992. They proposed the implementation of nanostructured semiconductors for the enhanced figure of merit. Furthermore, in 2004, the discovery of graphene accelerated the pace of thermoelectric research, as a whole new era of nanostructured materials was ready to be exposed for their novelty. Although a family of materials is available with unique properties, but not every material can be used for thermoelectric applications. Here, in the following sections, we will discuss how to choose materials for thermoelectric applications, followed by a brief introduction to some thermoelectric materials.

Criteria for the Selection of Thermoelectric Materials

As we saw in the previous section, the efficiency of a thermoelectric power generator depends on a dimensionless quantity called "Figure of Merit," which further indicates the qualitative performance of the materials chosen for the design of the device. We also observed that the application of the TE devices had been limited due to their low efficiency and, henceforth, a low FoM. Mathematically, the FoM for thermoelectric materials can be written as $zT = \left[\sigma S_B^2 T/(\kappa_e + \kappa_p)\right]$ where σ and S_B are the electrical conductivity and the Seebeck coefficient of the material, respectively. The factor $\kappa_{e/p}$ is the thermal conductivity due to electrons and phonons, respectively, while T is the temperature [10].

From the expression of FoM, it can be inferred that the material with high thermopower (Seebeck coefficient), and electrical conductivity with low lattice thermal conductivity will be the best choice for a thermoelectric material. Since the achievement of high FoM is mainly hindered due to the coupled relation of electronic thermal conductivity and electrical conductivity *via* Wiedmann-Franz law and cannot be tuned independently. The electrical conductivity and thermopower do depend on the carrier concentration with an inverse relation, hence we can conclude that the electrical conductivity and thermopower will be inversely proportional to each other, therefore increment in 'σ' will lead to a substantiated reduction in thermopower (S) as illustrated in Fig. (**3**).

A high carrier concentration is associated with high electrical conductivity (for metals), whereas a low carrier concentration is associated with a large thermopower (for insulators). The materials that lie between insulators and metals, however, display a fair balance between 'S' and 'σ'. As a result, the majority of effective thermoelectric materials are semiconductors that are severely doped and have carrier concentrations of between 10^{19}-10^{21} carriers/cm^3. Additionally, common ranges for the Seebeck coefficient and electrical conductivity are 150-250 μV/K and >10^3 Ohm^{-1}cm^{-1}, respectively, which are usually considered as a threshold for qualifying as a good choice for a thermoelectric material. For lattice thermal conductivity, the lowest value will be an obvious choice for good thermoelectric material.

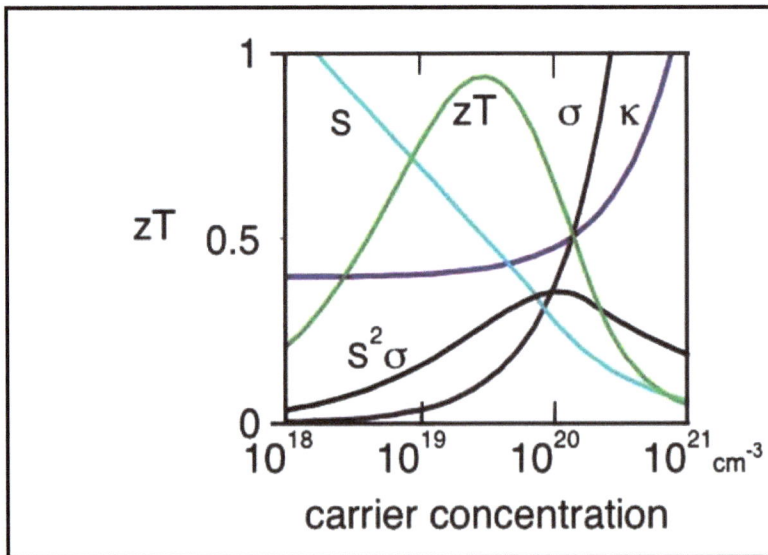

Fig. (3). The inter-coupling of the thermoelectric parameters [10].

Need of Nanostructuring

The performance of thermoelectric materials can be considerably improved by nanostructuring. Nanostructures such as nanowires, nanotubes, nanoforks, quantum dots, and superlattices outperform their bulk counterparts in terms of power factor and thermal conductivity. By combining the two ways, nanostructuring can improve thermoelectric performance. One method is to improve the power factor by boosting the DOS at the Fermi level *via* quantum confinement [11, 6, 12]. An alternative strategy is to use phonons, which scatter well at grain boundaries and have a longer mean free path than electrons [13, 14,

7], to reduce the thermal conductivity of the lattice. By boosting the scattering of long wavelength heat-carrying phonons in nanostructured thermoelectric materials, the phonon-electron scattering process reduces the lattice thermal conductivity. To achieve the maximum zT, nanostructured thermoelectric materials need a high doping concentration.

Low-energy minority charge carriers scatter more readily near grain boundaries than high-energy majority charge carriers, which results in a modest bipolar transition between the states [15, 16]. Due to this effect, the nanostructured thermoelectrics demonstrates a superior Seebeck coefficient than their bulk equivalents at comparable doping doses. By employing nanostructuring techniques, the FoM for thermoelectric materials is achieved in the range of beyond 3 [17, 18].

THE EMERGING CLASSES OF NANOMATERIALS FOR TE HARVESTING

There are several materials that have emerged as potential thermoelectric materials for high-performance thermoelectric devices. In this section, we will primarily discuss the most recent and emerging classes of layered materials which surge the thermoelectric performances in terms of high FoM. These materials can be classified as:

(a) Janus Materials

(b) Van-der-Waals Structures

(c) Chalcogenides Materials

(d) Organic nanomaterials

Janus Materials

The Janus materials are derived from the asymmetric functionalization of the central atoms. They are novel variants of two-dimensional (2D) materials that have gained the interest of researchers because of their unique features. Their characteristics differ significantly from those of standard 2D materials. The mirror asymmetry in the structure of the Janus 2D material gives it unique features [19, 20]. Due to their structure, stability, electrical characteristics, and other qualities, Janus transition metal dichalcogenides (TMDs) are an important family of materials. The name Janus is derived from the name of a Roman god who has two faces with extremely different properties. An example of the Janus material is shown in Fig. (4).

Fig. (4). Janus monolayer of Ge$_2$SeTe [23].

The first Janus material was MoSSe, which was successfully synthesized by Le and his co-worker through CVD techniques [21]. Since then, several different configurations of Janus materials have been proposed and investigated, such as MoSeTe, WSSe, WSeTe, Ge$_2$SeTe, HfSSe, and HfSTe [18, 22 - 24]. Owing to their structural asymmetry, these materials are found to be very promising to decouple the thermoelectric parameters such as high power factor with sufficiently low lattice thermal conductivity, which further increases the overall thermoelectric FoM to substantiate values in the range of 3-4.

Van-der Waals Structures

The van-der Waals structures are layered structures where more than one atomically thin layer, like graphene, is stacked in such a way that the interlayer is bounded by the weak vdW forces. These materials have tremendous potential for the next generation of nanoelectronics because they enable the development of high-performance, purpose-specific structures because of their distinctive interlayer coupling and breakdown of the inversion symmetry.

The several sets of vdW heterostructures are explored for numerous thermoelectric applications such as Graphene/h-BN vdW structures [26], MoS$_2$/h-BN [27], Graphene/MoS$_2$ [28], Sn/h-BN [29] *etc*. An example of the vdW structures with the combinations of hexagonal-Boron Nitride (h-BN) and monolayer antimonene (Sb), can be demonstrated in Fig. (**5**).

Chalcogenides Materials

Bi$_2$Te$_3$ and PbTe are the very first popular and widely used thermoelectric materials. In reality, materials based on bismuth and antimony chalcogenides, such as Bi$_2$Te$_3$, and Sb$_2$Te$_3$, have excellent thermoelectric characteristics. The compound exhibits a similar rhombohedral crystal structure. It's simpler to see it as a hexagonal cell with the chemical formula (X_2Y_3) repeated three times. A quintuple layer is formed by each repetition, which has the structure Y(1)-X-Y(2) X-Y(1). The anions' chemical states are represented by the numerals 1 and 2. For a long time, the highest figure of merit of Bi_2Te_3 has been about 1, which equates to an efficiency of roughly 5–7% in most applications. The van der Waals forces bind the quintuple layers together weakly, allowing them to be easily exfoliated. Thin films of p-type Sb_2Te_3 and n-type Bi_2Te_3 have been discovered with promising features such as a lower lattice thermal conductivity and a higher Seebeck coefficient [30]. These studies revealed that nanostructuring can effectively enhance the performance of the materials to be applied in thin film applications. Zinc antimony (Zn_4Sb_3) is one of the good thermoelectric materials owing to its poor capacity of heat conductivity and high efficiency. Zn_4Sb_3 reportedly operates most efficiently between 450 and 670 K, with a zT of around 1.3 [31]. The literature has also demonstrated that PbTe outperforms Bi_2Te_3 for high-temperature operation in automotive applications. Very recently, group-IV dichalcogenide Ge_2Se_2 monolayer has been explored for thermoelectric properties, and the results show a high value of zT of the order of 2.03 at 1000 K along with a low lattice thermal conductivity [18]. Materials based on group IV tellurides, such as PbTe, GeTe, and SnTe, are often used for power generation in the temperature range of 500 to 900 K. AgSbTe2 alloys have proven good performance for both p- and n-type semiconductors, with zT > 1.

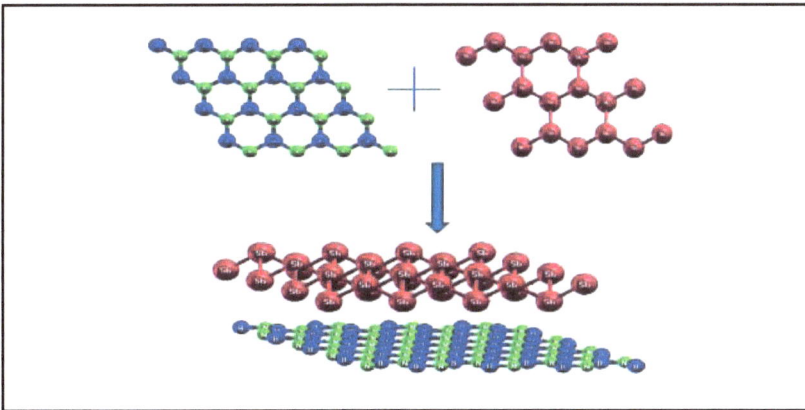

Fig. (5). Building Sb/hBN vdW structures [25].

The p-type alloy $(GeTe)_{0.85}(AgSbTe_2)_{0.15}$, in an instance, has a maximum zT > 1.2 and has been effectively applied to a durable TEG module [29]. Monolayer SnS and SnSe possess extremely low thermal conductivity (below 1 $Wm^{-1}K^{-1}$) with a high Seebeck coefficient, which makes them a promising material for thermoelectric application.

Organic Materials

The scientists also looked into organic thermoelectric materials as an alternative to inorganic thermoelectric due to their low cost, large-scale production methods, and unique mechanical flexibility. Organic thermoelectric materials and devices based on it have various applications in wearable's, human implants, automobile roofs and many more. These organic thermoelectric materials, when combined with inorganic nano-materials, improve their performance in terms of power factor and electrical conductivity. The nano-inclusion of PEDOT:PSS (CLEVIOS PH1000) with Bi_2Te_3 enhanced the power factor up to $130\mu W/mK^2$. Also, the nano-inclusions of graphene oxide, fullerenes, single wall CNTs, PbTe, Te nanowires, and polycarbonates ameliorate their thermoelectric performances. An n- or p-doped electron or hole transport in a polymer (insulating or conjugated) matrix can be facilitated by nanoinclusions through a hopping mechanism, but phonon scattering at the nanoparticle-polymer-nanoparticle interfaces prevents their efficient transmission and leads to low thermal conductivities. Therefore, despite the fact that their electrical conductivity, Seebeck coefficient, and power factor still need to be increased from their already stated values [32], nanocomposites are seen as prospective TE materials for thermoelectric power generation (used to generate TEGs).

SOLAR ENERGY HARVESTING

Sunlight is the major source of energy on the earth. Solar energy is a form of energy that relies on nuclear fusion at the core of the Sun. From a statistical report, an hour of solar radiation on earth produces 14 terawatt years of energy, which is nearly equal to the world's total annual energy consumption. The term solar energy harvesting refers to the approach of capturing solar energy and converting it into some other form of energy so that we can preserve it for later use.

Historical Perspective

The first historical evidence for the conversion of light into energy appeared in the name of the French scientist Edmund Becquerel in 1839. Later, in 1883, American inventor Charles Fritts proposed the first functioning solar cells made from selenium wafers [33]. Although it was the first major breakthrough, these solar panels ineffective, with an energy conversion rate of around 1%. In 1888 and 1901, two American scientists filed US patents mentioning solar cells, which further motivated the researchers to work on them. In 1905, Einstein introduced the photoelectric effect, which provides the baseline concept for the working of solar cells. Einstein was awarded the prestigious Nobel award in 1922 for his contribution to the photoelectric effect. The first high-power silicon solar photovoltaic cell was created by Gerald Pearson and his co-workers at Bell Laboratories in 1954 [34]. Just after decades, NASA launched the first solar PV array with a capacity of 470 Watts. In 1973, the first solar-powered building was erected at the University of Delaware, USA, followed by the establishment of NREL in 1977 [35]. In 1999, Germany launched a 100,000 solar roof program. The gradual research on optoelectronic materials further leads to the drastic reduction of the cost of PV modules during the last decades. The cost of PV modules fell from $5 per watt to $1 per watt in the USA and Europe, which further reduces to 0.7$ per watt around the year 2015. The rapid pace of PV research and investigation of new materials and techniques makes the solar cell a highly efficient and profound energy harvesting technique among other available approaches.

Advantages and Disadvantages of Solar Cells

Advantages

1. Clean, renewable, and sustainable energy sources.

2. Once installed, there are no recurring costs.

3. Environment-friendly and hence has no hazardous effects on the environment.

4. No moving or mechanical parts, so the maintenance cost and efforts are very less.

Disadvantages

1. High installation costs.

2. Required large area for installation to achieve a good level of efficiency.

3. Poor power conversion efficiency.

Classification of Solar Cells based on Materials

On the basis of the materials used, solar cells are categorized into three generations [36]. The first generations of solar cells largely used mono-crystalline or polycrystalline silicon wafers. These are called mono or polycrystalline based on the fact that the wafer slices are taken from a single crystalline structure or from multiple silicon crystals. Polycrystalline solar cells are a bit cheaper but also have low efficiency.

Further, in order to reduce the cost of silicon solar cells, inorganic materials have started to be used, such as CdTe, CIGS, and CdS, in the solar cells. The second-generation solar cell offers moderate efficiency at a comparatively low price. In the third generation of solar cells, novel approaches such as quantum dots, die-sensitized, perovskites, and organic materials are used. With the implications of novel nanomaterials and state-of-the-art techniques, third-generation solar cells pursue more efficient, more abundant and non-toxic materials with the durability of those particular systems. The detailed classification of solar-cells for their different generations can be summarized as in Fig. (**6**).

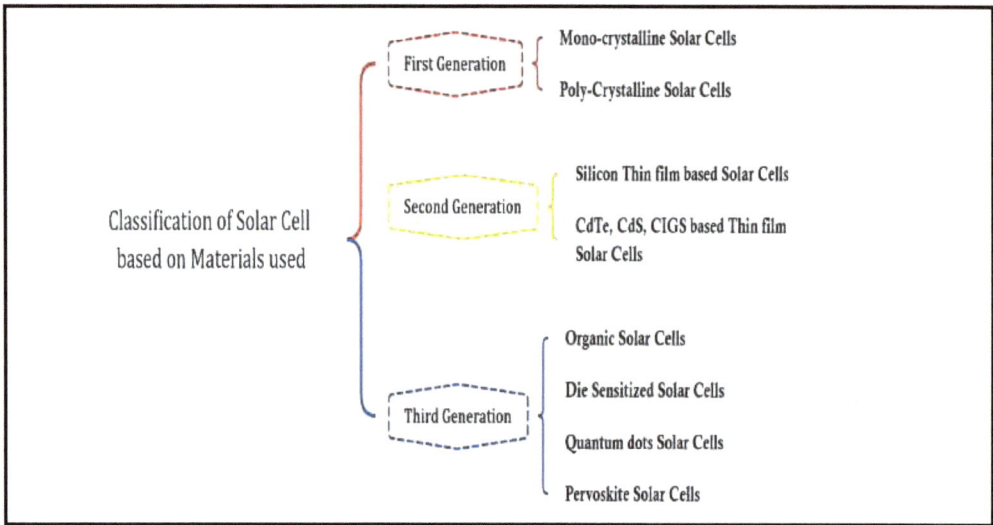

Fig. (6). Classification of solar cells based on materials used.

Choice of Material Selection

The criteria for the selection of nanomaterials for improved performance of solar cells are summarized in Fig. (**7**). The ideal materials to be used in solar cells should possess a proper and tunable bandgap. Basically, 1.0eV-1.8eV is an optimum value to harvest the maximum sunlight. The other parameters for the selection of suitable nanomaterials for solar cells are highly active for light-matter interactions, high conductivity, and high transparency, excellent flexibility, high surface area, good charge transport property, excellent stability, and, of course, cheap in cost [37].

Nanostructured Materials

In general, nanotechnology is a preeminent toolset for the most effective, sustainable energy conversion, storage, and conservation of energy. It enables a variety of techniques, such as the efficient processing of semiconductors into photovoltaic devices, the development of more efficient photo-catalysts for turning sunlight into chemical fuels, the creation of new materials and membranes for the separation needed in many energy applications, the conversion of chemical fuels into electrical energy, and increasing the power and energy density in batteries. So, in a simple sense, nanotechnology means we are dealing with the technology with some nanomaterials. In general, when we are using certain kind of light-absorbing materials, or electrodes/ transport layer materials, we basically try to make it as a composite like materials.

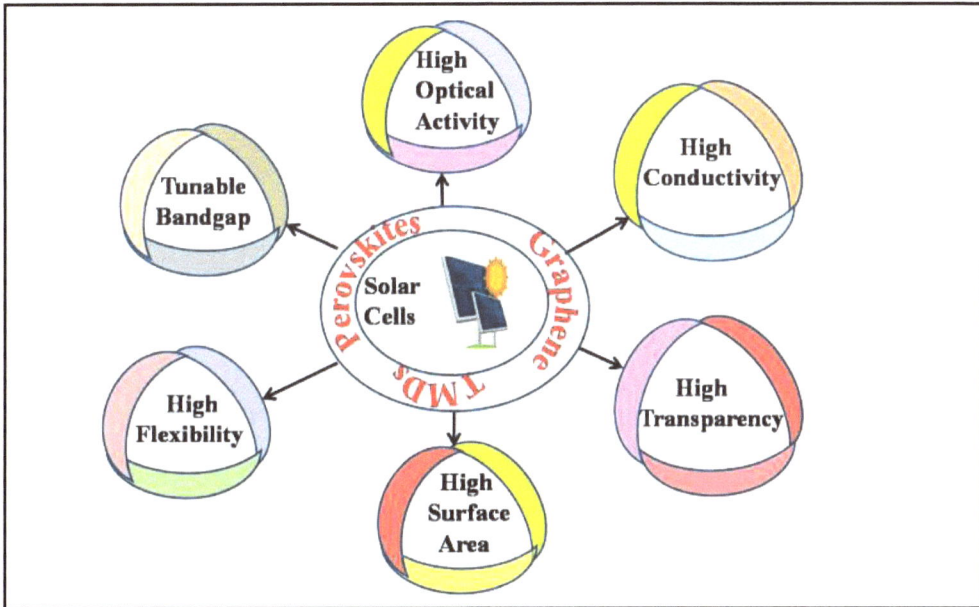

Fig. (7). Criteria for the selection of materials in solar cells.

So, in those composites, we are using certain kinds of filler materials which size into the nanometer range or maybe at the nanoscale. Since the last few decades, people have been tending to this because of the high surface-to-volume ratio. People use nano-particles, nano-fibers, carbon nano-tubes, nano-sheets, graphene, and quantum dots. So, these are all different types of examples from the nanotechnology point of view.

EMERGING MATERIALS FOR SOLAR CELLS

Janus Monolayers

After the synthesis of the first Janus monolayer (MoSSe) using the selenization of MoS_2, researchers are rigorously exploring their numerous potential applications. Recently, a few scientists demonstrated that the Janus monolayers based solar cells are outperforming in comparison to the other 2D, and bulk materials. Since, the structural asymmetry in the Janus monolayer leads to its extra-ordinary optical behavior, which further persuades its potential to be used in solar cells. The researchers from Nanjing University, China, proposed a Janus layer (Ga_2SeTe) based solar cells with a significantly high PCE of 19% [38]. In another work, Rajnish Chaurasia *et al.* at IIT-Jodhpur, India, proposed solar using an ultra-thin Janus monolayer of WSSe, which exhibited maximum power conversion efficiency in between 17-19% [39]. In the year 2021, I. Bouziani and his team demonstrated a solar* cell using Janus monolayer of Sn_2SSe, and $SnGeS_2$ as active materials and using the first-principle computations, they claimed the efficiency of proposed solar cells of the order of 27%, and 28%, respectively [40].

vdW heterostructures

Heterostructures built from atomically thin semiconductors are a potential new method for producing the thinnest and lightest photovoltaic solar cells on flexible substrates. Many 2D materials are stacked together and bounded with the weak van der Waals forces to form vdW heterostructures with desirable properties, depending on their stacking angles, number of layers and band alignment. Previous research reported the potential application of type-II band alignment in solar cells, because the relaxation in band offset-driven photogenerated carriers offers a significant photovoltaic effect and charge transfer across the junction [41]. The MoS2/WSe2 vdW structure offers type-II heterojunction, where the valence band states are spatially placed in WSe2 and the conduction band states are positioned in the MoS2 layer. Various other heterostructures also exhibit type-II band alignment including MoS2/WS2 [42], MoS2/black phosphorus [43], MoTe2/MoS2 [44], GaTe/MoS2 [45], MoSe2/WSe2 [46], MoS2/carbon nanotubes [47], MoS2/pentacene [48], MoS2/silicon [49], Sn/hBN [50, 51], and boost the efficiency of atomically thin solar cells.

Group-IV/VI Chalcogenides

Transition metal dichalcogenides (TMDCs) have been known for a long time as the most popular two-dimensional (2D) materials that are atomically thin, mechanically flexible, and have good semiconducting characteristics [52]. Recently, an emerging class of semiconducting chalcogenides (Group-IV/V chalcogenides) has drawn great consideration for their potential use in optoelectronics and photo devices. Up to 1nm thin group-IV chalcogenides have strong light-matter interactions with a low exciton binding energy and can absorb up to 10% of incident sunlight under the illumination of the full solar spectrum, which makes these materials more promising and appealing for photovoltaic devices. A group of researchers from Shandong University, China, obtained an external quantum efficiency of 10.27%, 30.32%, 25.43% and 22.01% for GeS, SnSe, GeSe, and SnS, monolayers-based solar cells. One more interesting work has been presented by Sumanyu Agarwal *et al.* [53], in which the authors reported a power conversion efficiency of 11% using the antimony chalcogenides $Sb_2(S, Se)_3$ based solar cells.

AIHP (All Inorganic Halide Perovskite)

All Inorganic Halide Perovskite (AIHP) has attracted immense interest for their prospective uses in solar cells due to their stabilities and high carrier mobilities compared to organic-inorganic perovskites [54]. These halide perovskites offer cost-effective fabrication of highly efficient optoelectronic devices. AIHPs are stable in various morphological states, such as single crystals, polycrystalline bulk films, and nano-crystal films. These morphologies and compositions in AIHP play a crucial role in determining the electronic and optical properties. The carrier mobility in $CsPbBr_3$ varies from 77.9 to an ultra-high value of 2000 cm^2 $V^{-1}s^{-1}$ when changing morphology from nano-sheet films to single crystals [55]. Similarly, polycrystalline $CsPbI_3$ film offers high mobility (60 cm^2 $V^{-1}s^{-1}$) compared to nano-crystal films (0.5 cm^2 $V^{-1}s^{-1}$). The photoluminescence quantum yield (PLQY) is highest for the $CsPbBr_3$ nano-crystal (95%) than for the $CsPbI_3$ and $CsPbCl_3$ nano-crystals, having photoluminescence of the order of 70% and 10%, respectively [56]. The cubic (α) phase of $CsPbI_3$ is considered to be the best inorganic perovskite for photovoltaic applications as it can absorb a wide range of solar spectrum due to its narrow band gap of 1.73 eV. Owing to its unique characteristics, AIHP provides potential candidates for solar cells and photovoltaic applications.

Besides the above-discussed emerging materials, there are numerous other approaches that have been developed to increase the efficiency of solar cells. While considering the overall efficiency of any harvesting system, the designing of devices are also very imperative in addition to the materials used.

DEVICES USED FOR ENERGY HARVESTING

Energy harvesting is an approach to collecting and storing energy from the environment (such as heat, light, vibrations, and radio waves), which is further used to power various devices such as sensors, implanted medical devices, and IoT (Internet of Things) devices. The effective implementation of energy harvesting devices leads to a genuinely wireless world in which physical interference-free functioning of equipment such as power cables, replacing batteries, charging batteries, filling fuel tanks, *etc.*, becomes obsolete. These devices require less maintenance, have a long life, and are especially helpful for powering electronics at remote locations. Based on the brief discussion about the energy harvesting techniques in the previous sections, we will now discuss the overview of the different nanogenerators in the subsequent sections. We will mainly discuss thermoelectric generators and solar cells as key devices for energy harvesting.

Thermoelectric Generators (TEG)

Thermoelectric generators (TEGs) are nanomaterials-based semiconducting devices which convert thermal energy into electrical energy without utilizing any mechanical components. The Seebeck effect, which asserts that the induced voltage is directly proportional to the temperature gradient between the two ends, is the basis for the TEGs' theory of converting thermal energy into electrical energy. The Seebeck coefficient, or thermopower, is another name for the proportionality constant. The temperature gradient, hot end, and cold end temperatures, as well as the thermoelectric figure of merit (zT) for the materials utilized, all have an impact on the TEG's efficiency.

The TEG device may consist of a single or more than one thermoelectric couple based on the power requirement or the area of application. The schematic diagram for a single thermoelectric couple-based TEG is shown in Fig. (**8**). The simplest TEG is made up of two N-type and P-type thermo-elements, also referred to as the TEG's legs, which are electrically coupled in series and thermally connected in parallel (as shown in Fig. (**8**)). The Seebeck coefficients of the N-type and P-type legs are oppositely positive and negative. Also, electrons are in excess in the N-type thermo-elements, while the P-type thermo-elements have an excessive number of holes. The two legs are connected together at one side using a metallic layer (shown in the figure with a clay color beneath the heated surface at the top). Now when the two ends are kept at different temperatures, due to the developed gradient in temperature, the holes are migrated from the hotter end to the colder end in the P-type legs, which further creates the charge density difference between two ends. This charge density difference will further repel electrons from the opposite ends (from cold to hot).

Similar phenomena will occur in the N-type leg, where the heat and charge carriers will be electrons instead of holes. Electrons will be migrated from the cold end to the hot end, which becomes negatively charged, as shown in Fig. (**8**). Now, if the hot end is connected to a metallic bar (electrically), and the cold end is connected to the load, a current will be generated and passed through the load. As the heat flows from the hot end to the cold end, the net current will be increased.

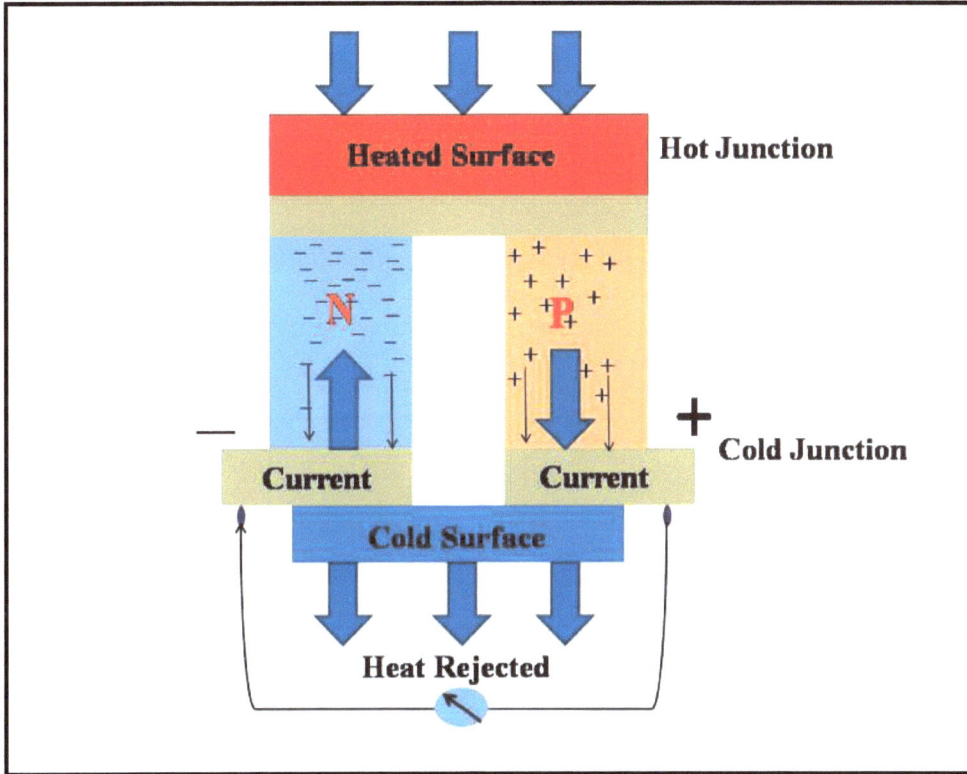

Fig. (8). Schematic for thermoelectric nanogenerators.

In order to generate higher power, wc generally cascaded the N-P pairs in series to the common heat source and heat sink. The efficiency of TEG is defined by the following relationship:

$$\eta = \left(\frac{\sqrt{1+zT}-1}{\sqrt{1+zT}+\frac{T_C}{T_H}} \right) \frac{\Delta T}{T_H} \tag{1}$$

Where, η the efficiency of TEG, zT is the FoM, T_c and T_h are the absolute temperatures of the cold end and hot end, respectively, and ΔT is the difference in temperature between the hot end and the cold end. The effectiveness of TEG varies linearly with the temperature gradient (ΔT). The greater the temperature difference, the greater the efficiency will be achieved.

If materials with zT = 15 exist, TEGs with the cold and hot end temperatures of T_c = 300 K and T_h = 400 K, respectively, will offer a TEG efficiency of about 15.8%. The biggest problem is to investigate the material with a high FoM.

Solar Cells

Construction and Working

Solar cells are basically used for harvesting sunlight. In the recent past, this device has become very popular owing to its low cost, maintenance-free production, and high efficiency. The simplest schematic for a planar perovskite solar cell is shown in Fig. (**9**).

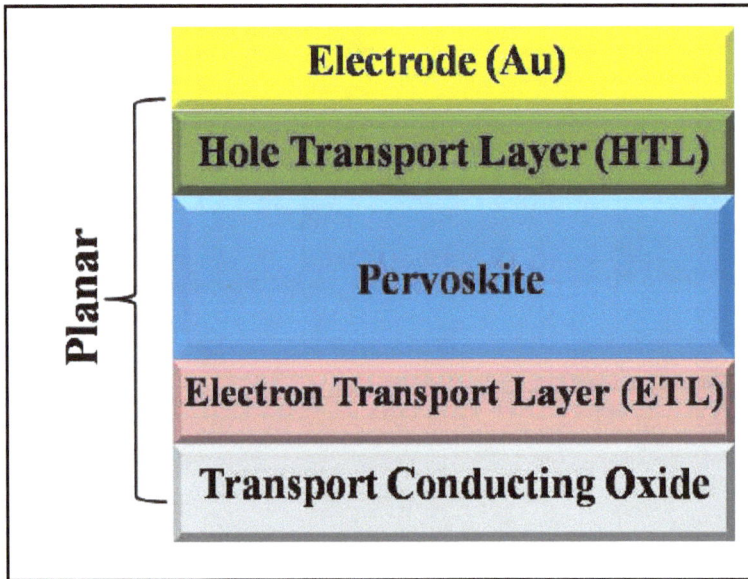

Fig. (9). Schematic for thermoelectric nanogenerators.

Solar cells basically work on the principle of photovoltaic. During the photovoltaic processes, under the illumination of sunlight on the absorbing materials (active layer), electron-hole pairs are generated. These generated pairs are also referred to as photo-generated carriers, and they are generally more energetic compared to incident photons. The generated photos carriers are further collected at the respective transport layers (ETL/HTL) adjacent to the central absorber/ active layer. The roles of the electron transport layer (ETL) and hole transport layer (HTL) are to collect the generated photo carriers and transfer them into the respective electrodes. Also, the ETL and HTL limit the perovskite crystal's crystal growth. The distinguishing features of ideal ETL/HTL materials are high carrier mobility, high transparency, inherent stability, appropriate band alignment, and electron affinity with low cost of manufacturing. A transparent conducting layer has been used as a substrate for the solar cell, which also works as a front electrode. The back electrode has been carefully designed with the proper matching of the work function in order to achieve high power conversion efficiency.

The Performance Parameters that Characterize the Solar Cells

A solar cell is quantitatively characterized through the estimation of the following performance parameters:

a. Short Circuit Current (I_{sc}): Short circuit current in solar cells is the maximum current withstand in the external circuit under no load condition. In other words, we may also say that the current flowing in the external circuit is subjected to the shorted contacts of the solar cell. The higher values of I_{sc} are favorable for a good solar cell.
b. Open-circuit voltage (V_{oc}): It can be referred to as the maximum voltage that can be drawn from the solar cell. More precisely, the open circuit voltage in a solar cell is the voltage produced by the device subjected to the proper illumination, when no current flows through the external circuit.
c. Fill Factor (FF): Fill factor can be described as the ratio of maximum electric power generated to the product of its open circuit voltage and short circuit current. It can be formulated as: $FF = \frac{V_{max}I_{max}}{J_{sc}V_{OC}}$
d. Power Conversion Efficiency (η): It is defined as the ratio of power generated to the incident light power. Mathematically,

$$\eta = \frac{P_{out}}{P_{in}} = \frac{J_{sc}V_{OC}FF}{I_{in}} \tag{2}$$

For a good solar cell, all four performance parameters are tried to keep as high as possible.

Nanomaterials-based Solar Cell Performance

Recently, a lot of research has been done to improve the performance of solar cells. Out of the several approaches, the use of nanomaterials and nanotechnology is the most enchanting. Recently, quantum dots have been employed by EPFL researchers to increase the effectiveness of perovskite solar cells. They have reported a significantly high PCE of the proposed solar cell at approximately

25.7% [57]. In another work, MIT scientists have devised a process to synthesize a larger sheet of graphene that is defect-free and can be used in lightweight solar cells [58]. Also, phosphorene nanosheets were employed by Flinders University researchers to construct low-temperature perovskite solar cells. It is much more effective than silicon solar cells with comparatively less expensive. Researchers from the Los Alamos National Laboratory have successfully used copper indium selenide sulphide quantum dots in solar cells [59]. Li D *et al*. [60] intriguing study offered anodic titanium oxide (ATO) nanotube-based dye-sensitized solar cells (DSSCs), which showed power conversion efficiencies of 2.9%, 3.9%, and fill factors of 0.51, 0.65 with and without bottom reductive doping treatment, respectively. Using single-walled carbon nanotubes, Zhang *et al*. achieved a solar cell with a conversion efficiency of 12.6% [61].

CONCLUSION

The scarcity of energy and the necessity to harness it from renewable sources are the most critical and challenging issues with the research community. In this chapter, we have discussed different aspects of nanomaterials and nano-devices for energy harvesting applications. More specifically, we have discussed nanotechnology and nanomaterials useful for the scavenging of thermal and solar energy. The criteria for the selection of the materials and their implications for thermoelectric and solar energy harvesting have also been discussed. In the later section of the chapter, we have explained the working mechanism and basic construction of thermoelectric generators and solar cells. Also, the chapter has summarized some emerging nanomaterials and state-of-the-art in the domain of energy harvesting at the nanoscale. This chapter will provide an overview to the reader about the applications of nanomaterials and nanodevices for energy harvesting applications.

LEARNING OBJECTIVES

• To understand the basic concept of nanomaterials with the global energy challenges.

• The key concept of green energy generation (different energy scavenging techniques).

• The introductory concept about the nano-devices used for energy harvesting.

• Role of emerging nano-materials (Janus monolayers, vdW structures, Group-IV (chalcogenides) in energy harvesting.

MULTIPLE CHOICE QUESTIONS

1. The term STM is an abbreviation for:

a. Scattering tunable microscope
b. Scanning tunable microscope
c. Scanning tunneling microscope
d. Scattering tunneling microscope

2. Nano materials refers to the material with one dimension less than the range of:

a. 1 nm
b. 50 nm
c. 100 nm
d. 0.1 nm

3. Which of the following not refers to non-renewable energy sources:

a. Fossil fuels
b. Coal
c. Solar cell
d. Petroleum

4. The first exfoliated 2D material is:

a. Graphene
b. Al_2O_3
c. Silicene
d. MoS_2

5. The efficiency of TEG has been derived by:

a. Seebeck
b. Charles Peltier
c. Edmund Altenkrich
d. William Thomson

6. For a good thermoelectric material, the optimum range of Seebeck coefficient is:

   a. 1500-2000 μV/K
   b. 150-200 μV/K
   c. 15-25 μV/K
   d. 1.5-2.5 μV/K

7. The thermoelectric figure of merit (FoM) does not depends on:

   a. Electrical conductivity
   b. Thermal conductivity
   c. Seebeck Co-efficient
   d. Peltier Co-efficient

8. In nano structured materials, the phonon-electron scattering mechanism:

   a. Will lower the thermal conductivity
   b. Will increase the thermal conductivity
   c. Will not affect the thermal conductivity
   d. None

9. Which of the following is not a Janus material:

   a. MoSeTe
   b. $CsPbI_3$
   c. WSSe
   d. Ge_2SeTe

10. The characteristics of Solar cell cannot be characterized by which of the following parameters:

   a. Short circuit current
   b. Open circuit current
   c. Fill-factor
   d. Power factor

11. The role of ETL/HTL in Solar cell is:

a. Generation of photons
b. Collection of charge carriers
c. Recombination of charge carriers
d. To provide stability in the structure

12. The ideal value of FF for a Solar cell is:

a. Zero
b. Unity
c. Infinity
d. None

13. The power conversion efficiency of a Solar cell does depends on:

a. Short Circuit Current (J_{sc})
b. Open Circuit Voltage (V_{oc})
c. Fill-Factor (FF)
d. All

14. TEG refers to:

a. Tunable electric generator
b. Thermal enhanced generator
c. Thermo-electric generator
d. Thermo effective generator

15. Which of the following statement is correct for a vdW structure?

a. It is a layered structure
b. It is a combination of two/more than two monolayers
c. The layers are bounded with covalent forces
d. Both (a) and (b)

ANSWER KEY

1. (c)

2. (c)

3. (c)

4. (a)

5. (c)

6. (b)

7. (d)

8. (a)

9. (b)

10. (d)

11. (b)

12. (b)

13. (d)

14. (c)

15. (b)

ACKNOWLEDGEMENT

The authors are thankful to the ministry of education, Govt. of India, and IIIT-Allahabad, India for providing the necessary financial support to carry out the research work.

REFERENCES

[1]   W.H. Hunt Jr, "Nanomaterials: Nomenclature, novelty, and necessity", *J. Miner. Met. Mater. Soc.,* vol. 56, no. 10, pp. 13-18, 2004.
[http://dx.doi.org/10.1007/s11837-004-0281-5]

[2]   R. Noebe, "Nanosystems molecular machinery, manufacturing, and computation", In: *Drexler in materials and manufacturing processes..* K. Eric, Ed., vol. 12, no. 1, p. 160-162, 1997.

[3]   E. Altenkirch, "Über den Nutzeffekt der Thermosäule", *Phys. Z.,* vol. 10, pp. 560-580, 1909.

[4]   E. Altenkirch, "Elektrothermische Kälteerzeugung und reversible elektrische Heizung", *Phys. Z.,* vol. 12, pp. 920-924, 1911.

[5]   M.V. Vedernikov, and E.K. Iordanishvili, "AF Ioffe and origin of modern semiconductor thermoelectric energy conversion", *17th Int. Conf. on Thermoelectrics,* vol. 1, pp. 37-42, 1998.

[6]   M.S. Dresselhaus, G. Chen, M.Y. Tang, R.G. Yang, H. Lee, D.Z. Wang, Z.F. Ren, J.P. Fleurial, and P. Gogna, "New directions for low-dimensional thermoelectric materials", *Adv. Mater.,* vol. 19, no. 8, pp. 1043-1053, 2007.

[http://dx.doi.org/10.1002/adma.200600527]

[7]   M.G. Kanatzidis, "Nanostructured thermoelectrics: The new paradigm?", *Chem. Mater.,* vol. 22, no. 3, pp. 648-659, 2010.
[http://dx.doi.org/10.1021/cm902195j]

[8]   Z. Chen, B. Ge, W. Li, S. Lin, J. Shen, Y. Chang, R. Hanus, G.J. Snyder, and Y. Pei, "Vacancy-induced dislocations within grains for high-performance PbSe thermoelectrics", *Nat. Commun.,* vol. 8, no. 1, p. 13828, 2017.
[http://dx.doi.org/10.1038/ncomms13828] [PMID: 28051063]

[9]   L.E. Bell, "Cooling, heating, generating power, and recovering waste heat with thermoelectric systems", *Science,* vol. 321, no. 5895, pp. 1457-1461, 2008.
[http://dx.doi.org/10.1126/science.1158899] [PMID: 18787160]

[10] G.J. Snyder, and E.S. Toberer, "Complex thermoelectric materials", *Nat. Mater.,* vol. 7, no. 2, pp. 105-114, 2008.
[http://dx.doi.org/10.1038/nmat2090] [PMID: 18219332]

[11] X. Shi, L. Chen, and C. Uher, "Recent advances in high-performance bulk thermoelectric materials", *Int. Mater. Rev.,* vol. 61, no. 6, pp. 379-415, 2016.
[http://dx.doi.org/10.1080/09506608.2016.1183075]

[12] L.D. Hicks, and M.S. Dresselhaus, "Effect of quantum-well structures on the thermoelectric figure of merit", *Phys. Rev. B Condens. Matter,* vol. 47, no. 19, pp. 12727-12731, 1993.
[http://dx.doi.org/10.1103/PhysRevB.47.12727] [PMID: 10005469]

[13] A.J. Minnich, M.S. Dresselhaus, Z.F. Ren, and G. Chen, "Bulk nanostructured thermoelectric materials: Current research and future prospects", *Energy Environ. Sci.,* vol. 2, no. 5, pp. 466-479, 2009.
[http://dx.doi.org/10.1039/b822664b]

[14] Y. Wang, H. Huang, and X. Ruan, "Decomposition of coherent and incoherent phonon conduction in superlattices and random multilayers", *Phys. Rev. B Condens. Matter Mater. Phys.,* vol. 90, no. 16, p. 165406, 2014.
[http://dx.doi.org/10.1103/PhysRevB.90.165406]

[15] Y. Lan, A.J. Minnich, G. Chen, and Z. Ren, "Enhancement of thermoelectric figure-of-merit by a bulk nanostructuring approach", *Adv. Funct. Mater.,* vol. 20, no. 3, pp. 357-376, 2010.
[http://dx.doi.org/10.1002/adfm.200901512]

[16] Z. Zamanipour, X. Shi, A.M. Dehkordi, J.S. Krasinski, and D. Vashaee, "The effect of synthesis parameters on transport properties of nanostructured bulk thermoelectric p-type silicon germanium alloy", *Phys. Status Solidi., A Appl. Mater. Sci.,* vol. 209, no. 10, pp. 2049-2058, 2012. [a].
[http://dx.doi.org/10.1002/pssa.201228102]

[17] S. Saini, A. Shrivastava, and S. Singh, "An optimum thermoelectric figure of merit using Ge 2 Se 2 monolayer: An ab-initio approach", *Physica E,* vol. 138, p. 115060, 2022.
[http://dx.doi.org/10.1016/j.physe.2021.115060]

[18] S. Saini, A. Shrivastava, A. Dixit, and S. Singh, "Ultra-low lattice thermal conductivity and high figure of merit for Janus MoSeTe monolayer: A peerless material for high temperature regime thermoelectric devices", *J. Mater. Sci.,* vol. 57, no. 13, pp. 7012-7022, 2022.
[http://dx.doi.org/10.1007/s10853-022-07065-3]

[19] A.Y. Lu, H. Zhu, J. Xiao, C.P. Chuu, Y. Han, M.H. Chiu, C.C. Cheng, C.W. Yang, K.H. Wei, Y. Yang, Y. Wang, D. Sokaras, D. Nordlund, P. Yang, D.A. Muller, M.Y. Chou, X. Zhang, and L.J. Li, "Janus monolayers of transition metal dichalcogenides", *Nat. Nanotechnol.,* vol. 12, no. 8, pp. 744-749, 2017.
[http://dx.doi.org/10.1038/nnano.2017.100] [PMID: 28507333]

[20] L. Zhang, Y. Xia, X. Li, L. Li, X. Fu, J. Cheng, and R. Pan, "Janus two-dimensional transition metal

dichalcogenides", *J. Appl. Phys.,* vol. 131, no. 23, p. 230902, 2022.
[http://dx.doi.org/10.1063/5.0095203]

[21]   R. Li, Y. Cheng, and W. Huang, "Recent progress of janus 2d transition metal chalcogenides: From theory to experiments", *Small,* vol. 14, no. 45, p. 1802091, 2018.
[http://dx.doi.org/10.1002/smll.201802091] [PMID: 30596407]

[22]   R. Chaurasiya, S. Tyagi, N. Singh, S. Auluck, and A. Dixit, "Enhancing thermoelectric properties of Janus WSSe monolayer by inducing strain mediated valley degeneracy", *J. Alloys Compd.,* vol. 855, p. 157304, 2021.
[http://dx.doi.org/10.1016/j.jallcom.2020.157304]

[23]   S. Saini, A. Shrivastava, and S. Singh, "A giant thermoelectric figure of merit and ultra-low lattice thermal conductivity using Janus Ge2SeTe monolayer: A first principle investigation", *Eur. Phys. J. Plus,* vol. 137, no. 7, p. 876, 2022.
[http://dx.doi.org/10.1140/epjp/s13360-022-02996-x]

[24]   D.M. Hoat, M. Naseri, N.N. Hieu, R. Ponce-Pérez, J.F. Rivas-Silva, T.V. Vu, and G.H. Cocoletzi, "A comprehensive investigation on electronic structure, optical and thermoelectric properties of the HfSSe Janus monolayer", *J. Phys. Chem. Solids,* vol. 144, p. 109490, 2020.
[http://dx.doi.org/10.1016/j.jpcs.2020.109490]

[25]   A.K. Geim, and I.V. Grigorieva, "Van der Waals heterostructures", *Nature,* vol. 499, no. 7459, pp. 419-425, 2013.
[http://dx.doi.org/10.1038/nature12385] [PMID: 23887427]

[26]   L.A. Algharagholy, Q. Al-Galiby, H.A. Marhoon, H. Sadeghi, H.M. Abduljalil, and C.J. Lambert, "Tuning thermoelectric properties of graphene/boron nitride heterostructures", *Nanotechnology,* vol. 26, no. 47, p. 475401, 2015.
[http://dx.doi.org/10.1088/0957-4484/26/47/475401]

[27]   J. Wu, Y. Liu, Y. Liu, Y. Cai, Y. Zhao, H.K. Ng, K. Watanabe, T. Taniguchi, G. Zhang, C.W. Qiu, D. Chi, A.H.C. Neto, J.T.L. Thong, K.P. Loh, and K. Hippalgaonkar, "Large enhancement of thermoelectric performance in MoS 2 / h -BN heterostructure due to vacancy-induced band hybridization", *Proc. Natl. Acad. Sci.,* vol. 117, no. 25, pp. 13929-13936, 2020.
[http://dx.doi.org/10.1073/pnas.2007495117] [PMID: 32522877]

[28]   J. Oh, Y. Kim, S. Chung, H. Kim, and J.G. Son, "Fabrication of a MoS 2 /Graphene nanoribbon heterojunction network for improved thermoelectric properties", *Adv. Mater. Interfaces,* vol. 6, no. 23, p. 1901333, 2019.
[http://dx.doi.org/10.1002/admi.201901333]

[29]   M.H. Rahman, M.S. Islam, M.S. Islam, E.H. Chowdhury, P. Bose, R. Jayan, and M.M. Islam, "Phonon thermal conductivity of the stanene/hBN van der Waals heterostructure", *Phys. Chem. Chem. Phys.,* vol. 23, no. 18, pp. 11028-11038, 2021.
[http://dx.doi.org/10.1039/D1CP00343G] [PMID: 33942827]

[30]   S. Twaha, J. Zhu, Y. Yan, and B. Li, "A comprehensive review of thermoelectric technology: Materials, applications, modelling and performance improvement", *Renew. Sustain. Energy Rev.,* vol. 65, pp. 698-726, 2016.
[http://dx.doi.org/10.1016/j.rser.2016.07.034]

[31]   G.J. Snyder, M. Christensen, E. Nishibori, T. Caillat, and B.B. Iversen, "Disordered zinc in Zn4Sb3 with phonon-glass and electron-crystal thermoelectric properties", *Nat. Mater.,* vol. 3, no. 7, pp. 458-463, 2004.
[http://dx.doi.org/10.1038/nmat1154] [PMID: 15220913]

[32]   Tzounis, L. "Organic Thermoelectrics and Thermoelectric Generators (TEGs)", In (Ed.), Advanced Thermoelectric Materials for Energy Harvesting Applications 2019.
[http://dx.doi.org/10.5772/intechopen.86946]

[33]   C. Fritts, "On the fritts selenium cell and batteries", *Van Nostrands Engineering Magazine,* vol. 32, pp.

388-395, 1885.

[34]   E.F. Kingsbury, and R.S. Ohl, "Photoelectric properties of ionically bombarded silicon", *Bell Syst. Tech. J.,* vol. 31, no. 4, pp. 802-815, 1952.
[http://dx.doi.org/10.1002/j.1538-7305.1952.tb01407.x]

[35]   W. Robert, "Introduction to the thin film photovoltaic" symposium commemorating the 25th anniversary of the institute of energy conversion at the university of delaware", *Progress in photovoltaics: Research and applications,* vol. 5, pp. 305-307, 1997.

[36]   M.A. Iqbal, M. Malik, W. Shahid, S.Z.U. Din, N. Anwar, M. Ikram, and F. Idrees, "Materials for photovoltaics: Overview, generations, recent advancements and future prospects", In: *Thin Films Photovoltaics.,* B. Zaidi, C. Shekhar, Eds., IntechOpen: London, 2022.
[http://dx.doi.org/10.5772/intechopen.101449]

[37]   R. Tala-Ighil, "Nanomaterials in solar cells", In: *Handbook of Nanoelectrochemistry.,* M. Aliofkhazraei, A. Makhlouf, Eds., Springer: Cham, 2015.
[http://dx.doi.org/10.1007/978-3-319-15207-3_26-1]

[38]   W. Jin, Hui. Guo, Junjun. Xue, Dunjun. Chen, Guofeng. Yang, Bin. Liu, Hai. Lu, and Youdou. Zheng, "Janus Ga2SeTe: A promising candidate for highly efficient solar cells",

[39]   R. Chaurasiya, G.K. Gupta, and A. Dixit, "Ultrathin janus wsse buffer layer for w(s/se)2 absorber based solar cells: A hybrid, DFT and macroscopic, simulation studies", *Sol. Energy Mater. Sol. Cells,* vol. 201, p. 110076, 2019.
[http://dx.doi.org/10.1016/j.solmat.2019.110076]

[40]   I. Bouziani, M. Kibbou, Z. Haman, N. Khossossi, I. Essaoudi, A. Ainane, and R. Ahuja, "Two-dimensional Janus Sn2SSe and SnGeS2 semiconductors as strong absorber candidates for photovoltaic solar cells: First principles computations", *Physica E,* vol. 134, p. 114900, 2021.
[http://dx.doi.org/10.1016/j.physe.2021.114900]

[41]   F. Marco, H. Florian, D. Lukas, and P. Dmitry, "Device physics of van der Waals heterojunction solar cells", *npj 2D Materials and Applications,* vol. 2, 2018.

[42]   N. Huo, J. Kang, Z. Wei, S.S. Li, J. Li, and S.H. Wei, "Novel and enhanced optoelectronic performances of multilayer MoS2-WS2 heterostructure transistors", *Adv. Funct. Mater.,* vol. 24, no. 44, pp. 7025-7031, 2014.
[http://dx.doi.org/10.1002/adfm.201401504]

[43]   Y. Deng, Z. Luo, N.J. Conrad, H. Liu, Y. Gong, S. Najmaei, P.M. Ajayan, J. Lou, X. Xu, and P.D. Ye, "Black phosphorus-monolayer MoS2 van der Waals heterojunction p-n diode", *ACS Nano,* vol. 8, no. 8, pp. 8292-8299, 2014.
[http://dx.doi.org/10.1021/nn5027388] [PMID: 25019534]

[44]   K. Zhang, T. Zhang, G. Cheng, T. Li, S. Wang, W. Wei, X. Zhou, W. Yu, Y. Sun, P. Wang, D. Zhang, C. Zeng, X. Wang, W. Hu, H.J. Fan, G. Shen, X. Chen, X. Duan, K. Chang, and N. Dai, "Interlayer transition and infrared photodetection in atomically thin type-II MoTe 2 /MoS2 van der Waals heterostructures", *ACS Nano,* vol. 10, no. 3, pp. 3852-3858, 2016.
[http://dx.doi.org/10.1021/acsnano.6b00980] [PMID: 26950255]

[45]   F. Wang, Z. Wang, K. Xu, F. Wang, Q. Wang, Y. Huang, L. Yin, and J. He, "Tunable GaTe-MoS2 van der Waals p-n junctions with novel optoelectronic performance", *Nano Lett.,* vol. 15, no. 11, pp. 7558-7566, 2015.
[http://dx.doi.org/10.1021/acs.nanolett.5b03291] [PMID: 26469092]

[46]   N. Flöry, A. Jain, P. Bharadwaj, M. Parzefall, T. Taniguchi, K. Watanabe, and L. Novotny, "A WSe 2 /MoSe 2 heterostructure photovoltaic device", *Appl. Phys. Lett.,* vol. 107, no. 12, p. 123106, 2015.
[http://dx.doi.org/10.1063/1.4931621]

[47]   D. Jariwala, V.K. Sangwan, C.C. Wu, P.L. Prabhumirashi, M.L. Geier, T.J. Marks, L.J. Lauhon, and M.C. Hersam, "Gate-tunable carbon nanotube–MoS 2 heterojunction p-n diode", *Proc. Natl. Acad.*

Sci., vol. 110, no. 45, pp. 18076-18080, 2013.
[http://dx.doi.org/10.1073/pnas.1317226110] [PMID: 24145425]

[48]   D. Jariwala, S.L. Howell, K.S. Chen, J. Kang, V.K. Sangwan, S.A. Filippone, R. Turrisi, T.J. Marks, L.J. Lauhon, and M.C. Hersam, "Hybrid, Gate-Tunable, van der Waals p-n heterojunctions from pentacene and MoS2", *Nano Lett.,* vol. 16, no. 1, pp. 497-503, 2016.
[http://dx.doi.org/10.1021/acs.nanolett.5b04141] [PMID: 26651229]

[49]   Tsai. Meng-Lin, Su. Sheng-Han, Chang. Jan-Kai, Tsai. Dung-Sheng, Chen. Chang-Hsiao, Wu. Chih-I, Li Lain-Jong , Chen. Lih-Juann, and He. Jr-Hau, "Monolayer MoS2 heterojunction solar cells", *ACS Nano,* vol. 8, pp. 8317-8322, 2014.

[50]   A. Shrivastava, S. Saini, and S. Singh, "Ab-Initio investigations of electronic and optical properties of Sn-hBN hetero-structure", *Physica B,* vol. 624, p. 413390, 2022.
[http://dx.doi.org/10.1016/j.physb.2021.413390]

[51]   A. Shrivastava, S. Saini, P. Kumar, and S. Singh, "A potential absorber for PHz electronics using Sn/h-BN Van der Waals structure: A hybrid DFT and macroscopic investigations", *Physica E,* vol. 144, p. 115423, 2022.
[http://dx.doi.org/10.1016/j.physe.2022.115423]

[52]   B. Radisavljevic, A. Radenovic, J. Brivio, V. Giacometti, and A. Kis, "Single-layer MoS2 transistors", *Nat. Nanotechnol.,* vol. 6, no. 3, pp. 147-150, 2011.
[http://dx.doi.org/10.1038/nnano.2010.279] [PMID: 21278752]

[53]   S. Agarwal, V. Nandal, H. Yadav, and K. Kumar, "Antimony chalcogenide-based thin film solar cells: Device engineering routes to boost the performance", *J. Appl. Phys.,* vol. 129, no. 20, p. 203101, 2021.
[http://dx.doi.org/10.1063/5.0047429]

[54]   J. Deng, J. Li, Z. Yang, and M. Wang, "All-inorganic lead halide perovskites: a promising choice for photovoltaics and detectors", *J. Mater. Chem. C Mater. Opt. Electron. Devices,* vol. 7, no. 40, pp. 12415-12440, 2019.
[http://dx.doi.org/10.1039/C9TC04164H]

[55]   J. Song, Q. Cui, J. Li, J. Xu, Y. Wang, L. Xu, J. Xue, Y. Dong, T. Tian, H. Sun, and H. Zeng, "Ultralarge all-inorganic perovskite bulk single crystal for high-performance visible-infrared dual-modal photodetectors", *Adv. Opt. Mater.,* vol. 5, no. 12, p. 1700157, 2017.
[http://dx.doi.org/10.1002/adom.201700157]

[56]   Q. Zeng, X. Zhang, C. Liu, T. Feng, Z. Chen, W. Zhang, W. Zheng, H. Zhang, and B. Yang, "Inorganic CsPbI 2 Br perovskite solar cells: The progress and perspective", *Sol. RRL,* vol. 3, no. 1, p. 1800239, 2019.
[http://dx.doi.org/10.1002/solr.201800239]

[57]   M. Kim, J. Jeong, H. Lu, T.K. Lee, F.T. Eickemeyer, Y. Liu, C. In-woo, S.J. Choi, Y. Jo, H-B. Kim, S-I. Mo, Y-K. Kim, H. Lee, N.G. An, S. Cho, W.R. Tress, S.M. Zakeeruddin, A. Hagfeldt, J.Y. Kim, M. Grätzel, and D.S. Kim, "Polymer-stabilized SnO2 quantum dot electron transporters for efficient perovskite solar cells", *Science,* p. 21, 2022.

[58]   M.M. Tavakoli, G. Azzellino, M. Hempel, A.Y. Lu, F.J. Martin-Martinez, J. Zhao, J. Yeo, T. Palacios, M.J. Buehler, and J. Kong, "Synergistic roll-to-roll transfer and doping of cvd-graphene using parylene for ambient-stable and ultra-lightweight photovoltaics", *Adv. Funct. Mater.,* vol. 30, no. 31, p. 2001924, 2020.
[http://dx.doi.org/10.1002/adfm.202001924]

[59]   H. McDaniel, N. Fuke, N.S. Makarov, J.M. Pietryga, and V.I. Klimov, "An integrated approach to realizing high-performance liquid-junction quantum dot sensitized solar cells", *Nat. Commun.,* vol. 4, no. 1, p. 2887, 2013.
[http://dx.doi.org/10.1038/ncomms3887] [PMID: 24322379]

[60]   D. Li, P.C. Chang, C.J. Chien, and J.G. Lu, "Applications of tunable TiO_2 nanotubes as nanotemplate and photovoltaic device", *Chem. Mater.,* vol. 22, no. 20, pp. 5707-5711, 2010.

[http://dx.doi.org/10.1021/cm101724t]

[61]   C. Chen, Y. Lu, E.S. Kong, Y. Zhang, and S.T. Lee, "Nanowelded carbon-nanotube-based solar microcells", *Small,* vol. 4, no. 9, pp. 1313-1318, 2008.
[http://dx.doi.org/10.1002/smll.200701309] [PMID: 18702123]

CHAPTER 14

Nanotechnological Advancement in Energy Harvesting and Energy Storage with Hybridization Potentiality

Shikha Kumari[1], Talapati Akhil Sai[1] and Koushik Dutta[1,*]

[1] PDPM Indian Institute of Information Technology, Design and Manufacturing, Jabalpur, India

Abstract: Decaying sources of non-renewable energy (fossil fuel) turned the research focus to other natural renewable resources. Among these, solar power is advantageous in terms of area and maintenance cost. However, the high installation cost of conventional solar cells restricts individual uses; alternatively, lightweight and flexible solar cells evolved. Among them, Dye-Sensitized Solar-Cell (DSSC) are inexpensive and considered nanotechnological advancement. Step-by-step improvisation of the photo-conversion efficiency has been discussed in light of nanoengineering on metal oxides. Simultaneously, the dependence of wavelength on the choice of dye has also been focused opting for a particular application field. Energy storage device (solid-state batteries and/or supercapacitors) is an inevitable part of solar-cell for ensured use at required time and space. With the help of nanotechnology, the major problems of storage efficiency are critically pointed out with possible way-out. In this connection, the adopted nanoengineering aspects are extensively discussed considering improvements in the battery capacity, cycle life, and charge and discharge cycles with the highest degree of safety. Linking with the nanostructures, the nanotubular array provides a higher specific surface area maximizing the performance for both the DSSCs and energy-storage devices, as anode material. Again, the unidirectionality of the carrier transport path enhances electron collection. The present endeavor includes such research instances probing towards the amalgamation of these two technologies to indicate the futuristic direction of the self-chargeable storage unit. The present scope is designed broadly in three sections, where the first section deals with the step-by-step improvement of DSSC with a prime focus on the oxide nanotube-based photoanode. The second section deliberates on the research trends for storage devices with the nanotube-based anode. In the last section, the unification of these two technologies within a single chip or area using a common anode is the main emphasis to enhance the utility and green approach for the future world.

Keywords: Energy harnessing, Energy storage, DSSC, Li-ion battery, Na-ion Battery, *In-situ* charging, In-built energy storage.

---

* **Corresponding author Koushik Dutta:** PDPM Indian Institute of Information Technology, Design and Manufacturing, Jabalpur, India; E-mails: koushikdutta@iiitdmj.ac.in, koushikdutta71187@gmail.com

PREAMBLE

The modern age is going towards automation to facilitate human efforts and vastly dependent on electric power consumption. Electric power generation, conventionally and dominantly, depends on fossil fuel (coal, petroleum, *etc.*), which is not either renewable or everlasting. Additionally, the adverse effect of conventional electric energy generation creates frowning for environmentalists. In this connection, solar power harnessing is the most popular among the ever-present and renewable energy forms, to address the concern. However, the climate condition restricts its continuous generation process, and the spatial distance is the main cause behind storing the energy. Hence, Energy Harvesting devices and Energy Storage devices are inevitable parts of renewable and green energy deployment. For this reason, the amalgamation of these two devices is necessary for a smarter future. Nanotechnological advancement arrowed towards such possibility within a single chip, eventually wearable and flexible form, providing the facility of mobility of the gadgets. In this proposed chapter, the cost-effective way of adjunction of energy harnessing and storage system will be discussed in light of nanostructural evolution for efficient device design with materialistic choice for futuristic applications.

ENERGY HARNESSING IN A COST-EFFECTIVE APPROACH

Restricted source of fossil fuel and their adverse effect through the conventional way of consumption wrinkles the thought process towards sustainable and renewable energy sources. Further, the advancement of nanotechnology thrusts us towards miniaturization and individual deployment of energy requirements. Through the connection, modern research is being focused on the cost-effective approach towards green technologies, which can be reformed towards individual deployment targeting large-scale effects on future generations. For this, solar power generation supported the integrated work with appreciable acceptability among all other renewable energy sources. This section will deal mainly with the cost-effective approach of solar power generation with the evolution strategies adopted by the scientific world.

Introduction

Energy consumption is an inevitable part of human society, and its consumption through different modes (transport, building, and/or individual gadgets) are in increasing pace as the population increases and technologies are invented [1]. In other words, no such fields exist that are powered solely by energy. We are far more reliant on fossil fuels for this purpose [2]. Generally, conventional energy resources, *i.e.*, crude oil and its derivatives, coal-related products, and natural gases, are identified as non-renewable energy sources and are together referred to as fossil fuels [3]. Compared to renewable energy sources like solar or wind, resources like coal and oil generally give us more energy [4]. According to the statistics, as of May 31, 2022, the total domestic coal output for the years 2021–22 was 137.85 MT, which is a 28.6% increase over the production of 104.83 MT during the same time the previous year [1 - 3]. In Fig. (**1**) the pie chart shows the percentage of generation of electricity by various resources in India in the year 2022 [1 - 3].

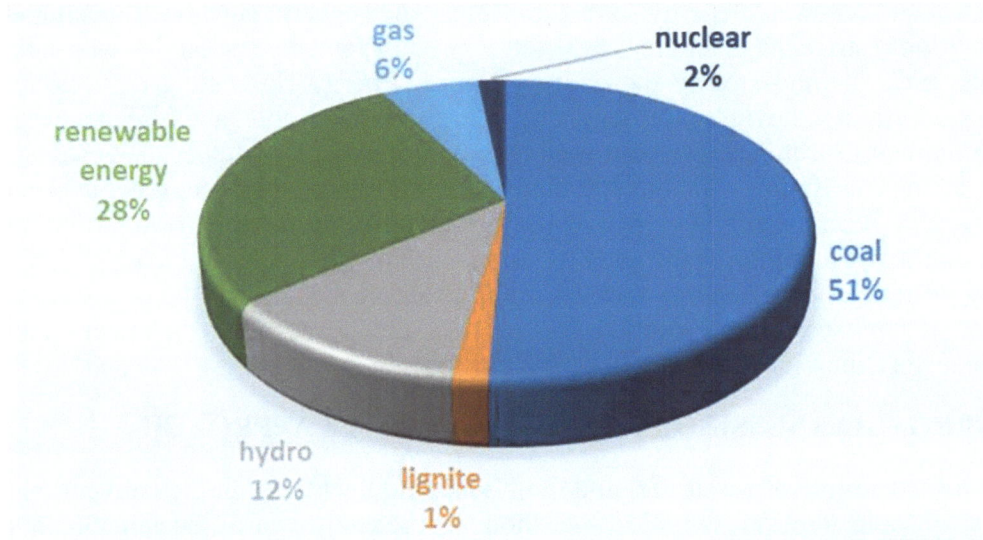

Fig. (1). Sources of electricity in India by installed Capacity 2022.

Non-renewable resources have many drawbacks in addition to their many advantages. The time-consuming nature of non-renewable energy is one of its main drawbacks [1 - 4]. It takes a long time to mine coal, look for oil, install oil drills, construct oil rigs, insert pipes to extract natural gas, and transport it [4, 5]. It also demands great human power as well as cost deployment [5]. As non-renewable energy sources like fossil fuels release chemicals like carbon monoxide, they can be harmful to the environment, and their harsh effects must affect human society in terms of health hazards [5]. A number of health concerns are more likely to affect employees who work in coal mines or oil drilling. As a result, there are numerous illnesses, accidents, and even fatalities [4, 5]. When burned, energy sources like coal, oil, and natural gas generate a lot of carbon dioxide. These substances are contributing to the ozone layer's quick demise. Sulphur oxide and other oxidants generated by the burning of fossil fuels change the precipitation to acid rain [4, 5]. The issues of global warming and climate change are still top-of-mind because fossil fuels emit a lot of greenhouse gases (GHGs), particularly carbon dioxide, into the atmosphere, which has an adverse effect on the environment [5, 6]. Also, non-renewable energy sources must form over billions of years; therefore, they are slowly but surely disappearing from the planet [3, 5]. It could be selfish to use non-renewable resources carelessly without considering the needs of future generations. In this regard, usage of renewable energy resources, like, geothermal, tidal, solar, wind, biomass, and biofuels, may possibly be considered to fulfill the rising demand for energy with an environment-friendly sustainable attire.

Being the most popular renewable energy resource, solar energy is often favored over other renewable energy sources since it can be deployed everywhere, whereas others face some limitations (may be artificially created) [5, 7]. It is also less noisy than wind energy [5, 7]. Its primary applications are in the fields of toys, watches, *etc.* They further utilize electrified fences. Additionally, it is utilized in the field of remote lighting systems. In the area of portable power supply, this might be used [8]. The portability and higher space density of these enable power generation in space for satellite operations [8].

Prior to delving into the technicalities of improving performance and lowering costs of solar energy harvesting schemes, it is important to have a basic understanding of the traditional solar cell and its performance parameters. The operation of solar cells and their generation-by-generation evolution towards cost-effectiveness are briefly described in the next subsection.

Basic Principle of Solar Cell

Solar energy is the natural source of all kinds of energy on the earth, and hence the free and abundant source can be channelized for energy harvesting and which is obtained by the Solar Cell [7, 8]. The solar cell is a special kind of semiconductor-based device that converts the photon into electrical ones [7]. Basically, photons can be described as the quantum particle of the electromagnetic field, including electromagnetic radiation such as light and radio waves [7, 8]. It is massless and travels at the speed of light in a vacuum [7]. The working of solar cells is related to the photovoltaic effect, which is the procedure of electric energy generation by absorbing the photon particles from light *via* a physio-chemical mechanism [7, 8].

By virtue of the PN junction, the voltage drop across the solar cell is limited to one (~0.6 V), and the current is a function of the area of the cell. Thus, upon illumination, one solar cell generates very little current, which in turn results in insufficient electrical power for operating modern electronic and electrical load/devices. Therefore, many of these solar cells are joined together to suffice the practical requirement. In this connection, the series connection of solar cells provides enhanced voltage, whereas the parallel connection is responsible for higher current. Interconnected solar cells are referred to as a 'PV module' (Fig. **2**) [9].

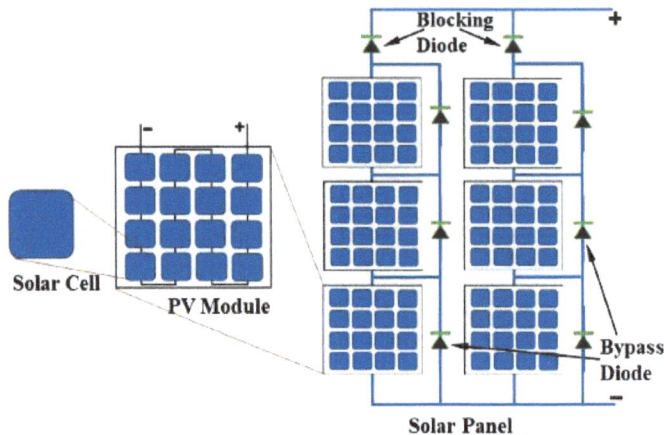

Fig. (2). Schematic view of Array deployment of solar cell module for commercial purposes with connection strategy.

For series interconnection, partial shadowing (low illumination) of the solar cell is the prime concern which results in lower power delivery from the PV module. To compensate for the effect, the bypass diode is connected to ensure the current level will not be restricted through the series connection of solar cells. On the other hand, a parallel connection is immune to such a problem, as the lower current is not deciding factor for such a connection. But, low voltage and high current combination lead to high resistive loss. Thus, the series connection of the solar cells with the bypass diode combination is the preferred one, in terms of batch fabrication and integration facility. The term 'solar panel' refers to a collection of several PV modules (as shown in Fig. (**2**)) [9]. Here, the blocking diode ensures higher resistance to faulty string (series connected modules) so that no power will deliver from the flawless string(s). Further, the bypass diode is connected in parallel to each module to protect from the partial shadowing effect.

The conventional or first-generation solar cell is basically a PN junction having a larger surface area [5, 9]. Without intruding on the conventional working mechanism of basic PN junction-based solar cells, the completeness of the evolution cannot be attained. For this purpose, the basic concepts can be discussed with the aid of Fig. (**3**) where the main focus will be concerned with the energy-band diagram of such a device.

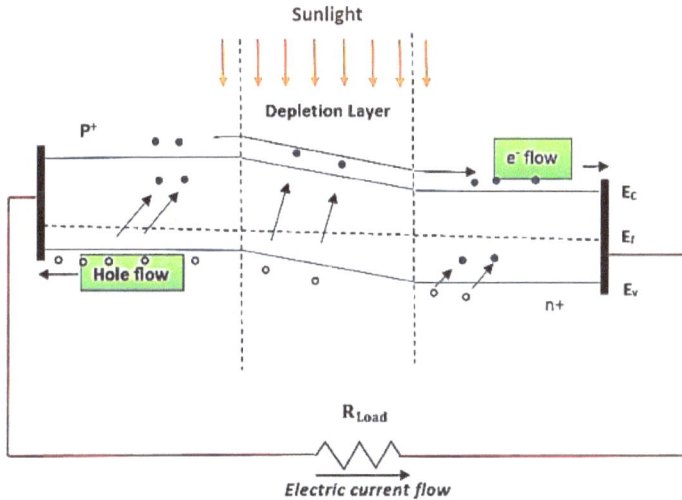

Fig. (3). Energy-band diagram of conventional solar cell working through electron-hole pair generation upon illumination.

A solar cell consists of a p-n junction with moderate doping concentration to achieve a large depletion surface [5, 9]. The p-type semiconductor is created by adding tetravalent elements, such as gallium, boron, *etc.*, as impurities, while the n-type semiconductor is created by adding group V elements as impurities to the intrinsic semiconductor [5, 9, 10]. Due to the concentration gradient (of charge carriers of different polarities), holes and electrons diffuse in the n-type and p-type semiconductors, respectively, forming a depletion zone owing to the balancing electric field [9, 10]. In a semiconductor diode, this region serves as a barrier that prevents electrons from moving from the n-part to the p-side. From the instance of sunlight exposure, absorption photon energy generates electron-hole pairs from the molecules at the depletion region, and due to the electric field (at the depletion region), the generated mobile carriers are collected to the corresponding electrode [10, 11]. Hence, current flows through external circuitry. However, all of the photon energy is not successfully converted to electrical energy, as some collisions and/or recombination generally occur [10, 11]. This defines the efficiency of the solar cell that what percentage of photon energy is converted to electrical one equation **(1)** [5, 9 - 11]. It is quite suggested that if a small reverse bias is applied to this p-n junction, the larger depletion area provides higher efficiency [5, 9].

$$\eta = \frac{\text{Generated Electric Power}}{\text{Incident Photon Power}} \times \mathbf{100\%} \qquad\qquad \textbf{(1)}$$

Solar Cell Evolution

The stepping stone towards solar cells can be related to the discovery of the photoconductivity of selenium in 1873, which was devised into the first solar cell, with $\eta < 1\%$, in 1883 [12]. As per the required applications, the efficiency did not address the practicability to address which silicon-based experiments were investigated under the umbrella of renowned Bell Labs [12]. Here, unintentional cracking probably acted as a PN junction and ignited the basis of the solar cell [12]. When it was exposed to light, a current started flowing through this sample, and it possesses $\eta \approx 1\%$. However, a particular scientific realization was achieved at a later time by adding various impurities in silicon [12, 13]. In this case, a silicon sample containing gallium as an impurity settled on arsenic and boron, resulting in a p-n junction. It was significantly more effective than any prior solar cells, turning sunlight's energy into electricity with a 6% efficacy. This empirical success instigated the aspiration for harnessing infinite energy toward the well-being of the civilization facing towards a green technological era.

Solar technologies are commonly grouped into "generations" based on how efficient they are [12]. The "first generation" of solar cells is made of crystalline silicon (c-Si) [12, 13]. However, because crystal Si is expensive and solar cells take a while to pay for themselves, it will be necessary to lower upfront costs before solar is more widely used [13]. Then second-generation solar cells entered the research scenario which had a different structure than the previous solar cell. These types of cells are based on thin films of amorphous, nanocrystalline, or polycrystalline semiconductors [12, 13]. Thin active layers enable simpler production by lowering manufacturing and module costs. They create the possibility for flexible and translucent solar cells [12 - 14]. These semiconductors are deposited on the substrates for mechanical support. The performance of the second-generation solar cells is discussed below one by one with the material perspective:

Copper Indium Gallium Selenide (CIGS): The best efficiency for a genuine thin-film solar cell has been attained using CIGS cells and is comparable to commercial crystalline silicon at 22.6% [14]. However, because indium is so rare, CIGS can be costly and challenging to manufacture. It is difficult to optimize CIGS cells due to the complex stoichiometry and many phases, which may limit future large-scale production.

Cadmium Telluride (CdTe): Semiconducting CdTe offers comparable efficiency to CIGS at 22.1% [14]. Additionally, due to the optimal band gap of 1.43 eV, it offers benefits, including efficient absorption and little energy losses. Low-temperature manufacturing of CdTe solar cells allows for flexible and economical production, and thus, this is preferred over complex manufacturing of Si-based solar cells [14]. However, the toxicity of cadmium and the rareness of tellurium limit general use [14]. These elements could indicate future problems with long-term, massive production.

Gallium Arsenide (GaAs): For the direct bandgap semiconductor, GaAs also possess near to optimum bandgap of 1.42 eV and offer the highest efficiency (28.8%) [14]. Technically, it is a thin-film cell deposited on GaAs wafer and hence, in terms of cost-effectiveness and complex fabrication, incurs the bottlenecks of c-Si cases [14]. However, for a highly efficient system, this material is preferred most. As an example, for operating satellites, this type of solar cell finds appreciable applicability [14].

Meanwhile, many researchers have attempted light harvesting concepts using Dye-Sensitized Solar Cells (DSSCs), a specific type of inexpensive solar (photoelectrochemical in nature) cell that effectively transforms different light spectra into electrical energy [15 - 24]. Considering the working principle and used material, this technology has been included in the third-generation solar cell due to cost-effectiveness, tunability through nanoengineering, flexibility, and its variety of wavelength (of light) absorption capability [16, 17]. Due to their material and straightforward structure, DSSCs offer a possible answer to corresponding issues of integration and cost-effectiveness [16, 17]. In the following sub-section, detailed evolution is discussed in an elaborate manner.

Prospectus of DSSC

The DSSC has emerged as one of the most promising alternatives to silicon solar cells in terms of affordability, maximum efficiency, and ease of product [15 - 19]. The DSSC is also known as Grätzel cell as per the innovator's name [17 - 19, 25]. For this creation, Michael Grätzel received the Millennium Technology Prize in 2010. The DSSC uses a wide band gap semiconductor and has a variety of appealing qualities, including being semi-flexible, semi-transparent, and easy to construct [18]. In a typical solar cell, silicon functions as both a photoelectron source and a generator of the electric field needed to separate the charges and produce a current. However, in DSSC, the mesoporous TiO_2 nanomaterial semiconductor transports charges, while photoelectrons are produced by photosensitizers or dyes [17 - 19]. This discussion is categorized through the basic principle behind DSSC operation followed by the research evolution towards state-of-arts to achieve higher efficiency to be compatible with the commercial c-Si-based solar cell.

Structure and working principle of DSSC

The structure of the Grätzel cell is such a way that makes it is different and advanced as compared to a conventional solar cell. On the top, there is a transparent FTO or fluoride-doped tin dioxide (SnO_2:F) on which a thin layer of a metal oxide semiconductor, having a wide energy band gap, is generally deposited, and the metal oxide is immersed into a mixture of a photosensitizer or dye and a solvent [19, 25]. The counter electrode (consisting of platinum or graphite) is used to seal the cell and complete the electric connection as a collector [17 - 19]. Generally, iodide electrolyte is used in between the photoanode and the counter electrode [15, 16]. In order to stop the electrolyte seepage, these two plates are linked and sealed together [15 - 19]. The following diagram in Fig. (**4**) shows the DSSC structure for a quick conception.

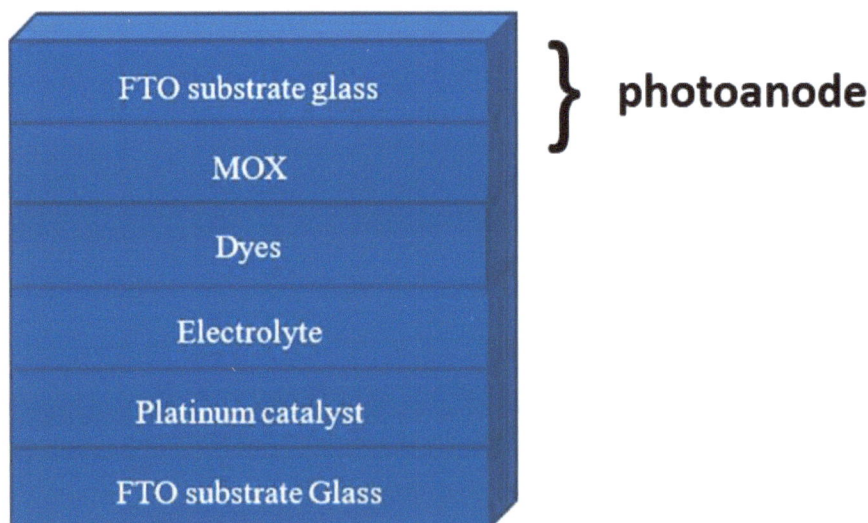

Fig. (4). Schematic view of DSSC structure showing different layers of generally used materials.

DSSC mainly consists of the following components:

(a) <u>Transparent and conductive substrate:</u> The Florine-doped tin oxide (FTO) thin film is a type of TCO, *i.e.*, transparent oxide coatings, which have attracted very much attention due to their high conductivity and optically transparent nature [15 - 19]. FTO has attracted great interest because of its thermal stability, chemical inertness, and also its high forbidden energy gap, and also their production cost is low. Generally, doping and the atomic ratio of F/Sn affect the structures, morphological, compositional, electrical, optical, and nanomechanical properties of the films. By doping and increasing F/Sn ratios, the resistivity of FTO can be decreased. FTO glass allows sunlight to enter the cell, for this, it should have very high transparency (> 90%) and be inert so that it cannot react with any chemicals in the cell. This serves as one electrode and provides mechanical strength to the DSSC.

(b) Working electrode: Although we have solar panels which employ silicon but are very expensive due to their complicated production process, which makes this technology not suitable for industrial and other commercial applications [13]. However, several thin films of organic solar cells have been invented recently that possess the advantages of being lightweight, flexible, and low module cost over conventional silicon solar cells [15 - 19]. In terms of cost-effectiveness, optimum efficiency, and easy fabrication, the DSSC, which is a subclass of thin-film organic solar cells, proved to be one of the most promising alternatives to silicon solar cells. Metal Oxide semiconductor (MOX) is used in DSSC instead of using silicon [15 - 19]. Various MOX (such as TiO_2, ZnO, SnO_2, *etc.*) has experimented with the DSSC application [15 - 25]. Among all of them, mesoporous TiO_2 is the most popular semiconductor for DSSC [15 - 19, 25]. It has the highest refractive index of any other material [26, 27]. Initially, TiO_2 nanoparticles (NPs) with a band gap of 3.0-3.2 eV were employed as the photoelectrode, which offers fast electron transfer and a large surface area [25]. The larger the surface area of TiO_2, the more dye molecules will be attached to it; more will be more absorption of the photon, and hence, higher photocurrent efficiency can be expected [25, 27]. But the bottleneck due to that grain boundaries, in the case of NPs, may lead to electron recombination, which results in a photocurrent loss and becomes a prominent loss for absorbing light at the near-infrared region [25, 26]. Hence, TiO_2 nanotubes (TNTs) have been experimented with to overcome the drawbacks of nanoparticles. TNT shows an amended effect due to light scattering, fast 1-dimensional electron transport, and low charge recombination with simple control geometry as compared to NPs [25 - 29]. In addition, the tubular structure of mesoporous TiO_2 provides a relatively large surface area, due to which more dye molecules can be adsorbed by the TiO_2 NT surface through inter and intra-nanotube spaces [25 - 29].

(c) Photosensitizer or Dye: This is the actual material responsible for photon absorption. There are varieties of dyes (ruthenium-based, porphyrin-based, *etc.*) that provide different options for choosing [20 - 24]. The redox flow is defined by the HOMO (highest occupied molecular orbital) and LUMO (lowest unoccupied molecular orbital) levels which in turn identify the absorbable wavelength [20 - 24]. The photo-generated electrons are transferred to the working electrode owes oxidation, whereas the charge compensation by reduction reaction has happened on the electrolyte side [18 - 20]. Hence, photo-to-electrochemical conversion is guided by the dye molecules within the DSSC.

(d) <u>Electrolyte:</u> Electrolytes play an important role in any electrochemical device. Redox couple, solvent, additives, ionic liquids, and cations are the five main components of electrolytes. Some properties should be possessed by an electrolyte [18, 19]. The properties are:

I. Redox couples should regenerate the oxidized dye successfully.
II. It should be stable for the long term, and durable electrically, chemically and thermally.
III. It should not be corrosive in nature. It should provide fast diffusion of charges so as to enhance conductivity.
IV. There should not be an overlay of absorption spectra with that of a dye.
V. It should be less viscous, with a high-dielectric constant and high-temperature tolerance.

(e) <u>Counter electrode:</u> The counter electrode collects of charges flowing from the photoanode through an external wire to it [23 - 25]. For successful operation, it is desirable that the reduction process rate at the counter electrode should be higher [23 - 25]. Platinum (Pt) electrode is a popular choice owing to its catalytic properties, high thermal stability, high charge transfer capability, and inertness, which results in high yield [30]. However, it is also very costly, so the replacement of Pt is much required [30, 31]. Thus, metal alloys and other carbon-based electrodes are preferred for use in DSSCs [30].

The working principle of DSSC can be understood in four phenomena: (i) absorption of light, (ii) electron injection, (iii) carrier transportation, and (iv) carrier collection, depicted schematically in Fig. (**5**).[30]. Firstly, the incident light (photon), with energy hv (h is Planck's constant and v is frequency), enters through FTO glass, and gets absorbed by the dye molecules and due to the photon-induced excitation, electrons of dye reached to LUMO, as per equation (**2**).

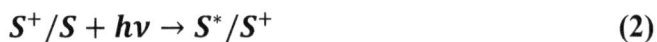

$$S^+/S + hv \rightarrow S^*/S^+ \tag{2}$$

These energized electrons move to the conduction band of TiO_2 equation (**3**). The oxidation of dye molecules is completed due to electron donation.

$$S^*/S^+ \rightarrow S^+/S + e^- \; (TiO_2) \tag{3}$$

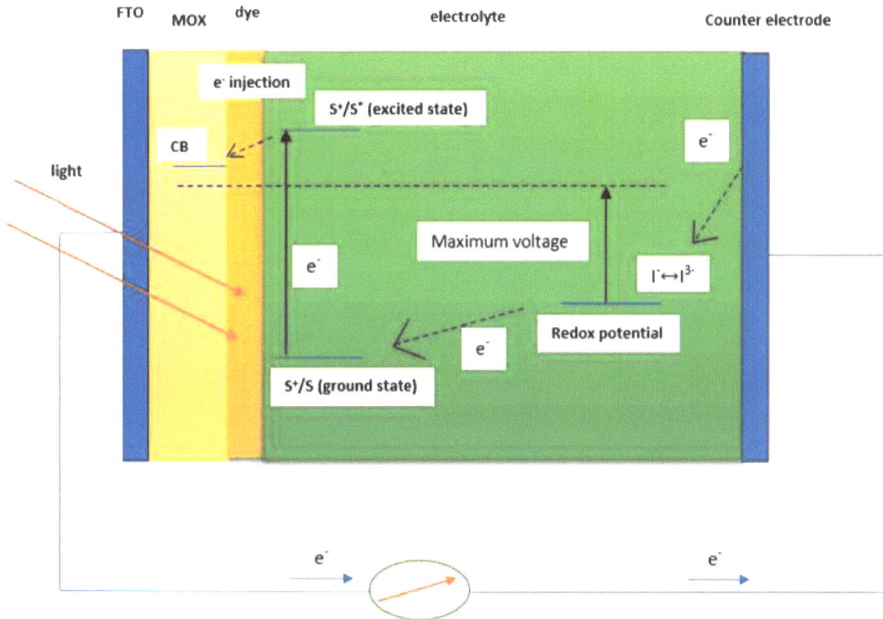

Fig. (5). Schematic view of DSSC Working principle.

There is an external wire connecting both electrodes, *i.e.*, the photoanode and the counter electrode (CE). The electron diffuses towards the back contact from the photoanode through the wire. The counter electrode releases the redox mediator ions (I^{-3}) in an electrolyte which in turn stabilizes the dye equation **(4)**.

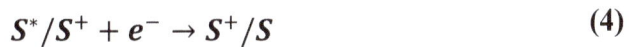

$$S^*/S^+ + e^- \rightarrow S^+/S \qquad \textbf{(4)}$$

Again, the redox mediator reduces to I^- as per equation **(5)**.

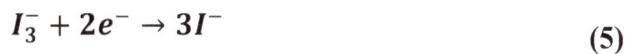

$$I_3^- + 2e^- \rightarrow 3I^- \qquad \textbf{(5)}$$

The performance of a dye-sensitized solar cell can be characterized by using a few parameters such as photo-conversion efficiency (PCE, %), short circuit current (J_{SC}, mAcm^{-2}), open circuit voltage (V_{OC}, V), maximum power output (P_{max}), overall efficiency (η, %), and fill factor (*FF*). A simple one-diode model of DSSC was reported by Murayama *et al.* [32] to estimate electrical behavior (empirical J-V characteristics) with the aid of fitting *via* Lambert W Function. Elaborate understanding can be inferred from the following discussion with the aid of Fig. **(6)**.

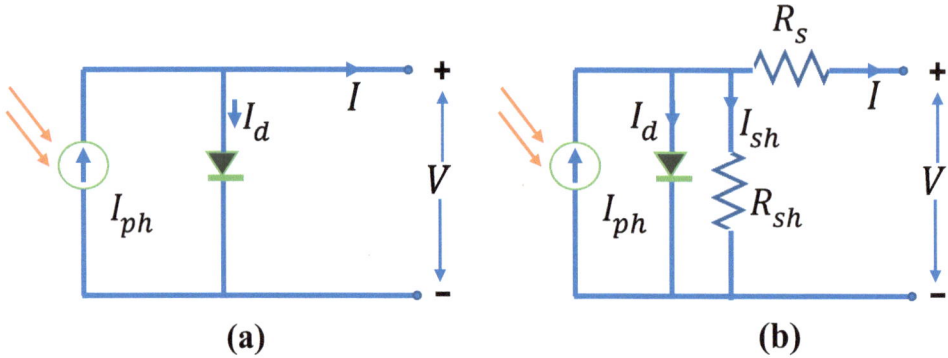

Fig. (6). Equivalent circuit of DSSC for **(a)** ideal condition and **(b)** with a practical aspect.

Ideally, the solar cell is a lossless device, and the equivalent circuit can be considered, as shown in Fig. (**6a**) (no resistance). The photogenerated current can be represented as a current source (I_{ph}) in parallel with a diode. From the circuit Fig. (**6a**), equation (**6**) can be derived with the standard diode current equation.

$$I = I_{ph} - I_d = I_{ph} - I_o \left(exp \left(\frac{qV}{nK_BT} \right) - 1 \right) \quad [A] \tag{6}$$

Where I_o is dark/reverse saturation current of the diode, q is the electronic charge, K_B is the Boltzmann constant, T is the absolute temperature, and n is the ideality factor of the diode. Under the non-illumination condition, $I_{ph}=0$, and the diode current is dominated by the reverse saturation current (I_o).

With practical consideration of solar cells, the loss elements are generally represented as resistances, *i.e.*, shunt resistance (R_{sh}) and series resistance (R_s), as depicted in Fig. (**6b**). R_{sh} is due to leakage across the PN junction around the edge, crystal defects, and impurities. Whereas, R_s is due to the bulk resistance of the semiconductor coupled with the bulk resistance of metal electrodes and contact resistance between the semiconductor and metal. Higher reverse bias resistance (R_{sh} here) ensures sufficient power delivery at the load and thus, the reverse bias condition of the diode is sustained. Otherwise, it acts as a sink for the current, and that will reduce power to the load. As per the circuit analysis of Fig. (**6b**), equation (**7**) can be written.

$$I = I_{ph} - I_d - I_{Rsh} \quad [A] \tag{7}$$

It is clear that the voltage drop across the R_{sh} will be equal to the summation of load voltage and voltage drop in R_s, which is again the voltage drop across the diode (*i.e.*, $V_d = V + IR_s$). Thus, modifying the diode equation, equation (7), can be rewritten as in equation (8).

$$I = I_{ph} - I_o\left(\exp\left(\frac{q(V+IR_s)}{nK_BT}\right) - 1\right) - \left(\frac{V+IR_s}{R_{sh}}\right) \quad [\text{A}] \tag{8}$$

Plotting of equation (8) results in the current (density) *vs.* voltage curve (Fig. 7), and it is generally characterized in terms of short circuit current density (J_{SC}), open circuit voltage (V), fill Factor (FF) as depicted below [32].

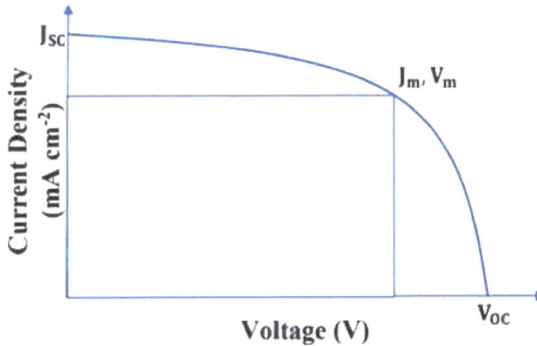

Fig. (7). J-V characteristic of DSSC/solar cell.

By definition, these parameters are described in the preceding points.

(a) <u>Current Density</u> (J_{SC}): This is the analytical point on the J-V curve, which is indicated by the amount of current flown between two electrodes (only due to photon absorption) when no potential difference is applied. As the external circuit is shorted, it is termed the short-circuited one. This factor depends upon various parameters, such as the incident photon flux on the exposed cell surface, the spectrum of the incident light, the effective active area, and the optical properties of the solar cell.

(b) <u>Open Circuit Voltage</u> (V_{OC}): If the two electrodes are in open circuit condition, *i.e.*, no path for current flow, the photon generated current will cause maximum potential drop across the Solar Cell. This is also an analytical measurement and extreme maxima. This defines the maximum deliverable voltage by the solar cell.

The above two points, however, are not practically measurable but define two extreme parts, which is essential to estimate usable range of the Solar Cell.

(c) <u>Fill Factor (FF)</u>: In regard to the previous discussion, to address the practicability of a Solar Cell, the fill factor defines the relatability between analytical and practical data. The fill factor describes the maximum power delivery ability of the corresponding solar cell Fig. (7). describes the expression of the Fill Factor mathematically (equation (9)).

$$FF = \frac{J_{max} \times V_{max}}{J_{sc} \times V_{oc}}$$
(9)

(d) <u>Cell Efficiency</u>: It is defined as the ratio of maximum output power to the incident photon power and mathematically expressed as in equation (10). Incident power is generally correlated by the photon energy. The incident power is generally measured in the laboratory environment by controlled radiation.

$$\eta \% = \frac{J_{sc} \times V_{oc} \times FF}{P_{incident}} \times 100 \%$$
(10)

Evolution

Historically, Grätzel invented the DSSC with TiO_2 nanoparticles as the active anode. The cell was illuminated with controlled sunlight exposure which yields 7.9% of efficiency [25]. It was investigated that efficiency decreased with increasing intensity. As a comparison, 83 Wm^{-2} light intensity provides 7.9% efficiency, whereas, for 750 Wm^{-2} light intensity, efficiency decreased to 7.12% [25]. Apart from TiO_2 nanoparticles, different semiconducting substrates/anodes were also investigated [25]. Dou *et al.* [20] proposed Bi_2Te_3 nanotubes for an anode of DSSC containing 0.3 mmol/L N719 dye which causes only 4.12% efficiency. Furthermore, ZnO nanotubes were also used as an active anode for DSSC. Here, only 1.6% efficiency resulted in and suggested the variation of nanostructures (from NPs to NTs) does not improve the efficiency of DSSC [17]. A DSSC fabricated by Bhande *et al.* [15] has obtained 1.1% only with SnO_2 NPs coated with N719 dye. However, the literature suggests the suitability of TiO_2 material for efficient DSSC.

To probe further, the research inventions focused mainly on two methodologies: (i) variations in the nanostructure [33], and (ii) treatment and/or decoration of TiO_2 anode [34]. As an example, Wang *et al.* [34] performed the influence of HCl pre-treatment of TiO_2 mesoporous film with N719 dye and observed an increase in efficiency to 10.5%.-On the other hand, a study on the effect of different nanostructures was reported by Ramakrishnan *et al.* [33]. Here, TiO_2 Nanoparticles, Nanoflakes, and Nanotubes were used as the active anode for fabricating DSSC. The comparison Table 1 is shown here for ready reference.

Table 1. Comparison table of efficiencies of DSSCs based on different nanostructures of TiO$_2$.

Sl. No.	Nanostructure	Pore Size	J_{sc} (mAcm^{-2})	V_{oc} (V)	FF	η (%)
1.	Nanoparticles	5	10.95	0.746	0.66	5.39
2.	Nanoflakes	7.8	12.90	0.728	0.68	6.39
3.	Nanotubes	7	13.75	0.748	0.70	7.20

From Table **1**, it can be observed that the TNTs are beneficial to obtain higher efficiency of DSSC, as this nanotubular structure possesses a higher specific surface area for favourable charge transfer phenomena over nanoparticle structure. The random electron transport, more grain boundaries, more resistances at the interfaces of grain boundaries, *etc.*, in the case of nanoparticles, are possible reasons behind such deteriorated performance [16, 19]. Interesting work was reported on the comparable performance of ZnO NTs and TiO$_2$ NTs, where both of the nanostructures were synthesized by sol-gel method and the corresponding DSSC was fabricated with 0.5 mmol/L N$_3$ in iodide electrolyte (0.5 M LiI, 0.05 M I$_2$) [18]. From this, it can also be concluded that the TNT array is a better candidate over ZnO NTs, due to its ordered alignment. The detailed concept of anode material for the DSSC is tabulated in Table **2**, as per the best of our knowledge, following the technical scenario of the recent world.

Table 2. Present Technological scenario on MOX-based DSSC focusing on the corresponding efficiency.

Sl. No.	Material	Synth. Method	Dye	Electrolyte	J_{sc} (mAcm^{-2})	V_{oc} (V)	FF	η (%)	Refs.
1.	SnO$_2$ NP	X	N719	NA	5.472	0.49	0.40	1.1	[15]
2.	TiO$_2$ NT Length = 5 μm = 10 μm = 15 μm = 20 μm	Anodic oxidation	N719	LiI + DMPII + I$_2$ + GuNCS + TBP in acetonitrile	2.02 3.02 3.95 5.2	0.6 0.69 0.71 0.70	0.42 0.49 0.55 0.54	0.50 1.03 1.54 1.96	[16]
3.	ZnO NT	Anodic oxidation method	N719	0.5 M LiI,0.05mmol I$_2$, 0.5 M TBP	3.3	0.739	0.64	1.6	[17]
4.	TiO$_2$ NT ZnO NT	Sol-gel method	N3	LiI + I$_2$	5.75 4.7	0.48 0.39	0.76 0.65	2.1 1.2	[18]
5.	TiO$_2$ NT	Anodic Oxidation	N719	LiI + DMPII + I$_2$ + TBP in acetonitrile	4.3 6.5	0.7 0.67	0.62-0.66	2.04 2.30	[19]

(Table 2) cont.....

Sl. No.	Material	Synth. Method	Dye	Electrolyte	J_{sc} (mAcm^{-2})	V_{oc} (V)	FF	η (%)	Refs.
6.	Bi$_2$Te$_3$ (wt%) NT 0.0 0.5 1.0 1.5 2.0 2.5	Anoidic Oxidation	0.3mmol/L N719	LiI + I$_2$ + TBP	8.189 10.403 11.040 11.767 9.957 9.178	0.656 0.645 0.639 0.637 0.623 0.611	0.552 0.558 0.561 0.570 0.551 0.537	2.96 3.74 3.96 4.27 3.41 3.01	[20]
7.	Deposited RGO nanosheets on TiO$_2$ NT BT-TiO$_2$ BT-TiO$_2$-RGO	Anodic Oxidation	N719	I- /I3- GTC in acetonitrile and valeronitrile	11.37 12.51	0.72 0.77	0.645 0.624	5.28 6.01	[21]
8.	TiO$_2$ NT (10, 20, 30 μm length)	Anodization	N3	NA	14.03 19.27 16.97	0.63 0.66 0.64	0.577 0.520 0.501	4.81 6.58 5.45	[22]
9.	TiO$_2$ NT (15, 22, 35, 30 μm length)	Anodic Oxidation	N719	LiI + DMPII + I$_2$ + GuNCS, + TBP in acetonitrile	10.26 11.68 12.84 13.12	0.790 0.783 0.779 0.762	0.723 0.701 0.703 0.661	5.9 6.4 6.6 7.0	[23]
10.	TiO$_2$ NT	Anodic Oxidation	N719	BMIM-I + I$_2$ + GTC +TBP in acetonitrile and valeronitrile	15.64	0.77	0.62	7.75	[24]

Improvisation from Efficiency Perspective

From Table **2**, it is observable that TiO$_2$ Nanotubes (NTs) do not provide the same results (efficiency of DSSC) at each instance. It is due to the physical arrangements of the NTs on the anode contact. The NT grown *via* the Hydrothermal or sol-gel method, the vertical alignment is less appreciable with respect to that grown *via* the electrochemical anodization route. This is because of the application of the electric field lines, while synthesizing, the directivity of the oxide growth is well retained.

Now, the different morphological parameters are crucial for defining the ultimate efficiency of DSSC, as shown in Table **2**, as an example, the reduced graphene oxide decoration of the TiO$_2$ enhanced the Photocurrent Efficiency (from 5.28% -

6.01%). In this verge, the catalytic effect of novel metal NPs has also been reported indicating the rate at the vicinity of catalytic sites.

It is also indicated that a large surface area and high crystallinity are preferable for the photoanodes of DSSC to ensure the large absorptance to dye and good conductivity of electrons. Hence, TiO_2 NTs are preferred over the other nanostructures. However, in the case of TiO_2 NT-based photoanode, limited PCE is resulted in, owing to the recombination phenomena between dye and TiO_2. This limitation is possibly mitigated by introducing self-catalytic sites within the TiO_2 NT morphology. It is interesting to note that electrochemical anodization provides the flexibility to nanoengineer the TiO_2 film. The variations in the water content during anodization allows the variation in the intrinsic stoichiometry of TiO_2 NTs. It was also observed the DSSC performances is dependent on different nanotube lengths, which is the result of different anodizing time. Simultaneously, the electrolyte and dye combination have a great impact on DSSC performances as well as the choice of light wavelength. For this reason, the versability of DSSC can be addressed with sunny weather, cloudy environments, and even with Neon-based lights.

Constraints and Design Considerations

A detailed literature review is presented in this chapter; starting from the basics of Solar Cell. The chapter describes the working principle of DSSC in a step-by-step manner. It has been discussed how a solar cell works on the photovoltaic effect categorizing into different generations of solar cell, indicating corresponding prospects and constraints. First-generation (Si-based) Solar cells and their production consist of many (and) expensive processes. To reduce the cost of material and production process of solar cells, thin-layer film solar cells also have been invented, such as CIGS, CdTe, and GaAs. They perform better and also give better efficiency. CIGS, CdTe, and GaAs give efficiency up to 22.6%, 22.1%, and 28.8%, respectively. But CIGS and GaAs are quite costly, while in CdTe, cadmium is toxic, and tellurium is rarely available. In this regard, DSSC has been proven to be the better choice among all. As photoanode is the heart of the DSSC, which can exist in different nanostructures such as nanoparticles, nanoflakes, nanosheets, nanotubes, *etc.* Although the nanoparticulate photoanode gives better efficiency, there is a risk of charge (electron-hole) recombination at the boundary of the particles which limits the device's performance. The present technological scenario suggests that TiO_2 NT grown by the electrochemical anodization method is the most suitable candidate for DSSC. The morphological variation (tube length, pore diameter, wall thickness) of the cell has a pivotal impact on DSSC efficiency. The length of a nanotubular semiconductor up to a certain extent is beneficial, but a very long nanotubular structure gives poor efficiency, and also the device becomes weaker. Moreover, higher dye concentration is undesirable as the charges diffuse easily resulting lower PCE. The implementation of self-catalytic sites may possibly be a solution, which is yet to be investigated. However, the optimistic approach to improve the DSSC, the anode material will be nanoengineered.

ENERGY STORAGE: FEASIBLE CRITERIA TOWARDS RENEWABLE ENERGIES

Renewable energy sources possess intrinsic intermittent virtue in terms of power generationhence, energy storage is indispensable for efficient and successful operations [3, 5, 7]. In the case of solar energy harnessing, compulsory barring of required photon energy (from sunlight) at monsoon and at the night. Besides, the wind energy is completely dependent on monsoon and air direction, having limitations at high wind velocity [3, 5, 7]. Tidal wave, a great replacement to energy resources, can also be considered distance-limiting due to selective sea-shore condition [3, 5, 7]. Hence, energy storage and power management, throughout the globe, is important for renewable energy production to cope with the environmental condition and sustainability issues [3, 5, 7]. The main requirement of the energy storage device may be identified as the high-power delivering capability with a long discharge cycle [3, 5, 7]. Other desired qualities must include high charging-discharging cyclability, ease of maintenance, portability, cost-effective production and flexibility to adopt new technological trends [3, 5, 7]. All these aspects are very important to go for a good storage device, and this is only possible with the good convolution of renewables with energy storage might give to a new Industrial revolution as this 21st century also already started its glorious way towards commercial Electrical vehicles (EVs) and many portable electronics [35]. Fig. (**8**) represents the different schemes of anode materials for lithium ion batteries (LIBs) their usage around the globe, with the life-time and efficiency comparison.

Introduction

In a large variety of Batteries, from the basic button batteries to the megawatt loading applications, the efficacy is almost around 90 percent except for the high-power densities. Primitive battery (Volta's cell) had alternating discs of zinc and copper separated by cardboard in a Brine solution. The Daniel cell, in 1836, evolved as a dry cell and had two electrolytes continued to the Lechlanche cell development [36, 37]. In the dry cell, an alkaline electrolyte, with a Zinc anode and Magnesium oxide cathode, was utilized. Until now, discussed battery types are all one-time use, *i.e.*, non-rechargeable. Coming to the rechargeable battery (secondary battery) era, the lead acid battery takes the dominant position, for power backup in industrial areas, owing to its high-power delivery capability [35 - 37]. To the contrary, the portability of nickel-cadmium batteries makes them be used in toys [37]. It is the cell with redox material as the base, nickel, and separators are the main components. Generally, to obtain ~3-4 V (required for the operation of toys), the series connection of one cell, producing 1.2 V, is the main approach [36, 37]. A schematic representation is shown in Fig. (**8**) for the ready concept.

Further, with concern on the toxic effect of cadmium, the evolvement of nickel metal hydride (NiMH) has been observed [36, 37]. However, the integration in portable electronic gadgets (like laptop, mobile phones, *etc.*) continued with the invention of the LIBs owing to their high energy density and thus addressing light-weight, portable and appreciable operation in electronic regimes [35, 36].

To discuss progress with portable batteries of the modern era, it will be relevant to explore the basic battery operation, in short. At the anode, oxidation takes place whereas, the reduction process is followed in the cathode during the charging cycle and; hence the batteries are also termed redox flow one [36 - 38]. Alternatively, at the time of discharging, the reverse phenomenon is observed. As an example of a Ni-Zn battery, equation (**11**) represents the oxidation process at the cathode during charging operation [39].

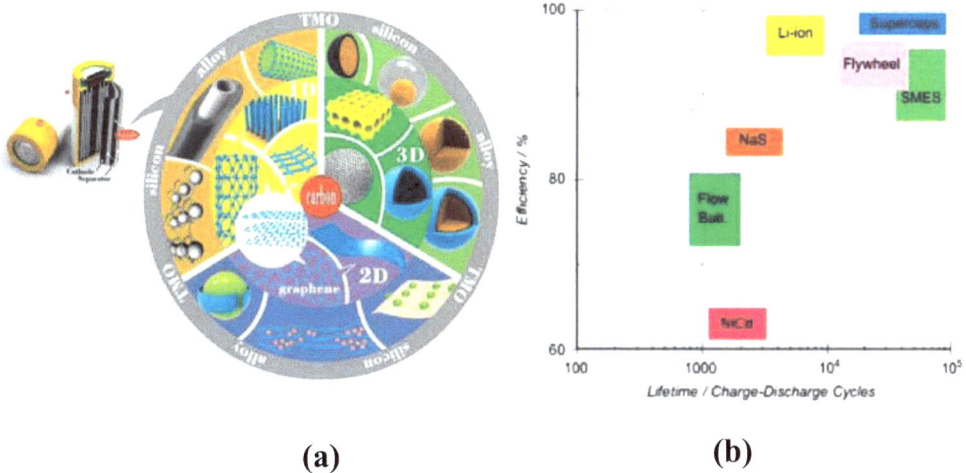

(a)   **(b)**

Fig. (8). (a) Schematic diagram of recent anode materials of Lithium Ion Battery and **(b)** Comparative of efficiency and life-time of different batteries; with permission from [36].

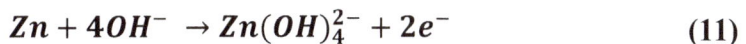

$$Zn + 4OH^- \rightarrow Zn(OH)_4^{2-} + 2e^- \qquad (11)$$

Coming to the reduction, (gaining of electrons) is shown in Eequation **(12)** for the discharge cycle of LIB [36].

$$Li_{1-x}CoO_2 + xLi^+ + xe^- \rightarrow LiCoO_2 \tag{12}$$

From equations **(11 & 12)**, it can be elucidated that the cathode material serves as either source of electrons for electrolytes during charging or acts as an electron collector from an external circuit at discharging. On the other hand, the anode material is of immense importance to add the feature of ion storage into its lattice void [36, 37, 40]. Porosity, conductivity, voltage match, high durability, purity, less weight, less current density (<0.5 Am^{-2}), cheaper and readily available is the main criteria for choosing the anode material [40]. For a clearer reference, a comparison chart is summarized in Table **3**, comprising different primary (non-rechargeable) batteries and their figure-of-merits. In contrast, Table **4** provides the performance information of secondary batteries in a comparative manner.

Table 3. Comparison table of primary batteries [41].

Sl. No.	Battery	Anode	Cathode	Max.voltage (V)	Specific Energy (W/Kg)	Comments
1	Zinc- Chloride	Zn	Cl	1.5	150	Heavyduty battery
2	Zinc-Manganese dioxide	Zn	MnO_2	1.5	400	Comparable energy density
3	Lithium-Manganese dioxide	Li	MnO_2	3	800	Extra duty and expensive
4	Silver oxide	Zn	Ag_2O	1.85	450	Commercial button batteries

Table 4. Comparison table for different rechargeable batteries [42].

Sl. No.	Battery	Voltage (V)	Energy Density (W/Kg)	Cycle Life	Self-Discharge
1	Lead-Acid	2	30	200-300	Low
2	Nickel-Cadmium	1.2	45-75	1500	Moderate
3	Nickel-Metal Hydride	1.25	60-120	300-500	High
4	Lithium-Ion	3.6	110-160	500-1000	Very low

The battery technology, till now, can be considered to have various material aspects, as observed from Table **3** and **4**. These also further developed in the present technical scenario, majorly expertized in the Lithium-ion based route to address the most recent technological boom towards EV technology.

Prospects and Constraints of Commercial One

The present commercial Li-based storage device has a lot of research background [40, 41]. LIB comes under the secondary and high voltage capacity battery [41]. Mainly, the metallic lithium is converted to deal with the storage purpose of battery, but the tendency of the dendrite formation (shown in Fig. (9)) causes easy short-circuits, confining the direct use of Li metal [41]. Also, in terms of safety, the high reactivity of Li with the environment and inadequate cycle durability with inherent thermal runaway limits the direct use of the same. Hence, different materials are being researched for safer and more efficient battery behavior. To enhance the knowledge about the respective compositions and combinations, in search of an ideal and advantageous material, Table **5** may be used as a point of reference.

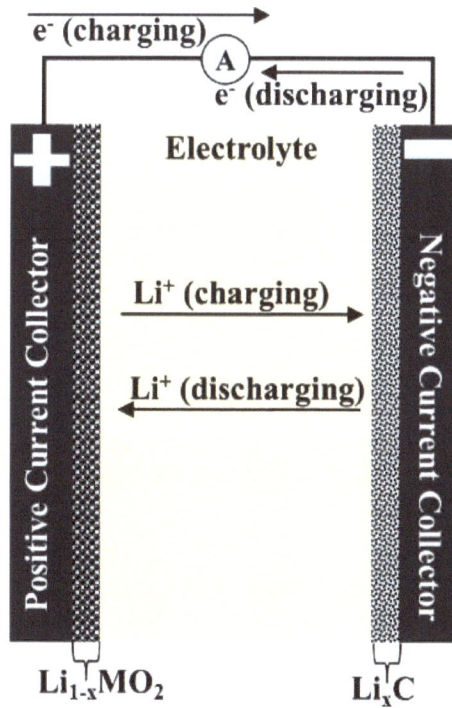

Fig. (9). Redox reaction flow within Lithium-Ion cell (schematic view).

Table 5. Different parts of LIB architecture [40, 41].

Sl. No.	Components	Process	Material
1	Anode	Lithium-ion enter the anode when charging state and leave the anode in discharging state	Carbon-based, CNTs, Graphite, *etc.*

(Table 5) cont.....

Sl. No.	Components	Process	Material
2	Cathode	Lithium-ion leaves the cathode when charging state and enters the cathode in discharging state	Lithium metal oxide based
3	Electrolyte	Transport of lithium ions but not electrons between anode to cathode	Lithium salts and organic solvents
4	Separator	For insulation to prevent short circuits and have pores to travel Li^+ ions but not electrons	Porous membranes

The commercial LIB is, generally, a non-aqueous secondary battery with transitional metal (M) oxides with $LiMO_2$ as a positive electrode and the carbonaceous as the negative electrode, as shown schematically in Fig. (**9**) [35].

The cathodic reaction will be as in equation (**12**), only replacing the Co with M for generalized consideration and represented in equation (**13**) showing a two-way redox reaction [35, 36]. The obtained Li^+ ions are stored in an anode following the compensating reaction as per equation (**14**).

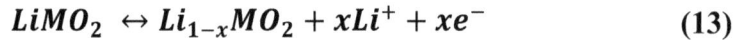

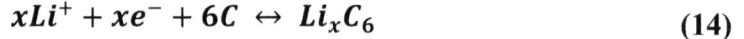

$$LiMO_2 \leftrightarrow Li_{1-x}MO_2 + xLi^+ + xe^- \qquad (13)$$

$$xLi^+ + xe^- + 6C \leftrightarrow Li_xC_6 \qquad (14)$$

The movement of Li^+ ions occurs between the anode and cathode *via* the electrolyte, as shown in Fig. (**9**). The commonly and most popularly used electrolyte is $LiPF_6$ dissolved in organic solvent (ethylene carbonate) [35, 36]. Normally, lithium cobalt oxide will be the cathode material due to its high energy density. When its charging cycle, a voltage is applied, the lithium ions from lithium metal oxide diffuse across to the carbon anode (intercalation). These electrons provide the necessary charges for the insertion half-reaction [35, 36]. While discharging, the whole scenario is again reversed: electrons move from carbon to lithium metal oxide, which now acts as a positive terminal. This commercial grade LIBs, carbon/graphite anode, is generally preferred due to the best Van-der Waals forces which were hexagonally bonded. The forces were stronger between the two concurrent sheets of carbon on the same sheet.

In other words, there can be one lithium ion for every six carbon atoms in the graphite sheet because lithium ions can only unite on every second hexagon of carbon atoms [40, 41, 43]. This relation can be modelled to incur the theoretical capacity of 372 mAh/g for fully intercalated LIB [43]. To the contrary, if other metal is used instead of Li, higher theoretical capacity (*e.g.*, Al: 993mAh/g, Sn: 994 mAh/g, Sb: 536 mAh/g, *etc.*) can be achieved as these metal ions require a smaller number of carbon atoms for intercalation purpose. But this intends to form alloys (through pulverization) at the anode side and in turn rechargeability will be lost. Concerning the cathode materials, there is a lot of choices of materials, possessing higher capacity and chemically stable compounds, long cycle life, low cost, good electronic conductivity, as desired [43]. However, these are almost all Lithiated metal oxides or Lithiated metal phosphates [40, 41, 43, 44]. Compounds discussed will be Lithium Cobalt Oxide (LCO), Lithium Nickel-Manganese-Cobalt Oxide (LMCO), Lithium Manganese Oxide (LMO), Lithium Iron Phosphate (LFP), *etc.* The properties of such materials are also summarized in Table **6** to understand their potentiality.

Table 6. Potentiality of cathode materials [44].

Sl.No.	Properties	LCO	LMCO	LMO	LFP
1	Voltage	3.9	3.6	3.7	3.4
2	Specific energy	155	200	100	160
3	Specific power	1 C	10 C	10 C,40 C	35 C
4	Cycle life	500	1200	1000	1500
5	Charge limit	4.20 V	4.20 V	4.20 V	3.60 V
6	Thermal runaway	150	210	250	270
7	Safety	Poor	Good	Average	Very good
8	Cost	High	Moderate	Moderate	Low

The LCO (Lithium Cobalt Oxide; $LiCoO_2$), cathode material with the carbon anode, is identified as the first commercial LIB [40, 41, 43]. LCO has a layered structure with cubic closely packed oxygen in CoO_2 forming a hexagonal lattice. It has a high capacity of 155 mAh/g with good conducting performance, is easily manufactured, and is insensitive to moisture. But the main concern with cobalt is an expensive material making it non-suitable for individual usage [40, 41, 43]. As per Table **6**, it has a power density of 1 C, which indicates a fully charged battery for a capacity of 1 hour. It is to be noted at this point that a higher value of 'C' indicates a lesser time for capacity delivery. As an example, 30 C indicates battery operation for 2 minutes, whereas 2 C is meant for 30 min battery operation with full capacity. Likewise, the power-density for different components is explained. However, in spite of the good performance of LCO-based LIB, it suffers from heating issues at highly charged conditions [40, 41, 43]. So, for these, safety and lifetime is the main trade-off. On the other hand, Lithium Manganese Oxide (LMO; $LiMn_2O_4$) is cost-effective for available Li-based batteries [45]. It has a good number of cycle life (higher cyclability); but the existing lacuna is its lower capacity at higher temperature. In the case of Lithium Iron Phosphate (LFP;

LiFePO$_4$) electrolyte, phosphates are introduced to achieve good stability in charging and in short circuit conditions, hence, it withstands high-temperature conditions. It has the highest Thermal runaway suggesting an advantage for the LIB applications. It not only withstands the Thermal probability, but also oxidation and the acidic environment of the battery. Non-toxic in nature with 160 mAh/g and an average voltage of about 3.4 V is also utilized in the EV technology. LMCO or Lithium mixed Nickel Manganese Cobalt oxide, which is a complex compound, is another potential candidate for LIB cathode as its synthesis is accomplished using modified Mixed-hydroxide reacting with a Lithium salt in the air at 750 °C [40, 45]. LMCO is based mainly on a structure similar to the LCO with less cost and high thermal stability. Initial capacity of the 200 mAh/g and its high stability maintains its popularity as cathode material. LFP based LIBs were very promising as phosphates can withstand some high values of energies without heating concern and hence suitable for EVs battery [35]. Also, LFP is comparably safer for environment. It is to be noted that, higher capacity and higher voltage is necessary for the more energy. However, the nanotechnological aspects cannot be related in cathodic material whereas the scope is extended towards anode owing to the requirement of pore, surface area and lattice intercalation of ions.

Hence, to enhance the battery performance, the research concentration focuses on the anode materials owing to nanotechnology adaptability. Allotrope of carbon, Carbon Nanotubes (CNT), is being considered as potential anode material for LIB owing to its cylindrical structure (having a higher number of hexagons), 600mAh/g reversible capacitances and increased up to 1000 mAh/g, theoretically. The alloy-formed metallic-coated CNT will have higher advantages [35, 41, 45]. Going on with the discussion of the LIB compositions, the requirement of achieving an appreciable advancement in storage technology can be addressed by the nanostructured anode materials (especially transition metal oxide based) simultaneously with the prior discussed phosphate-cathode. In the next section, the Nanostructured Transition metal oxides and their effect on energy storage performances are precisely discussed.

Research Art towards Cost-Effective and Safe Storage

Transition metal oxides are considered for the LIBs as they are compounds comprising oxygen atoms possessing semiconductive properties and utilized for catalytic effect. In conventional LIB, many metal oxides, in a bulk structure, are in use for generic purposes. The advent of nanoscience generated the freedom to nanoengineer the morphology of metal oxides and for the sake of superior electrochemical properties [46]. The nanotechnology provides the facile base of surface chemistry of the oxides, having higher capacity values and being more stable compared to bulk counterparts]. Moreover, mismatching in the capacity of anode and cathode material causes capacity penalty and as a result, the dendrite formation (insulation failure) is evidenced [35, 45 - 47]. Thus, compatibility between two electrodes has to be maintained. Further, Thick electrodes lead to less efficiency as more material is utilized, which in turn increases the cost. Thus, nanostructured transition metal oxides are preferred as anode material. For this purpose, various transitional metal oxides comprising their theoretical capacity are presented in Fig. (**10**).

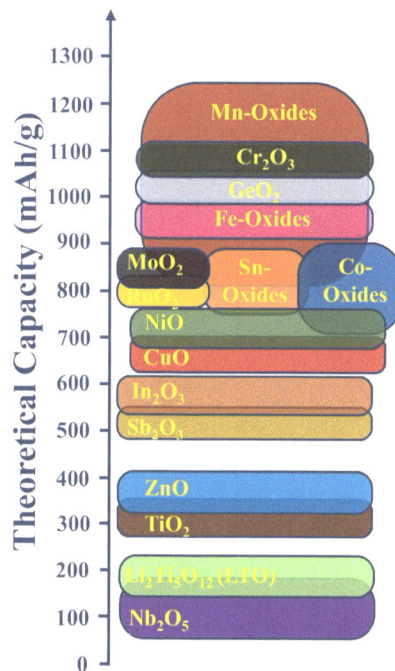

Fig. (10). Theoretical Capacities of LIBs comprising of different Metal oxide anode [35, 45 - 47].

The reaction mechanism on an anode can be identified into three main groups: (A) Intercalation, (B) Conversion and (C) Alloying [47]. Intercalation indicates capacity retention; for example, CNTs possess high capacitive retention after intercalated with the Li ions (discussed earlier). Alloying of the materials will result in large crystallographic stress and volume expansion tending to weak cycle life. To the contrary, the conversion anodes having a future treatable material. Combining strengths of different materials such as alloys, carbons, transition metals to prepare anodes for a highly recommended electrochemical behavior. For the respective transitional metal oxides M_xO_y type we consider M with several materials like Nb, Ti, Zn, Sb, In, Cu, Ni, Mo, Co, Ru, Fe, Cr, Mn, *etc.* metal oxides, in which few were related and compared for nanostructured counterpart for LIB. Table **7** shows the corresponding advantages and disadvantages of different types of material and the same suggest the suitability of metal oxides as anode of LIB.

Table 7. Advantages and disadvantages of Transitional Metal Oxides based anode for solid-state battery [47].

Sl. No.	Material	Advantages	Disadvantages
1	Carbon	Good electronic conductivity, hierarchical design, cost-effective and very highly available.	Very low specific capacity and having enormous safety issues
2	Alloys	Specific capacity of 400 to 2300 mAh/g	Less volume change and electronic conductivity
3	Silicon	Very high specific capacitance of 3579 mAh/g, easily available and clean	Very large volume change which is around 300%
4	Transitional Metal Oxides	High Specific capacity compared to Alloys up to 1000 mAh/g	Low coloumbic efficiency and large potential hysteresis

Effects of Nanoengineering

It is desirable that, there should be very little volume change and no significant degradation of original crystal structure, when cations are reversibly stored in the anode material, which subsequently results in high cyclability and good capacity retention. In the case of the transitional metal oxides, Titanium (Ti) serves as the redox active center, in most traditional insertion-type metal oxide anodes (TiO_2, $Li_4Ti_5O_{12}$, *etc.*) [46, 47]. Upon Lithiation, Ti is reduced to Ti^{3+} and then reoxidized to Ti^{4+} when the alkaline metal cations are subsequently deinserted. The usage of these materials is largely restricted to high-power rather than high-energy applications. For this, cycle stability is more crucial than energy density due to the comparatively high mass of these compounds and relatively low Lithium absorption [46, 47]. Alloy-type materials of Si, Sn, Ge, Zn (like, $Li_{15}Si_4$) have fast lithium capacities. However, the substantial volume fluctuations reduce the life-cycle of such electrodes. As a consequence, the electrolyte is constantly coming into contact with new surfaces, causing a continual corrosion process. Reversible cycling of TMOs which is advantageously helpful in the formation of Li_2O, possesses capabilities of about 1000 mAh/g for Fe_2O_3 [46, 47]. For the past 20 years, it was helpful to maintain and develop such high-capacity and durable transition metal oxides. Since transition metals normally show the metallic state, conversion reactions commonly show higher capacities. Considering cobalt oxide, Co_3O_4, reduction of Co^{2+} in Li^+ forms Co^0 and Li_2O, achieving overall capacity will be around 890 mAh/g whereas $LiCoO_2$ anode has 273 mAh/g [46, 47]. The alkaline metals, like Li, Na, affect more on the conversion reaction. During the first Lithiation, a Solid electrolyte interface is generally formed on the surface of the electrode, which is a product of utilized electrode. The effect of different nanostructures is discussed considering different TMOs in the following paragraphs.

Nano α-Fe_2O_3 possesses a specific capacitance of about 1006 mAh/g with a voltage range 1.5 V to 4 V. Further, α-Fe_2O_3 nanotubes with a discharge capacity of 1415 mAh/g [46, 47]; whereas α-Fe_2O_3 nanoflakes possess better reversibility. It is also reported that α-Fe_2O_3 nanostructures-based anode with discharge current density of 100 mA/g of negligible capacity delays up to 80 cycles for the voltage ranging from 0.005 to 3.0 volts. Apart from this, Fe-based hollow structures, microboxes and cylinders provide high specific capacity. Due to its low cost, environment-friendly, and abundant richness in chemistry making Iron oxide is a promising anode. However, for lithium storage cells, problems will be faced mainly in the concept of voltage hysteresis which is the main cause and having low storage efficiency for iron oxides.

These have great redox reactions and deliver about 716 mAh/g specific capacity [46, 47]. Different nanostructures (nanowires, nanotubes, *etc.*) and composites of the same are possessing substantial properties as anode material. Table **8** summarized the battery performance comprising of cobalt oxide-based anode. These have high reversibility and use large amounts of cobalt [42, 46 - 48]. However, the issues related to abundance and toxicity restrict high usage. The Manganese oxides will be having a prominent usage in the field of chemistry, as it is most abundant on the earth surface, compared to the prior discussed. Manganese oxide-based anode provides high theoretical capacitance; but is mostly affected by poor cyclability as a result of large volume change. When it is used in LIB, the capacity is 830 mAh/g after 300 cycles at 500 mA/g current. However, low electronic conductivity is a prime problem to use and probably so it is less realized in battery-related researches.

On the other hand, Cu_2O and CuO anode for LIB offers a high reverse capacity of 375 and 674 mAh/g, respectively. Further, spherical Cu_2O-CuO-TiO_2 nanocages reduce volumetric capacity exhibiting 700 mAh/g at 50 mA/g over 80 cycles [46]. Cu-based graphite nanorod is responsible for stable cycle performance. But the lower capacity retention is a prime lacuna of this material. NiO, which is a p-type wide bandgap semiconductor, became an attractive anode for LIBs due to its theoretical capacity of 718 mAh/g. 2D nanosheets and nanoparticles of NiO are vastly used as anode material in LIBs. NiO generally is intermediate material in the aspects of its variables in the conversion type anodes. However, the comparably expensive nature can be compensated with the non-toxic consideration. Besides, TiO_2-based anode is very much studied in recent days. TiO_2 nanotubes, having a higher degree of freedom to be tuned through different morphological aspects, are mostly preferred. Tunability of TiO_2 NT can easily be achieved for the confinement of 1D or 2D electron transportation, which develops the main reason to have efficient Li ion transport [42, 46 - 48].

Table 8. Comparison table of parametric effects in a battery [42, 48].

Sl. No.	Cause of Ageing	Mechanism	Effect
1	High temperature (more than 35 °C)	Electrolyte decomposition	SEI growth Micropore clogging etc
2	High temperature (more than 35 °C)	Electrode decomposition	Precipitation
3	Low temperature (less than 35 °C)	Lithium plating	Dendrite growth
4	High C rates	Lithium plating	Dendrite growth
5	Over-discharge	Current collector corrosion	Loss of conductivity
6	Overcharge	Electrode decomposition	Gas evolution
7	Safety	Poor	Good
8	Cost	High	Moderate

TMOs such as the Cr_2O_3, Nb_2O_3, and MoO_x exhibit higher theoretical capacity values. Cr_2O_3 for instance provides a capacity of about 1058 mAh/g as an anode and the lowest lithiation voltage (0.2 V). To the best of my knowledge, the highest empirical charge capacity was reported well below 900mAh/g. Thin film electrodes have intrinsic limitations with the volumetric and gravimetric capacities; whereas for thicker films, the performance is faded out. Coming to Molybdenum oxide (MoO_3) these have the theoretical capacitances of 1117 mAh/g. With high-capacity values, it is also known that after several cycles there is a high discharge on the storage. Other MoO_2 salts give cycle discharge capacitance values of 838 mAh/g and also 700 mAh/g after 30 cycles [46, 47]. Table **9** provides an overview of such investigations as a representative entry.

Technological Evolution

The technological scenario must concentrate on desired ideal parameters of solid-state batteries. Mainly, the thermal runaway, self-discharge and effects of physis of materials used (as anode, cathode, and electrolyte) draw the research emphasis for further improvement aiming for higher performance in lower volume. Table **9** listed the corresponding effect of different parameters for real-field applications.

Table 9. Comparison table for battery performances for the nanostructured TMO anode [46].

Sl. No.	Material	Specific Capacity (mAh/g)	Specific Current (mA/g)	Capacity Retention/ Number of Cycles
1	α- Fe_2O_3 nanotubes	510	100	36%/100
2	α- Fe_2O_3 nanoflakes	680	65	83%/80
3	Hierarchical Fe_2O_3 Microboxes	945	200	80%/30
4	Fe_3O_4 @N-C	1063	1000	98.4%/1000
5	Fe_3O_4-C nanospindles	530	462	70%/80
6	CoO-Co-C	800	44.5	87.5%/80
7	Co_3O_4 nanotubes	500	50	58.8%/100
8	Co_3O_4/ Graphene	935	50	>100%/30
9	Co_3O_4 NW arrays	700	111	NA
10	Ti-CoO@C	1108	200	100%/150
11	MnO@C/N	837	500	100%/300
12	MnO@Mn_3O_4/NPCF	1500	200	100%/300
13	MnO_2@CNT	500	50	NA
14	Mn_3O_4/RGO	900	40	NA
15	Mn_3O_4	800	30	92%/40
16	Pure CuO nanoparticles	210	2000	90%/40
17	CuO nanostructures	560	150	100%/50
18	Co_3O_4/CuO	1191	200	90.9%/200
19	CuO-CNT	650		100%/100
20	CuO/MCMB	400	50	77%/50
21	NiO nanowall	638	895	98%/85
22	NiONS/Graphene	1000	50	90%/50
23	CuO@NiO Spheres	1061	100	87%/200
24	NiO/Ni/Graphene	962	2000	100%/1000
25	TiO_2, Anatase	168	NA	NA
26	TiO_2-B	335	NA	NA
27	$Li_4Ti_5O_{12}$	175	NA	NA
28	$TiNb_2O_7$	387	NA	NA
29	$Ti_2Nb_{10}O_{29}$	396	NA	NA

In short, the LCO cathode is toxic and of high cost whereas the ideal one should be non-toxic and cheaper. Again, nickel-based cathode suffers from issues due to ageing and causes capacity fade. Popular LFP cathode, in spite of its cost-effectiveness and high thermal stability, suffers from thermal runaway. Additionally, for the active material of anode, in battery applications, the effect of different bulk materials and later nanostructured materials were thoroughly discussed. For graphite anode, preferred due to its safe operation, the high volumetric change and non-ability for full recycling extend the research world for further probe. Besides the physical properties the materials should withstand low-volume change and high mechanical strains are considered a major technological concern [42, 46]. In the case of the electrolyte, it should be an organic solvent for the higher output efficiency. The solid electrolyte interface layer is to be observed as it helps further battery performance. In battery ageing, operating temperature, cut-off voltage should have to be noticed frequently as these changes deteriorate the final operation. Moreover, lithium plating or dendrite formation is a burning that should be avoided as far as possible. When Carbon type electrodes are used as anode it is observed that the highest dendrite formation for the short circuit. Dendrite formation can be avoided in the case of the nanotubular structure of TMO with defined crystalline providing an acceptable performance [42, 46].

In search of a cost-effective yet highly-efficient charge storage module, it can be summarized that nanostructured TMO-based anode provides an acceptable condition of low volume change, non-toxic and longer lifetime. Moreover, the nanoengineering possibilities *via* tuning different synthesis parameters is a possible way to improve the battery performances, simultaneously, mitigating the concerned issue due to the different properties of the nanostructures.

IN-SITU ENERGY HARVEST AND STORAGE

As per the preceding sections, the evolution of the DSSC in the field of energy harnessing and the same for energy storage devices was discussed elucidating the nanotechnological impact. In the present section, the integration of these technologies will be addressed, correlating the nitty gritty of the amalgamation process with probable way-outs.

Introduction

Consciousness towards the environment in conjunction with perishable storage of fossil fuels thrusts the research world into the newer technological advancement from which smart grid and the electric vehicle are emerged as the utmost applicable field [3, 5, 7, 35]. The prime motto of such modern technologies is sustainability, renewability and huge deployment [3, 5, 7, 35, 49, 50]. The background research on renewable and sustainable energy harvesting flourished to its best by the invention of solar cells; whereas the huge deployment is yet to be addressed following cost-effective approaches [49, 50]. In addition, the storage of the generated energy, *via* renewable resources, is of vital importance as the electrical energy can be utilized at times of need [49 - 51]. As per the present state of technology, silicon-based photovoltaic (PV) cell is used for solar power generation and a battery, which is not mandatorily solid-state one, stores the generated energy [50 - 52]. On the other hand, solid-state batteries are generally used to store the charge, not compulsorily from renewable energy sources, which are being used for carbon-emission-free (green) processes [49 - 52].

However, the charging process is dependent on the conventional electrical power consumption and hence these approaches (for EV and/or smart grid) can be considered as pseudo-green technology [50 - 52]. On the verge of fully green techniques, integration of the energy storage device with the energy harnessing one is a novel step in retaining sustainability and renewability.

The initial approach of such integration was reported with a separate PV module (Si-based) with Li-ion batteries (connected in series) confirming the potential applications of sustainable and green criteria for a better future [53]. Systematic studies of this scientific report, containing the relations and circuitry detail, paved the path towards mathematical modeling for further research in this particular field. Further, the inflexibility and higher installation cost of the crystalline Si-based PV cell, the incorporation with the storage system possesses design constraints for huge deployment targeting individual usage [49, 51, 53]. As an alternative, other than the crystalline Si, amorphous Si or other materials (CdTe) provides a cost-effective solution at the cost of less efficiency, which in turn affects the packing density for successful operation [49, 51]. However, the separate modules, for energy harvesting and energy storage system, restrict the compactness of the final product to some extent. The compactness of such a system (with the self-charging condition) ensures the adaptability in manifold from low dimensions (as smart watches) to the larger scale (such as electric vehicles) [50 - 52] Here also the appliance voltage of the power source became an important concern for further practical applications [53]. In that consideration, Photo-Electrochemical (PEC) power source is a suitable alternative to address the high voltage specification [53]. Dye-Sensitized Solar Cell (DSSC) is one of the choices under the PEC category to directly charge the storage element, as the direct charging from a conventional PV cell is responsible for higher charging time and less capacity towards Li-ion batteries [50 - 53].

This section is further organized to discuss the design restriction for the integration of energy harvesting technology with the energy storage system along with the degree of freedom to do so. The research trend and future scope of such integration will also be discussed in the following sub-section to indicate smart futuristic green technology for individual deployment. The following discussion can also be considered as a tutorial part for further research as through these sub-sections research paths and lacunas are tried to be explored.

Design Restrictions and Liberty

The sub-section will mainly deal with the integration of renewable energy harvesters with the energy storage system aiming to point out the prospects and constraints for designing such. Broadly, this discussion is categorized into three categories; namely, (i) Static Integration, (ii) Mobile Integration, and (iii) Wearable Integration, as per Fig. (**11**), following the used charging mechanism from renewable energy.

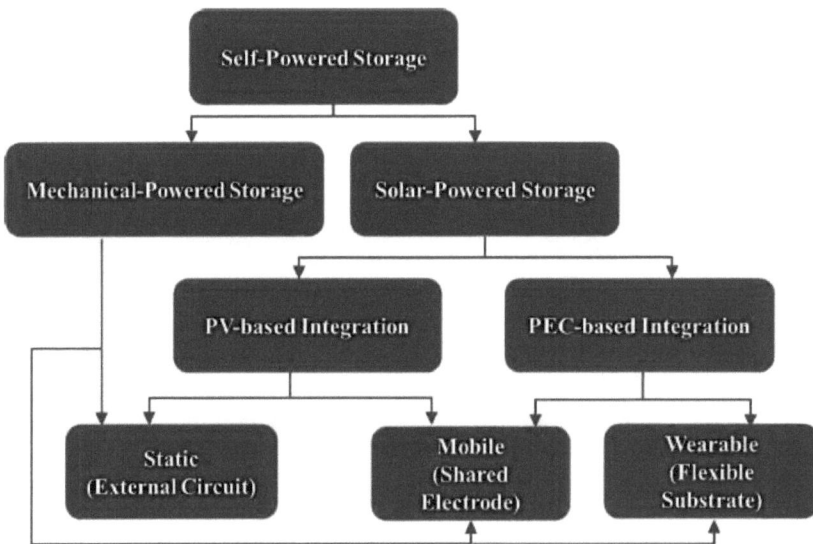

Fig. (11). A crude categorization of the self-powered energy storage system.

Static Integration

As per the conventional design consideration, the energy harvester (renewable) and storage element are kept in distance, as the generated energy is generally not always the same and mandatorily not at the usable space [49, 51]. In this case, the energy harnessing unit is static by position whereas the chargeable unit is moved to the required point of use. Specific mechanical energy sources, like electromagnetic generators, windmills, tidal sources, *etc.*, are mandatorily static due to their large and heavy design [51]. Thus, charging automation, especially for accessed mobility, is somehow restricted. Hence, the solar energy source dominates the amalgamation method towards the self-charging criterion.

The potential design has been adopted for PV cells for integration with the battery system [51]. PV cells are also categorized into four: (A) Crystalline Silicon-based, (B) Amorphous Si and Other semiconducting thin film-based, (C) Perovskite-based, and (D) Organic Semiconductor based energy harvester [50, 51]. It has been pointed out that, the integration design of the former one results in extended charging time with comparatively less storage capacity, to compensate for which the introduction of supercapacitors is evident from scientific literatures [49 - 52]. Three electrode configuration is generally incorporated for These kinds of designs (PV-supercapacitor-battery) [50 - 52]. PV cells generally operate the supercapacitor and in turn, convert it into electrochemical energy to charge the electrochemical battery devices [52].

Thus, as per the design consideration, the involvement of the supercapacitor, develops the size constraints for desired integration. However, the design approach mitigates the mobility issue to some extent [49 - 53]. Moreover, from the cost analysis viewpoint, such kind of integration demands a higher one owing to the high initial cost for conventional PV cells and additional cost for supercapacitor design and efforts. For these reasons, the individual deployments of such are restricted in a significant manner. To cope with this, the (B) category of PV cells comes into the scenario.

Considering the PV cell with amorphous Si and/or semiconducting material, the design perceptibly provides a cost-effective approach for energy harvesting devices [51]. However, here also, the requirement of the supercapacitor cannot be neglected and hence persists in the same design constraints as in the former case [50 - 52]. Similar design constraints are evidenced in the case of perovskite solar cells, as the electron-hole pair generation in semiconducting material is the dominant principle [49, 51, 53]. Further, organic solar cells are also used for battery integration, especially due to the ease of fabrication [50 - 52]. But the overall efficiency is quite lower. On the contrary, perovskite solar cell-based integration of energy storage systems leads to higher overall efficiency as this kind of solar cell provides facile high output voltage [51]. The mandatory conversion of solar energy into different energy forms prior to store/charge the battery restraints the incorporation of the supercapacitor interface. It is rightly pointed out that, solar energy to electrochemical energy conversion makes the favorable conditions for charging solid-state batteries (which is principally electrochemical device) [49, 51]. Continuing with this design strategy, the requirement of supercapacitor interface is however eliminated and paved the path for in-situ charging of battery availing mobility. This is elaborately discussed in the following points.

Mobile Integration

The motion of humans is also one kind of renewable energy that can be integrated through the deployment of nanogenerators [49, 51, 53]. Piezoelectric crystals are used for such motion to electric signal generation which in turn charges the lithium-ion battery, integrated into a coin cell [51]. Polarization of the PVDF (poly-vinylidene fluoride) film has an impact on electrical leakage of the battery and use of the film restricts the battery performance as well [51]. To enhance the battery performance, sacrificial porosity was adopted as a solution [51]. Moreover, the packaging is another important concern to enhance the sensitivity of the piezoelectric [51]. However, the design restrictions owing to its piezo effect limited the applicability with respect to that solar cells [51, 54].

The chief principle of direct solar energy to electrochemical energy conversion is dominated in the case of DSSC and hence a huge research effort is reported for the amalgamating purpose for charging electrochemical batteries [51, 54]. The PEC cell relies on the photo to the electrochemical conversion of solar energy *via* redox flow reaction, which in turn matches with the same redox flow rate within the storage element [51, 54]. This type of integration possesses a shared electrode and hence the volume density became higher for this case [51, 54 - 56]. From the cost-effective and tunable fabrication point of view, DSSC possesses an immense interest in such integrated storage design as the same electrochemical principle is shared for both [51, 54, 55]. The design is mainly focused on the integration of DSSC with electrochemical storage in such a manner that common materials can be used for both of the devices [51, 54, 55, 57]. For ready reference, a schematic of such integrated devices is shown in Fig. (**12**). From the scheme, it can be stated that the particular design increases the volume density of charges and the components are to be lightweight.

In the case of DSSC, dye molecules get sensitized by illumination and release the electrons to the photoanode (a detail mechanism was discussed in the first section) [54, 55]. Integration with the battery (EES in Fig. (**12**)) shares the external connection with this photoanode, from where the photogenerated electrons are conducted to the electrolyte of the battery. The electrochemical phenomena will be processed due to the flow of these electrons, within the electrolyte of the battery. By this procedure, electrochemical energy storage from solar energy is being feasible [55, 57]. While discharging, the load is connected across the electrodes of the battery and here electrochemical energy is converted into electrical energy [55, 57].

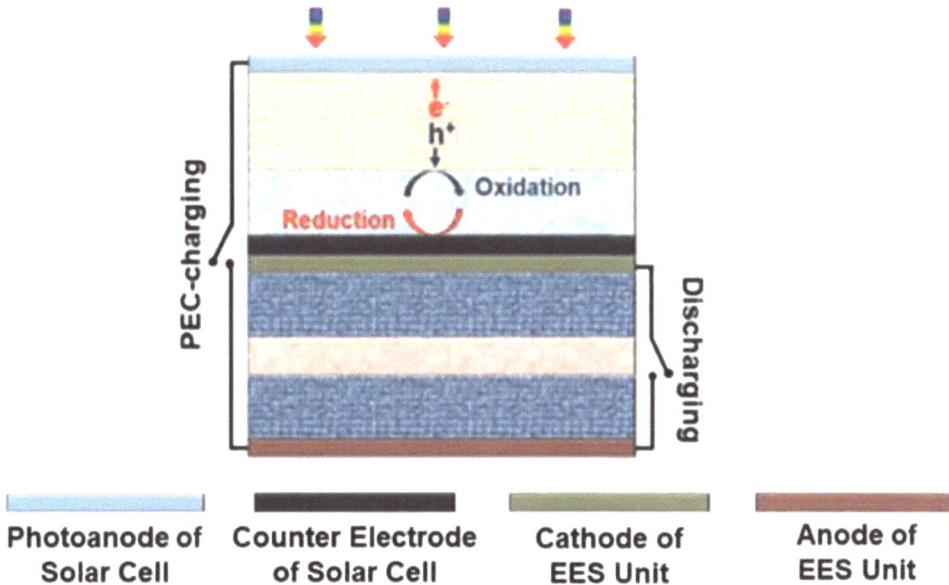

Fig. (12). Three electrode configurations for PEC (DSSC integration with Electrochemical Energy Storage (EES) system; with permission from [57].

It should be relevant to point out that, some novel works have also been reported on the monolithic integration of Si-based PV cell and Li-ion battery (solid-state), which provides good applicability for an electronic device (like smartphone, smartwatch, *etc.*) charging [57]. Here also, the low voltage specification of Si-based PV cells lacks the futuristic applications of electric vehicles [41, 56, 57].

Sticking with DSSC-based integration, the initial attempts with a double electrode and aqueous electrolyte system showed up with lesser efficiency for compatible applications [55 - 57]. However, the invention of ion-storage batteries gears up the design consideration as both of them share the same electrochemical principle [56]. In this connection, it is worth to be noted that, the proper choice of electrolytes and/or dyes of DSSC enable energy harvesting even from artificial light (from lamps) and hence the storage integration may be efficient enough for energy-effective households [56]. The facile synthesis of titania nanotubes, *via* electrochemical anodization, on both sides of titanium metal foil, also makes feasible conditions of desired integration [28, 29, 55, 57]. As discussed in earlier sections, the Titania nanotube array is a suitable candidate for both of the technologies (as battery anode and photoanode of DSSC), grown oxide at either side of Ti foil is used for either purpose. Fabrication and design of such are reported by Guo *et al.* [55] possessing an acceptable efficiency.

The particular design has lots of flexibility in choosing the dye and electrolyte combination so that the working illumination can be altered [55 - 57]. Further, the use of a flexible substrate makes suitability towards its incorporation in different applicable fields [57]. However, the cost element for the dyes is a considerable bottleneck for DSSC design and hence semiconducting nanocrystals, as photo-sensitizer, are of new research interest [57]. The working principle of these nanocrystals prevails over the quantum dot mechanism, which generally possesses a comparably high extinction coefficient [57]. Additionally, its tunable band-gap energy opens the research facility for further increase in the efficiency of the DSSC [55, 57].

Wearable Integration

This is a special type of mobile integration technique, where the battery, as well as the energy harvester, are generally with a flexible and wearable substrate (preferably fabric) [49, 51]. Here, DSSCs with different fabrics and flexible substrates were investigated thoroughly for amalgamation with the battery on the same substrate [49, 51]. In this condition, the triboelectric nanogenerator proved its potential as it supports electrostatic energy generation [49, 51]. In such designs, suitable power management is mandatory state-of-art, for which series connection with DSSC (favoring wearability) extended the performance efficiency of the integrated battery [49, 51, 57]. One should be noted that to achieve wearability, besides flexibility, the fibre shape device structure is generally preferred [49, 51]. This condition imposes further design constraints which either limit the fabrication freedom or tend for adding cost-element. However, the wearability of the integrated device suffices the power solutions for small-scale devices whereas, for large-scale technology, *e.g.*, electric vehicles, it is still needed some mobile integration so that dependency on fossil fuel can be reduced in a significant manner.

Particular integration design technique owes corresponding intrinsic features and drawbacks stating the development trend in the research-oriented fields. As per the desired need for recent applications, an electric vehicle is of the most recent and tempting research field, for which the self-powered battery is an inevitable part of fully answering the green technology. Hence, the following subsection describes the trends and future for further amalgamation of energy harvesting devices with storage elements.

AMALGAMATION: TRENDS AND FUTURE

With an objective of self-powered storage elements for electric vehicles, the conjunction of DSSC and electrochemical storage is popular research adoption, as it provides a higher degree of freedom for choosing materials with compatible electrodes to achieve appreciable performance with good stability and cyclability [51, 54, 55]. In this verge, two-electrode mode and three-electrode mode is a different circuital approaches for integrating these applications [57, 58]. A former one is simple but suffers from higher resistance due to the barrier owing to the oxide layer and diffusion of the ions in the DSSC [57]. On the other hand, the three-electrode configuration offers a separate path of battery discharge (towards load) without hampering or being hampered by the physical contents of the energy harvesting module [55, 57]. This type of configuration is the modern trend for integrated battery modules of interest.

Maintaining a similar approach, the usage of a three-electrode configuration moderates the problem of high voltage requirements for electric vehicle applications [55, 57]. However, overall efficiency is the dominant figure of merit. To achieve a high value of the same, material compatibility, long-term feasibility and mechanical stability have to be considered in a significant manner [55]. The trends of such research also considered some power management schemes for Maximum-Power-Point-Tracking (MPPT) to ensure maximum power delivery from the solar module to the storage element [56]. Material compatibility is a suitable alternative to cut down the requirement of such a power management scheme [55].

In these circumstances, facile double side growth of TiO_2 nanotube array, *via* electrochemical route, became a popular trend for integration in three-electrode modes [27, 29]. Here, the underlying Ti foil served as a shared electrode between DSSC and the battery. The subsequent electrochemical conversion of solar energy through a nanotube array is enhanced due to the higher available surface for dye-to-oxide interface (electron transfer) [56, 57]. The ion storage is also subject to TiO_2 (of the other side of the metal)-electrolyte interface, as the transfer of electrons from DSSC builds the favorable condition for ion insertion (lithiation) [55]. For the discharging path, the sandwiched Ti metal separates the operation from that of DSSC.

It is also indicated that, to enhance the battery performance as well as the energy harvesting efficiency, the catalytic effect is considered as one of the primordial conditions [51, 54, 57]. Hence, doping of oxide and/or decoration of the same is one of the popular schemes to enhance the performance of both device technologies [51, 54, 57]. The grapheme-reduced graphene-oxide layer shows extraordinary efficacy to achieve so [57]. However, the synthesis process still be vigorous and with lesser repeatability [54, 57]. This may be addressed with self-catalytic sites within the oxide itself. In this approach, the anodization route provides the flexibility to tune synthesis parameters to nanoengineer with the grown oxide layer [59]. Instigated by the fact of catalytic properties of nanocrystals, even of noble metals, morphological tuning of the TiO_2 nanotubes is one of the way-outs for achieving self-catalytic sites through a single process step, which by virtue holds the repeatable nanostructure and its distribution [57]. However, room temperature synthesis of such restricts the accessibility of the surface area and catalytic sites as the inadvertent coverage of oxide resulted in [60]. To avoid this, the application of higher anodization temperature is suggested and the catalytic properties were correlated with gas sensing applications [59].

The self-catalytic TiO_2 nanotubes, grown on both sides of Ti, may be considered as the future of DSSC in conjunction with Li/Na ion batteries. This type of integration is space-saving and has a large volume density of power. Thus, the futuristic application for electric vehicles may be addressed. Further, the power generation does not depend only the day time charging but also have the capability to charge at night time, if the array of such integrated device can be developed with various combination of dye and electrolyte. The self-catalytic sites on the oxide morphology itself will be responsible for enhancing the separate efficiency of each module and shared electrode mode probably suffices the overall efficacy. However, this kind of research is yet to be concluded with methodical and scientific intervention and successful implementation will open up a new generation of fully green technology in the transportation business.

SUMMARY AND CONCLUSION

Sustainable and green energy efficient system with a cost-effective approach delivers a mean of individual deployment of such to revolutionize the eco-friendly and pollution-free future custom. This scope can be identified as a base material for understanding the possibilities of a self-powered battery storage system in an integrated manner so that portability with appreciable operation can be obtained through an environment-friendly attire. The nanotechnological advancement enhanced the possibilities for cost-effective yet efficient material fabrication, tuning with a nanoengineering perspective, and its deployment in different devices, especially in solar cells and storage elements. Moreover, engineered usage and sharing of the same nanomaterial for both, evolved the path for efficient integration with a space-efficient energy scheme. For this purpose, an overview of conventional solar cell theory is discussed in the first section to have a ready reference for the trending technology of DSSC. A detailed review of the present state-of-art will be helpful to understand the materialistic approach for designing such with an insight into nanotechnology. The basic working principle and current trends of the storage element are elaborated with specific examples to understand the parts, design and prospects of an energy storage device *via* a path of its evolution. A combination of these two technologies is of recent demand to achieve fully green technology for the transportation routine as a part of the responsibility to mother nature. The design criteria with different conditions have been discussed with an envision of an integrated solar-chargeable battery device, indicating the research suitability of DSSC with a solid-state battery.

The present attempt may also be considered as the basic reference for future research to give an effort for the betterment of integrated battery storage with solar power for mitigating the charging issue of EV technology. Successful development of such a storage system genuinely solves the problem of charging stations for EVs, and the dependency on fossil fuels will be cut down in a significant manner. The self-powered solid-state batteries, as a single device, also be a potential entity for individual deployment if cost-effectiveness is achieved *via* incorporating nanoengineering techniques for the shared electrode.

LEARNING OBJECTIVES

• An overview of the DSSC working principle.

• Evolution of DSSC through different nanostructures.

• Concept of solid-state battery in the light of nanostructured electrode(s).

• Brief of the present technological scenario.

• Amalgamation of energy harnessing and energy storage devices using common nanostructures.

• Wearable and flexible self-charging system with solar power.

• Futuristic research overview to mitigate the challenges by virtue of nanostructuring of the common electrode.

MULTIPLE CHOICE QUESTIONS

1. Choose the most cheapest technology:

a. Single crystalline Si-based solar cell
b. Polycrystalline Si-based solar cell
c. Tidal power generator
d. Wind mill

2. Nanotubular structure is benefic in DSSC due to:

a. High power delivering capacity

b. Corrosion resistance
c. 1 dimenstional electron transporattion path:
d. None of the above

3. Aggregation of dye results in:

a. Less efficicency in DSSC performance
b. Higher longibility
c. Efficient DSSC
d. Lesser oxidation time

4. Main requirements for energy storage device may be identified with:

a. High power delivering capacity
b. Long distance cycle
c. Small in size
d. All the above

5. Solution used in primitive battery *i.e* volta's cell is:

a. Alkalline
b. Organic
c. Binge
d. Phosphorous

6. In Rechargable Batteries, the Self discharge of LIB will be:

a. High
b. Low
c. Unpredictable
d. Negligible

7. Commercial LIB the material type used for positive electrode in Li-base:

a. Metal sulphate
b. Metal nitride
c. Metal oxides
d. Metallics

8. LFP stands for:

a. Lithium Fusion Process
b. Lithium Iron Phosphate
c. Lead Ferro Phosphate
d. Lead Iron Phosphate

9. Reaction mechanism on anode can be identified into:

a. Intercalation
b. Conversion
c. Alloying
d. All of these

10. _____ is considered as potential anode for LIB:

a. Carbon nanotubes (CNT)
b. Tetra-hedral calcinogens
c. Hydro thermal metal nitrides
d. Sodium composite

11. The value of specific capacity of TiO_2 anatase phase material:

a. 120 mAh/s
b. 168 mAh/s
c. 200 mAh/s
d. 500 mAh/s

12. To enhance wearability of device which nnostructure is preferred:

a. Nanowire
b. Nanoparticle
c. Nanosphere
d. Nanotube

13. Main advantage of three-electrode configuration:

a. Low i/p impedance
b. High o/p impedance
c. High o/p voltage

d. High o/p current

14. How the higher reversability of a chargeable battery can be presumed from cyclic voltammogram measurement?

a. Cathodic and anodic peak at same potential.
b. Cathodic and Anodic peak at different potential.
c. Number of ocuurance of any of the peaks at corresponding potential.
d. Cannot be presumed.

15. Nanotubular morphology with rupture may increase the electrochemical reaction rate (applicable to both DSSC and battery) due to:

a. Higher surface space for the reaction.
b. Higher free energy from the material.
c. High absorption of electron in ruptured portion.
d. Faster electron transport.

ANSWER KEY

1. (b)

2. (c)

3. (a)

4. (d)

5. (c)

6. (d)

7. (c)

8. (b)

9. (d)

10. (a)

11. (b)

12. (a)

13. (c)

14. (a)

15. (c)

ACKNOWLEDGEMENT

S. Kumari and Talapati A.S. are thankful to MHRD, Govt. of India, for their MTech scholarship.

REFERENCES

[1]   A. George, D. Moline, and J. Wagner, "A mobile energy harvesting autowinder – Build and test", *Renew. Energy,* vol. 146, pp. 2659-2667, 2020.
[http://dx.doi.org/10.1016/j.renene.2019.08.106]

[2]   G.A. Lenferna, "Can we equitably manage the end of the fossil fuel era?", *Energy Research & Social Science,* vol. 35, pp. 217-223, 2018.
[http://dx.doi.org/10.1016/j.erss.2017.11.007]

[3]   M.A. Destek, and A. Sinha, "Renewable, non-renewable energy consumption, economic growth, trade openness and ecological footprint: Evidence from organisation for economic Co-operation and development countries", *J. Clean. Prod.,* vol. 242, p. 118537, 2020.
[http://dx.doi.org/10.1016/j.jclepro.2019.118537]

[4]   L. Tripathi, A.K. Mishra, A.K. Dubey, C.B. Tripathi, and P. Baredar, "Renewable energy: An overview on its contribution in current energy scenario of India", *Renew. Sustain. Energy Rev.,* vol. 60, pp. 226-233, 2016.
[http://dx.doi.org/10.1016/j.rser.2016.01.047]

[5]   I. Dincer, *Renewable energy and sustainable development: A crucial review.,* vol. 4, pp. 157-175, 2000.
[http://dx.doi.org/10.1016/S1364-0321(99)00011-8]

[6]   S. Kim, and B.E. Dale, "Environmental aspects of ethanol derived from no-tilled corn grain: nonrenewable energy consumption and greenhouse gas emissions", *Biomass Bioenergy,* vol. 28, no. 5, pp. 475-489, 2005.
[http://dx.doi.org/10.1016/j.biombioe.2004.11.005]

[7]   S. Mekhilef, R. Saidur, and A. Safari, "A review on solar energy use in industries", *Renew. Sustain. Energy Rev.,* vol. 15, no. 4, pp. 1777-1790, 2011.
[http://dx.doi.org/10.1016/j.rser.2010.12.018]

[8]   J. Blanco, S. Malato, P. Fernández-Ibañez, D. Alarcón, W. Gernjak, and M.I. Maldonado, "Review of feasible solar energy applications to water processes", *Renew. Sustain. Energy Rev.,* vol. 13, no. 6-7, pp. 1437-1445, 2009.
[http://dx.doi.org/10.1016/j.rser.2008.08.016]

[9]   Y.B. Park, H. Im, M. Im, and Y.K. Choi, "Self-cleaning effect of highly water-repellent microshell structures for solar cell applications", *J. Mater. Chem.,* vol. 21, no. 3, pp. 633-636, 2011.
[http://dx.doi.org/10.1039/C0JM02463E]

[10] G. Dong, F. Liu, J. Liu, H. Zhang, and M. Zhu, "Realization of radial p-n junction silicon nanowire solar cell based on low-temperature and shallow phosphorus doping", *Nanoscale Res. Lett.,* vol. 8, no. 1, p. 544, 2013.
[http://dx.doi.org/10.1186/1556-276X-8-544] [PMID: 24369781]

[11] F.A. Lindholm, and Sah Chih-Tang, "Fundamental electronic mechanisms limiting the performance of solar cells", *IEEE Trans. Electron Dev.,* vol. 24, no. 4, pp. 299-304, 1977.
[http://dx.doi.org/10.1109/T-ED.1977.18733]

[12]   B.C. Brusso, "A brief history of the energy conversion of light history", *IEEE Ind. Appl. Mag.,* vol. 25, no. 4, pp. 8-13, 2019.
[http://dx.doi.org/10.1109/MIAS.2019.2908804]

[13]   L.M. Fraas, "Low-cost solar electric power, Low-Cost Sol", *Electr. Power,* vol. 9783319075, pp. 1-181, 2014.
[http://dx.doi.org/10.1007/978-3-319-07530-3]

[14]   V. Fthenakis, "Sustainability of photovoltaics: The case for thin-film solar cells", *Renew. Sustain. Energy Rev.,* vol. 13, no. 9, pp. 2746-2750, 2009.
[http://dx.doi.org/10.1016/j.rser.2009.05.001]

[15]   S.S. Bhande, G.A. Taur, A.V. Shaikh, O-S. Joo, M-M. Sung, R.S. Mane, A.V. Ghule, and S-H. Han, "Structural analysis and dye-sensitized solar cell application of electrodeposited tin oxide nanoparticles", *Mater. Lett.,* vol. 79, pp. 29-31, 2012.
[http://dx.doi.org/10.1016/j.matlet.2012.03.074]

[16]   Z. Yi, Y. Zeng, H. Wu, X. Chen, Y. Fan, H. Yang, Y. Tang, Y. Yi, J. Wang, and P. Wu, "Synthesis, surface properties, crystal structure and dye-sensitized solar cell performance of TiO_2 nanotube arrays anodized under different parameters", *Results Phys.,* vol. 15, p. 102609, 2019.
[http://dx.doi.org/10.1016/j.rinp.2019.102609]

[17]   A.B.F. Martinson, J.W. Elam, J.T. Hupp, and M.J. Pellin, "ZnO nanotube based dye-sensitized solar cells", *Nano Lett.,* vol. 7, no. 8, pp. 2183-2187, 2007.
[http://dx.doi.org/10.1021/nl070160+] [PMID: 17602535]

[18]   Z. Liu, C. Liu, J. Ya, and E. Lei, "Controlled synthesis of ZnO and TiO_2 nanotubes by chemical method and their application in dye-sensitized solar cells", *Renew. Energy,* vol. 36, no. 4, pp. 1177-1181, 2011.
[http://dx.doi.org/10.1016/j.renene.2010.09.019]

[19]   D. Luo, B. Liu, A. Fujishima, and K. Nakata, "TiO_2 Nanotube Arrays Formed on Ti Meshes with Periodically Arranged Holes for Flexible Dye-Sensitized Solar Cells", *ACS Appl. Nano Mater.,* vol. 2, no. 6, pp. 3943-3950, 2019.
[http://dx.doi.org/10.1021/acsanm.9b00849]

[20]   Y. Dou, F. Wu, L. Fang, G. Liu, C. Mao, K. Wan, and M. Zhou, "Enhanced performance of dye-sensitized solar cell using Bi2Te3 nanotube/ZnO nanoparticle composite photoanode by the synergistic effect of photovoltaic and thermoelectric conversion", *J. Power Sources,* vol. 307, pp. 181-189, 2016.
[http://dx.doi.org/10.1016/j.jpowsour.2015.12.113]

[21]   X. Luan, L. Chen, J. Zhang, G. Qu, J.C. Flake, and Y. Wang, "Electrophoretic deposition of reduced graphene oxide nanosheets on TiO2 nanotube arrays for dye-sensitized solar cells", *Electrochim. Acta,* vol. 111, pp. 216-222, 2013.
[http://dx.doi.org/10.1016/j.electacta.2013.08.016]

[22]   T.H. Meen, Y.T. Jhuo, S.M. Chao, N.Y. Lin, L.W. Ji, J.K. Tsai, T.C. Wu, W.R. Chen, W. Water, and C.J. Huang, "Effect of TiO2 nanotubes with TiCl4 treatment on the photoelectrode of dye-sensitized solar cells", *Nanoscale Res. Lett.,* vol. 7, no. 1, p. 579, 2012.
[http://dx.doi.org/10.1186/1556-276X-7-579] [PMID: 23092158]

[23]   Z. Seidalilir, R. Malekfar, H.P. Wu, J.W. Shiu, and E.W.G. Diau, "High-performance and stable gel-state dye-sensitized solar cells using anodic tio 2 nanotube arrays and polymer-based gel electrolytes", *ACS Appl. Mater. Interfaces,* vol. 7, no. 23, pp. 12731-12739, 2015.
[http://dx.doi.org/10.1021/acsami.5b01519] [PMID: 25984747]

[24]   M. Ye, X. Xin, C. Lin, and Z. Lin, "High efficiency dye-sensitized solar cells based on hierarchically structured nanotubes", *Nano Lett.,* vol. 11, no. 8, pp. 3214-3220, 2011.
[http://dx.doi.org/10.1021/nl2014845] [PMID: 21728278]

[25]   P. Roy, D. Kim, K. Lee, and P. Schmuki, "TiO 2 nanotubes and their application in dye-sensitized solar cells", *Nanoscale,* vol. 2, pp. 45-59, 2010.
[http://dx.doi.org/10.1039/B9NR00131J]

[26]   S. Wang, X. Zhou, X. Xiao, Y. Fang, and Y. Lin, "An increase in conversion efficiency of dye-sensitized solar cells using bamboo-type TiO$_2$ nanotube arrays", *Electrochim. Acta,* vol. 116, pp. 26-30, 2014.
[http://dx.doi.org/10.1016/j.electacta.2013.11.011]

[27]   A. Hazra, K. Dutta, B. Bhowmik, V. Manjuladevi, R.K. Gupta, P.P. Chattopadhyay, and P. Bhattacharyya, "Structural and optical characterizations of electrochemically grown connected and free-standing TiO$_2$ nanotube array", *J. Electron. Mater.,* vol. 43, no. 9, pp. 3229-3235, 2014.
[http://dx.doi.org/10.1007/s11664-014-3183-5]

[28]   B. Bhowmik, K. Dutta, and P. Bhattacharyya, "An efficient room temperature ethanol sensor device based on p-n Homojunction of TiO2 nanostructures", *IEEE Trans. Electron Dev.,* vol. 66, no. 2, pp. 1063-1068, 2019.
[http://dx.doi.org/10.1109/TED.2018.2885360]

[29]   A. Hazra, B. Bhowmik, K. Dutta, P.P. Chattopadhyay, and P. Bhattacharyya, "Stoichiometry, length, and wall thickness optimization of TiO$_2$ nanotube array for efficient alcohol sensing", *ACS Appl. Mater. Interfaces,* vol. 7, no. 18, pp. 9336-9348, 2015.
[http://dx.doi.org/10.1021/acsami.5b01785] [PMID: 25918822]

[30]   K. Sharma, V. Sharma, and S.S. Sharma, "Dye-sensitized solar cells: Fundamentals and current status", *Nanoscale Res. Lett.,* vol. 13, no. 1, p. 381, 2018.
[http://dx.doi.org/10.1186/s11671-018-2760-6] [PMID: 30488132]

[31]   D.W. Zhang, X.D. Li, H.B. Li, S. Chen, Z. Sun, X.J. Yin, and S.M. Huang, "Graphene-based counter electrode for dye-sensitized solar cells", *Carbon,* vol. 49, no. 15, pp. 5382-5388, 2011.
[http://dx.doi.org/10.1016/j.carbon.2011.08.005]

[32]   M. Murayama, and T. Mori, "Evaluation of treatment effects for high-performance dye-sensitized solar cells using equivalent circuit analysis", *Thin Solid Films,* vol. 509, no. 1-2, pp. 123-126, 2006.
[http://dx.doi.org/10.1016/j.tsf.2005.09.145]

[33]   V. Madurai Ramakrishnan, S. Pitchaiya, N. Muthukumarasamy, K. Kvamme, G. Rajesh, S. Agilan, A. Pugazhendhi, and D. Velauthapillai, "Performance of TiO$_2$ nanoparticles synthesized by microwave and solvothermal methods as photoanode in dye-sensitized solar cells (DSSC)", *Int. J. Hydrogen Energy,* vol. 45, no. 51, pp. 27036-27046, 2020.
[http://dx.doi.org/10.1016/j.ijhydene.2020.07.018]

[34]   Z.S. Wang, T. Yamaguchi, H. Sugihara, and H. Arakawa, "Significant efficiency improvement of the black dye-sensitized solar cell through protonation of TiO$_2$ films", *Langmuir,* vol. 21, no. 10, pp. 4272-4276, 2005.
[http://dx.doi.org/10.1021/la050134w] [PMID: 16032834]

[35]   X. Chen, W. Shen, T.T. Vo, Z. Cao, and A. Kapoor, "An overview of lithium-ion batteries for electric vehicles", *10th Int. Power Energy Conf. IPEC 2012,* pp. 230-235, 2012.
[http://dx.doi.org/10.1109/ASSCC.2012.6523269]

[36]   W. Qi, J.G. Shapter, Q. Wu, T. Yin, G. Gao, and D. Cui, "Nanostructured anode materials for lithium-ion batteries: Principle, recent progress and future perspectives", *J. Mater. Chem. A Mater. Energy Sustain.,* vol. 5, no. 37, pp. 19521-19540, 2017.
[http://dx.doi.org/10.1039/C7TA05283A]

[37]   A. Poullikkas, "A comparative overview of large-scale battery systems for electricity storage", *Renew. Sustain. Energy Rev.,* vol. 27, pp. 778-788, 2013.
[http://dx.doi.org/10.1016/j.rser.2013.07.017]

[38]   Y. Wang, P. He, and H. Zhou, "Li-redox flow batteries based on hybrid electrolytes: At the cross road

between Li-ion and redox flow batteries", *Adv. Energy Mater.,* vol. 2, no. 7, pp. 770-779, 2012.
[http://dx.doi.org/10.1002/aenm.201200100]

[39]   S.S. Zhang, "Problems and their origins of Ni-rich layered oxide cathode materials", *Energy Storage Mater.,* vol. 24, no. August, pp. 247-254, 2020.
[http://dx.doi.org/10.1016/j.ensm.2019.08.013]

[40]   Y. Wang, and G. Cao, "Developments in nanostructured cathode materials for high-performance lithium-ion batteries", *Adv. Mater.,* vol. 20, no. 12, pp. 2251-2269, 2008.
[http://dx.doi.org/10.1002/adma.200702242]

[41]   S. Seki, Y. Kobayashi, H. Miyashiro, Y. Mita, and T. Iwahori, "Fabrication of high-voltage, high-capacity all-solid-state lithium polymer secondary batteries by application of the polymer electrolyte/inorganic electrolyte composite concept", *Chem. Mater.,* vol. 17, no. 8, pp. 2041-2045, 2005.
[http://dx.doi.org/10.1021/cm047846c]

[42]   B. Yu, T. Tao, S. Mateti, S. Lu, and Y. Chen, "Nanoflake Arrays of Lithiophilic Metal Oxides for the Ultra-Stable Anodes of Lithium-Metal Batteries", *Adv. Funct. Mater.,* vol. 28, no. 36, p. 1803023, 2018.
[http://dx.doi.org/10.1002/adfm.201803023]

[43]   Q. Yuan, and F. Ding, "How a zigzag carbon nanotube grows", *Angew. Chem. Int. Ed.,* vol. 54, no. 20, pp. 5924-5928, 2015.
[http://dx.doi.org/10.1002/anie.201500477] [PMID: 25766145]

[44]   M.Y. Saïdi, J. Barker, H. Huang, J.L. Swoyer, and G. Adamson, "Performance characteristics of lithium vanadium phosphate as a cathode material for lithium-ion batteries", *J. Power Sources,* vol. 119-121, pp. 266-272, 2003.
[http://dx.doi.org/10.1016/S0378-7753(03)00245-3]

[45]   M. Bakierska, M. Świętosławski, K. Chudzik, M. Lis, and M. Molenda, *Enhancing the lithium ion diffusivity in $LiMn_2O_{4-y}S_y$ cathode materials through potassium doping,* 2018.
[http://dx.doi.org/10.1016/j.ssi.2018.01.014]

[46]   S. Fang, D. Bresser, and S. Passerini, "Transition metal oxide anodes for electrochemical energy storage in lithium- and sodium-ion batteries", *Adv. Energy Mater.,* vol. 10, no. 1, p. 1902485, 2020.
[http://dx.doi.org/10.1002/aenm.201902485]

[47]   Y. Hou, Y. Cheng, T. Hobson, and J. Liu, "Design and synthesis of hierarchical MnO_2 nanospheres/carbon nanotubes/conducting polymer ternary composite for high performance electrochemical electrodes", *Nano Lett.,* vol. 10, no. 7, pp. 2727-2733, 2010.
[http://dx.doi.org/10.1021/nl101723g] [PMID: 20586479]

[48]   A.M. Omer, "Energy, environment and sustainable development", *Renew. Sustain. Energy Rev.,* vol. 12, no. 9, pp. 2265-2300, 2008.
[http://dx.doi.org/10.1016/j.rser.2007.05.001]

[49]   D. Devadiga, M. Selvakumar, P. Shetty, and M.S. Santosh, "The integration of flexible dye-sensitized solar cells and storage devices towards wearable self-charging power systems: A review", *Renewable and Sustainable Energy Reviews,* vol. 159, p. 11252, 2022.
[http://dx.doi.org/10.1016/j.rser.2022.112252]

[50]   B. Luo, D. Ye, and L. Wang, "Recent progress on integrated energy conversion and storage systems", *Adv. Sci.,* vol. 4, no. 9, p. 1700104, 2017.
[http://dx.doi.org/10.1002/advs.201700104] [PMID: 28932673]

[51]   X. Pu, W. Hu, and Z.L. Wang, "Toward wearable self-charging power systems: The integration of energy-harvesting and storage devices", *Small,* vol. 14, no. 1, p. 1702817, 2018.
[http://dx.doi.org/10.1002/smll.201702817] [PMID: 29194960]

[52]   L. Fagiolari, M. Sampò, A. Lamberti, J. Amici, C. Francia, S. Bodoardo, and F. Bella, "Integrated

energy conversion and storage devices: Interfacing solar cells, batteries and supercapacitors", *Energy Storage Mater.,* vol. 51, no. June, pp. 400-434, 2022.
[http://dx.doi.org/10.1016/j.ensm.2022.06.051]

[53]   T.L. Gibson, and N.A. Kelly, "Solar photovoltaic charging of lithium-ion batteries", *J. Power Sources,* vol. 195, no. 12, pp. 3928-3932, 2010.
[http://dx.doi.org/10.1016/j.jpowsour.2009.12.082]

[54]   P. Chen, G.R. Li, T.T. Li, and X.P. Gao, "Solar-Driven Rechargeable Lithium–Sulfur Battery", *Adv. Sci.,* vol. 6, no. 15, p. 1900620, 2019.
[http://dx.doi.org/10.1002/advs.201900620] [PMID: 31406674]

[55]   W. Guo, X. Xue, S. Wang, C. Lin, and Z.L. Wang, "An integrated power pack of dye-sensitized solar cell and Li battery based on double-sided TiO2 nanotube arrays", *Nano Lett.,* vol. 12, no. 5, pp. 2520-2523, 2012.
[http://dx.doi.org/10.1021/nl3007159] [PMID: 22519631]

[56]   P. Poulose, and P. Sreejaya, "Indoor light harvesting using dye sensitized solar cell", *2018 Int. CET Conf. Control. Commun. Comput. IC4 2018,* pp. 152-156, 2018.
[http://dx.doi.org/10.1109/CETIC4.2018.8530924]

[57]   L. Wang, L. Wen, Y. Tong, S. Wang, X. Hou, X. An, S.X. Dou, and J. Liang, "Photo-rechargeable batteries and supercapacitors: Critical roles of carbon-based functional materials", *Carbon Energy,* vol. 3, no. 2, pp. 225-252, 2021.
[http://dx.doi.org/10.1002/cey2.105]

[58]   A. Gurung, and Q. Qiao, "Solar charging batteries: Advances, challenges, and opportunities", *Joule,* vol. 2, no. 7, pp. 1217-1230, 2018.
[http://dx.doi.org/10.1016/j.joule.2018.04.006]

[59]   K. Dutta, P.P. Chattopadhyay, and P. Bhattacharyya, "Voltage controlled rupturing of TiO_2 nanotubes for gas sensor device applications: Correlation with surface and edge energy", *IEEE Trans. Electron Dev.,* vol. 63, no. 12, pp. 4933-4938, 2016.
[http://dx.doi.org/10.1109/TED.2016.2620560]

[60]   J.H. Lim, and J. Choi, "Titanium oxide nanowires originating from anodically grown nanotubes: The bamboo-splitting model", *Small,* vol. 3, no. 9, pp. 1504-1507, 2007.
[http://dx.doi.org/10.1002/smll.200700114] [PMID: 17647256]

SUBJECT INDEX

www.ingramcontent.com/pod-product-compliance
Lightning Source LLC
Chambersburg PA
CBHW050758220326
41598CB00006B/57